Lecture Notes in Mechanical Engineering

Lecture Notes in Mechanical Engineering (LNME) publishes the latest developments in Mechanical Engineering—quickly, informally and with high quality. Original research or contributions reported in proceedings and post-proceedings represents the core of LNME. Volumes published in LNME embrace all aspects, subfields and new challenges of mechanical engineering.

To submit a proposal or request further information, please contact the Springer Editor of your location:

Europe, USA, Africa: Leontina Di Cecco at Leontina.dicecco@springer.com
China: Ella Zhang at ella.zhang@cn.springernature.com
India, Rest of Asia, Australia, New Zealand: Swati Meherishi at swati.meherishi@springer.com

Topics in the series include:

- Engineering Design
- Machinery and Machine Elements
- Mechanical Structures and Stress Analysis
- Automotive Engineering
- Engine Technology
- Aerospace Technology and Astronautics
- Nanotechnology and Microengineering
- Control, Robotics, Mechatronics
- MEMS
- Theoretical and Applied Mechanics
- Dynamical Systems, Control
- Fluid Mechanics
- Engineering Thermodynamics, Heat and Mass Transfer
- Manufacturing Engineering and Smart Manufacturing
- Precision Engineering, Instrumentation, Measurement
- Materials Engineering
- Tribology and Surface Technology

Indexed by SCOPUS, EI Compendex, and INSPEC.

All books published in the series are evaluated by Web of Science for the Conference Proceedings Citation Index (CPCI).

To submit a proposal for a monograph, please check our Springer Tracts in Mechanical Engineering at https://link.springer.com/bookseries/11693.

Bartosz Gapiński · Olaf Ciszak · Vitalii Ivanov ·
José Machado · Anna Karwasz
Editors

Advances in Manufacturing V

Volume 1 - Mechanical Engineering: Factory of the Future

Springer

Editors
Bartosz Gapiński
Faculty of Mechanical Engineering
Poznań University of Technology
Poznań, Poland

Olaf Ciszak
Faculty of Mechanical Engineering
Poznań University of Technology
Poznań, Poland

Vitalii Ivanov
Department of Manufacturing Engineering, Machines and Tools
Sumy State University
Sumy, Ukraine

José Machado
Mechanical Engineering Department
University of Minho
Guimarães, Portugal

Anna Karwasz
Faculty of Mechanical Enginering
Poznań University of Technology
Poznań, Poland

ISSN 2195-4356 ISSN 2195-4364 (electronic)
Lecture Notes in Mechanical Engineering
ISBN 978-3-032-22942-7 ISBN 978-3-032-22943-4 (eBook)
https://doi.org/10.1007/978-3-032-22943-4

This Springer imprint is published by the registered company Springer Nature Switzerland AG
The registered company address is: Gewerbestrasse 11, 6330 Cham, Switzerland

Preface

This volume of Lecture Notes in Mechanical Engineering gathers accepted papers presented at the 9th International Scientific-Technical Conference MANUFACTURING 2026, held in Poznan, Poland, on May 19–21, 2026. The conference was hosted by the Faculty of Mechanical Engineering, Poznan University of Technology, Poland.

The primary objective of the MANUFACTURING 2026 was to showcase cutting-edge advancements across the broad spectrum of mechanical engineering, providing a dynamic forum for the exchange of innovative ideas and expert opinions. Focused on the theme of the Factory of the Future, the conference serves as a vital platform for dialogue on industrial evolution and science-industry synergy. Its global reach continues to expand, as evidenced by the consistently increasing participation from international research centers. Furthermore, the event fosters the integration of academia with the economic sector, encouraging collaboration with industry leaders and management professionals interested in implementing research results into business practice. The scope encompassed the following areas:

- Mechanical Engineering
- Production Engineering
- Quality Engineering
- Measurement and Control Systems
- Biomedical Engineering

The global significance of the conference is reflected in the diversity of the submissions. The Organizing Committee received 219 contributions from 37 countries. To ensure the highest academic standards, all papers underwent a rigorous double-blind peer review process. Each manuscript was evaluated by at least three independent international experts, with a strict limit of five assignments per reviewer to guarantee depth and quality of feedback. Following this comprehensive appraisal, 120 papers were accepted for publication (an acceptance rate of 55%), representing the work of 347 authors from 25 countries. A total of 109 articles were published in the proceedings Advances in Manufacturing V, while 11 selected papers appeared in the journal Management and Production Engineering Review, issued under the auspices of the Committee on Production Engineering of the Polish Academy of Sciences and the Polish Society for Production Management.

The review process focused on originality of paper, significance of presented results, comprehensibility and quality of the presentation, relevance for this conference and overall impression. Furthermore, reviewers meticulously assessed the timeliness of the research, the adequacy of the literature review, methodological rigor, and the clarity of conclusions, as well as the quality of English. Feedback provided to authors included specific recommendations for improving methodology, data presentation, and editorial quality. Acceptance required at least two positive reviews and a full resolution of all reviewers' concerns. This international effort involved experts from 30 countries across

Europe, North America, South Asia, and North Africa. Additionally, all manuscripts were checked using anti-plagiarism software, which allowed the similarity index to be less than 15%.

The book **Advances in Manufacturing V** is organized into five volumes that correspond to the main conference disciplines mentioned above.

The **Advances in Manufacturing V - Volume 1 - Mechanical Engineering: Factory of the Future** volume focuses on the paradigm shift toward the Factory of the Future, gathering cutting-edge research and technologies designed to bridge the gap between theoretical excellence and industrial application. The content is structured into five thematic parts, reflecting the multidisciplinary nature of modern mechanical engineering.

- Production & Testing: Research in the field of advanced technologies covering materials and their properties, manufacturing and processing, methods of analysis, modelling, optimisation and monitoring.
- Design: Covers the advanced design supported by numerical simulations and describes engineering solutions used in industrial practice.
- Robotics: Addresses the use of modern technologies such as robotics, autonomous systems, AI and digital tools to increase the efficiency, precision and safety of processes, both in industry and in crisis management, while supporting human-machine collaboration.
- Green Technology: Addresses the issue of sustainable development and the impact of industrial activity on the environment, providing practical recommendations on how to reduce emissions from human industrial activity.
- Internet of Things: Describes advanced industrial and digital technologies, Industry 4.0, IoT systems, data acquisition, predictive diagnostics and communication in an industrial environment, while monitoring and optimising processes based on modern methods and digital tools.

Comprising 28 chapters prepared by 98 authors from 11 countries, this volume serves as a roadmap for the effective implementation of Industry 4.0 and 5.0 solutions, fostering a vital synergy between science and the global economy.

We wish to express our sincere gratitude to the members of the Scientific and Program Committees for their dedication and expertise during the review process. We also extend our thanks to the authors, sponsors, and all contributors whose commitment was instrumental in making MANUFACTURING 2026 a resounding success.

May 2026

Bartosz Gapiński
Olaf Ciszak
Vitalii Ivanov
José Machado
Anna Karwasz

Organization

Steering Committee

General Chair

Adam Hamrol	Poznan University of Technology, Poland

Chairs

Bartosz Gapiński	Poznan University of Technology, Poland
Magdalena Hryb	Poznan University of Technology, Poland
Justyna Trojanowska	Poznan University of Technology, Poland

Conference Founders

†Zenobia Weiss	Poznan University of Technology, Poland
Adam Hamrol	Poznan University of Technology, Poland

Scientific Committee

Katarzyna Antosz	Rzeszow University of Technology, Poland
Christopher Brown	Worcester Polytechnic Institute, USA
Anna Burduk	Wrocław University of Science and Technology, Poland
Suthep Butdee	Rajamangala University of Technology Krungthep, Thailand
Marcin Butlewski	Poznan University of Technology, Poland
Olaf Ciszak	Poznan University of Technology, Poland
Reggie Davidrajuh	University of Stavanger, Norway
Ewa Dostatni	Poznan University of Technology, Poland
Jan Duda	Cracow University of Technology, Poland
Mosè Gallo	University of Naples Federico II, Italy
Bartosz Gapiński	Poznan University of Technology, Poland
Józef Gawlik	Cracow University of Technology, Poland
Adam Gąska	Cracow University of Technology, Poland

Arkadiusz Gola	Lublin University of Technology, Poland
Filip Górski	Poznan University of Technology, Poland
Marta Grabowska	Poznan University of Technology, Poland
Adam Hamrol	Poznan University of Technology, Poland
Magdalena Hryb	Poznan University of Technology, Poland
Jozef Husár	Technical University of Košice, Slovakia
Aminul Islam	Technical University of Denmark, Denmark
Vitalii Ivanov	Sumy State University, Ukraine
Michał Jakubowicz	Poznan University of Technology, Poland
Andrzej Jardzioch	West Pomeranian University of Technology in Szczecin, Poland
Sabahudin Jasarevic	University of Zenica, Bosnia & Herzegovina
Małgorzata Jasiulewicz-Kaczmarek	Poznan University of Technology, Poland
Anna Karwasz	Poznan University of Technology, Poland
Sławomir Kłos	Poznan University of Technology, Poland
Lucia Knapčíková	Technical University of Košice, Slovakia
Agnieszka Kujawińska	Poznan University of Technology, Poland
Vijaya Kumar Manupati	National Institute of Industrial Engineering, India
Stanisław Legutko	Poznan University of Technology, Poland
Sławomir Luściński	Kielce University of Technology, Poland
José Machado	University of Minho, Portugal
Damjan Maletic	University of Maribor, Slovenia
Dariusz Mikołajewski	Kazimierz Wielki University, Poland
Piotr Moncarz	Stanford University, USA
Magdalena Niemczewska-Wójcik	Cracow University of Technology, Poland
Przemysław Niewiadomski	University of Zielona Góra, Poland
Razvan Ioan Pacurar	Technical University of Cluj-Napoca, Romania
Piotr Paczos	Poznan University of Technology, Poland
Mirosław Pajor	Maritime University of Szczecin, Poland
Justyna Patalas-Maliszewska	University of Zielona Góra, Poland
Ivan Pavlenko	Sumy State University, Ukraine
Erwin Rauch	Free University of Bozen-Bolzano, Italy
Paulina Rewers	Poznan University of Technology, Poland
Izabela Rojek	Kazimierz Wielki University, Poland
Michał Rychlik	Poznan University of Technology, Poland
Krzysztof Santarek	Warsaw University of Technology, Poland
Sebastian Skoczypiec	Cracow University of Technology, Poland
Bożena Skołud	Silesian University of Technology, Poland
Jerzy Andrzej Sładek	Cracow University of Technology, Poland
Beata Starzyńska	Poznan University of Technology, Poland

Liangyan Tao	Nanjing University of Aeronautics and Astronautics, China
Anna Timofiejczuk	Silesian University of Technology, Poland
Justyna Trojanowska	Poznan University of Technology, Poland
Wiesław Urban	Bialystok University of Technology, Poland
Leonilde Varela	University of Minho, Portugal
Nikola Vitković	University of Niš, Serbia
Michał Wieczorowski	Poznan University of Technology, Poland
Ewa Więcek-Janka	Poznan University of Technology, Poland
Szymon Wojciechowski	Poznan University of Technology, Poland
Ralf Woll	BTU Cottbus-Senftenberg, Germany
Naiming Xie	Nanjing University of Aeronautics and Astronautics, China
Jiehan Zhou	Shandong University of Science and Technology, China

Program Committee

Adam Gąska	Cracow University of Technology, Poland
Adam Hamrol	Poznan University of Technology, Poland
Agnieszka Kaczmarek-Pawelska	University of Zielona Góra, Poland
Agnieszka Stachowiak	Poznan University of Technology, Poland
Alejandro Pereira	University of Vigo, Portugal
Aleksandar Miltenovic	University of Nis, Serbia
Alice Chlupová	Czech Academy of Sciences, Czech Republic
Alzbeta Sapietova	University of Žilina, Slovakia
Ancuta Pacurar	Technical University of Cluj-Napoca, Romania
André Guimarães	University of Beira Interior, Portugal
Andrzej Jardzioch	West Pomeranian University of Technology in Szczecin, Poland
Angelika Hofer	Free University of Bozen-Bolzano, Italy
Anna Borucka	Military University of Technology Warsaw, Poland
Anna Burduk	Wroclaw University of Science and Technology, Poland
Anna Karwasz	Poznan University of Technology, Poland
Anna M. Deptuła	Opole University of Technology, Poland
Anna Saniuk	University of Zielona Góra, Poland
Arkadiusz Gola	Lublin University of Technology, Poland
Bartosz Gapinski	Poznan University of Technology, Poland
Bogdan Dybała	Wroclaw University of Science and Technology, Poland

Olha Kalman	Technical University of Košice, Slovakia
Paulina Rewers	Poznan University of Technology, Poland
Paweł Buń	Poznan University of Technology, Poland
Paweł Szczygieł	Kielce University of Technology, Poland
Paweł Zmarzły	Kielce University of Technology, Poland
Pedro Torres	Polytechnic Institute of Castelo Branco, Portugal
Piotr Gąska	AGH University of Krakow, Poland
Piotr Kotlarz	Kazimierz Wielki University, Poland
Piotr Trojanowski	West Pomeranian University of Technology in Szczecin, Poland
Piotr Ziembicki	University of Zielona Góra, Poland
Przemysław Zawadzki	Poznan University of Technology, Poland
Radosław Wichniarek	Poznan University of Technology, Poland
Răzvan Păcurar	Technical University of Cluj-Napoca, Romania
Roman Kielec	University of Zielona Góra, Poland
Sandra Grabowska	Silesian University of Technology, Poland
Sebastian Saniuk	University of Zielona Góra, Poland
Sebastian Skoczypiec	Cracow University of Technology, Poland
Sergej Hloch	Technical University of Košice, Slovakia
Sławomir Kłos	University of Zielona Góra, Poland
Stanislaw Legutko	Poznan University of Technology, Poland
Stella Hrehova	Technical University of Košice, Slovakia
Sven Maricic	Juraj Dobrila University of Pula, Croatia
Sylvia Kusmierczak	University of Jan Evangelista in Ústí nad Labem, Czech Republic
Szymon Wojciechowski	Poznan University of Technology, Poland
Thanh Tran	RMIT University, Vietnam
Tomasz Bednarek	Kazimierz Wielki University, Poland
Vijaya Kumar Manupati	NITIE Mumbai, India
Vitalii Ivanov	Sumy State University, Ukraine
Wiesław Kuczko	Poznan University of Technology, Poland
Wiktor Szot	Kielce University of Technology, Poland
Wojciech Babirecki	University of Zielona Góra, Poland
Wojciech Macek	Gdańsk University of Technology, Poland
Yuliia Denysenko	Sumy State University, Ukraine
Zbigniew Wiśniewski	Lodz University of Technology, Poland
Zuzana Mitaľová	Technical University of Košice, Slovakia

Special Sessions and Special Session Organizing Committee

Efficiency-Driven Manufacturing: Tools and Trends in the Era of Industry 4.0

Justyna Trojanowska	Poznan University of Technology, Poland
Leonilde Varela	University of Minho, Portugal
Vijaya Kumar Manupati	NITIE Mumbai, India
Thanh Tran	RMIT University, Vietnam
Jozef Husár	Technical University of Košice, Slovakia

Intersecting Quality Management and Mechanical Design for Manufacturing Sustainability

Florin Popişter	Technical University of Cluj-Napoca, Romania
Diana Cristina Dragomir	Technical University of Cluj-Napoca, Romania
Corina Julieta Birleanu	Technical University of Cluj-Napoca, Romania
Mircea Claudiu Fulea	Technical University of Cluj-Napoca, Romania

Human Factors in Manufacturing

Marcin Butlewski	Poznan University of Technology, Poland
Beata Mrugalska	Poznan University of Technology, Poland
Wiktoria Czernecka	Poznan University of Technology, Poland
Fryderyk Wachowiak	Poznan University of Technology, Poland
Krzysztof Hankiewicz	Poznan University of Technology, Poland

Circularity in Advanced Production Systems: Materials, Methods, and Digital Transformation

Marek Kočisko	Technical University of Košice, Slovakia
Lucia Knapčikova	Technical University of Košice, Slovakia
Jakub Kaščak	Technical University of Košice, Slovakia
Jozef Török	Technical University of Košice, Slovakia

Smart Manufacturing and Sustainability: Interdisciplinary Approaches in the Era of Industry 4.0/5.0

Radosław Maruda	University of Zielona Góra, Poland
Sebastian Saniuk	University of Zielona Góra, Poland
Katarzyna Skrzypek	University of Zielona Góra, Poland
Daria Moskwa-Bęczkowska	Kielce University of Technology, Poland

Karol Dąbrowski	University of Zielona Góra, Poland

From Additive Manufacturing to Healthcare Innovation through Digital Transformation in Biomedical Engineering

Filip Górski	Poznan University of Technology, Poland
Răzvan Păcurar	Technical University of Cluj-Napoca, Romania
Nikola Vitkovic	University of Niš, Serbia
Michał Rychlik	Poznan, University of Technology, Poland

Intelligent Adaptive Systems in Integrated Production Management Processes

Mariusz Piechowski	Merito WSB University in Poznan, Poland
Małgorzata Jasiulewicz-Kaczmarek	Poznan University of Technology, Poland
Katarzyna Antosz	Rzeszow University of Technology, Poland
Ryszard Wyczółkowski	Silesian University of Technology, Poland
Arkadiusz Gola	Lublin University of Technology, Poland

Intelligent Solutions for Industry of the Future

Izabela Rojek	Kazimierz Wielki University in Bydgoszcz, Poland
Ewa Dostatni	Poznan University of Technology, Poland
Dariusz Mikołajewski	Kazimierz Wielki University in Bydgoszcz, Poland
Nataša Náprstková	University of Jan Evangelista in Ústí nad Labem, Czech Republic

Grey Systems Modeling for Engineering Solutions and Business Optimization

Magdalena Hryb	Poznan University of Technology, Poland
Agnieszka Kujawińska	Poznan University of Technology, Poland
Naiming Xie	Nanjing University of Aeronautics and Astronautics, China
Liangyan Tao	Nanjing University of Aeronautics and Astronautics, China
Rafał Mierzwiak	Poznan University of Technology, Poland

Intelligent Methods Supporting Manufacturing Systems Efficiency

Anna Burduk	Wroclaw University of Science and Technology, Poland
Damian Krenczyk	Silesian University of Technology, Poland
Dagmara Łapczyńska	Wroclaw University of Science and Technology, Poland
Kamil Krot	Wroclaw University of Science and Technology, Poland

Contents

Production and Testing

Production and Testing

Comparative Characteristics of Milling Zones for Polymer Composites and Aluminum Alloy

Krzysztof Ciecieląg(✉), Agnieszka Skoczylas, and Jakub Matuszak

Lublin University of Technology, Faculty of Mechanical Engineering, Department of Production Engineering, Nadbystrzycka 36, 20-618 Lublin, Poland
k.ciecielag@pollub.pl

Abstract. Fiber reinforced plastics are frequently used due to their specific properties. However, their application requires machining processes such as milling. These materials are often used as an alternative to aluminum alloys. The novelty in this article are the comparative results of study on the milling of polymer composites and aluminum alloy in three cutting zones. According to the research presented in the article, it was determined that changing the milling parameters such as feed per tooth and cutting speed increases the maximum feed forces in each analyzed zone. The maximum values of 580 N were achieved for the aluminum alloy in the first zone at a cutting speed of 600 m/min. In the case of surface quality analysis, an increase in feed per tooth resulted in an increase in the Ra parameter in each zone, with the highest values occurring in the first zone. An increase in cutting speed resulted in a decrease in Ra values, with the lowest values of 0.40 μm being achieved in the last zone for aluminum alloy at a cutting speed of 600 m/min.

Keywords: Milling · Roughness · Polymer composites · Feed force

1 Introduction

Milling is the methods of shaping polymer composites [1] and aluminum alloys [2]. The method is particularly common in the aerospace, automotive, and defense industries. It allows for obtaining the correct dimensions and shapes of elements and the appropriate surface quality [3, 4]. In the process of composite machining, two most common milling operations can be distinguished: face milling and side milling. Face milling involves cutting the material mainly with the face edges of the milling cutter, perpendicular to the machined surface [5, 6]. It is mainly used for milling pockets and face surfaces, but also for making holes with complex geometries [7] and preparing surfaces for bonding processes. Side milling is a process in which the material is machined using the side edges of the tool. This technique is most often used for shaping and finishing the edges of components after forming [8]. Side milling allows burrs and undefined fibers to be removed and the desired shape and dimensions to be obtained. In the process of milling polymer composites, defects occur due to the interaction of the tool with the material [9]. The key to limiting their formation is to optimize the basic machining parameters that characterize milling. In practice, the following parameter ranges are used: feed per

B. Gapiński et al. (Eds.): MANUFACTURING 2026, LNME, pp. 3–12, 2026.
https://doi.org/10.1007/978-3-032-22943-4_1

tooth 0.01–0.6 mm/tooth [4, 10], cutting speed 50–600 m/min [11, 12] and depth of cut 0.1–4 mm [6, 13]. Different values have different effects on the process [14]. In the process of milling polymer composites, it is very important to select the right tools, such as inserts with polycrystalline diamond [15]. Milling polymer composites requires the use of appropriately selected techniques [16], which include co-rotational and counter-rotational milling [17]. Milling cutters with diamond coatings allow for a smooth surface (Ra 1–2 μm) to be obtained at low feed per tooth (0.05–0.1 mm/tooth) and cutting depths (0.5–1 mm) [18]. The effects of polymer composite machining also depend on the kind of tool [19]. Due to the roughness of the surface, cutters with diamond inserts are most commonly used [20].

Removal machining of aluminum is not a difficult challenge, but the specific properties of aluminum alloys, such as high plasticity, low hardness, and tendency to adhere, require the use of specific machining [21]. The selection of appropriate tools, strategies, and machining parameters is crucial for cost-effective and high-quality production [22]. The most commonly used method of machining aluminum alloys is milling [23]. Milling makes it possible to produce complex shapes that require high precision and accuracy. When milling aluminum alloys, the main obstacles are problems with chip removal, susceptibility to overheating, build-up on the cutting edge, and deterioration of surface quality [24]. So machining aluminum alloys, cutters made of high-speed steel, cemented carbides, polycrystalline diamond or CVD and PVD coated cutters are most commonly used. The appropriate geometry and material of the tool directly affect its wear and, consequently, the increase in cutting forces and deterioration of surface quality. If the rake angle is the greater, the lower the cutting forces, but the greater the rake angle, the smaller the tooth angle, which weakens the tool. Depending on the specific alloy, rake angles from $+10^\circ$ to $+30^\circ$ are recommended. The use of appropriate geometry, well-chosen milling cutter materials, and a suitable coating can improve surface quality and reduce burr formation by up to 40% [25]. The fundamental machining parameters determine the quality of the finished parts and the wear on the tools. [26]. Feed per tooth plays an significant impact in the surface. The higher this parameter, the greater the surface roughness [27]. The specific value depends on the type of alloy; softer alloys can be machined with higher feed per tooth, while harder alloys can be machined with lower feed. Similar to feed per tooth, cutting speed also significantly affects surface quality. Typically, as cutting speed increases, roughness decreases [28]. An increase in this parameter is associated with an increase in temperature, which in aluminum alloys is a key phenomenon that must be kept under control, as the melting point of aluminum is 660 °C. This is a low value that is very easy to reach, and once exceeded, the workpiece material begins to build up on the tool. Therefore, it is very important to select the appropriate value for this parameter. The depth of cut also affects the temperature increase; the greater the depth, the higher the temperature. To avoid unnecessary temperature increases that can cause buildup on the tooth, low values for this parameter are usually used [27]. The surface quality after machining aluminum alloys can also be improved using various techniques such as burnishing [29]. Heat dissipation minimizes or completely eliminates build-up on the tool edge, which is particularly useful in the case of aluminum alloys [30]. The use of low cutting speeds, low feed per tooth, and greater cutting depths makes the chips thicker and therefore more brittle and with a smaller rounding radius, which makes them easier to remove [28].

This article analyzes feed force in relation to milling process parameters and surface roughness in three cutting zones. This is a novelty in this work, as it presents the results of testing two alternative materials in specific machining areas where the machining results differ. In addition, the results of tests on fiber reinforced plastics were compared, and an aluminum alloy was also tested for comparison.

2 Methodology

The machining operations were conducted on the AVIA-VMC 800 HS. The machine tool uses Heidenhein iTNC 530 control. Polymer composites based on EGL/EP 3200–120 glass fiber prepreg fabrics (designated as GFRP) and HexPly AG193PW/3501/6SRC41 carbon fiber prepreg fabrics (designated as CFRP) were used for the tests. Aluminum alloy EN-AW-2024 T351 was also used for comparative analysis. The face milling process was done using a cutter with a diameter of ϕ 12 mm. A SECO cutting tool was used, made of a body (R217.69-1212 0-06-2AN) and an insert (XOEX060204FR PCD05). During milling, the cutting forces will be measured using the main element of the system, i.e., a Kistler 3D dynamometer (marked as 9257B) connected to a Kistler amplifier (marked as 5070) that converts the electrical charge into proportional voltages. DynoWare software (marked as 2825A) and Dynoware data acquisition card (marked as 5697A) will be used to record the signals. The data acquisition system is used to connect and control charge amplifiers and works with a PC running the universal DynoWare software. A scheme of the test setup and analysis of the results is presented in Fig. 1.

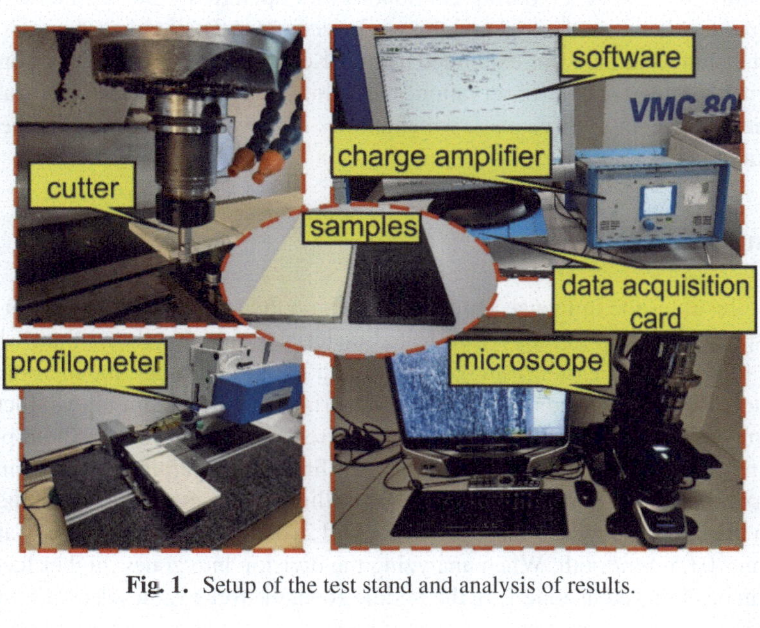

Fig. 1. Setup of the test stand and analysis of results.

In order to compare the feed force and surface roughness values, the same set of parameters was used for all tested materials, as shown in Table 1. The cutting depth for all tests was 1 mm..

Table 1. Milling parameters.

No.	f_z [mm/tooth]	v_c [m/min]
1	0.05	300
2	0.1	300
3	0.2	300
4	0.4	300
5	0.6	300
6	0.2	50
7	0.2	200
8	0.2	400
9	0.2	600

Hommel-Etamic T8000RC 120–140 used in measure, allowed the parameters of the 2D and 3D geometric structure to be determined. An area measuring 4.8 mm x 4.8 mm with an elementary section length of 0.8 mm was designated for measurement. Roughness was examined in three zones: at the first point of contact between the tool and the workpiece (zone 1), during continuous tool operation (zone 2), and at the last point of contact between the tool and the workpiece (zone 3).

The machined surface was examined using a Keyence VHX-5000 microscope. This device enables microscopic observation, recording, and measurement. The table has automatic adjustment in three axes, which allows for precise positioning of the object under examination.

3 Results and Discussion

Figure 2 illustrates the influence of feed per tooth on the feed force values F_f at a fixed cutting speed $v_c = 300$ m/min for all tested materials. The plot shows that, when the feed per tooth was increased from 0.05 mm/tooth to 0.4 mm/tooth, the measured feed forces increased for all kinds of composite materials and aluminum alloys. Increasing the parameter to 0.6 mm/tooth resulted in an increase in feed forces for both polymer composites and aluminum alloy. According to the literature, a parameter of 0.6 mm/tooth is the maximum value for this parameter when milling composites. The lowest measured values in the first zone were 90.2 N for GFRP and 210.4 N for aluminum alloy at a feed per tooth of 0.05 mm/tooth. When analyzing the division into zones, higher feed force values can be observed in zone 1. In the middle zone, the force values decrease by about 10–15%.

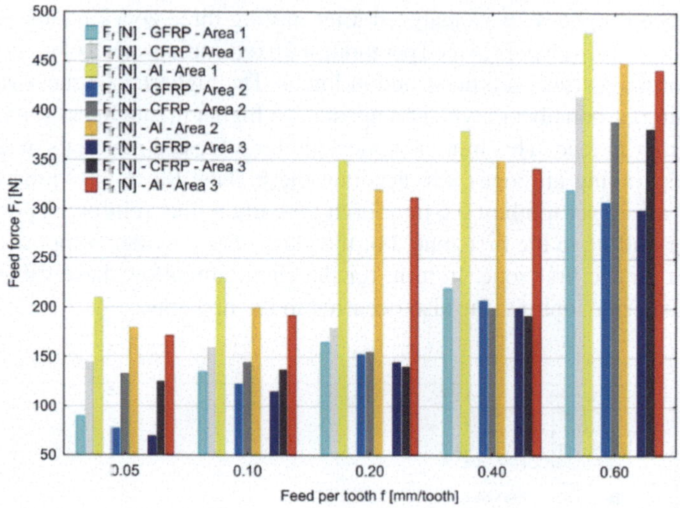

Fig. 2. Maximum force values depending on feed per tooth in three cutting zones.

Figure 3 illustrates the influence of cutting speed on feed forces for a fixed feed per tooth of $f_z = 0.2$ mm/tooth. The lowest value for GFRP is 120.3 N for $v_c = 50$ m/min, and the highest is 580 N for aluminum alloy for $v_c = 600$ m/min. For carbon fiber reinforced plastics, the lowest value is 160 N for $v_c = 50$ m/min. When analyzing the other cutting zones, a decrease in the maximum feed forces can be observed. It can be explained by the stabilization of the cutting process. However, the upward trend in maximum feed force values is still maintained, i.e., as the cutting speed increases, the force values increase.

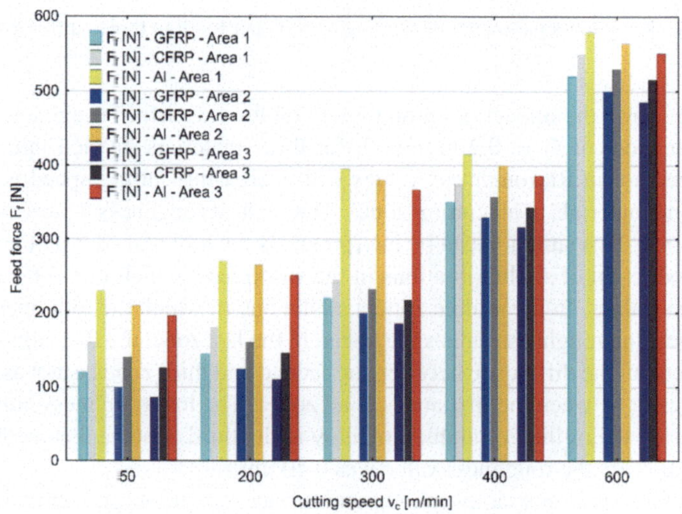

Fig. 3. Maximum force values depending on cutting speed in three cutting zones.

The surface roughness was analyzed after milling three zones for the three tested materials, along with a change in feed per tooth at a fixed cutting speed of $v_c = 300$ m/min. The results of the research are presented in Fig. 4. The highest roughness values occur after machining carbon fiber polymer composites. A higher in roughness values is noticeable with higher in feed. This trend is noticeable for all tested materials in all analyzed cutting zones. Taking all zones into account, the highest Ra = 3.03 μm values were achieved for a feed per tooth of 0.6 mm/tooth for carbon fiber reinforced polymer composite during milling in the first zone. For the glass fiber polymer composite, the value was 2.02 μm for the first zone. In turn, for the aluminum alloy, these values were the lowest, but their maximum value also occurred in the first zone.

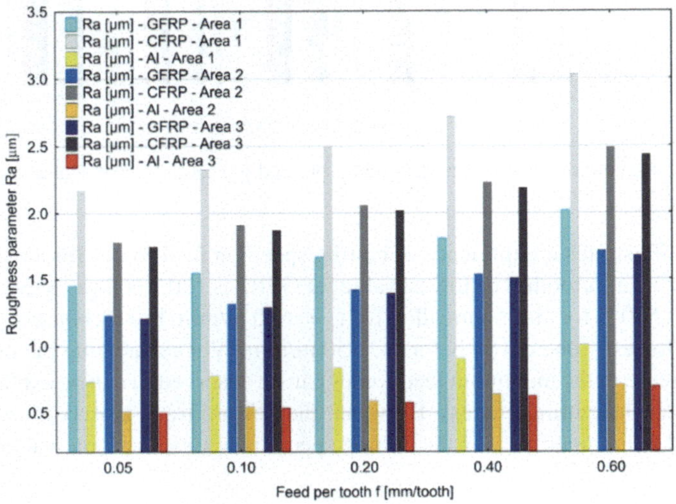

Fig. 4. Roughness parameter depending on feed per tooth in three cutting zones.

Figure 5 shows the impact of cutting speed on the Ra surface roughness parameter at a feed per tooth of $f_z = 0.2$ mm/tooth for three materials divided into machining zones. A decrease in Ra roughness values with increasing cutting speed is noticeable for each of the materials in all cutting zones. The highest roughness values are observed in the first zone. This situation can be interpreted as the tool's impact on the workpiece and the introduction of slight vibrations in the workpiece, which translates into higher Ra roughness values. In the middle zone, i.e., the one in which the tool operates at full circumference, the roughness values are lower. In the last zone, the Ra roughness values are still lower, but the difference between the second and third zones is not as significant as the difference between the first and second zones. The low roughness numbers were recorded during the milling of aluminum alloy in the third zone, and at a cutting speed of $v_c = 600$ m/min, the roughness value was 0.40 μm.

The paper also presents sample photos and surface topographies. Figure 6a shows the microstructure of CFRP after milling at 100x magnification. The characteristic structure of CFRP is visible – carbon fibers in a polymer matrix, which immediately suggests that

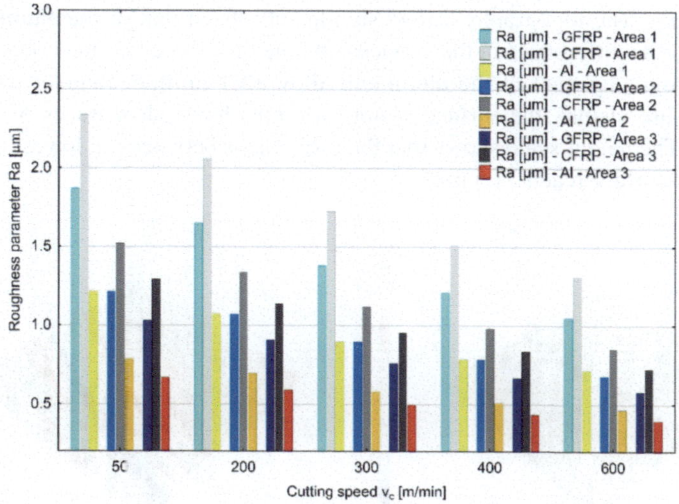

Fig. 5. Roughness parameter depending on cutting speed in three cutting zones.

it is a non-uniform material. The arrangement of the prepregs, alternating at an angle of 90°, is also visible. Figure 6b shows the microstructure of the tested aluminum alloy after milling at 100x magnification. It is characterized by elongated, parallel machining marks caused by the cutting tool's impact on the material surface. It is a material with a uniform structure. The structure also shows grain boundaries and local clusters of small surface discontinuities, which may be the result of microcracks or detached fragments of material caused by the impact of the milling cutter tooth. The image clearly shows that this is a material with a uniform structure.

Fig. 6. Microstructure after milling of CFRP (a) and aluminum alloy (b) at 100x magnification.

Figure 7a shows a map of the CFRP topography after milling. The surface is significantly more irregular than that of the aluminum alloy (Fig. 7b), resulting from the heterogeneous structure of the composite. There are noticeable higher points above 30 µm and several deeper grooves, but these are not regular features. The range of values reaches 35 µm. The topography differs significantly from that of the aluminum alloy, which shows the differences in the surfaces obtained for these two materials. Figure 7b shows a topography map of the aluminum alloy after milling. Regular traces of the cutting tool are visible, the surface is not even but shows clear traces of subtractive machining. The height scale shows that the differences between the lowest and highest points of the surface reach 21.5 µm.

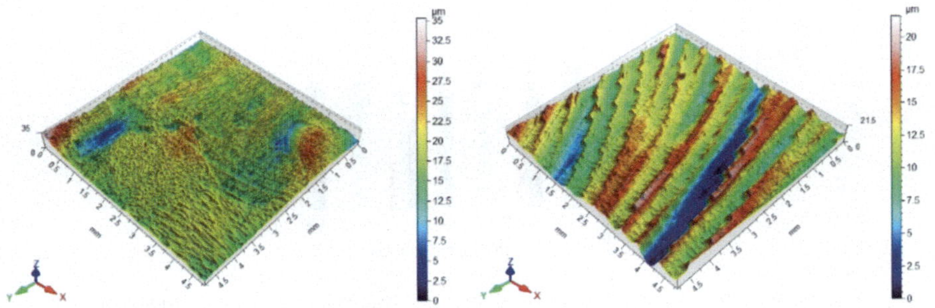

Fig. 7. 3D Topography after milling of CFRP (a) and aluminum alloy (b) at 100x magnification.

4 Conclusions

The article presents an analysis of machinability indices for milling GFRP, CFRP, and aluminum alloy. The materials were milled with a 12 mm diameter milling cutter, in which two inserts were mounted.

The results of the analyses allow the following conclusions to be presented:

1. The highest cutting forces, depending on the feed per tooth and cutting speed, occur in the first cutting zone, while in the last zone the maximum forces are lowest.
2. An increase in feed per tooth causes an increase, while an increase in cutting speed causes a decrease in roughness values for all tested materials. The highest roughness values occur in the tool entry zone for all tested materials, which is related to the disruption of the fiber structure during initial contact with the tool in the case of composites, while in the exit zone for aluminum alloy, the roughness parameters are noticeably lowest, which may be due to the smoothing effect of the tool and the plasticity of the material.
3. Analysis of machinability indicators allows for real optimization of the process through the selection of appropriate parameters.

The research results and their analysis indicate the need to continue work in the field of machining polymer composites and aluminum alloys. These materials are used in many industries and have specific properties. Future work will focus on studying the cutting tools. The authors also intend to perform surface analysis using a scanning microscope.

References

1. Vinayagamoorthy, R.: A review on the machining of fiber-reinforced polymeric laminates. J. Reinf. Plast. Compos. **37**, 49–59 (2018). https://doi.org/10.1177/0731684417731530
2. Matuszak, J., Kłonica, M., Zagórski, I.: Measurements of forces and selected surface layer properties of AW-7075 aluminum alloy used in the aviation industry after abrasive machining. Materials. **12**, 3707 (2019). https://doi.org/10.3390/ma12223707
3. Caggiano, A.: Machining of fibre reinforced plastic composite materials. Materials. **11**, 442 (2018). https://doi.org/10.3390/ma11030442
4. Yuanyushkin, A.S., Rychkov, D.A., Lobanov, D.V.: Surface quality of the fiberglass composite material after milling. Appl. Mech. Mater. **682**, 183–187 (2014). https://doi.org/10.4028/www.scientific.net/AMM.682.183
5. Kiliçkap, E., Yardimeden, A., Çelik, Y.H.: Investigation of experimental study of end milling of CFRP composite. Sci. Eng. Compos. Mater. **22**, 89–95 (2015). https://doi.org/10.1515/secm-2013-0143
6. Azmi, A.I., Lin, R.J.T., Bhattacharyya, D.: Machinability study of glass fibre-reinforced polymer composites during end milling. Int. J. Adv. Manuf. Technol. **64**, 247–261 (2013). https://doi.org/10.1007/s00170-012-4006-6
7. Davim, J.P., Reis, P.: Damage and dimensional precision on milling carbon fiber-reinforced plastics using design experiments. J. Mater. Process. Technol. **160**, 160–167 (2005). https://doi.org/10.1016/j.jmatprotec.2004.06.003
8. Newman, B., Creighton, C., Henderson, L.C., Stojcevski, F.: A review of milled carbon fibres in composite materials. Compos. Part A Appl. Sci. Manuf. **163**, 107249 (2022). https://doi.org/10.1016/j.compositesa.2022.107249
9. Kumar, D., Gururaja, S.: Machining damage and surface integrity evaluation during milling of UD-CFRP laminates: dry vs. cryogenic. Compos. Struct. **247**, 112504 (2020). https://doi.org/10.1016/j.compstruct.2020.112504
10. Karpat, Y., Polat, N.: Mechanistic force modeling for milling of carbon fiber reinforced polymers with double helix tools. CIRP Ann. **62**, 95–98 (2013). https://doi.org/10.1016/j.cirp.2013.03.105
11. Ghidossi, P., El Mansori, M., Pierron, F.: Edge machining effects on the failure of polymer matrix composite coupons. Compos. Part A Appl. Sci. Manuf. **35**, 989–999 (2004). https://doi.org/10.1016/j.compositesa.2004.01.015
12. Teicher, U., Rosenbaum, T., Nestler, A., Brosius, A.: Characterization of the surface roughness of milled carbon fiber reinforced plastic structures. Procedia CIRP. **66**, 199–203 (2017). https://doi.org/10.1016/j.procir.2017.03.282
13. Hintze, W., Hartmann, D.: Modeling of delamination during milling of unidirectional CFRP. Procedia CIRP. **8**, 444–449 (2013). https://doi.org/10.1016/j.procir.2013.06.131
14. Ciecieląg, K., Zaleski, K., Kęcik, K.: Effect of milling parameters on the formation of surface defects in polymer composites. Mater. Sci. **57**, 882–893 (2022). https://doi.org/10.1007/s11003-022-00622-w
15. Ciecieląg, K.: Analysis of the surface layer and feed force after milling polymer composites with coated and uncoated tools. Adv. Sci. Technol. Res. J. **17**, 70–78 (2023). https://doi.org/10.12913/22998624/168832
16. Ciecieląg, K.: Research on the main and side surfaces after face milling of polymer composites. Adv. Sci. Technol. Res. J. **19**, 17–29 (2025). https://doi.org/10.12913/22998624/208651
17. Palanikumar, K., Karunamoorthy, L., Karthikeyan, R.: Assessment of factors influencing surface roughness on the machining of glass fiber-reinforced polymer composites. Mater. Des. **27**, 862–871 (2006). https://doi.org/10.1016/j.matdes.2005.03.011

18. Nurhaniza, M., Ariffin, M.K.A.M., Mustapha, F., Baharudin, B.T.H.T.: Analyzing the effect of machining parameters setting to the surface roughness during end milling of CFRP-aluminium composite laminates. Int. J. Manuf. Eng. **2016**, 1–9 (2016). https://doi.org/10.1155/2016/4680380
19. Doluk, E., Rudawska, A.: Effect of machining settings and tool geometry on surface quality after machining of Al/CFRP sandwich structures. Adv. Sci. Technol. Res. J. **16**, 22–33 (2022). https://doi.org/10.12913/22998624/147787
20. Kuo, C., Liu, J., Chang, T., Ko, S.: The effects of cutting conditions and tool geometry on mechanics, tool wear and machined surface integrity when routing CFRP composites. J. Manuf. Process. **64**, 113–129 (2021). https://doi.org/10.1016/j.jmapro.2021.01.011
21. Zaleski, K., Skoczylas, A., Ciecielag, K.: The investigations of the surface layer properties of C45 steel after plasma cutting and centrifugal shot peening. In: Królczyk, G.M., Niesłony, P., Królczyk, J. (eds.) Industrial Measurements in Machining, pp. 172–185. Springer International Publishing, Cham (2020). https://doi.org/10.1007/978-3-030-49910-5_16
22. Matuszak, J., Zaleski, K., Ciecieląg, K., Skoczylas, A.: Analysis of the effectiveness of removing surface defects by brushing. Materials. **15**, 7833 (2022). https://doi.org/10.3390/ma15217833
23. Unnikrishna Pillai, J., Sanghrajka, I., Shunmugavel, M., Muthuramalingam, T., Goldberg, M., Littlefair, G.: Optimisation of multiple response characteristics on end milling of aluminium alloy using Taguchi-Grey relational approach. Measurement. **124**, 291–298 (2018). https://doi.org/10.1016/j.measurement.2018.04.052
24. Matuszak, J.: Comparative analysis of the effect of machining with wire and ceramic brushes on selected properties of the surface layer of EN AW-7075 aluminium alloy. Adv. Sci. Technol. Res. J. **16**, 50–56 (2022). https://doi.org/10.12913/22998624/146211
25. Pimenov, D.Y., et al.: Review of improvement of machinability and surface integrity in machining on aluminum alloys. Int. J. Adv. Manuf. Technol. **129**, 4743–4779 (2023). https://doi.org/10.1007/s00170-023-12630-4
26. Matuszak, J., Kawalec, A., Gdula, M.: Analysis of the deburring efficiency of EN-AW 7075 aluminum alloy parts with complex geometric shapes considering the tool path strategy during multi-axis brushing. Materials. **17**, 6267 (2024). https://doi.org/10.3390/ma17246267
27. Srivastava, V.S., Gupta, T.K., Srivastava, A.K., Chauhan, S., Chauhan, P.K.: Effects of cutting parameters on aluminium alloys-a review. Mater. Today Proc. **47**, 3823–3827 (2021). https://doi.org/10.1016/j.matpr.2021.03.091
28. Rahman, M.A., et al.: Influence of feed rate response (FRR) on chip formation in Micro and macro machining of Al alloy. Meta. **11**, 159 (2021). https://doi.org/10.3390/met11010159
29. Skoczylas, A.: Influence of centrifugal shot peening parameters on the impact force and surface roughness of EN-AW 2024 aluminum alloy elements. Adv. Sci. Technol. Res. J. **15**, 71–78 (2021). https://doi.org/10.12913/22998624/130511
30. Teti, R.: Machining of composite materials. CIRP Ann. **51**, 611–634 (2002). https://doi.org/10.1016/S0007-8506(07)61703-X

The Fatigue Life of Ball Burnished Objects Immersed in Oil

Agnieszka Skoczylas(✉), Jakub Matuszak, Krzysztof Ciecieląg, and Kazimierz Zaleski

Lublin University of Technology, Mechanical Department, Nadbystrzycka str. 36, 20-618 Lublin, Poland

a.skoczylas@pollub.pl

Abstract. When designing machine and equipment components, one of the requirements is to ensure sufficient durability of these components. A process that increases the strength and fatigue life (FL) of machine components is ball burnishing. The subject of the research were of C45 steel samples, which were subjected to a ball burnishing (BB). The samples were pre-ground and then burnished using the following parameters: force F = 400 N, burnishing feed f = 0.1 mm/rev. and speed v_b = 0.62 m/min. As a result of BB, the surface topography changed in relation to the pre-treatment. The 3D surface roughness parameters are lower, with a maximum reduction of 82%. After BB, an increase in microhardness was achieved, with a hardened layer thickness of 50 μm. Compressive residual stresses (CRS) occur in the surface layer after BB, whose were 202% greater than after pretreatment. The fatigue life of ball burnished samples is 13% higher for oil than for air, and 8% higher for ground samples. The use of oil as a medium for fatigue testing limits crack propagation.

Keywords: Ball Burnishing · Surface Layer Properties · Fatigue Life

1 Introduction

One of the most important requirements for machines and devices is that they maintain sufficient durability. Machine components are very often exposed to variable loads, which is why their strength and fatigue life are investigated. It is particularly important to ensure the adequate fatigue strength of aircraft components to guarantee passenger safety and, at the same time, take into account the weight requirements of these components. These requirements justify the need for research involving the selection of materials and technology for aircraft components, which can be made from composite materials [1–4]. A serious problem is to ensure the required durability of aircraft engine compressor blades which are exposed to cyclic aerodynamic loads during flight [5]. Materials used in the automotive and railway industries are also subjected to fatigue testing, as components made from these materials are susceptible to variable loads during operation [6, 7]. Examples of components exposed to variable forces, widely used in means of transport and various machines and devices, include shafts and screw connections, whose fatigue strength was studied in [8–10].

B. Gapiński et al. (Eds.): MANUFACTURING 2026, LNME, pp. 13–25, 2026.
https://doi.org/10.1007/978-3-032-22943-4_2

Components subjected to variable loads during operation operate in different environments. Therefore, to obtain reliable results, it is necessary to conduct fatigue tests in the same environments. Many studies report the results of strength and fatigue life tests conducted on samples exposed to air [11–13]. Fatigue tests conducted in corrosive environments showed that these environments had an adverse effect on the fatigue life of the tested samples [14, 15]. Numerous machine and equipment components operate in an oil environment. The fatigue testing of AZ91D and AM50 magnesium alloys in air, mineral oil and gear oil conducted by Eliezer et al. showed that the service life of both alloys was longer in mineral oil than in gear oil, with the AZ91D alloy showing a greater sensitivity to the environment [16]. Omidvar and Ghorbanpoor found that lubricating oil seepage could accelerate the development of fatigue cracks in rolling-sliding systems [17]. The shape of the oil supply channel also affects lubrication conditions. The work of Wang et al. found that the fatigue life of diesel engine pistons depended on the shape and position of the cooling channel through which the oil would flow [18]. Xie et al. investigated the causes of fatigue damage in a valve in a paraffin oil cracking plant due to vibrations generated by a pump [19]. Li et al. compared the results of fatigue life tests for helical gears after honing, shaving and polishing, operated under mixed lubrication conditions [20]. Kagami et al. studies the fatigue properties of JIS SCM415 samples in diesel fuel and also in air, finding that deterioration of diesel fuel quality led to reduced fatigue strength [21].

The fatigue properties of machine components can be improved by modifying their surface layer [22, 23]. A frequently used method to shape properties of surface and thereby improve the fatigue life of workpieces is burnishing [24–26]. In terms of fatigue properties, a positive effect of burnishing is reduced surface roughness, the formation of CRS, and increased hardness of the surface layer [27–29]. A study [30] demonstrated that the condition of the surface layer of stainless steel after slide burnishing could be assessed in a non-destructive way using the average lifetime of positrons. Research [31] showed that after burnishing, the surface layer exhibited better properties than after milling and polishing. Ball burnishing made on the CNC machines is successfully used in injection molds, dies or stamping dies production. The ball burnishing is often made after milling, which allows for the deformation of micro-irregularities after the pre-treatment, which is very important in the production of moulds [31]. Depending on the manner in which the tool affects the surface of the workpiece, can be distinguished slide burnishing [32, 33] and ball burnishing [34, 35]. It is also possible to use slide roller burnishing [36]. The surface layer properties depend largely on the technological parameters of burnishing, primarily burnishing force and the feed [37, 38]. Achieving a good surface quality after burnishing requires the use of highly rigid tools [39]. Burnishing can be carried out either dry or with the use of a processing fluid. A study [40] found that the type of liquid used in the burnishing process affected the surface texture, microhardness and free surface energy. Research presented in [41] indicated that the use of variable feed in multi-pass burnishing allowed for low surface roughness while maintaining good efficiency. Another study [42] showed that surface roughness and microhardness of the surface layer after BB depended on the angle between the direction of the traces left by the preceding machining and the direction of burnishing.

The studies analyzed in the literature review shows that the burnishing process shapes properties of the surface layer, which has a positive effect on the fatigue life of machined components. Prior research has also showed that fatigue properties depend on the medium in which the test object is tested. In the analysed works, the investigations into the effect of burnishing elements made on C45 steel on strength and fatigue life were conducted in air. Considering that machine components exposed to variable loads often operate in an oil environment, this paper presents the results of a study aimed at comparing the effect of BB conducted in air and in oil on the fatigue life of tested elements.

2 Research Methodology

The research was conducted on samples of C45 steel. This material is used for manufacturing shafts, spindles, discs, and levers. C45 steel is quite difficult to weld due to the occurrence of cracks in the heat-affected zone. The material has the following properties: $R_e = 430$ MPa, $R_m = 740$ MPa, hardness: approx. 225 HV. The C45 steel samples were in the form of rings with the following dimensions: external $D_e = 56$ mm, internal $D_i = 50$ mm, width $b = 10$ mm. Grinding (G), as pre-treatment, was performed on a cylindrical grinder using a Norton Saint-Gobain electrocorundum grinding wheel of 8A60M7VS3 type, having the dimensions 400x50 mm. The following grinding parameters were used: $a_p = 0.01$ mm and $v_g = 35$ m/s. In the next stage, the thin-walled C45 steel samples were subjected to ball burnishing (BB). A universal lathe C11/MB made in Bulgaria was used to BB. The thin-walled rings were attached on a mandrel that rotated with a speed of *n*, while the burnishing tool performed a feed *f*. The tool used for processing was a burnisher consisting of a mechanism for exerting pressure and. a silicon nitride Si_3N_4 with a diameter of $d_k = 8$ mm. During BB, the following technological parameters were used: force $F = 400$ N, feed $f = 0.1$ mm/rev, speed $v_b = 0.62$ m/s. BB was performed in the presence of synthetic oil with the trade name Mobile Vactra TM Oil No. 2. These parameters are dedicated to ball burning of C45 steel. Figure 1 shows the stations and equipment used in this study.

EN ISO 25178-2 2022 specifies the conditions for measuring surface topography and 3D surface roughness parameters. A T8000 RC120–400 device was used for measurements. Surface roughness was measured along the cylindrical outer surface of the specimens. A 3x3 mm area was scanned, for which the following parameters were then determined: Sa, Sz, Sp, Sv, and Smr at cut-off levels of $c = 35\%$ and $c = 60\%$. Microhardness was measured on angled sections after standard machining.. The use of angled sections enables microhardness measurements at smaller depths. The Vickers method was employed with a LM 700at tester. The indenter load was 50 g (HV0.05). A portable Theta-Theta EDGE diffractometer was used to measure the residual stresses by use d X-ray diffraction method. The source of the XRD beam was a chromium lamp. A filter made from V and a 0.5 mm collimator were used during the measurements. All measurements were performed in one direction, and the exposure time of the sample surface to the XRD beam was 30 s. The schematic diagram of a rig for testing the fatigue life of ring-shaped samples was presented on Fig. 2 During the fatigue life tests, the samples were subjected deformation in order to reduce their diameter. Before the tests, the

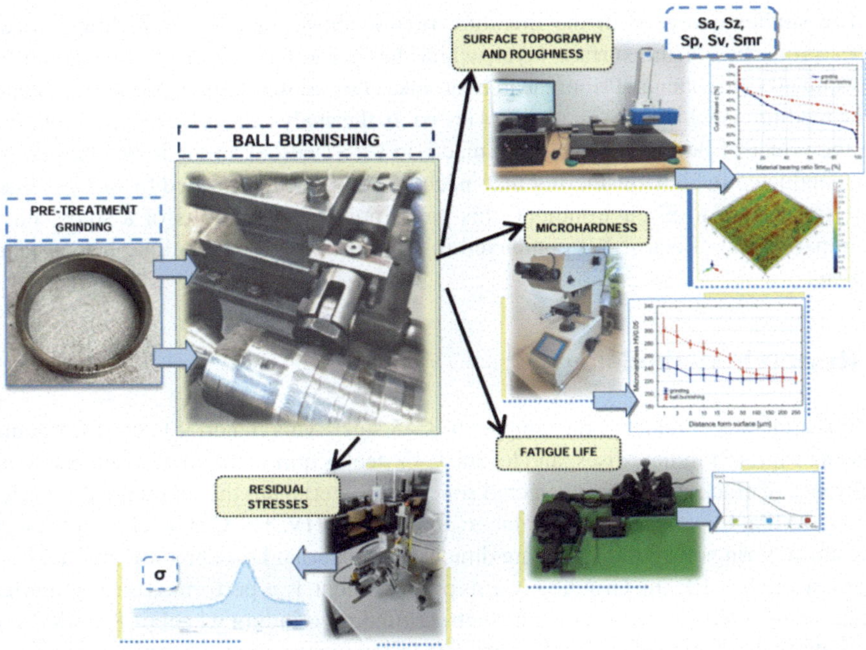

Fig. 1. Stations and equipment used in the study

specimens were cut twice along the generatrix to remove a fragment thereof. A fixed holder (1) and a movable holder (3) were used to attach a sample (2). After fixing the sample (2), a disc (7), in which a pin (6) is located at a distance c from the axis, was set in rotational motion. A connecting rod (5), mounted on the pin (6), was moved. The displacement of the connecting rod (5) was transferred to the sample (2) by means of a lever (4) and the movable holder (3), which caused cyclic deformation of the sample. Fatigue life was measured by the number of deformations of the sample until it broke. Fatigue life tests were performed for two media: air and oil. During fatigue life testing in oil, an additional component was used in the system, namely a bath (8) in which the sample subjected to cyclic loading was immersed.

A statistical analysis was performed for the fatigue life results. In the statistical analysis, was compared between two groups from independent samples. The fatigue life of ground and burnished specimens was compared when fatigue tests were made in air and oil (the effect of the type of pre-treatment was examined), and the fatigue life of ground samples and ball burnished samples for which fatigue tests were performed in air and oil (the type of medium was examined) was compared. The first step in the statistical analysis was normality of the distribution using the Shapiro-Wilk test. If the distribution of the variable under study was normal, a parametric test was used. Next, the equality of variance was analysed (using Levene's test), which influenced the choice of the test to verifying the assumption of equality of means (T-student test or Cochran-Cox adjustment). If the distribution under study was not normal, a non-parametric test was used (U Manna-Whitney test). Results and discussion

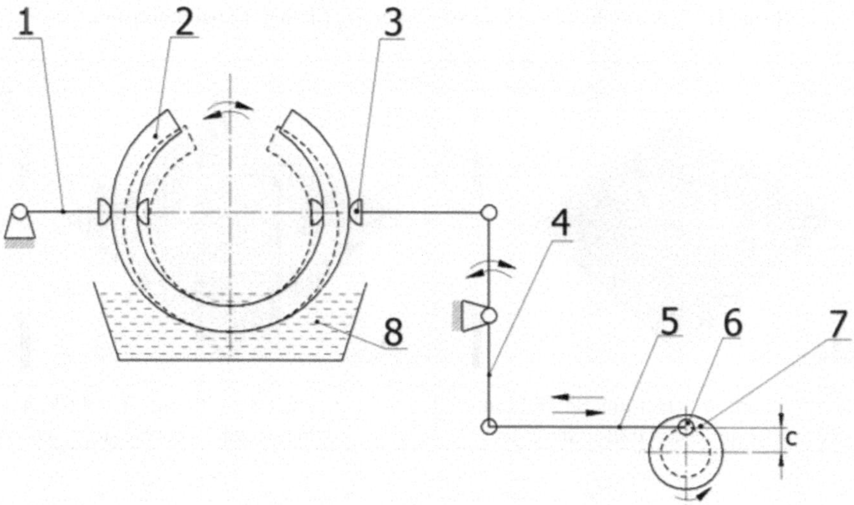

Fig. 2. Design of a rig for testing the fatigue life of ring samples (1 – fixed holder, 2 – sample, 3 – movable holder, 4 – lever, 5 – connecting rod, 6 – pin, 7 – disc, 8 – bath)

2.1 Surface Topography and Roughness

Surface topography after pre-treatment (grinding) (Table 1a) and after ball burnishing (Table 1b) was presented. A unidirectional, straight-line of micro-irregularities are visible on the ground surface. The resulting micro-irregularities are evenly spaced. The ratio of peaks is greater than that of valleys in the maximum height of micro-irregularities. The peaks of micro-irregularities are sharp. Areas with a higher degree of deformation (marked with a black circle) are sometimes visible in the scanned area, and they are most likely due to pre-treatment instability. After ball burnishing, the surface topography is rebuilt. The peaks of micro-irregularities are flattened, their height is reduced, but they are still visible on the surface. The burnished surface also shows the presence of considerably deformed areas (marked with a black circle), which confirms the occurrence of intense deformation. After ball burnishing, the skewness coefficient (Ssk parameter) has a negative value, which means that the material is concentrated near the peaks of the profile. The Ssk coefficient with negative value indicates a greater ability of the material to transfer contact loads across surfaces and lower tribological wear.

The material bearing curve, also known as the Abbott-Firestone curve (AFC curve), reflects changes in micro-irregularities (Fig. 3). After pre-treatment, a degressive-progressive curve with a large angle of inclination is obtained. Ball burnishing does not change the AFC curve; it remains a degressive-progressive curve, only its angle of inclination changes. After ball burnishing, the AFC curve is flattened and the Smr parameter is increased. For the ball burnished sample, Smr at the cut-off level c = 60% is $Smr_{(c=60\%)} = 94.65\%$, while for the pre-treated sample it is $Smr_{(c=60\%)} = 47.71\%$.

Figures 4 and 5 show the parameters Sa, Sz (Fig. 4) and Sp, Sv (Fig. 5) after pre-treatment and ball burnishing. The parameters are lower after BB than after pre-treatment. The Sa parameter had the greatest changes in the analysed 3D parameters. Its value is

Table 1. Topography of the C45 steel after a) grinding, b) ball burnishing

a)	b)
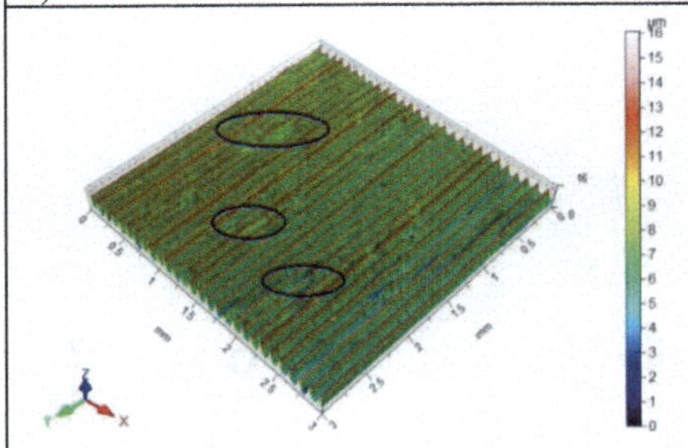	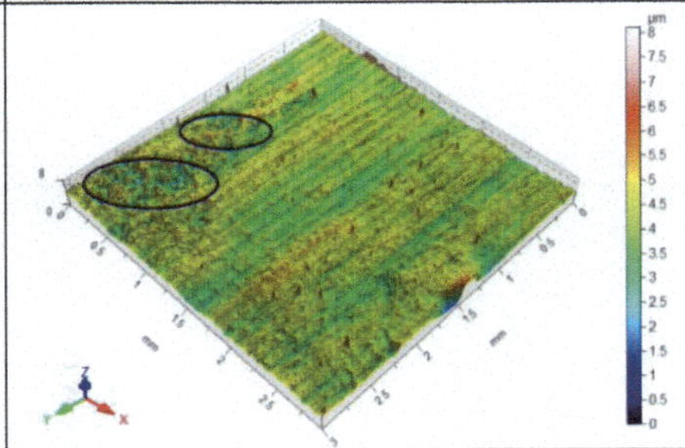
Sa = 2.34 μm; Sz = 16.1 μm; Sp =9.15 μm, Sv = 6.91 μm; Ssk = 0.615; Sku = 2.41	Sa = 0.46 μm; Sz = 8.10 μm; Sp = 4.03 μm, Sv = 4.06 μm; Ssk = -0.49; Sku = 5.11

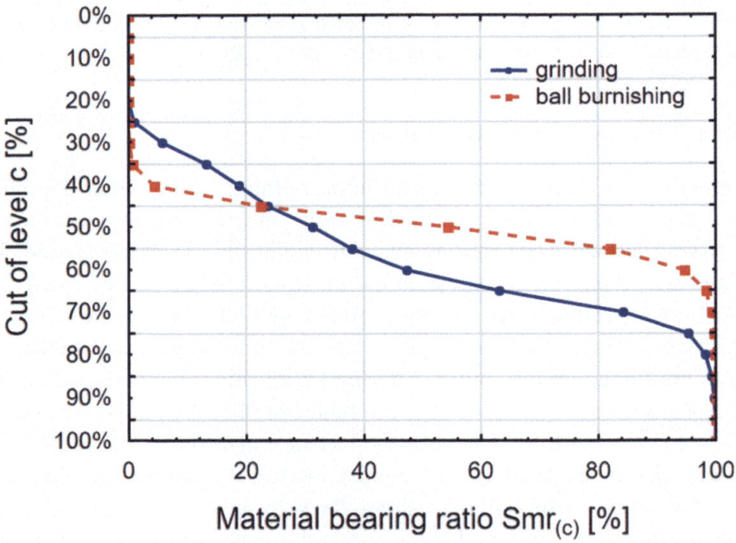

grinding

$Smr_{(c=35\%)} = 13.18\%$

$Smr_{(c=60\%)} = 47.71\%$

ball burnishing

$Smr_{(c=35\%)} = 0.63\%$

$Smr_{(c=60\%)} = 94.65\%$

Fig. 3. AFC curve after grinding and BB

82% lower after ball burnishing than after pre-treatment. As a result of the contact between the ball and the machined surface, the micro-irregularities induced by grinding are deformed, and the maximum surface height is reduced (the Sz is 49% lower than after grinding). After ball burnishing, one can observe the predominance of valleys (the Sv is 9% higher than the Sp). After pre-treatment, the surface is dominated by peaks, and the Sp is 31% higher than the Sv. After ball burnishing, the Sp parameter is 57% lower and

Sv is 39% lower compared to their pre-treatment values. The micro-irregularities pattern (surface topography) obtained after BB is like in our earlier work [28], with the values of 3D parameters being higher, but lower than reported in [43], where the ASI 1045 steel was ball burnished using variable burnishing speed, feed and burnishing force.

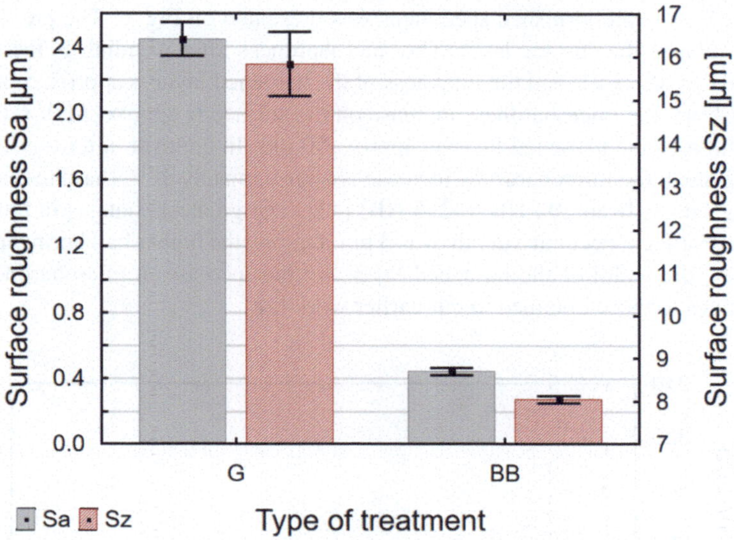

Fig. 4. Parameters Sa and Sz after pre-treatment and ball burnishing

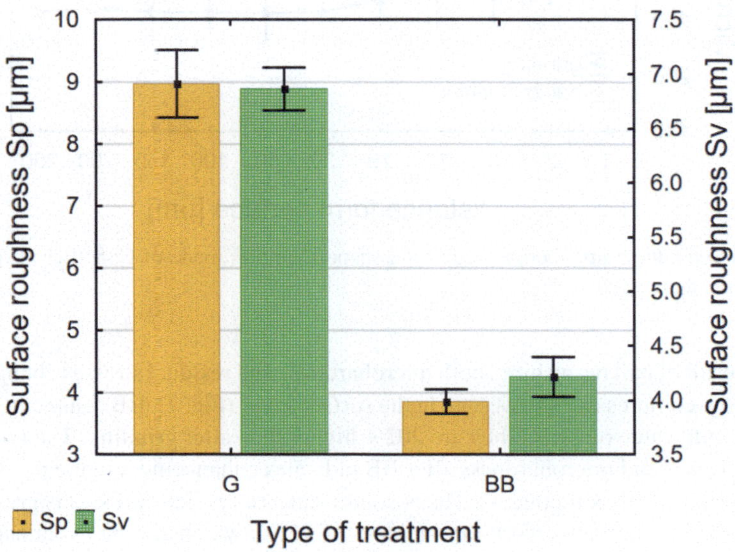

Fig. 5. Parameters Sp and Sv after pre-treatment and ball burnishing

2.2 Microhardness and Residual Stresses

The effect of the occurring contact between the ball and the surface, the material undergoes deformation. This causes microstructural changes and promotes an increase in the dislocation density, which leads to changes in the microhardness of the surface layer. The distribution of microhardness as a function of the distance from the surface for ground (pre-treated) and ball burnished specimens was presented on Fig. 6. The pre-treatment itself strengthens the surface layer. The microhardness after grinding is increased to approx. ΔHV 0.05 = 21, and the thickness of the hardened layer is approx. 5 μm. After ball burnishing, the microhardness increase on the surface is approx. ΔHV 0.05 $\approx$ 54, and the width of the hardened layer is approx. 50 μm. In contrast, after the BB of the AISI 1045 steel, the microhardness increase is approximately 14% (hardness after ball burnishing ranges from 202 HB to 236 HB) [43]. The greatest changes in the surface layer microhardness occur at a depth of 1–3 μm. Despite the higher ball burnishing force F = 400 N, the width of the hardened layer and the increase in microhardness have similar values to those obtained in our earlier work [28].

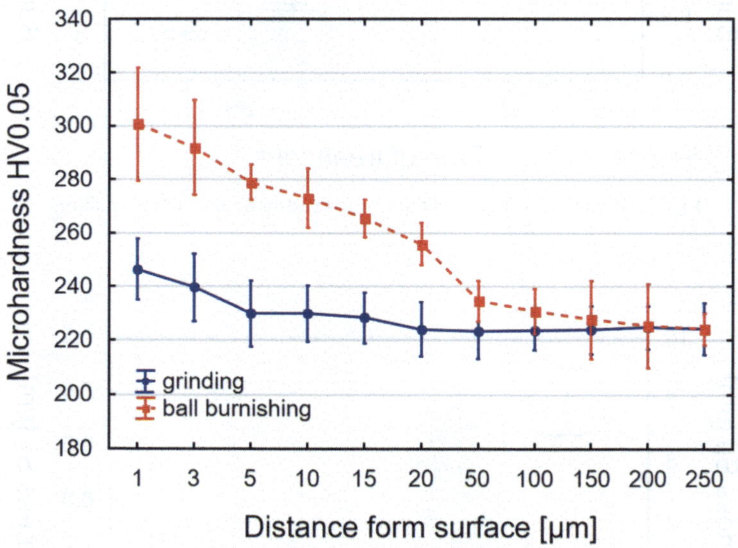

Fig. 6. Microhardness distribution in the ring-shaped samples made of C45 after G (grinding) and ball burnishing (BB)

As a result of ball burnishing, both microhardness and residual stresses change. After the grinding, compressive stresses are in the surface layer (Fig. 7). BB causes an increase in the absolute value of CRS. They are 202% higher than after grinding. The increase in residual stresses and microhardness after BB indicates enhancement in the performance properties of the C45 components. The absolute value of residual stresses is greater than that reported in a previous work (an increase of 11%), which can be explained by the use of a bigger ball burnishing force than in [28].

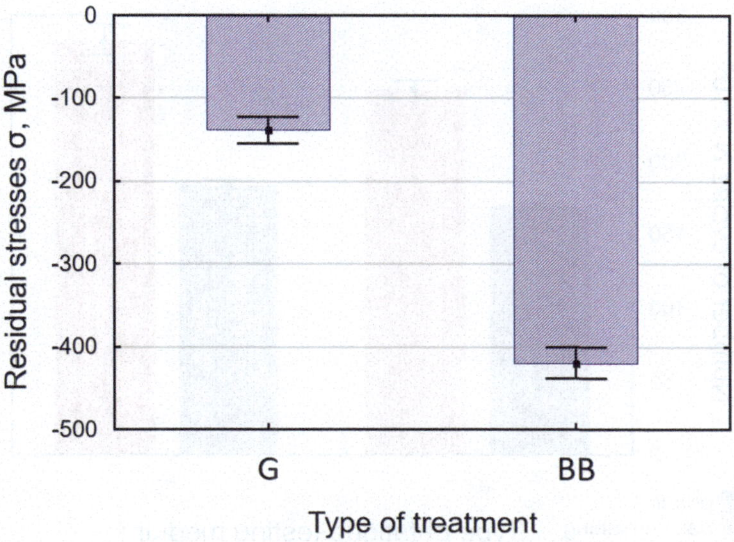

Fig. 7. Surface residual stresses for the ring-shaped samples made of C45 after G (grinding) and ball burnishing (BB)

2.3 Fatigue Life

On Fig. 8 bee seen the impact of the testing environment on the FL of samples after pre-treatment and ball burnishing. Whatever the fatigue testing medium, the strength of the ball burnished samples is bigger than that of the ground samples. For fatigue tests performed in air, the strength of ball the burnished samples is 46.3% higher, while for tests in oil, it is 53.3% higher than the values for the pre-treated samples. The higher number of cycles to failure for the ball burnished specimens is related to their microhardness, the depth of microhardness changes, and residual stresses. For the ball burnished samples, these indicators were larger than those after grinding. Regarding the medium versus fatigue life relationship, the ground and burnished samples showed a longer fatigue life for oil than for air, i.e it was 8% longer for the ground samples and 13% longer for the ball burnished samples. The fatigue life differences can be explained by the fact that Mobil Vactra No. 2 oil prevents the propagation of fatigue cracks. During fatigue testing, the oil flows into the crack, lubricating the resulting fracture and thus slowing down the propagation of the crack in the sample. Similar results to those obtained in our study were obtained in [16], but they differ from those reported in [21], where a decreased fatigue life was obtained along with a reduction in diesel oil quality.

In all cases, the analysed dependent variable (fatigue life) has a normal distribution and its variances are homogeneous. This allowed the use of the Student's t-test to examine the The analysis (Table 2) shows that the mean fatigue life values for ground and ball-burnished samples after fatigue tests performed in oil and air differ significantly.

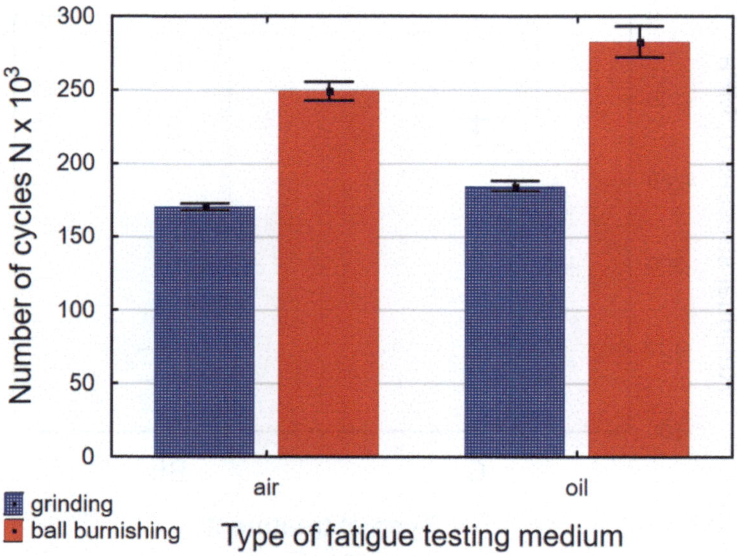

Fig. 8. Influence of the research environment on the fatigue life of C45 ring samples after grinding (G) and ball burnishing (BB)

Table 2. Comparison of mean values using the T-student test (p – probability, α – significance level)

Parameter	Variable factor	p	Conclusions
Fatigue life – air	treatment	0.00000	$p < \alpha$
Fatigue life – oil	treatment	0.00000	$p < \alpha$
Fatigue life – grinding samples	medium	0.00010	$p < \alpha$
Fatigue life – ball burnished samples	medium	0.00025	$p < \alpha$

3 Conclusion

Ring specimens from C45, which were used in the study, were subjected to pre-treatment (grinding) first and then to ball burnishing. The properties of the surface layer were examined, and then the samples were subjected to fatigue testing using two different media (air and oil). The next conclusions summarize the results of the work performed on C45 specimens in terms of the parameters applied in the experiment.:

- ball burnishing changes the ratio of peaks and valleys in the maximum height of micro-irregularities, the Sa, Sz, Sp and Sv are lower than after the pre-treatment (up to 82% lower),
- after ball burnishing, the AFC curve still is a degressive-progressive; only its angle of inclination changes,

- ball burnishing causes an increase in microhardness (strengthening degree e = 8–22%), and the maximum thickness of the strengthened layer is approx. 50 μm,
- after BB, CRS occur in the surface layer and their absolute value is 202% greater than after grinding,
- the fatigue life is greater by 13% (ball burnished samples) and 8% (ground samples) for oil than for air, and the use of oil as a fatigue testing medium constrains the propagation of cracks.

References

1. Spak, B., Kästner, M.: Numerical and experimental fatigue strength evaluation of thin walled aluminum components for aerospace application manufactured by roll forming. Procedia Struct. Integr. **57**, 445–451 (2024)
2. Fan, J., Wang, Z., Liu, C., Shi, D.A.: Tensile properties-related fatigue strength predicted machine learning framework for alloys used in aerospace. Eng. Fract. Mech. **301**, 110057 (2024)
3. Ciecieląg, K.: Evaluation of machinability indicators in milling of thin-walled composite structures. Adv. Sci. Technol. Res. J. **18**(1), 202–212 (2024)
4. Ciecieląg, K.: Influence of the length of components from polymer composite on selected machinability indicators in the circumferential milling process. Adv. Sci. Technol. Res. J. **14**(4), 229–239 (2020)
5. Sun, S. et al.: Predicting high-cycle fatigue strength and three-dimensional fatigue crack growth in simulated compressor blade by phase-field model. Aerosp. Sci. Technol. **159**, 110009 (2025)
6. Chaurasiya, R., Maji, P., Mukhopadhyay, G.: High cycle fatigue behaviour of advanced high strength steel sheet (HS1000) in automotive application. Mater. Today Proc., available online (2023)
7. Zhao, H., Han, R.-P., Gao, J.-W., Zhu, S.-P., Han, J.: Fatigue performance evaluation of high-strength railway axles subjected to different surface defects. Alex. Eng. J. **127**, 66–74 (2025)
8. Vetter, S., Leidich, E., Becker, N., Schlecht, B., Hasse, A.: Probabilistic method to estimate the scatter of the fatigue strength of shafts in the HCF region. Procedia Struct. Integr. **37**, 746–754 (2022)
9. Guo, S., Ji, X., Wang, G., Liu, X., Liu, H., Lei, H.: A novel structural stress method and stress intensity factor solutions for fatigue analysis of high-strength bolted connections in bolted spherical joints. Eng. Fract. Mech. **321**, 111138 (2025)
10. Jiao, J., Du, H., Deng, J., Chen, P., Lu, G.: Experimental and theoretical investigation of the high-cycle fatigue failure mechanism of M24 high-strength bolts. J. Constr. Steel Res. **230**, 109560 (2025)
11. Matuszak, J., Zaleski, K., Zyśko, A.: Investigation of the impact of high-speed machining in the milling process of titanium alloy on tool wear, surface layer properties, and fatigue life of the machined object. Materials. **16**, 5361 (2023)
12. Wang, P., et al.: The highest fatigue strength for steels. Acta Mater. **289**, 120888 (2025)
13. Zaleski, K.: The effect of vibratory and rotational shot peening and wear on fatigue life of steel. Eksploat. Niezawodn.– Maint. Reliab. **19**(1), 102–107 (2017)
14. Khan, M.A., Singh, S.S.K., Abdullah, S., Arifin, A., Bashir, M.: Fatigue life characterisation of API X65 steel pipeline for internal vibrational loads under sea water condition. Appl. Ocean Res. **155**, 104454 (2025)

15. Otto, J.L., et al.: Fatigue performance and failure mechanisms of vacuum brazed AISI 304L joints with experimental nickel-based filler metals in air and corrosive environments. Int. J. Fatigue. **199**, 109082 (2025)
16. Eliezer, A., Medlinsky, O., Haddad, J., Ben-Hamu, G.: Corrosion fatigue behavior of magnesium alloys under oil environments. Mater. Sci. Eng. A. **477**(1-2), 129–136 (2008)
17. Omidvar, B., Ghorbanpoor, A.: The role of oil seepage in fatigue crack growth of lubricated wearing systems. Eng. Fract. Mech. **60**(2), 239–250 (1998)
18. Wang, B., Wang, D., Lei, J., Deng, X., Liu, Y., Yang, H.: Study on low-cycle fatigue life of diesel engine piston considering oil cooling gallery structure. Appl. Therm. Eng. **264**(1), 125382 (2025)
19. Xie, H., et al.: Failure analysis of weld cracking in wax oil cracking unit. Eng. Fail. Anal. **174**(1), 109467 (2025)
20. Li, R., Ma, J., Zhang, X., Shi, X., Sun, W., Lu, X.: Modeling and analysis of relative fatigue life under 3D mixed lubrication in marine helical gears. Tribol. Int. **211**, 110834 (2025)
21. Kagami, S., Ueno, A., Miyakawa, S., Miyamoto, N., Ishibashi, N., Guennec, B.: Fatigue properties of high-strength steel in diesel oil. Procedia Struct. Integr. **2**, 1738–1745 (2016)
22. Schneller, W., et al.: Fatigue strength assessment of additively manufactured metallic structures considering bulk and surface layer characteristics. Addit. Manuf. **40**, 101930 (2021)
23. Sayid, A.A., El-Kashif, E., Adly, M.A., Morsy, M.A., Abdelkawy, A.: Fatigue behavior of surfaced C45 steel. J. Eng. Appl. Sci. **68**, 16 (2021)
24. Zaleski, K., Skoczylas, A.: Effect of slide burnishing on the surface layer and fatigue life of titanium alloy parts. Adv. Mater. Sci. **19**(4), 35–45 (2019)
25. Zielecki, W., Bucior, M., Trzepiecinski, T., Ochał, K.: Effect of slide burnishing of shoulder fillets on the fatigue strength of X19NiCrMo4 steel shafts. Int. J. Adv. Manuf. Technol. **106**, 2583–2593 (2020)
26. Maximov, J., Duncheva, G.: Effects of diamond burnishing on surface integrity, fatigue, wear, and corrosion of metal components—review and perspectives. Int. J. Adv. Manuf. Technol. **139**, 4233–4267 (2025)
27. Kułakowska, A., Bohdal, Ł.: Surface characterization of carbon steel after rolling burnishing treatment. Meta. **14**, 31 (2024)
28. Skoczylas, A., Zaleski, K.: Study on the surface layer properties and fatigue life of a workpiece machined by centrifugal shot peening and burnishing. Materials. **15**, 6677 (2022)
29. Labuda, W., Wieczorska, A.: Possibility of using the hydrostatic burnishing process under marine conditions. Adv. Sci. Technol. Res. J. **18**(6), 189–205 (2024)
30. Skoczylas, A., Zaleski, K., Matuszak, J., Ciecieląg, K., Zaleski, R., Gorgol, M.: Influence of slide burnishing parameters on the surface layer properties of stainless steel and mean positron lifetime. Materials. **15**, 8131 (2022)
31. Kalisz, J., Zak, K., Wojciechowski, S., Gupta, M.K., Krolczyk, G.M.: Technological and tribological aspects of milling-burnishing process of complex surfaces. Tribol. Int. **155**, 106770 (2021)
32. Ichkova, M., Anastasov, K., Daskalova, P.: Governing factors – surface integrity correlations and optimizations of diamond burnishing process of chromium-nickel austenitic stainless steels of the 18/8 type. Int. J. Innov. Res. Sci. Stud. **8**(3), 4136–4149 (2025)
33. Korzynski, M., Dudek, K., Korzynska, K.: Effect of slide diamond burnishing on the surface layer of valve stems and the durability of the stem-graphite seal friction pair. Appl. Sci. **13**, 6392 (2023)
34. Kowalik, M., Trzepieciński, T., Kukiełka, L., Paszta, P., Maciąg, P., Legutko, S.: Experimental and numerical analysis of the depth of the strengthened layer on shafts resulting from roller burnishing with roller braking moment. Materials. **14**, 5844 (2021)
35. Swirad, S., Pawlus, P.: Effect of ball burnishing on fretting at elevated temperatures. Materials. **17**, 5960 (2024)

36. Duncheva, G., Maximov, J., Anchev, A., Dunchev, V., Anastasov, K., Argirov, Y.: Effect of roller burnishing and slide roller burnishing on fatigue strength of AISI 304 steel: comparative analysis. Meta. **14**, 710 (2024)
37. Tugay, I.O., Hosseinzadeh, A., Yapici, G.G.: Hardness and wear resistance of roller burnished 316L stainless steel. Mater. Today Proc. **47**, 2405–2409 (2021)
38. Dyl, T., Rydz, D., Szarek, A., Stradomski, G., Fik, J., Opydo, M.: The influence of slide burnishing on the technological quality of X2CrNiMo17-12-2 steel. Materials. **17**, 3403 (2024)
39. Maheshwari, A.S., Gawande, R.R.: Influence of specially designed high – stiffness ball burnishing tool on surface quality of titanium alloy. Mater. Today Proc. **4**, 1405–1413 (2017)
40. Skoczylas, A., Kłonica, M.: Selected properties of the surface layer of C45 steel samples after slide burnishing. Materials. **16**, 6513 (2023)
41. Kebede, F.T., Felho, C., Sztankovics, I.: Improving surface roughness of 42CrMo4 low alloy steel shafts by applying varying feed in the multi-pass slide burnishing process. Appl. Sci. **15**, 9063 (2025)
42. Skoczylas, A., Matuszak, J., Ciecieląg, K., Zaleski, K.: Analysis of selected surface layer properties after ball burnishing of samples cut with a laser parallel and perpendicular to the rolling direction. In: Gapiński, B., Ciszak, O., Ivanov, V., Machado, J.M. (eds.) Advances in Manufacturing IV. Volume 1 - Mechanical Engineering: Digitalization, Sustainability and Industry Applications Lecture Notes in Mechanical Engineering, pp. 138–164. Springer Nature Switzerland AG (2024)
43. Saldaña-Robles, A., Plascencia-Mora, H., Aguilera-Gómez, E., Saldaña-Robles, A., Marquez-Herrera, A., Diosdado-De la Peña, J.A.: Influence of ball-burnishing on roughness, hardness and corrosion resistance of AISI 1045 steel. Surf. Coat. Technol. **339**, 191–198 (2018)

The Influence of Controlled Atmosphere Brazing Process Parameters of the Charge Air Cooler on the Brazing Degree and the Amount of Flux Residues

Sławomir A. Nadolny[1,2(✉)], Adam Hamrol[1], and Michał Rogalewicz[1]

[1] Faculty of Mechanical Engineering, Poznan University of Technology, Poznań, Poland
nadolny.slawomir@gmail.com

[2] MAHLE Thermal and Fluid Systems Poland Sp. z.o.o., Ostrów Wielkopolski, Poland

Abstract. This article presents optimization studies of the controlled atmosphere brazing process used for aluminum charge air coolers. The aim of the research was to reduce the amount of flux residues while maintaining the required brazing degree, including the presence and quality of brazed joints. The study focused on two key process parameters: maximum brazing temperature and flux grammage. Their influence on the process outcome was evaluated using a full factorial experiment design. The results showed that increasing the maximum brazing temperature to 606 °C significantly improves the flowability of both the flux and the filler metal, enabling proper joint formation even at reduced flux grammage. Conversely, increasing the flux grammage above 3 g/m^2 does not significantly affect the brazing degree, while leading to a noticeable increase in amount of flux residues. Based on the findings, the optimal process parameters were defined as flux grammage of 3 g/m^2 and maximum brazing temperature of 606 °C. These settings allow for an up to 35% reduction in amount of flux residues and a 24% improvement in brazing degree. The results confirm the potential of technological improvement of the process through parameter control.

Keywords: Aluminum Heat Exchangers · Flux Grammage · Flux Level · Brazeability · RLM-DIC Method

1 Introduction

Heat exchangers play a key role in efficient thermal energy management across various industrial sectors [1], including the automotive industry, where they significantly affect vehicle performance, durability, and design [2]. Their application contributes to reduced energy consumption, improved cooling system efficiency, and compliance with increasingly stringent environmental regulations, such as CO_2 emission standards [3]. A specific example of aluminum heat exchangers is the charge air cooler used in turbocharged internal combustion engines. Its function is to cool the compressed air before it enters the combustion chamber, thereby increasing the oxygen concentration in the intake air, enhancing the combustion process. This leads to reduced fuel consumption, lower exhaust emissions, and increased engine power output [4].

B. Gapiński et al. (Eds.): MANUFACTURING 2026, LNME, pp. 26–41, 2026.
https://doi.org/10.1007/978-3-032-22943-4_3

The subject of this study is the technological process of aluminum charge air coolers used in the automotive industry. For their reliability, it is crucial to achieve durability and leak-tightness during the controlled atmosphere brazing process carried out in a continuous furnace [5]. This is achieved through the formation of brazed joints which are formed between components by a filler metal that has a lower melting point than the base [6]. The quality of brazed joints largely depends on the flowability of the filler metal, that is, its ability in the liquid state to fill the gaps between components [7]. Flowability is influenced, among other factors, by the amount of flux used in the brazing process. Flux is a substance that facilitates brazing by cleaning the surfaces and improving the wettability of the filler metal [8].

The amount of flux, measured as flux grammage (mass per unit surface area of the components), must be adapted to the specific requirements of the heat exchanger. Increasing the flux grammage can improve joint quality by compensating for unfavorable brazing conditions, such as suboptimal temperature, complex joint geometry, or surface contamination [5, 9]. However, the use of flux results in residues remaining after the brazing process, which may negatively affect the thermal efficiency of the heat exchanger and lead to galvanic corrosion [10, 11].

The aim of this study was to determine the influence of two key brazing process parameters (maximum brazing temperature and flux grammage) on the brazing degree (the extent to which all intended joints are formed and meet quality requirements) and the amount of flux residues after controlled atmosphere brazing. The goal was to find a balance between ensuring high-quality brazed joints and minimizing flux residues. The results are application-oriented and can be used to optimize the technological process of charge air coolers under industrial conditions, contributing to improved product quality, reduced operational costs, and lower environmental impact.

2 Research Problem

The primary metric for assessing the quality of the controlled atmosphere brazing process in aluminum charge air coolers is the brazing degree. It defines what portion of the surface or volume of joints between components has been effectively filled with filler metal, which is affected by the material brazeability characteristics [12]. A high brazing degree indicates durable, leak-tight, and homogeneous connections between components (Fig. 1).

A low brazing degree may lead to leakage, weakened mechanical structure, and reduced thermal conductivity. The quality characteristics of brazed joints are defined by the destructive test of brazed joints standard PN-EN 12797:2002 [13].

To achieve the appropriate brazing degree and joint quality, it is necessary to reach a brazing temperature within the range of 590–610 °C [14]. Excessive temperatures may cause partial melting of the base material and erosion, while simultaneously increasing energy consumption costs [5, 15].

To ensure a sufficient brazing degree, the use of a potassium fluoroaluminate-based flux is essential, as it enables the dispersion of the aluminum oxide layer from the surface during heating, which would otherwise prevent wetting by the molten filler metal [16]. As a result, the molten filler metal can adhere to the aluminum base material, forming

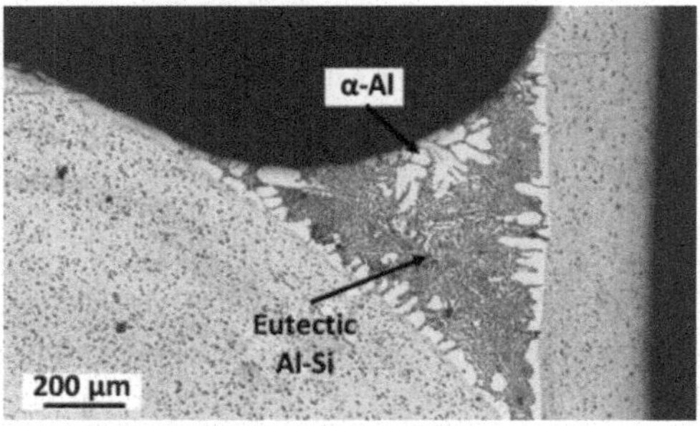

Fig. 1. Brazed joint microstructure [5]

a durable joint. Deficiency of flux leads to incomplete joints and leakage [17]. The minimum flux grammage (mass of flux per unit surface area of the component) required to disperse the aluminum oxide layer is estimated at 2 g/m^2 [18]. However, considering the potential presence of moisture, lubricants, oxygen particles, or magnesium content, 5 g/m^2 is typically used, and in special cases even up to 10 g/m^2 [5].

Increasing flux grammage helps compensate for suboptimal process conditions, such as unfavorable atmosphere, excessively low or high brazing temperature, complex joint geometry, mismatched capillary gaps, or surface contamination [5, 9].

Flux grammage in range of 10–20 g/m^2 has been linked to decrease in corrosion resistance of the joint [19]. It is due to the flux residues deposit on the surface of the components after the controlled atmosphere brazing process [8, 20, 21]. These residues appear as square-shaped potassium fluoroaluminate plates ranging from 10 to 20 μm in size and as magnesium fluoride needles [22] (Fig. 2). The exact structure of flux residues depends on various factors, such as the brazing atmosphere (presence of oxygen and moisture), heating profile (temperature changes over time), and the elemental composition of the aluminum alloys used [15, 18]. However, based on the available literature, the influence of brazing process parameters on the structure of flux residues has not been systematically described to date.

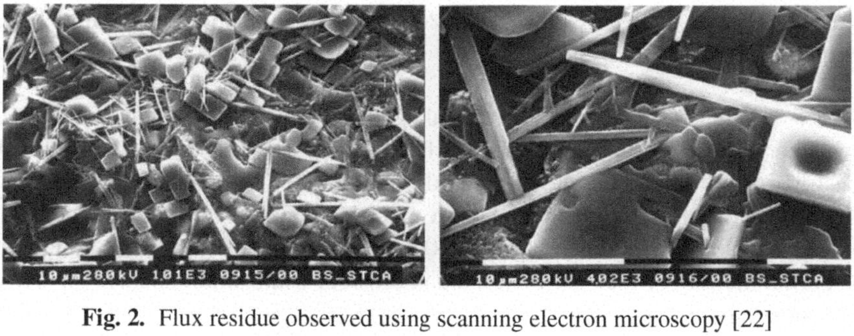

Fig. 2. Flux residue observed using scanning electron microscopy [22]

Flux residues represent a form of contamination, the negative effects of which are gaining increasing importance in the automotive industry as the dimensions of heat exchangers continue to decrease [23]. The negative effects of flux residues can be categorized as follows:

- Influence on performance: flux residue particles act as an additional insulating barrier. This reduces the effective heat exchange surface area, resulting in decreased thermal efficiency of the charge air cooler [10].
- Influence on coolant flow: the presence of flux residue particles may lead to the formation of deposits that block the flow channels of the coolant [8, 10, 11, 24, 25]. This causes an undesirable pressure drop in the cooling system due to disrupted coolant flow [26].
- Influence on corrosion: excessive flux residues can destabilize the coolant, leading to precipitation of its protective components [18], such as nitrites, nitrates, molybdates, and silicates [11]. This promotes the accumulation of flux-related deposits and the formation of corrosion sites [10].
- Influence on chemical properties: flux residue may react negatively with coolants, especially those containing nanoadditives. These reactions can lead to gelation, increased viscosity, and reduced flowability of the coolant [10, 23, 25]. Additional effects may include coolant discoloration, ammonia formation, pH increase, and accelerated degradation of inorganic corrosion inhibitors [11].

The importance of limiting solid flux residue particles is closely tied to the properties of the coolants used, particularly those containing additives such as nanoparticles [24]. In 2009, Kanjirakat et al. [27] linked increased coolant viscosity to particle aggregation in the cooling system. In 2014, Zeinali Heris et al. [28] summarized studies on nanofluids, emphasizing the importance of parameters such as particle volume, morphology, shape, and size. Pourfarhang et al. [29] reported a 3–18% increase in intermolecular friction in ethylene glycol depending on solid particle concentration ranging from 0.1% to 0.8%. In 2015, Li et al. [30] observed that nanofluid viscosity significantly affects pressure drop in cooling systems, leading to pumping difficulties and reduced processability. They also noted that research on nanofluid rheological properties remains incomplete. In 2017, Akbari et al. [31] recorded a 116% increase in dynamic viscosity of ethylene glycol when solid particle concentration rose from 0.1% to 3%. In 2021, Huy et al. [32] described the role of corrosion inhibitor additives in improving the durability of cooling system components. In 2025, Pundkar et al. [33] highlighted the importance of nanofluids in cooling systems, stressing the need to ensure compatibility with the coolants used.

Despite extensive research on the effects of flux residues, systematic optimization studies examining the interaction between brazing temperature and flux grammage on the amount of flux residues formed during controlled atmosphere brazing remain limited.

3 Methodology

Controlled atmosphere brazing experiments were carried out in a continuous furnace under production-relevant conditions. The furnace was configured for a family of similarly sized heat exchangers and operated without direct modification of set-point temperatures to avoid long stabilization times and additional operational costs. Consequently, brazing conditions were adjusted indirectly through furnace load organization, as described in the following sections.

Two key process parameters were selected based on process relevance and industrial applicability: flux grammage and maximum brazing temperature.

Flux grammage was defined as the mass of flux applied per unit area to the inner surfaces of the charge air cooler covers. Three grammage levels were investigated: 1 g/m^2, 3 g/m^2, and 5 g/m^2, where 5 g/m^2 represents the typical flux application level used in the current production process. The lower grammage levels of 1 g/m^2 and 3 g/m^2 were intentionally prepared for the present study to assess the process sensitivity to flux reduction under controlled brazing conditions (Fig. 3). Visible linear contact patterns (highlighted in yellow) correspond to mechanical contact between covers and adjacent components during brazing.

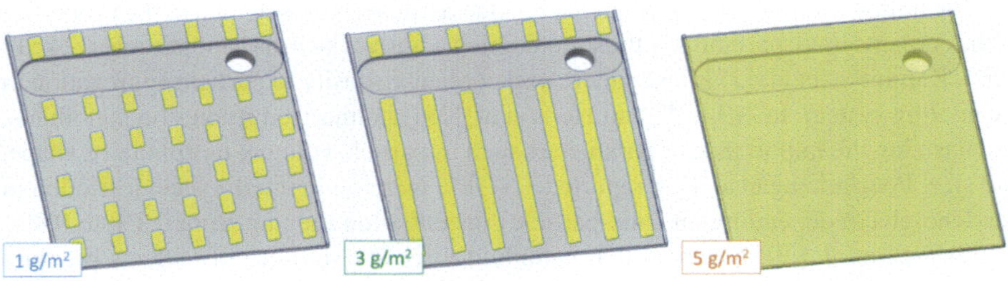

Fig. 3. Flux layer distribution (yellow) on covers of charge air cooler depending on its grammage

Maximum brazing temperature was constrained by the furnace configuration and by literature-reported effective brazing windows for aluminum alloys (~590–610 °C) [14]. Within this technological range, three peak temperatures were achieved: 593 °C, 602 °C, and 606 °C, where 602 °C represents the typical level used in the current production process. The upper temperature of 606 °C was selected as the highest reproducible peak temperature achievable under current continuous furnace constraints while remaining below the onset of base-material erosion and excessive filler metal depletion reported in the literature above 610 °C. Temperatures above this level were therefore not considered, as they are known to increase metallurgical risk without providing proportional improvements in braze joint quality. Peak temperatures were controlled indirectly through adjustments to the thermal mass and spatial distribution of the furnace load. Due to significant differences in thermal conductivity and heat capacity between aluminum components and stainless steel brazing frames (fixtures used to maintain dimensional accuracy and joint alignment during brazing [34]), distinct temperature profiles can be obtained by altering brazing frame usage and load spacing along the conveyor belt of the continuous furnace [35].

The lowest peak temperature (593 °C) was achieved by omitting brazing frames, thereby reducing the thermal mass of the furnace load. Conversely, the highest peak temperature (606 °C) was obtained by increasing the spacing between charge air coolers mounted in brazing frames, which reduced the effective thermal mass per unit length and allowed a greater proportion of the supplied heat to be absorbed by individual components (Fig. 4).

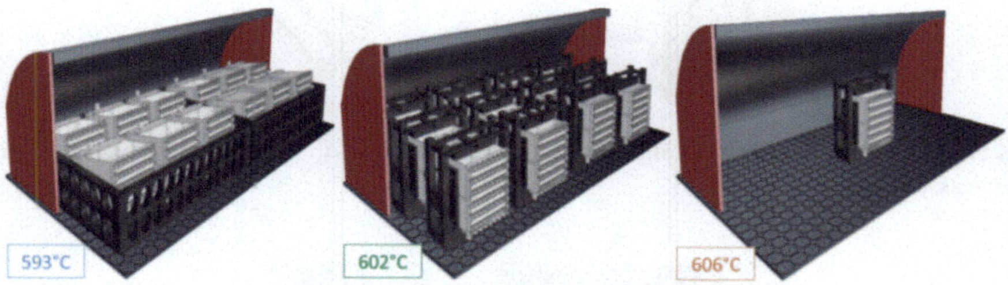

Fig. 4. Arrangement of charge air coolers in a continuous furnace used to control the achievable maximum brazing temperature

Two key response variables were defined to quantitatively assess the influence of the brazing process parameters investigated: brazing degree and amount of flux residues.

The brazing degree was selected as a quantitative measure of brazed joint quality and evaluated using optical microscopy on metallographically prepared cross-sections. Specimen preparation followed the guidelines of ASTM E3-11 (2017) [36] and PN-EN 12799:2003 [37], which specify standardized procedures for metallographic sample preparation and classify destructive and non-destructive testing methods, including requirements for joint sampling and testing equipment. Samples were sectioned using a band saw from selected regions of the charge air cooler covers, distinguishing between the upper (U) and bottom (B) covers relative to the continuous furnace conveyor (Fig. 5a). For brazing degree assessment, the joint region between the covers and the tubes of the heat-exchange packet was selected (Fig. 5b). The brazing degree was defined as the percentage of successfully formed joints between the tubes and covers along the perimeter of coolant-channel connections, relative to the total number of potential joint locations.

Due to the flux residues being present on the surface, samples for residue analysis could not undergo further processing after cutting. This caused high surface roughness, which had to be considered during optical measurements. Each flux residue specimen contained three regions located between tube cross-sections, from which six measurement points were selected for image acquisition and analysis (Fig. 6).

The amount of flux residues was measured using an original RLM-DIC method proposed by the author [38]. This technique is based on reflected light microscopy (RLM), enabling observation of opaque samples by analyzing light reflected from their surfaces. When combined with differential interference contrast (DIC), it enables visualization of transparent flux residue agglomerates due to differences in reflectivity and surface topography (Fig. 7a). Image processing of the RLM-DIC output enables the isolation of flux residue agglomerates (Fig. 7b), followed by calculation of their surface coverage using a color-based histogram that accounts for the corresponding pixels. The result is

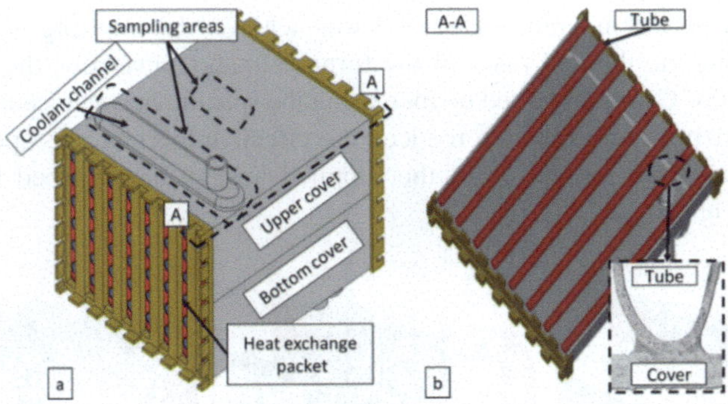

Fig. 5. Charged air cooler: a) Design and sampling area, b) A-A cross section - area of brazing degree assessment

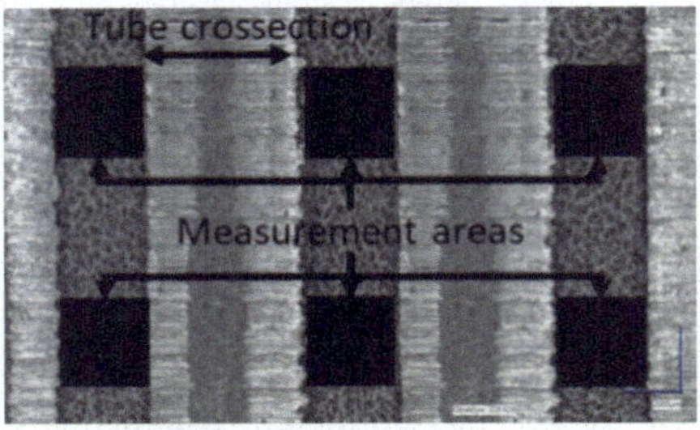

Fig. 6. Designated flux residue measurement areas on the surface of the specimen

expressed using a relative metric [Px%]. It should be noted that this method quantifies surface coverage of flux residues rather than their individual thickness or mass, which may influence interpretation of the results.

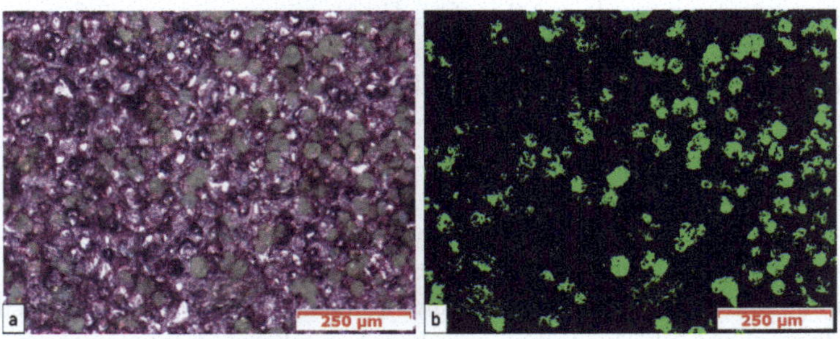

Fig. 7. Image processing using the RLM-DIC method: (a) flux residues visible as yellow regions before processing; (b) isolated flux residue areas after processing

Based on the defined process factors and response variables, a full factorial 3^2 experimental design was implemented to evaluate their combined effects on brazed joint quality and flux residues formation. In total, 64 independently manufactured charge air coolers were produced according to the experimental matrix (Fig. 8).

Flux grammage		1 [g/m^2]			3 [g/m^2]			5 [g/m^2]		
Batch no.		1	2	3	1	2	3	1	2	3
Max temp.	593 [°C]	4 pc.	3 pc.	3 pc.	3 pc.	4 pc.	3 pc.	1 pc.	1 pc.	1 pc.
	602 [°C]	2 pc.	3 pc.	3 pc.	3 pc.	2 pc.	3 pc.	1 pc.	1 pc.	0 pc.
	606 [°C]	4 pc.	3 pc.	3 pc.	3 pc.	4 pc.	3 pc.	1 pc.	1 pc.	1 pc.

Fig. 8. Experimental matrix of the full factorial 3^2 experimental design with flux grammage and maximum brazing temperature as the investigated factors

For each brazing-temperature condition, three independent production batches were conducted, and all flux grammage levels were applied within each batch. The parameter combination corresponding to the current process reference parameters was included with limited replication and treated as a comparative benchmark.

Brazing degree was evaluated for all 64 specimens. For flux residue measurements, a reduced subset of specimens was used. Following manufacturing, one charge air cooler from each experimental group was allocated to additional tests unrelated to the present study, which might affect the surface chemical properties. As a result, residue analysis was conducted on 55 specimens. The removal was systematic and evenly distributed across all factor combinations and therefore is not expected to bias the evaluation of factor effects.

Owing to unequal replication across factor combinations, the realized design deviates from a fully balanced factorial experiment. Consequently, statistical evaluation was performed using a general linear modeling (GLM) framework. Fixed effects included maximum brazing temperature and flux grammage as the primary process parameters, and additionally the cover side of the charge air cooler (upper and bottom), which was treated as a separate factor in the statistical analysis. Production batch was initially considered as a potential blocking factor in the statistical analysis to account for variability between independent production runs.

Model adequacy was assessed using coefficients of determination and the standard error of regression. Statistical interpretation emphasized effect direction, magnitude, and practical relevance under production-relevant constraints, with formal hypothesis testing and p-values reported where appropriate.

4 Results

The analysis began with a general linear model (GLM) that included batch number as a factor to assess its effect on the brazing degree. The initial model had a coefficient of determination (R^2) of 40.79%, indicating that the model explained a substantial portion of the variability in brazing degree. However, the adjusted coefficient of determination, which accounts for the number of predictors in the model, was lower at 33.25%. This decrease reflects the penalty applied for including batch number as a predictor, which was found to be statistically non-significant (p-value = 0.997).

Subsequently, the analysis was repeated without the batch number factor. The final model included three fixed factors: flux grammage, maximum brazing temperature, and cover side (upper and bottom). Separate GLM models were fitted for brazing degree and amount of flux residues, following the same fixed-effect structure.

GLM model for brazing degree demonstrated moderate explanatory power ($R^2 = 44.32\%$, adjusted $R^2 = 40.10\%$, predicted $R^2 = 36.61\%$, standard error of the regression $(S) = 0.0952$), which is acceptable given the complexity of the process.

Main effects analysis (Fig. 9) highlights the relative influence of the investigated factors on brazing degree. Analysis of variance confirmed that all three factors are statistically significant (maximum brazing temperature: $p = 0.000$, flux grammage: $p = 0.004$ and cover side: $p = 0.001$). Temperature has the strongest effect, increasing from approximately 85.5% at 593 °C to 97.5% at 602 °C, followed by a smaller gain to about 99.5% at 606 °C, indicating that most improvement occurs within the lower part of the investigated range. Flux grammage also shows a positive trend with the increase from ~91% at 1 g/m^2 to ~95% at 3 g/m^2, while the change from 3 g/m^2 to 5 g/m^2 is about half as large, reaching ~97%, suggesting a gradual rather than abrupt saturation. Cover side exerts a comparable influence, with the bottom cover reaching ~98% compared to ~91% for the upper cover. These patterns confirm that temperature is the primary driver of brazing quality, while flux grammage and cover side provide secondary but significant contributions.

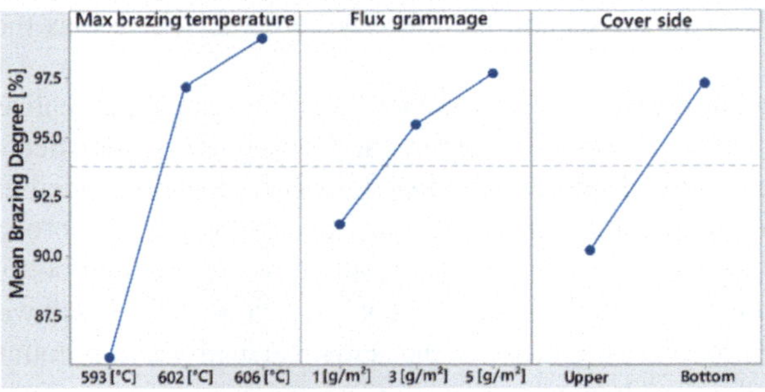

Fig. 9. Main effects plot of brazing temperature, flux grammage, and cover side on brazing degree (GLM marginal means)

The interval plot (Fig. 10) provides additional insight into factors interactions, variability and uncertainty. It is based on GLM-estimated marginal means with 95% confidence intervals. At lower temperatures (593 °C), brazing degree was highest when flux grammage reached 5 g/m^2, whereas at 606 °C even 1 g/m^2 produced values comparable to or exceeding those at higher grammage levels. This crossover behavior indicates that optimal flux grammage is temperature-dependent, and process parameters cannot be optimized independently.

Flux residue analysis was based on 55 specimens, as noted earlier. The general linear model demonstrated strong explanatory power ($R^2 = 69.62\%$, adjusted $R^2 = 66.91\%$, predicted $R^2 = 65.02\%$, standard error of the regression $(S) = 1.3586$), indicating that the investigated factors explain most of the variability in amount of flux residues. This

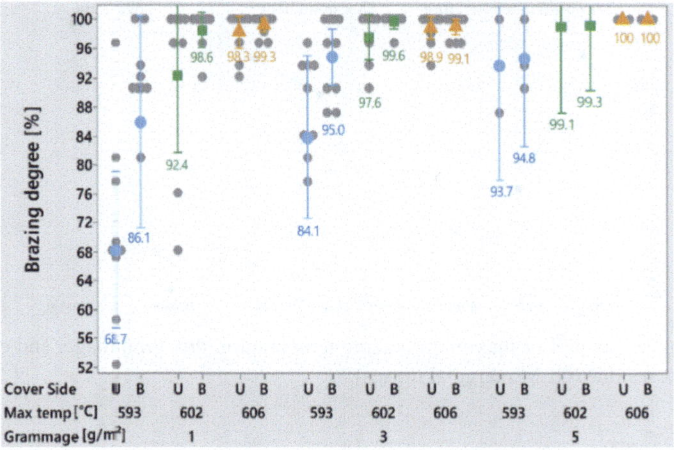

Fig. 10. Interval plot of brazing degree across brazing temperature, flux grammage, and cover side (data labels indicate GLM marginal means)

reflects the higher sensitivity of this response to selected process parameters in contrast to the brazing degree.

Main effects analysis (Fig. 11) highlights the relative influence of the investigated factors on amount of flux residues. Analysis of variance confirmed that all three factors are statistically significant (maximum brazing temperature: $p = 0.000$, flux grammage: $p = 0.013$ and cover side: $p = 0.001$). Consistently, temperature shows the strongest effect, with amount of flux residues decreasing as temperature increases, although the reduction is not uniform. The largest drop occurs between 593 °C and 602 °C (decrease of 1.3 pp.), with an additional smaller reduction to 606 °C (decrease of 0.6 pp.). Flux grammage shows a stepwise increase, with a rise from 1 g/m^2 to 3 g/m^2 (increase of 1.5 pp.) and smaller gain to 5 g/m^2 (increase of 0.4 pp.), suggesting saturation. Cover side exhibits the most pronounced contrast, with the bottom cover averaging about 3 pp. higher amount of flux residues than the upper cover, which suggests a gravitational effect on distribution during brazing.

The interval plot (Fig. 12) illustrates that the amount of flux residues generally decreases with increasing brazing temperature, but the trend varies across flux grammage levels and cover sides. At 593 °C and 1 g/m^2, residue levels are disproportionately high (~11.7 Px%–12.9 Px%), which may be explained by internal leakage resulting from insufficient brazing degree. At 606 °C, the difference between amount of flux residue means for grammage levels 3 g/m^2 and 5 g/m^2 decreases from 11.7 Px% to 10.8 Px% for the entire charge air cooler (mean value averaged over the upper and bottom covers). This may indicate that elevated brazing temperature affects flux residue morphology, such as the formation of thicker agglomerates, which cannot be distinguished using the RLM-DIC method. Variation between upper and bottom covers remains substantial, ranging from approximately 10% to 58%, which again suggests gravitational flux migration driven by flowability.

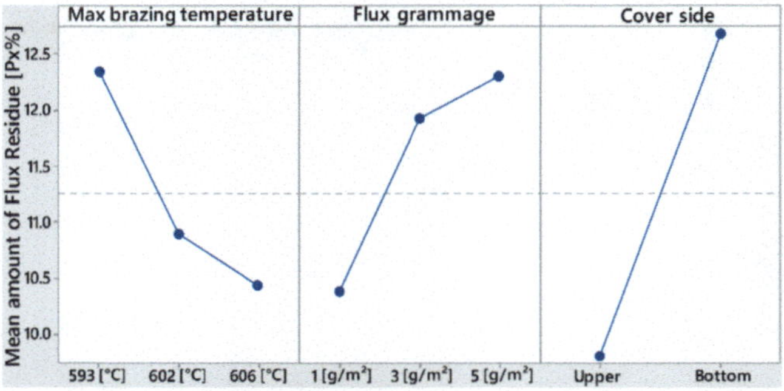

Fig. 11. Main effects plot of maximum brazing temperature, flux grammage, and cover side on amount of flux residue (GLM marginal means)

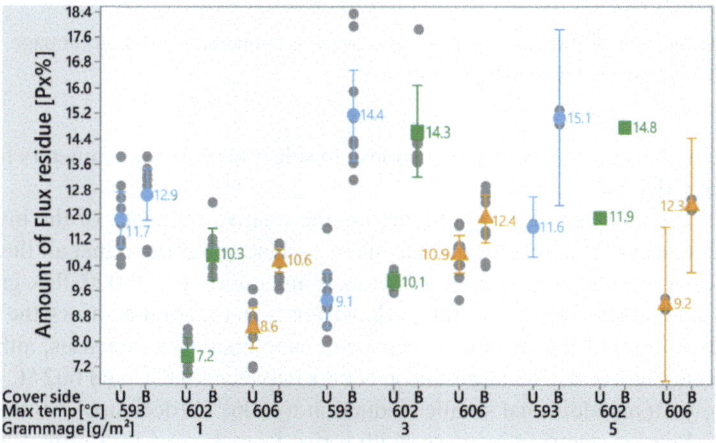

Fig. 12. Interval plot of amount of flux residues across brazing temperature, flux grammage, and cover side (data labels indicate GLM marginal means)

A one-way ANOVA analysis was performed to examine the relationship between brazing degree and amount of flux residue. Results showed no statistically significant dependence: the p-value (0.385) exceeded the 0.05 threshold, the F-value was low (1.08) and R^2 values were minimal ($R^2 = 17.44\%$, $R^2(\text{adj}) = 1.28\%$). Thus, within the studied range, brazing degree does not explain variability in residues amount.

Therefore, minimizing flux residues requires considering the combined effects of heating profile, flux grammage, and cover side rather than relying on a single parameter. Interpreting the observed trends requires acknowledging that the RLM-DIC method measures surface coverage only, not thickness or mass.

5 Discussion

The influence of the studied factors was compared to the central values: flux grammage of 3 g/m^2 and maximum brazing temperature of 602 °C (Fig. 13).

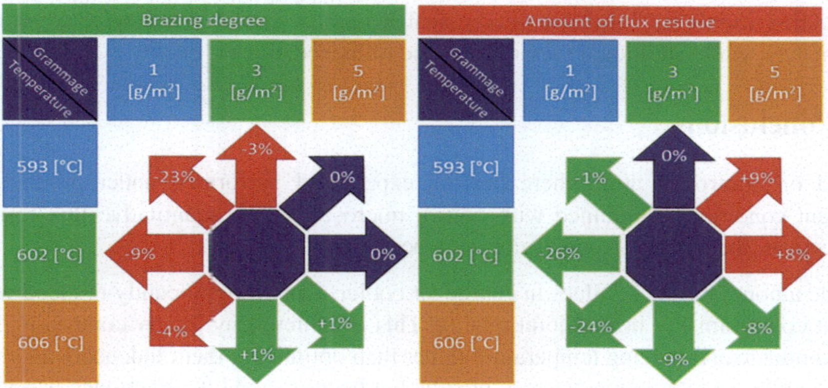

Fig. 13. Comparison of brazing degree and amount of flux residue depending on maximum brazing temperature and flux grammage

Although the maximum brazing temperature shows the strongest effect on both brazing degree and the amount of flux residues, when combined with production constraints, the initial flux grammage becomes the key parameter for process optimization. Increasing the grammage above 3 g/m^2 does not notably improve brazing degree (up to 3 pp.), while it clearly increases the amount of flux residues (by 1 to 9 pp.), indicating a practical saturation effect for flux application.

An increase in maximum brazing temperature above 602 °C improves the flowability of the flux and filler metal, resulting in a 1–5 pp. increase in brazing degree. Although brazing temperatures close to 600 °C are sufficient to achieve acceptable joint formation, the present results demonstrate that a moderate increase to 606 °C improves process robustness non-linearly, enabling effective joint formation even at reduced flux grammage levels. Since the literature places the practical upper limit at ~610 °C [14], with temperatures above this point showing diminishing returns and increased risks of erosion and filler depletion, selecting 606 °C is supported as a safe and effective upper boundary for industrial optimization.

At the same time, elevated brazing temperature reduces amount of flux residues by 2–16 pp. Comparison of the extreme values within the investigated parameter range indicates that the brazing degree can be improved by up to 24 pp., and the amount of flux residues reduced by up to 35%, depending on the parameter combination.

The full factorial design revealed a synergistic interaction between temperature and flux grammage. Increasing the maximum brazing temperature to 606 °C allows a reduction in flux grammage without compromising brazing degree, thereby reducing residue-related risks to thermal performance, flow stability, and corrosion resistance.

Increasing the flux grammage beyond 3 g/m^2 did not yield proportional benefits in brazing degree, further confirming the presence of a saturation point beyond which additional flux increases residue formation without improving process performance.

Despite the clear trends observed, several limitations should be noted. The study was conducted under laboratory and semi-industrial conditions, which may not fully reflect production variability. Furthermore, the analysis focused only on two selected process parameters, while additional factors may also influence brazing quality.

6 Conclusions

Based on controlled atmosphere brazing experiment performed under production-relevant conditions, combined with optical microscopy and quantitative flux residue assessment, the following conclusions can be drawn:

- The amount of flux residues in charge air coolers can be significantly reduced without compromising brazed joint quality. This is achieved by jointly controlling flux grammage and brazing temperature rather than optimizing them independently.
- A significant interaction between brazing temperature and flux grammage was identified. Increasing the maximum brazing temperature enhances flux and filler metal flowability, enabling effective joint formation even at reduced flux grammage levels.
- Optimal process conditions within the investigated range were identified as a flux grammage of 3 g/m^2 combined with a maximum brazing temperature of 606 °C.
  Compared to current reference parameters, these conditions allow reducing flux residues by up to 35% or improving brazing degree by up to 24 pp., depending on priority.
- Increasing flux grammage beyond 3 g/m^2 does not improve brazing degree proportionally but markedly increases flux residues, assuming other parameters remain within specification. This indicates the presence of a practical saturation limit.
- Brazing degree and amount of flux residues show no statistical dependence, highlighting the need for multi-objective optimization.
- The identified process parameters can be implemented without modifying existing furnace infrastructure, improving product quality and reducing cleaning needs, but at the cost of reduced capacity and higher energy use.

The RLM-DIC method used for flux residue assessment measures surface coverage rather than thickness or mass. Therefore, the apparent reduction at higher brazing temperatures may reflect changes in residue morphology and distribution rather than an actual decrease in total residue amount. This limitation should be considered when interpreting the results and planning future improvements in residue quantification.

Future work should focus on the standardization of flux residue measurement techniques and on the inclusion of additional process parameters, such as brazing time, atmosphere composition, and geometry of both braze frames and heat exchangers, to further enhance the robustness and generality of the proposed optimization strategy.

Acknowledgments. The research was conducted as a part of the Applied Doctorate Program of the Ministry of Education and Science implemented in the years 2021–2025 (Agreement No. DWD/(PP)RU00019673).

References

1. Huminic, G., Huminic, A.: Application of nanofluids in heat exchangers: a review. Renew. Sust. Energ. Rev. **16**(8), 5625–5638 (2012)

2. Syed, Z.A., Wagner, J.R.: Modeling and control of a multiple-heat-exchanger thermal management system for conventional and hybrid electric vehicles. Designs. **7**(1), 19 (2023)
3. European Union: Regulation (EU) 2019/851 of the European Parliament and of the Council. https://eur-lex.europa.eu/legal-content/EN/TXT/?uri=CELEX:32023R0851. Last accessed 01 Aug 2025
4. Nguyen-Schäfer, H.: Turbocharging concepts. In: Rotordynamics of Automotive Turbochargers, pp. 1–16. Springer, Heidelberg (2012)
5. Zhao, H., Woods, R.: Controlled atmosphere brazing of aluminum. In: Advances in Brazing Science Technology and Applications, pp. 280–322. Woodhead Publishing Series in Welding and Other Joining Technologies, Pleasanton (2013)
6. Lathabai, S.: Joining of aluminium and its alloys. In: Fundamentals of Aluminium Metallurgy, pp. 607–654. Woodhead Publishing (2011)
7. Swidersky, H.-W.: Aluminium brazing with non-corrosive fluxes state of the art and trends in NOCOLOK® flux technology. In: 6th International Conference on Brazing, High Temperature Brazing and Diffusion Bonding, pp. 1–6, Aachen (2001)
8. Mirski, Z., Pabian, J.: Modern trends in production of brazed heat exchangers for automotive industry. Weld. Technol. Rev. **89**(8) (2017)
9. Claydon, D.G., Sugihara, A.: Brazing aluminum automotive heat exchanger assemblies using a non-corrosive flux process. SAE Tech. Pap. **830021** (1983)
10. Wojdat, T., Winnicki, M., Mirski, Z., Żuk, A.: An innovative method of applying fluxes using the low-pressure cold gas spraying method. Weld. Technol. Rev. **91**(10), 17–24 (2019)
11. Alverson, F.C., Ranger, M., DeBaun, H.J.: The effects of residual controlled atmospheric brazing flux on engine coolants. In: STP 1556: Engine Coolants and Related Technologies, pp. 175–194. ASTM International, West Conshohocken (2014)
12. Winiowski, A.: Brazeability – the definitions and examples. Weld. Technol. Rev. **82**(9), 58–62 (2010)
13. Polish Committee for Standardization: PN-EN 12797:2002. Brazing — Destructive testing of brazed joints. PKN, Warsaw (2002)
14. Mirski, Z., Pabian, J., Wojdat, T., Hejna, J.: Significance of the brazing gap in the brazing of aluminium heat exchangers for automotive industry. Weld. Technol. Rev. **92**(4), 7–14 (2020)
15. Zähr, J., Ullrich, H.-J. Oswald, S., Türpe, M., Füssel, U.: Analyses about the influence of the natural oxide layer of aluminium on the brazeability in a shielding gas furnace. Weld. World. **57**, 449–455 (2013)
16. Solvay Special Chemicals: NOCOLOK® Flux Brazed Aluminum Heat Exchangers For the Refrigeration and Air Conditioning Industry. https://www.solvay.com/sites/g/files/srpend221/files/tridion/documents/NOCOLOK_Flux-Brazed-Aluminum-Heat-Exchangers.pdf. Last accessed 13 Jun 2025
17. Leszczyńska-Madej, B., Richert, M., Wąsik, A., Szafron, A.: Analysis of the microstructure and selected properties of the Aluminium alloys used in automotive air-conditioning systems. Meta. **8**(1), 10 (2018)
18. Bourgeais, H., Demarcq, C.: Cleaning methods for flux pollution measurement in automotive coolant loop components. In: CONAT 2016 International Congress of Automotive, Brasov (2016)
19. Gao, Z., Qin, Z., Lu, Q.: Controlled atmosphere brazing of 3003 aluminum alloy using low-melting-point filler metal fabricated by melt-spinning technology. Materials. **15**(17), 6080 (2022)
20. Wittebrood, A.J.: Process for fluxless brazing of aluminium and brazing sheet for use therein. Eur. Patent. **EP2382087B1** (2017)
21. Choi, S.U., Eastman, J.A.: Enhancing thermal conductivity of fluids with nanoparticles. In: ASME International Mechanical Engineering Congress & Exposition, San Francisco (1995)

22. Gershun, A.V., Woyciesjes, P.M.: Engine coolants and coolant system corrosion. In: Corrosion: Environments and Industries, vol. 13C, pp. 531–537. ASM International (2006)
23. Kakaç, S., Liu, H., Pramuanjaroenkij, A.: Heat Exchangers: Selection, Rating, and Thermal Design, 3rd edn. CRC Press, Boca Raton (2012)
24. Zaldivar, J.G., Swidersky, H., Pelsemaeker, G.D.: Study of the compatibility of the CAB flux. In: Proceedings of the 2007 vehicle thermal management conference, Nottingham (2007)
25. Garcia, J., Massoulier, C., Faille, P.: Brazeability of Aluminum Alloys Containing Magnesium by CAB Process Using Cesium Flux. SAE Tech. Pap. (2001)
26. Gray, A., Swidersky, H.W., Lauzon, D.C.: The Morphology and Structure of Post-Braze Flux Residues. http://www.aluminium-brazing.com/sponsor/nocolok/Files/PDFs/ Morphology_Paper.pdf. Last accessed 17 Sept 2025
27. Cocian, G., Popa, C.: Measurement method for determining the residual flux deposits in a heat exchanger from a controlled atmosphere brazing line. Studia Universitatis Babes-Bolyai Chemia. **60**(1), 187–193 (2015)
28. Jin, H.: Development of an aluminum brazing sheet product with barrier layer for high-performance automotive heat exchangers. Metall. Mater. Trans. A. **52**, 1409–1426 (2021)
29. Kanjirakat, A., Kabelac, S., Sundararajan, T., Das, S.K.: Rheological and flow characteristics of nanofluids: influence of electroviscous effects and particle agglomeration. J. Appl. Phys. **106**(034909), 1–7 (2009)
30. Zeinali Heris, S., Shokrgozar, M., Poorpharhang, S., Shanbedi, M., Noie, S.H.: Experimental study of heat transfer of a car radiator with CuO/ethylene glycol-water as a coolant. J. Dispers. Sci. Technol. **35**(5), 677–684 (2014)
31. Pourfarhang, S., Heris, S.Z., Motahare, S., Kahani, M.: Pressure drop and thermal performance of CuO/ethylene glycol (60%)-water (40%) nanofluid in car radiator. Korean J. Chem. Eng. **32**(4), 609–616 (2014)
32. Li, X., Zou, C., Wang, T., Lei, X.: Rheological behavior of ethylene glycol-based SiC nanofluids. Int. J. Heat Mass Transf. **84**, 925–930 (2015)
33. Akbari, M., Afrand, M., Arshi, A., Karimipour, A.: An experimental study on rheological behavior of ethylene glycol-based nanofluid: proposing a new correlation as a function of silica concentration and temperature. J. Mol. Liq. **233**, 352–357 (2017)
34. Huy, V.V., Quang, H.D., Katolik, A., Dan, N.T., Gaidar, S.M.: The development of corrosion inhibitor used in the automotive coolant. Russ. J. Appl. Chem. **94**, 1577–1584 (2021)
35. Pundkar, K.A., Chaudhari, S.S., Daryapurkar, A.S., Himte, R.: Overall impact on thermal and tribological performance of air conditioning system with nanoparticle application: a critical review. Int. J. Air-Cond. Refrig. **33**, 15 (2025)
36. Nadolny, S., Rogalewicz, M., Hamrol, A.: The effect of continuous furnace load organization on temperature distribution during controlled atmosphere brazing process. Bull. Pol. Acad. Sci. Tech. Sci. **73**(3), 1–9 (2025)
37. Nadolny, S., Rogalewicz, M.: Transition of controlled atmosphere brazing jig for aluminum heat exchangers from spring-loaded to fixed-dimension. In: Advances in Manufacturing IV, MANUFACTURING 2024, Poznan (2024)
38. ASTM International: ASTM E3-11 (2017), Standard Guide for Preparation of Metallographic Specimens, ASTM International, West Conshohocken (2017)
39. Polish Committee for Standardization: PN-EN 12799:2003. Brazing — Non-destructive testing of brazed joints. PKN, Warsaw (2003)
40. Nadolny, S., Hamrol, A., Rogalewicz, M., Piasecki, A.: Measurement methods for flux residue quantity after controlled atmosphere brazing of aluminum coolers. Manag. Prod. Eng. Rev. **14**(3), 156–163 (2023)

41. Hamrol, A.: Quality Management with Examples (in Polish). PWN Scientific Publishers, Warsaw (2013)
42. Koleňák, R., Augustin R., Ruža, V., Nerádová, M.: The influence of high-temperature brazing upon indicators of material brazeability. Mater. Eng. **17**(1), 17–22 (2010)

An AI-Driven Slicing Framework for Optimising Energy Consumption and Printing Time in Wire Arc Additive Manufacturing

Elena Manoli[1], Giulio Mattera[1(✉)], Thao Le Van[2], Mosè Gallo[1], Van Canh Nguyen[2], and Luigi Nele[1]

[1] Department of Chemical, Materials and Industrial Production Engineering, University of Naples Federico II, Naples, Italy
giulio.mattera@unina.it

[2] Advanced Technology Center, Le Quy Don Technical University, Hanoi, Vietnam

Abstract. The pursuit of sustainability in manufacturing often involves balancing conflicting goals, such as minimising production time or reducing energy consumption to meet environmental demands. Additive Manufacturing (AM) processes, as Wire Arc Additive Manufacturing (WAAM), face particular challenges in this regard because of their complexity and sensitivity to process parameters. Nevertheless, Artificial Intelligence (AI) provides valuable support for process optimisation planning, as it enables the application of advanced parameter search algorithms in conjunction with ML-based models to maximise given reward functions. In this regard, this study proposes a data-driven framework that integrates AI models into slicing software to predict layer geometry and specific energy consumption (SEC), thus enabling multi-objective optimisation of WAAM parameters and consequent generation of the path planning strategy. By incorporating both process parameters and scanning strategy variables into a geometry-specific optimisation framework around the slicer software, the user can achieve component-specific planning optimisation with minimal intervention. A prototype production of an Invar 36 alloy impeller showcases the potential of the framework to support sustainable and scalable production planning, aligned with modern green manufacturing paradigms, particularly when dealing with complex geometries produced via WAAM.

Keywords: Wire Arc Additive Manufacturing (WAAM) · Artificial Intelligence · Slicer Software · Multi-objective Optimisation · Sustainable Manufacturing

1 Introduction

Manufacturing industries frequently face the challenge of simultaneously optimising conflicting objectives, primarily minimising production time and energy consumption. Minimising processing time is typically critical in high-volume manufacturing contexts, where enhanced productivity directly translates into economic benefits through faster

B. Gapiński et al. (Eds.): MANUFACTURING 2026, LNME, pp. 42–55, 2026.
https://doi.org/10.1007/978-3-032-22943-4_4

return on investment and reduced unit production costs. In contrast, minimising energy consumption becomes increasingly prioritised in scenarios that involve limited production volumes, high customisation, or stringent environmental regulations. In these contexts, energy reduction not only lowers operational costs but also aligns production processes with sustainability objectives, environmental compliance, and corporate responsibility strategies. The selection of an optimal balance between processing time and energy consumption is a non-trivial optimisation task, influenced heavily by application context, corporate strategies, and associated economic and environmental sustainability goals. Thus, robust decision-support methodologies are required to systematically identify and reconcile these complex trade-offs.

In the context of the Industry 4.0 transformation paradigm, Additive Manufacturing (AM) has become increasingly adopted as a production technology [1, 2]. It is commonly employed to fabricate small to medium batches of both prototypes and high-value components [3]. Recent studies have highlighted its broader potential, particularly in improving the sustainability of manufactured products [4]. AM can contribute to reducing carbon footprint emissions and achieving significant energy savings [2]. Consequently, as with other manufacturing processes, nowadays it is possible to pursue two objectives simultaneously, also with AM: reducing processing time and minimising the energy consumption per unit produced.

Among the various AM technologies, Wire Arc Additive Manufacturing (WAAM) is generally preferred for producing medium-to-large components, which would be difficult to fabricate using other AM technologies [5]. This is largely driven by physical constraints, such as the limited build volumes of powder-based systems, and by economic factors, including the high costs of equipment and feedstock [6]. The WAAM process, shown in Fig. 1, leverages Gas Metal Arc Welding (GMAW) technology to fabricate metallic components in a layer-by-layer approach.

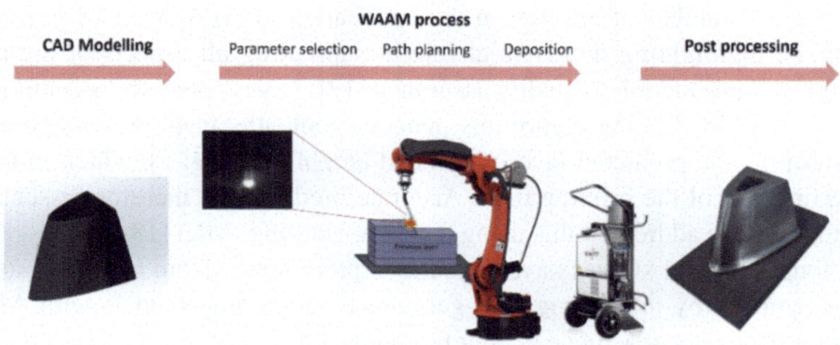

Fig. 1. The WAAM process begins with the creation of a CAD model. The slicing software then defines the process parameters and selects the infill deposition strategy. Once the build is completed, the component undergoes machining to meet the final dimensional and surface-quality requirements.

Before printing, an optimised scanning strategy must be defined, which involves selecting the most appropriate robotic path. This strategy can be tuned either to reduce processing time or to minimise the energy consumed during deposition. Although this issue is less problematic in other AM processes, in WAAM both objectives are strongly affected by process-parameter selection [7, 8] which remains one of the most critical challenges [9, 10].

Indeed, it is well-established that parameters such as Welding Speed (WS) and Wire Feed Speed (WFS) significantly affect the geometric characteristics of the deposited material. An increase in WS typically results in narrower bead widths, requiring more passes to fill the same geometry. Conversely, higher WFS tends to increase layer height, thereby reducing the total number of layers needed to build a component of equivalent height. These geometric variations, in turn, impact the process's energy efficiency [11]. For instance, higher WFS increases welding current [12–14], consequently, arc energy demand. However, greater travel speed reduces total deposition time, potentially offsetting overall energy consumption. As a result, identifying the optimal trade-off between these parameters is non-trivial, as it depends largely on the component's geometry, as well as considerations related to process stability, defect formation, and final part quality [15]. In this context, AI-based tools emerge as a promising solution to the problem of finding optimal solutions in the presence of conflicting objectives [16].

Therefore, this work presents a framework that integrates artificial intelligence into a slicer specifically developed for WAAM, enabling a new combination of sustainability and productivity objectives while generating an offline toolpath optimised according to a product-centric paradigm. Section 2 outlines the state of the art, the research gap, and the motivation for the proposed approach. Section 3 introduces the framework, and Sect. 4 illustrates its potential through the application to a prototypical component.

2 Background and Work Proposal

2.1 State of the Art

As of today, the number of articles on the application of AI in WAAM is increasing exponentially, highlighting its essential role in supporting all aspects of the process, from design and production to quality assurance [17]. Layer geometry is central to path planning in WAAM. Slicing algorithms generate toolpaths from process parameters, which determine the predicted layer width and height (see Fig. 2), which in turn also enables estimation of the printing time. Accurate prediction is therefore essential, and several studies have addressed this using machine learning (ML) [18–20].

Additionally, recent studies have begun to explore how AI can be used to optimise process parameters by integrating surrogate machine-learning models with AI-driven optimisation strategies to achieve greater precision.

For example, Tomaz et al. [21] used an artificial neural network (ANN) to model the welding process and a genetic algorithm to optimise parameters with the goal of maximising weld joint penetration in a gas tungsten arc welding (GTAW) application. In the context of WAAM, similar strategies have been explored. Van Thao et al. [22] used polynomial regression combined with response surface methodology to minimise heat input while maximising layer width and height, with the dual aim of reducing heat accumulation and improving build efficiency. Similar works explored the possibility of reducing energy consumption [23], and energy efficiency, assuring process stability [24]. Despite

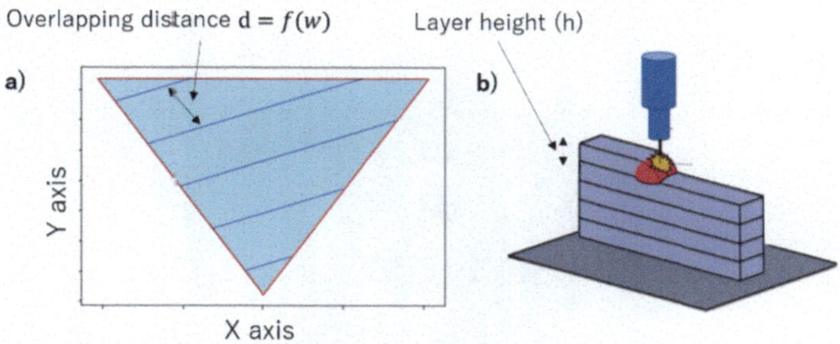

Fig. 2. Influence of layer geometry on build time: (a) Effect of layer width on the time required for a single layer; (b) Relationship between layer height and the number of layers required, based on component height.

the amount of research in this direction, current approaches remain fragmented and lack a unified, systematic framework necessary for effective industrial implementation.

2.2 Research Gap and Work Proposal

In light of the above, it is clear that AI methods are not only beneficial but increasingly essential for informed decision-making in WAAM. However, the current literature still lacks a comprehensive framework capable of integrating recent advances in surrogate machine-learning models and AI-based optimisation directly into a unified slicer environment. Existing studies on AI for WAAM path planning have primarily focused on interpolating waypoints to generate trajectories or on improving classical raster or contour strategies [25–27]. While these contributions advance toolpath generation, they overlook an equally critical aspect: the systematic selection of process parameters, which remains central to industrial applicability and standardisation of the parts [28].

Most prior works address isolated elements of the workflow, resulting in fragmented and often disconnected solutions. Consolidating these developments offers a significant opportunity to embed AI-driven decision-making directly within the slicer software, thereby enabling an intelligent, adaptive, and truly integrated approach to WAAM process planning.

Through a case study, this work demonstrates how AI can be used to simulate build time and energy consumption as functions of both process parameters and component geometry. Starting from the CAD model, appropriate surrogate models enable the prediction of key outcomes linked to the sustainability–productivity trade-off, namely specific energy consumption and printing time. These predictions can then be coupled with an optimisation algorithm capable of balancing both objectives for a given geometry, thereby guiding parameter selection within a certification-oriented and production-centric framework. To illustrate the proposed approach, the component considered is an impeller designed for cryogenic applications and manufactured in Invar 36, as shown in Fig. 3. This part provides a promising yet demanding test case for WAAM due to its geometric complexity, the need for dimensional stability at low temperatures, and the high cost and machining difficulty associated with Invar 36 [29].

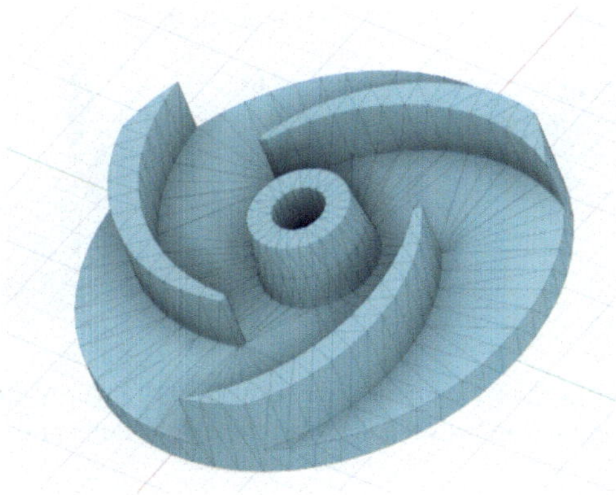

Fig. 3. The CAD model of the impeller made of Invar 36 considered in this work

3 The Proposed AI-Integrated Slicer Optimisation Framework

As outlined above, layer geometry plays a critical role in determining build time in WAAM. Specifically, layer height (h) defines the number of layers (L) needed to build a part of height H (usually L = H/h + 1), while layer width (w) affects overlap and the number of deposition passes needed to fill it [30]. To minimise build time, WFS and moderate WS are typically preferred to maximise deposition per pass. However, increasing these parameters also raises the heat input and overall power consumption. In fact, while reducing build time, faster deposition can lead to higher Specific Energy Consumption (SEC) and increased Global Warming Potential (GWP) due to elevated energy demand [31, 33]. The proposed framework is shown in Fig. 4 and consists of two main components:

- A predictive model to estimate the geometric characteristics of the deposited layer based on selected process parameters (e.g., travel speed, wire feed speed, current, etc.);
- An optimization module that uses the output of the predictive model (i.e., layer geometry) alongside a user-defined deposition strategy (e.g., infill angle, number of contour passes, infill percentage, etc.) to determine the optimal combination of parameters according to given objectives.

In the following, the specific components of the proposed framework are defined. The subsequent section presents a case study that illustrates its application and validates the solution to the productivity versus energy-efficiency problem introduced earlier.

3.1 Surrogate ML Model

As discussed in the literature background, the predictive or surrogate model can be implemented using various ML techniques (e.g., neural networks, support vector machines, random forest, etc.), with the selection depending on the nature of the available data, the material used, and the process characteristics. The framework is conceived as modular,

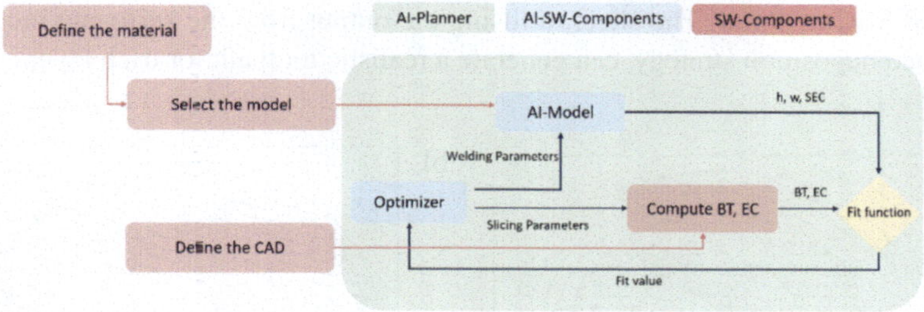

Fig. 4. A flow-diagram of the proposed AI-based WAAM process planning model

meaning that the user can train and load the most suitable ML model for their specific case.

The workflow for implementing the predictive model is illustrated as follows. Once the material is selected, an experimental campaign must be conducted. Specifically, N combinations of process parameters should be tested by depositing multi-layer walls (at least 10 layers) to account for heat accumulation effects and their impact on layer geometry. The average values of layer geometry can be obtained by measuring the total height and width of the deposited walls. These measurements can be performed using a laser scanner or through macrostructural analysis, either layer-by-layer or at the end of the deposition process.

While layer geometry is essential for estimating the build time, the multi-objective nature of the problem also requires evaluating power consumption. Therefore, the proposed framework necessitates the measurement of welding current and voltage, which are used to compute the power consumption as defined in [34]. By dividing the PAI by the welding speed, the SEC can be calculated $SEC = \frac{1}{WS*T}\sum V_t \cdot I_t$.

The model outputs are the predicted layer geometry and SEC, while the inputs consist of the and, in the case of WAAM, typically include wire feed speed (WFS), arc voltage, travel speed (WS), and contact tip to workpiece distance (CTWD).

3.2 Optimiser

Once the input parameters are defined, the proposed framework operates in a fully automated manner. A data-driven optimisation algorithm (e.g., Genetic Algorithm or Particle Swarm Optimisation) generates the set of process parameters (e.g., travel speed, wire feed speed, current) and defines the deposition strategy. The deposition strategy, which represents another set of parameters that the optimiser must determine (see Fig. 5), includes additional outputs such as:

- Activate the contour passes: It is managed by a discrete variable (ON/OFF).
- Activate infill: It is managed by a discrete variable (ON/OFF).
- Generation of the number of contour repetition: An integer between 1–10 (K).
- Selection of the infill angle (α): An integer value, associated to [0, 22, 44, 66, 90] degrees.

The process parameters are passed to the surrogate model, which outputs the expected geometric dimensions of the deposited layer—namely, height (h) and width (w) and the estimated SEC in J/mm. The slicer, utilising the output from the surrogate model and the defined deposition strategy, can generate a realistic toolpath for each layer.

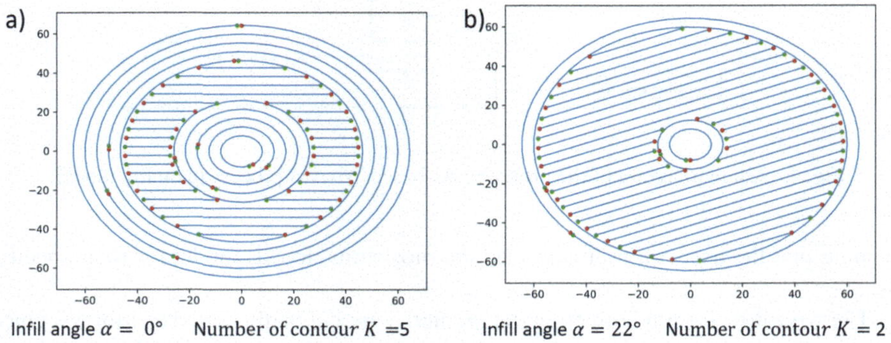

Fig. 5. Example of different possible printing strategies for the impeller under study

This toolpath is determined by computing the travel distances once the waypoints are known. Let D_{travel} represent the total travel distance. Given the WS and the intra-layer cooling time ($T_{cooling}$), the total build time (T_{build}), can be estimated as shown in Eq. 1.

$$T_{build} \approx \frac{D_{travel}}{WS} + T_{cooling} \tag{1}$$

Additionally, the total energy consumption (E_{tot}),) can be estimated by knowing the specific energy consumption (SEC) per unit length of deposition, as shown in Eq. 2.

$$E_{tot} \approx SEC \cdot D_{travel} \tag{2}$$

Thus, the slicer computes both the total build time and energy consumption based on the travel distances and process parameters. In this first exploratory study, we assumed that all layers are produced using the same strategy. Hence, the metrics for each layer are aggregated to compute the overall production metrics.

3.3 Fitness Function Design

In the proposed framework, the optimisation process is driven by a metaheuristic algorithm which aims to maximise the fitness function, balancing two key objectives:

- Minimising the building time T_{build}
- Minimising the total energy consumption E_{tot}.

The multi-objective fitness function is defined in Eq. 3, where the parameter β is a user-defined weight that regulates the trade-off between build time and energy consumption. Its value determines the primary objective of the optimisation: values of $\beta > 0.5$ give more emphasis to minimising the energy consumption, while values below 0.5 prioritise minimum processing time, therefore fast production. In the proposed framework,

β is constrained within the range $0.05 < \beta < 0.95$ to ensure that both criteria are always considered to some extent.

$$f = (1 - \beta) \cdot \frac{1}{T_{build}} + \beta \cdot \frac{1}{E_{tot}} \tag{3}$$

This is the only decision required from the user, in addition to selecting the material. The optimisation loop continues until a stopping criterion is met (e.g., maximum iterations or convergence), iteratively refining both process and slicing parameters.

4 Preliminary Demonstration via Case Study

4.1 Experimental Procedure and Surrogate Model Development

The experimental activities were carried out using a robotic WAAM cell comprising an Omron Viper s850 robotic arm integrated with a Miller Phoenix 456 welding power source, as shown in Fig. 6(a). A total of 15 walls (see Fig. 6(c)) were deposited using Invar 36 wire with a diameter of 1.2 mm, each produced under different combinations of process parameters and by varying the printing direction between layers, as shown in Fig. 6(b). The parameters varied were wire feed speed (WFS), reference welding voltage, and welding speed (WS). During each deposition trial, welding voltage and current signals were recorded at a sampling rate of 5 kHz using a National Instruments 6361 DAQ system, enabling the computation of power consumption. After deposition, each wall was measured to determine the average layer height and width. These values were obtained by measuring the total height and width of each wall and dividing by 10, corresponding to the number of deposited layers.

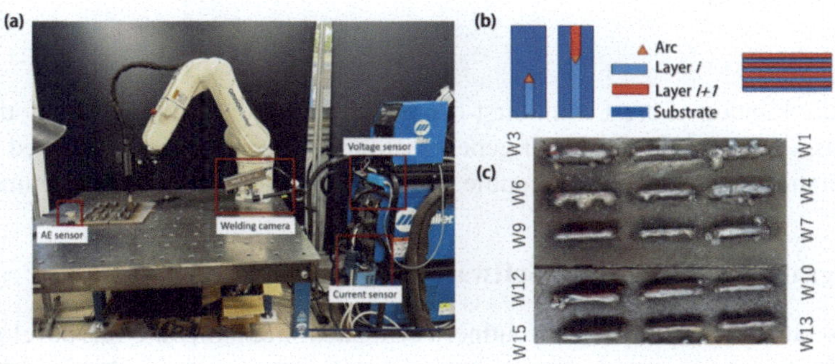

Fig. 6. (a) WAAM system used in this study. Summary of the experimental campaign, highlighting (b) the deposition strategy employed and (c) the appearance of the deposited walls.

Once the dataset is collected during the experimental setup phase, an AI-based model must be chosen, trained and then implemented to predict the optimal process parameters based on the CAD model. To assess the predictive performance of various regression strategies applied to WAAM process parameters, four models were tested: Linear Regression (LR), second-order Polynomial Regression (PR), and Random Forest

Regressor, both with and without polynomial feature expansion. Given the limited size of the dataset, consisting of 15 parameter combinations, models were evaluated using Leave-One-Out cross-validation (LOO) to assess generalisation ability, and the mean absolute error (MAE) percentage was used to measure in-sample accuracy, as shown in Table 1. Finally, the best-performing model was trained on the entire dataset for final predictions. Table 1 shows that the LR model achieved the lowest cross-validation error (MAE% = 10.89%), indicating a good generalisation. However, the second-order PR model, while significantly reducing the training MAE (4.67%), suffered from overfitting, with the highest cross-validation MAE (17.89%) and standard deviation (14.47%). In contrast, Random Forest models showed more stable and reliable performance. With or without polynomial expansion, the models achieved nearly identical MAE% on the full dataset (4.45%–4.62%) and a consistent cross-validation MAE% of approximately 12.7–12.8%. This confirms that ensemble-based regressors are better suited for capturing non-linear dependencies in WAAM process data without the need for explicit polynomial transformations, especially in data-limited scenarios.

Table 1. Comparison of regression models for predicting WAAM performance indicators.

Model	MAE% (LOO Mean)	MAE% (LOO Std)	MAE% (Full Dataset)
Linear Regression	10.89	6.58	8.12
Second-order Polynomial Regression	17.89	14.47	4.67
Random Forest	12.68	7.95	4.45
Random Forest (polynomial features)	12.80	8.76	4.62

Overall, Random Forest with first-degree polynomial features offered the best compromise between accuracy and generalisation, and was therefore selected for the implementation of the predictive module within the proposed optimisation framework.

4.2 Integration with the Slicer Software

Prior to executing the optimisation routine, a dedicated slicer software was developed, as illustrated in Fig. 7a. Upon selecting the material and process type—in this case, Invar 36 and short-circuit GMAW—the surrogate model described in the previous section, namely the Random Forest Regression, is automatically loaded. Two sets of process parameters were investigated to quantify, using the CAD model, the influence of parameter selection on the fitness function. The objective was to assess the sensitivity of the problem to variations in process settings once the CAD geometry, path strategy, infill angle of 22°, and a single contour repetition were fixed (see Fig. 7). In particular, *Set 1* employed a WFS of 6 m/min, a welding voltage of 21 V, and a WS of 4.5 mm/s; while *Set 2* used a WFS of 4 m/min, a voltage of 22.5 V, and a WS of 3.2 mm/s. Although the two parameter sets appear similar from a process perspective, selecting between them is not straightforward. Set 1 employs higher power and higher speed, while Set 2 uses lower power and lower speed, resulting in distinct layer geometries. Set 1 produces layers of 2.1 mm in height and 6.7 mm in width, whereas Set 2 yields 2.0 mm and 9.9 mm, respectively. Despite producing the same number of layers, one configuration offers narrower beads with a longer processing time, and the other wider beads with potentially reduced speed. This makes it difficult to identify the preferred option without dedicated decision-support tools. Therefore, for the two parameter sets, the total build time T_{build} and total energy consumption E_{tot} were computed, as described in the optimisation strategy section, assuming a fixed inter-layer cooling time of 60 s.

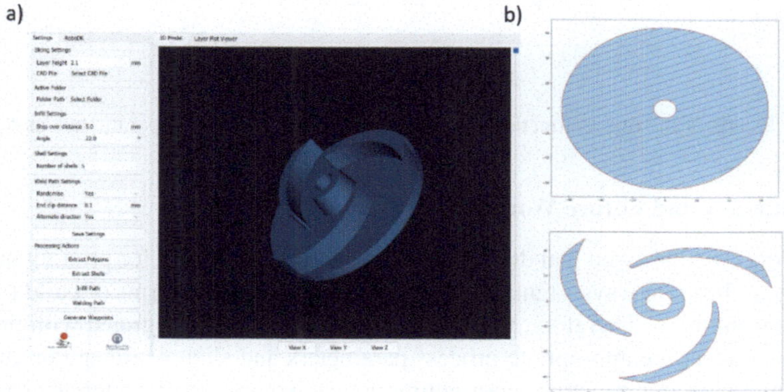

Fig. 7. (a) The GUI of the slicer (b) An example of the planned path based on process parameters Set 1.

The simulation environment integrated within the slicer demonstrated that the two parameter sets yielded markedly different results in terms of build time and energy consumption, highlighting the critical influence of process parameters on manufacturing performance. *Set 1* completed the build in 1.5 h, consuming 77.36 MJ of energy, while *Set 2* required 3.24 h but consumed only 13 MJ of energy.

Figure 8 illustrates how the fitness value varies with the weighting factor β which adjusts the relative importance of the total build time T_{build} and total energy consumption E_{tot} in the objective function. As shown, when $\beta \to 0$, the fitness is dominated by the inverse of build time, favouring faster deposition strategies such as Set 1. Conversely, for $\beta \to 1$, the fitness reflects predominantly the inverse of energy consumption, thereby favouring Set 2, which is significantly more energy-efficient despite the longer build time. The plot clearly highlights the trade-off enabled by tuning β, allowing the user to select the optimal parameter set depending on whether the manufacturing priority is time efficiency or energy conservation. Moreover, it is interesting to observe that for a given β^* value (around 0.9), both process parameter sets yield the same results.

This behaviour reinforces the importance of carefully selecting β, as it prevents ambiguous choices and ensures that the final decision accurately reflects the intended balance between the conflicting objectives of productivity and sustainability.

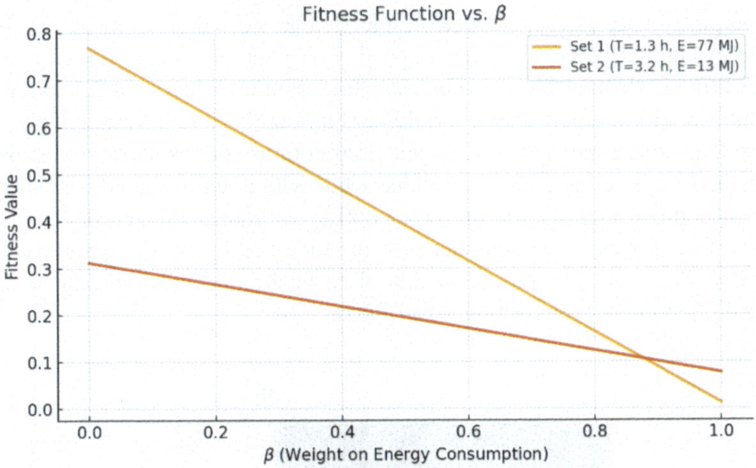

Fig. 8. Trends of the fitness function based on the two sets of process parameters as β varies

4.3 Summary and Future Works

In this work, we discussed that different process parameters allow for printing the same geometry in different ways, significantly affecting the output in terms of build time and energy consumption. Therefore, we discussed how AI can be integrated with the slicer software to select the appropriate process parameters and slicing strategy on the basis of a defined objective function representing the decision maker's preferences in terms of process performance.

As a result, we proposed a conceptual AI-based framework to support process planning optimisation. We defined a fitness function and, in the case of two sets of process parameters, we showcased how this function gets different values, also as a consequence of the weighting parameter β, i.e., decision maker's preferences. The results confirm that the process planning problem of WAAM processes is a complex multi-objective optimisation problem with not always univocal solutions (see Fig. 8).

This initial study has many possible ways to proceed further. An important one is selecting the best optimisation method from those available in the literature for the Optimiser module (refer to optimiser in Fig. 4). This module could, in principle, leverage different metaheuristics to explore the process parameter space. Thus, a crucial point in implementing the proposed framework is to select an appropriate optimization algorithm, or their combination. This selection could be done by directly comparing the perfomance of variuos algorithms, or by initially screening them according to defined criteria. Future research may focus on applying the optimisation procedure to different CAD geometries, identifying the most appropriate strategy for each case, showcasing the application of the framework. In this work, the problem has been formally defined and its inherent complexity has been demonstrated. Moreover, in showcasing the proposed framework, we applied the same slicing approach to all layers (e.g., without changing the number of contours or the infill angle for successive layers). Therefore, a natural evolution of this framework involves the possibility of modifying the slicing strategy at the individual layer level, allowing for fine-tuning according to the component's geometry. However, in that regard, to avoid undesirable effects on the microstructure of the deposited material, an important constraint could be the consistency of process parameters along the Z-direction (i.e., between layers), which could compromise the mechanical properties of the component. Moreover, it is possible to vary the infill strategy, such as the infill angle or the number of contour repetitions, layer by layer, in order to improve surface quality or reduce local time/energy consumption. Finally, we did not consider any quality aspects or mechanical properties in the fitness function, leaving open the possibility for additional information integration into this function.

5 Conclusions

In this work, we acknowledge the challenge of balancing processing time and energy consumption in Wire Arc Additive Manufacturing, a metal Additive Manufacturing (AM) technique. As a response to this challenge, a conceptual AI-based slicer software framework is proposed. By integrating AI-driven process modelling into the slicer environment, the framework demonstrates how data-driven tools can support more informed decision-making, using a nickel-based alloy impeller as a representative case study. Specifically, we proposed a framework for AI-driven process parameter optimisation that, through a single tunable hyperparameter, adapts to different geometries to minimise either build time or energy usage, providing as output the welding process parameters and the optimal path planning strategy. Finally, we discussed current limitations and outlined future directions, including the development of a fully integrated AI-based optimiser.

References

1. Ahmadi, M., Rahmatabadi, D., Karimi, A., et al.: The role of additive manufacturing in the age of sustainable manufacturing 4.0. In: Sustainable Manufacturing in Industry 4.0, pp. 57–78. Springer Nature Singapore, Singapore (2023)
2. Kokare, S., Oliveira, J.P., Godina, R.: A LCA and LCC analysis of pure subtractive manufacturing, wire arc additive manufacturing, and selective laser melting approaches. J. Manuf. Process. **101**, 67–85 (2023). https://doi.org/10.1016/j.jmapro.2023.05.102
3. Priarone, P.C., Pagone, E., Martina, F., et al.: Multi-criteria environmental and economic impact assessment of wire arc additive manufacturing. CIRP Ann. **69**, 37–40 (2020)
4. Gao, H., Li, H., Shao, D., et al.: Towards quality controllable strategies in wire-arc directed energy deposition. Int. J. Extreme Manuf. **7**(4), 042004 (2025)
5. Xiong, J., Zhang, G., Hu, J., Wu, L.: Bead geometry prediction for robotic GMAW-based rapid manufacturing through a neural network and a second-order regression analysis. J. Intell. Manuf. **25**, 157–163 (2014)
6. Ding, D., He, F., Yuan, L., et al.: The first step towards intelligent wire arc additive manufacturing: an automatic bead modelling system using machine learning through industrial information integration. J. Ind. Inf. Integr. **23**, 100218 (2021)
7. Petrik, J., Bambach, M.: RLTube: reinforcement learning based deposition path planner for thin-walled bent tubes with optionally varying diameter manufactured by wire-arc additive manufacturing. Manuf. Lett. **40**, 31–36 (2024). https://doi.org/10.1016/J.MFGLET.2024.01.007

8. Sideris, I., Petrik, J., Bambach, M.: Too hot to print, too slow to handle; finding optimal path characteristics for WAAM. Manuf. Lett. **41**, 879–890 (2024). https://doi.org/10.1016/J.MFGLET.2024.09.108
9. Mattera, G., Pan, Z., Nele, L., Laghi, V.: Reducing energy consumption of pulsed-gas metal arc additive manufacturing through machine learning algorithms. J. Manuf. Process. **156**, 13–28 (2025)
10. Doodman Tipi, A.: The study on the drop detachment for automatic pipeline GMAW system: free flight mode. Int. J. Adv. Manuf. Technol. **50**, 137–147 (2010). https://doi.org/10.1007/s00170-010-2515-8
11. Doodman Tipi, A.R., Sani, S.H., Pariz, N.: Improving the dynamic metal transfer model of gas metal arc welding (GMAW) process. Int. J. Adv. Manuf. Technol. **76**, 657–668 (2015)
12. Zhao, S., Qiu, X., Burnett, I., et al.: A lumped-parameter model for sound generation in gas metal arc welding. Mech. Syst. Signal Process. **147**, 107085 (2021)
13. Wu, B., Pan, Z., Ding, D., et al.: A review of the wire arc additive manufacturing of metals: properties, defects and quality improvement. J. Manuf. Process. **35**, 127–139 (2018)
14. Mattera, G., Chozaki, S.P., Norrish, J.: Advances in machine learning for parameters optimisation and in-situ monitoring of wire arc additive manufacturing. Weld. World. (2025). https://doi.org/10.1007/s40194-025-02200-5
15. He, F., Yuan, L., Mu, H., et al.: Research and application of artificial intelligence techniques for wire arc additive manufacturing: a state-of-the-art review. Robot. Comput. Integr. Manuf. **82**, 102525 (2023). https://doi.org/10.1016/j.rcim.2023.102525
16. Manikandan, N., Arumugam, M.: Optimizing cold metal transfer-wire arc additive manufacturing parameters for enhanced mechanical properties and microstructure of ER5356 aluminum alloy using artificial neural network and response surface methodology. J. Mater. Eng. Perform. (2024). https://doi.org/10.1007/s11665-024-10403-y
17. Krishnaveni, S., Kunchala, B.R., Gamini, S., Ch Anilkumar, T.: Machine learning-based bead modeling of wire arc additive manufacturing (WAAM) using an industrial robot. Mater. Today Proc. (2023). https://doi.org/10.1016/j.matpr.2023.04.534
18. Šket, K., Brezočnik, M., Karner, T., et al.: Predictive modelling of weld bead geometry in wire arc additive manufacturing. J. Manuf. Mater. Proc. **9**, 67 (2025). https://doi.org/10.3390/jmmp9020067
19. Tomaz, I.D.V., Colaço, F.H.G., Sarfraz, S., et al.: Investigations on quality characteristics in gas tungsten arc welding process using artificial neural network integrated with genetic algorithm. Int. J. Adv. Manuf. Technol. **113**, 3569–3583 (2021). https://doi.org/10.1007/s00170-021-06846-5
20. Le, V.T., Mai, D.S., Doan, T.K., Paris, H.: Wire and arc additive manufacturing of 308L stainless steel components: optimization of processing parameters and material properties. Eng. Sci. Technol. Int. J. **24**, 1015–1026 (2021). https://doi.org/10.1016/j.jestch.2021.01.009
21. Le Van, T., Dinh, D.M., Bui, M.C., et al.: Optimizing process stability and energy efficiency in wire arc additive manufacturing of nickel-based superalloys using machine learning. Int. J. Adv. Manuf. Technol. (2025). https://doi.org/10.1007/s00170-025-15748-9
22. Mattera, G., Caggiano, A., Nele, L.: Energy efficiency optimisation in wire arc additive manufacturing of invar 36 alloy via intelligent data-driven techniques. Int. J. Precis. Eng. Manuf.-Green Technol. (2025). https://doi.org/10.1007/s40684-025-00705-4
23. Ferreira, R.P., Schubert, E., Scotti, A.: Exploring multi-armed bandit (MAB) as an AI tool for optimising GMA-WAAM path planning. J. Manuf. Mater. Proc. **8**, 99 (2024). https://doi.org/10.3390/jmmp8030099
24. Ferreira, R.P., Schubert, E., Scotti, A.: Reducing computational time in pixel-based path planning for GMA-DED by using multi-armed bandit reinforcement learning algorithm. J. Manuf. Mater. Proc. **9**, 107 (2025). https://doi.org/10.3390/jmmp9040107

25. Petrik, J., Bambach, M.: Reinforcement learning and optimization based path planning for thin-walled structures in wire arc additive manufacturing. J. Manuf. Process. **93**, 75–89 (2023). https://doi.org/10.1016/j.jmapro.2023.03.013
26. Rosli, N.A., Alkahari, M.R., bin, A.M.F., et al.: Review on effect of heat input for wire arc additive manufacturing process. J. Mater. Res. Technol. **11**, 2127–2145 (2021). https://doi.org/10.1016/j.jmrt.2021.02.002
27. Mattera, G., Manoli, E., Nele, L.: Study on the printing stability of invar 36 alloy under different process parameter conditions in gas metal arc additive manufacturing. Mater. Sci. Addit. Manuf. **4**, 025220046 (2025) https://doi.org/10.36922/MSAM025220046
28. Ding, D., Pan, Z., Cuiuri, D., Li, H.: A multi-bead overlapping model for robotic wire and arc additive manufacturing (WAAM). Robot. Comput. Integr. Manuf. **31**, 101–110 (2015)
29. Henckell, P., Gierth, M., Ali, Y., et al.: Reduction of energy input in wire arc additive manufacturing (WAAM) with gas metal arc welding (GMAW). Materials. **13**, 2491 (2020). https://doi.org/10.3390/ma13112491
30. Liu, Z., Zhao, Y., Wang, Q., et al.: Modeling and assessment of carbon emissions in additive-subtractive integrated hybrid manufacturing based on energy and material analysis. Int. J. Precis. Eng. Manuf.-Green Technol. (2024). https://doi.org/10.1007/s40684-023-00588-3
31. Joseph, A., Harwig, D., Farson, D.F., Richardson, R.: Measurement and calculation of arc power and heat transfer efficiency in pulsed gas metal arc welding. Sci. Technol. Weld. Join. **8**, 400–406 (2003). https://doi.org/10.1179/136217103225005642

Multi-modal Performance Analysis of Harmonic Gearboxes Manufactured Through 3D-Printing Technology Using PLA and PETG

Horea-Ştefan Goia[1], Corina Birleanu[1], Florin Popişter[1](✉), Friedemann Schaber[2], Mihai Dragomir[1], and Paul Ciudin[1]

[1] Technical University of Cluj-Napoca, Cluj-Napoca, Romania
Florin.Popister@muri.utcluj.ro
[2] University of Northampton, Northampton, UK

Abstract. The study aims to evaluate the acoustic, thermal, and tribological performance of two geometrically identical harmonic gearboxes manufactured from Polylactic Acid (PLA) and Polyethylene Terephthalate Glycol (PETG) through Fused Filament Fabrication (FFF), with the objective of determining how material choice affects dynamic stability and long-term dimensional integrity. A custom experimental setup was developed to enable simultaneous, no-load endurance testing under controlled laboratory conditions, incorporating sound pressure level (SPL) measurements, thermal imaging, and high-resolution wear analysis of the Flexspline teeth. Each gearbox was subjected to ten cycles of 30-min operation, during which acoustic emissions and peak surface temperatures were recorded. Subsequently, a final wear analysis was conducted. The resulting dataset reveals material-dependent trends in both acoustic and thermal behaviour, with each polymer exhibiting distinct SPL evolution patterns and heat accumulation characteristics. Correlation between thermal rise and acoustic instability indicates that frictional processes at the tooth interface evolve differently for PLA and PETG over repeated cycling. Wear examination confirms significant geometric alterations and wear debris formation, consistent with the expected tribological degradation mechanisms in polymer–polymer contact. These findings underscore the significance of material selection in the design of additively manufactured harmonic drives, particularly for low-load, precision, and noise-sensitive applications, where subtle thermal and tribological effects can influence operational consistency and functional longevity.

Keywords: Harmonic Drive · Fused Filament Fabrication · Tribology · Acoustic Emission · Temperature · Wear

1 Introduction

1.1 Harmonic Drive Technology and Applications

The demand for high torque, precise angular positioning, and minimal power-to-weight ratios in contemporary robotics, aerospace mechanisms, and precision machine tools has solidified the Harmonic Drive, also known as strain wave gearing, as a critical speed

B. Gapiński et al. (Eds.): MANUFACTURING 2026, LNME, pp. 56–70, 2026.
https://doi.org/10.1007/978-3-032-22943-4_5

reduction technology. The system's foundational principle involves the elastic deformation of a thin-walled gear, the Flexspline, via an input Wave Generator to sequentially engage with a fixed Circular Spline [1–3]. This design yields distinct advantages, notably its near-zero backlash, compact coaxial configuration, and ability to achieve high reduction ratios in a single stage [1, 4]. The complex operation, involving continuous flexing, significant tooth-to-tooth friction, and cyclic loading on the thin-walled components, renders the system highly sensitive to material properties, dimensional accuracy, and thermal management [5, 6]. Investigations into novel materials for these components are essential to expand the applicability of this technology.

1.2 Additive Manufacturing Context

The emergence of Additive Manufacturing (AM), particularly the Fused Filament Fabrication (FFF) technique, offers a compelling solution for rapid prototyping and fabricating specialized, complex components. FFF allows for the creation of intricate geometries, such as the thin, flexible walls and fine teeth of the Flexspline, without the constraints and costs associated with traditional injection moulding or machining [6, 7]. For polymer gears, FFF presents an avenue for reducing part mass and enhancing customization, making it particularly attractive for low-cost, low-load applications in mechatronics and robotics, especially at the prototyping stage [8].

The FFF fabrication technique introduces unique challenges that impact the final mechanical performance. The layer-by-layer deposition process results in material anisotropy and inter-layer porosity, which lead to reduced strength and different mechanical properties compared to bulk material [9]. Furthermore, the resulting surface roughness is typically higher than that of moulded or machined gears [4, 10].

1.3 Tribological Considerations and Material Properties

The service life and efficiency of a polymer gear system, even under low or no-load conditions, are predominantly dictated by its tribological performance and resistance to thermal or mechanical degradation [11]. In a harmonic drive, the Flexspline teeth undergo continuous cyclic radial and tangential motion against the Circular Spline, resulting in a complex kinematic field characterized by a combination of rolling and sliding contact, that generate friction head, which is critical for polymer components [12, 13].

For FFF-printed polymers, the rough surface topography inherent to the process influences the initial running-in period, exacerbating wear mechanisms such as the abrasion, and accelerating temperature rise at the meshing interfaces, altering mechanical and tribological parameters, which are critical for wear analysis [14, 15]. The complexity of analysing the degradation under continuous mechanical stress and temperature fluctuation is also highlighted by research into the thermo-mechanical behaviour of components under dynamic load, where friction and internal heat generation significantly modify the stress state [16].

Unlike traditional metallic systems where wear governs failure, polymer gears frequently fail due to thermal softening and plastic deformation well before tooth fracture occurs [13]. The maximum operating temperature is constrained by the material's Glass Transition Temperature (T_g) or Heat Deflection Temperature (HDT), exceeding this threshold rapidly reduces the polymer's elastic modulus and yield strength, leading to rapid tooth profile deviation and failure of the kinematic precision [17].

This study focuses on a direct comparison between the PLA and PETG materials, two of the most widely adopted FFF polymers. PLA is often favoured for its rigidity and wide accessibility, being however susceptible to thermal creep and deformation even at moderate operating temperatures [18]. Conversely, PETG is known for its superior ductility and impact resistance, suggesting a potentially greater capacity for sustained kinetic energy dissipation and greater resistance to thermally induced degradation [19]. Additionally, FFF-manufactured gears are susceptible to interlayer delamination and structural anisotropy [9, 11]. Furthermore, the molecular structure differences between the two polymers may influence their surface quality and, consequently, the coefficient of friction and wear particle generation during use, impacting overall gearbox noise and internal temperatures [20].

1.4 Rationale and Research Gap

Despite the growing utilization of FFF polymers for functional gear applications, a significant gap remains in the quantitative understanding of their endurance and stability under dynamic operational conditions, particularly in high-contact-ratio systems like the harmonic drive [11]. Previous studies have largely focused on static mechanical properties or highly loaded gear tests, where wear and structural failure are dominant [11, 21]. In applications characterized by high cycle counts and low torque, the primary indicators of component degradation are often subtle shifts in the dynamic signature, specifically thermal or acoustic emissions, as well as wear markers that can be identified and observed through analytic methods such as high-resolution wear analysis [14].

Acoustic measurement serves as a highly sensitive non-destructive indicator of both the initial state, and the progressive degradation of the 3D printed surfaces [22]. Variations in SPL reflect changes in the gearbox's internal condition, such as increased tooth profile error or the onset of wear debris accumulation, types of tribological failures [23]. By conducting a direct comparison between PLA and PETG, this study aims to isolate the material-specific influence on the gearbox's acoustic stability, as well as the effects of internal friction. This isolation is critical because acoustic data is capable of reflecting the interplay between the material's damping characteristics and the dimensional or structural changes occurring due to frictional heating [23].

The two chosen materials, PLA and PETG, represent divergent thermal and mechanical profiles within the accessible FFF material range. Their selection allows for a direct comparative assessment of how a material's inherent properties dictate performance stability under identical dynamic stress conditions [19, 20].

Therefore, the rationale for this work is to provide specific, quantifiable comparative data on three key performance indicators: acoustic stability, thermal behaviour, and long-term dimensional fidelity. This study directly addresses the need for empirical data to inform material selection for FFF-printed harmonic drive components, focusing on the dynamic and dimensional precision, essential for low-load applications where noise, temperature and subtle structural changes are significant failure modes [11].

1.5 Research Aim and Novelty

The primary research aim of this study is to perform a rigorous, comparative analysis of two FFF 3D-printed harmonic drive gearboxes fabricated from PLA and. Specifically, the investigation seeks to quantify and contrast the materials' response across three key performance metrics: acoustic emissions, thermal stability, and mechanical/ structural degradation.

This work offers two distinct contributions to the field of additive manufacturing for mechanical systems. Firstly, by focusing on no-load running, the methodology effectively isolates the influence of internal frictional losses and thermal effects as the singular drivers of component degradation. Secondly, the use of acoustic monitoring coupled with pre- and post-test high-resolution wear analysis of the Flexspline teeth provides a novel, multi-modal diagnostic approach to correlate the dynamic acoustic signature with the resulting permanent dimensional change. This comparative data set, specifically detailing the endurance behaviour of common FFF materials in the demanding harmonic drive geometry, serves as a crucial resource for engineers selecting materials for precision, low-torque, and even noise-sensitive applications.

Furthermore, the present paper aims to bring a contribution to the current state of the art by analyzing FFF-printed polymers under a regime of continuous elastic deformation. While the extant literature predominantly addresses the tribological performance of rigid-body gearing, the present study investigates the interaction between the performance of 3D-printed gearbox components and the high-frequency flexing cycles fundamental to strain wave kinematics, aiming to provide a predictive performance framework essential for the deployment of lightweight actuators. Consequently, this work serves as a guide for evaluating the feasibility of functional FFF specific components in precision.

2 Materials and Methods

The present chapter outlines the methodology employed for the comparative endurance testing of the two FFF-printed harmonic drive gearboxes. The section details the custom-built experimental setup, the two harmonic gearboxes tested, the specific parameters used in the fabrication of the gearboxes, as well as the data acquisition techniques and methodology employed.

2.1 The Experimental Setup

The comparative endurance testing was conducted using a custom-designed experimental test bench, depicted in Fig. 1. The setup was specifically designed to facilitate concurrent, identical dynamic testing of both the PLA and PETG harmonic drive gearboxes while minimizing external influences on acoustic measurements.

The core of the dynamic system consisted of two identical harmonic gearboxes acted by two NEMA 23 stepper motors. The use of stepper motors, connected to the input shafts of the gearboxes using elastic couplings, ensured precise control over the input speed and consistency across both test subjects. The elastic couplings were used as a mean of dampening minor angular misalignments, minimizing parasitic vibrations that

Fig. 1. The experimental test bench.

could artificially inflate acoustic readings or induce uneven wear. Both the motors and the gearboxes were rigidly fixed to the test stand base using custom 3Dprinted supports.

To ensure the integrity of the acoustic measurements, the entire test stand was placed upon a 10-mm-thick foam sheet to isolate the setup from external vibrations. Furthermore, all measurements were conducted while the stand was enclosed within a removable foam box, fulfilling the role of an acoustic enclosure, isolating the gearboxes from ambient acoustic noise.

2.2 The Harmonic Drive Reducer

The two geometrically identical harmonic drive gearboxes, offering a reduction ratio of 50:1, were fabricated using the FFF process. The difference between the two gearboxes was the material used, PLA, respectively PETG, labelled as seen in Fig. 1.

The design featured a standard harmonic drive topology, comprising a fixed Circular Spline, presenting a denture composed of 102 teeth, a deformable Flexspline, containing a denture of 100 teeth, while bearings were used to minimize unwanted friction. During the assembly process, the gear teeth interfaces were lubricated using a multi-purpose grease [24]. The lubricant was applied to replicate standard industrial operating conditions, to minimize interfacial friction, dissipate heat, and prevent premature abrasive wear, being selected for its good mechanical stability, and ability to maintain performance over a wide operating temperature range, thus ensuring consistent tribological conditions throughout the endurance test. An equal quantity of 10 g of lubricant was applied to the denture interfaces and distributed uniformly. Subsequently, the reducer was rotated 20 times to ensure uniform distribution of the lubricant.

The nature of the trials mandated a no-load condition, meaning that no external torque or resistance was applied to the exit flange of either gearbox. This isolation ensured that all frictional heat generation, sound profile and subsequent mechanical degradation were solely attributable to the internal tribological interactions, at tooth level, between the flexspline and circular spline during operation.

2.3 The Fused Filament Fabrication of the Harmonic Drive Reducers

The component parts of both the PLA and PETG harmonic drive gearboxes were fabricated using the same FFF equipment, specifically an Ender 5 Max 3D printer. Utilizing a single machine minimizes systematic variability between the two material groups, ensuring that any difference between the two gearboxes would be related to the polymeric material employed within the fabrication process.

The parameters selected for the FFF process were optimized to balance dimensional accuracy and mechanical strength. These parameters varied between the PLA and PETG materials to account for their distinct thermal properties and specific fabrication requirements [25, 26]. Relevant 3D printing parameters are summarized in Table 1.

Table 1. FFF 3D Printing Parameters for the PLA and PETG Harmonic Gearbox Components.

Parameter	PLA Gearbox Parts	PETG Gearbox Parts
Nozzle Diameter (mm)	0.4	0.4
Layer Height (mm)	0.16	0.16
Infill Density (%)	100	100
Nozzle Temperature (°C)	220	230
Bed Temperature (°C)	60	70
Printing Speed (mm/s)	400	400

Following the additive manufacturing process, the resulting parts were not subjected to any post-fabrication chemical or thermal treatments, but only the necessary removal of support structures and any plastic residue inherent to the FFF process.

2.4 Laboratory Requirements and Standards

The experimental testing was conducted under strict laboratory conditions to ensure the validity and comparability of the acquired data, particularly concerning the acoustic and thermal measurements. Acoustic emission readings were performed following the principles of relevant standards, considering ISO 16032:2024 [27], while thermal measurements respected the metrological principles for temperature measurement instruments, classified under ICS 17.200.20 [28], ensuring the accuracy of the collected data.

2.5 Experimental Procedure

The experimental procedure was structured into two phases: an initial Preliminary Phase for establishing baseline data, and a repetitive Cyclic Endurance Phase for tracking material degradation over time. The entire process was designed to ensure that the acoustic and thermal data were collected under strictly identical operating conditions for both gearboxes. The subsequent steps conducted during the trial and the data gathering process of the practical work are described below. All measurements were strictly conducted considering the proper stabilisation of a PWBKA 4 A1 thermal imaging camera reading before registering the temperature, as well as conducting sound profile measurements for a minimum of 60 s, for each sound profile or thermal value.

During the **Preliminary Phase**, the gearboxes were subjected to a series of baseline measurements to establish the initial state of the test subjects and the ambient environment. This included the process of capturing high quality images of the initial tooth profiles of each harmonic gearbox.

1. **Initial Dimensional Assessment:** High-resolution images of the teeth on both Flexsplines of the gearboxes are acquired, serving as the baseline data for the final high resolution wear analysis used to quantify post-test mechanical degradation.
2. **Instrumentation Setup:** The PCE-322A Sound Level Meter is positioned within the acoustic enclosure (Fig. 1), at an equal distance between both gearboxes.
3. **Initial Thermal Stabilization:** The gearboxes are allowed to acclimatize to the controlled laboratory environment. The initial resting temperature of each gearbox surface is measured using a PWBKA 4 A1 thermal imaging camera and recorded.
4. **Ambient Noise Measurement:** The SPL of the enclosure is measured with the assembly being stationary, for the predetermined time duration.
5. **Initial Operational Sound Measurement:** The motors are briefly started, the sound pressure level emitted by each gearbox being measured separately at the trial initiation, providing the initial acoustic baseline for the newly assembled system.

The main phase of the experiment involved the systematic repetition of a standardized 30-min operational cycle, designed to induce thermal and mechanical stress on the components, which was consistently followed by a controlled cool-down period. The process of thermal cycling was executed for a total of 10 complete cycles for both harmonic gearboxes.

1. **Pre-Cycle Temperature Check:** The temperature of each gearbox is measured and recorded, confirming that both units were acclimatized to the ambiental temperature.
2. **Acoustic Isolation:** The foam enclosure is closed.
3. **Dynamic Run:** The NEMA 23 stepper motors are started, running the gearboxes simultaneously at an input speed of 100 Revolutions Per Minute (RPM), for an exact duration of 30 min.
4. **Peak Temperature Measurement:** Immediately following the 30-min run, the enclosure is opened, and the maximum operational surface temperature of each gearbox is measured sequentially using the thermal camera. The enclosure is then immediately closed again to minimize thermal exchange.
5. **Acoustic Measurement (PLA Harmonic Drive):** The motor driving the PLA gearbox is maintained running while the PETG motor is briefly powered off. The SPL emitted exclusively by the PLA gearbox is measured and registered.
6. **Acoustic Measurement (PETG Harmonic Drive):** The motor driving the PLA gearbox is powered off, and the PETG motor is restarted. The SPL emitted exclusively by the PETG gearbox is measured and registered.
7. **Cool-down Phase:** Both motors are powered off, and the gearboxes are allowed to cool down to the initial temperature.
8. **Cool-down Confirmation:** The temperature of both gearboxes is measured, confirming that the temperature reached the initial value before the next cycle.

9. **Repetition:** Steps 1 through 8 are repeated for the subsequent 10 cycles.

3 Results and Discussion

3.1 Thermal Behaviour Results and Analysis

Thermal measurements were collected at the end of each 30-min operational cycle to characterise the heat generation behaviour of the two 3D-printed harmonic gearboxes. Figure 2 illustrates representative thermal images obtained after Cycle 10, highlighting the higher surface temperature consistently recorded on the PLA reducer compared to the PETG unit. This difference is maintained throughout the test and reflects the lower thermal resistance and heat dissipation capability of PLA.

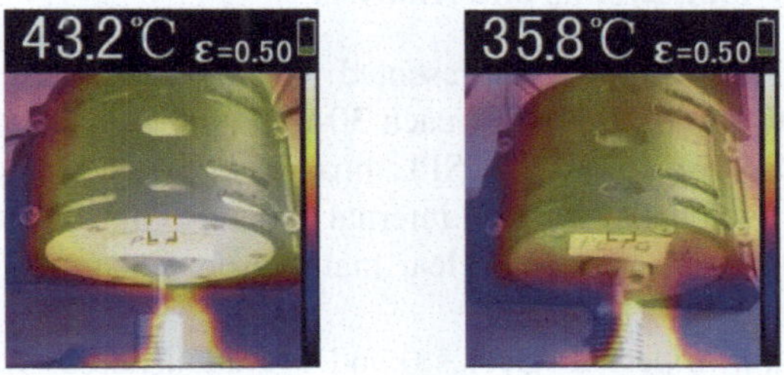

(a) (b)

Fig. 2. Thermal Imaging measurements (a) the PLA Harmonic Reducer and (b) the PETG Harmonic Reducer – example from cycle 10.

The complete temperature evolution for all ten cycles is presented in Fig. 3, where both reducers exhibit distinct material-dependent trends. The initial temperature decreases between Cycles 2 and 3 is characteristic of the running-in stage, during which asperity smoothing reduces friction at the tooth interface. A subsequent rise around Cycles 4–5 indicates the transition to a more conformal contact, where increased sliding interaction temporarily elevates heat generation, especially in the PLA gearbox.

The marked temperature reduction observed between Cycles 6–8 suggests the onset of a secondary stabilization regime, in which continued sliding yields smoother tooth flanks and reduced friction. A renewed temperature increase in Cycle 9, more pronounced in the PLA reducer, points to the beginning of structural degradation and geometric deviation of the internal components, which amplifies sliding irregularities. The final temperature drop in Cycle 10 likely reflects partial loss of structural integrity of the Flexspline. As internal damage progresses, the resulted reduced contact pressure and diminished load transfer between the dentures of the Flexspline and the Circular Spline, lead to lower frictional energy generation despite continued motion.

Overall, the thermal results show that PLA experiences stronger temperature fluctuations, while PETG demonstrates more stable and moderate thermal behaviour.

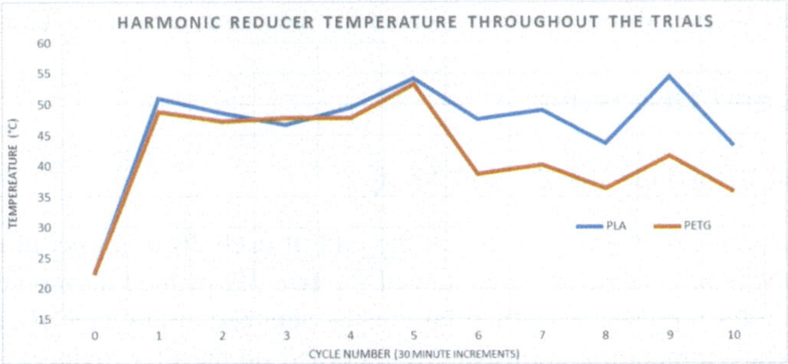

Fig. 3. The Graphical representation of the measured temperature trends.

3.2 Acoustic Behaviour Results and Analysis

The initial baseline measurement is presented in Table 2, as well as ten subsequent recordings corresponding to the end of each 30-min operational cycle. These acoustic traces characterise the evolution of the SPL throughout the endurance test and provide insight into the frictional behaviour and internal dynamic changes occurring within the two harmonic gearboxes, revealing a clear material-dependent difference in acoustic performance.

The initial running-in period (Cycles 0 and 1) established a critical divergence in acoustic behavior. While both gearboxes exhibited high fluctuation in the SPL reading during the Cycle 0 baseline measurement, indicative of FFF-induced surface imperfections, the PLA unit registered a slightly lower average SPL, approximately 74 to 75 dBA, compared to the PETG unit, approximately 77 to 78 dBA. This relationship was immediately inverted in the first operational cycle. The PLA gearbox experienced a rapid and significant escalation in noise, with measured peaks soaring to 80 to 83 dBA. Conversely, the PETG unit maintained a more constrained SPL range, stabilizing at 75 to 77 dBA during this initial phase. This suggests the immediate thermo-mechanical stresses of the first cycle had a substantially greater and more disruptive effect on the kinematic precision of the PLA component.

During the transient adjustment phase (Cycles 2 and 3), the SPL of the PLA unit dropped, temporarily converging with the PETG noise profile, which had quickly settled into a stable range of 72 to 74 dBA. However, the long-term endurance performance (Cycles 4 through 10) solidified the acoustic performance gap. The PLA gearbox consistently operated at a significantly higher and highly unstable acoustic level. From Cycle 4 onward, the PLA SPL regularly spanned 78 dBA to 85 dBA. This sustained, highly erratic acoustic signature indicates a persistent and amplified source of dynamic excitation. In strong contrast, the PETG gearbox demonstrated superior acoustic stability, maintaining a constrained SPL profile, predominantly between 74 dBA and 77 dBA, across the entire endurance phase. The stability of the PETG acoustic signature, characterized by both a lower average SPL and less extreme peak-to-peak variation, confirms its greater resilience to the combined thermal and frictional stresses inherent to the no-load harmonic drive operation.

Table 2. Sound Pressure Level measurements for the Baseline and the 10 Cycles.

Cycle Number	Sound Pressure Level (SPL) Measurement PLA PETG
0 (Baseline)	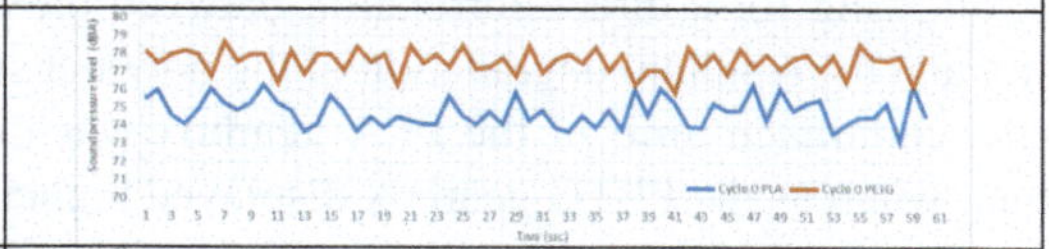
1	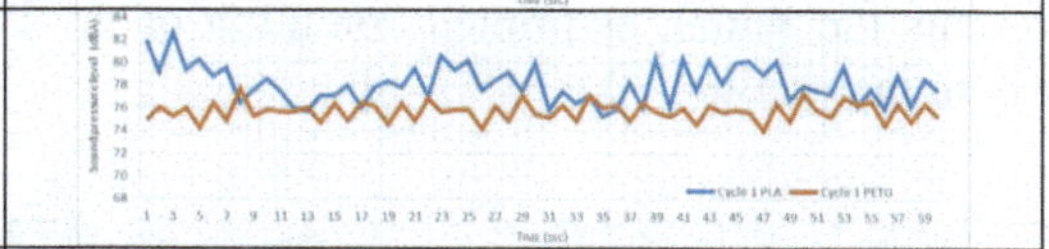
2	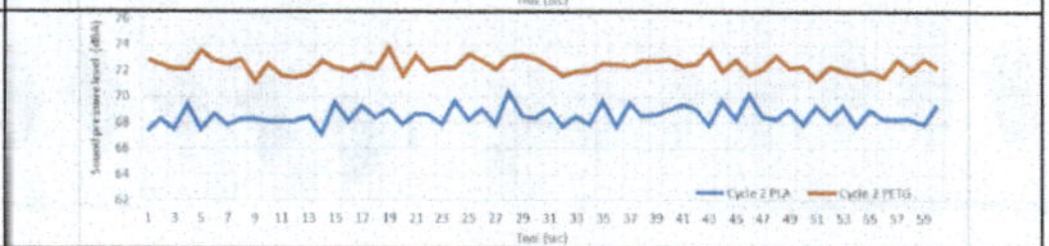
3	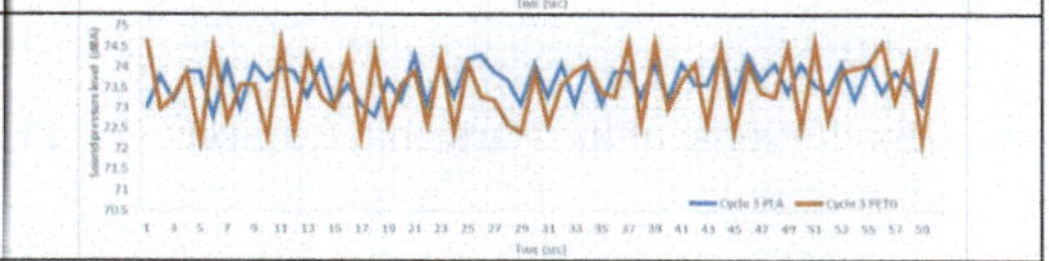
4	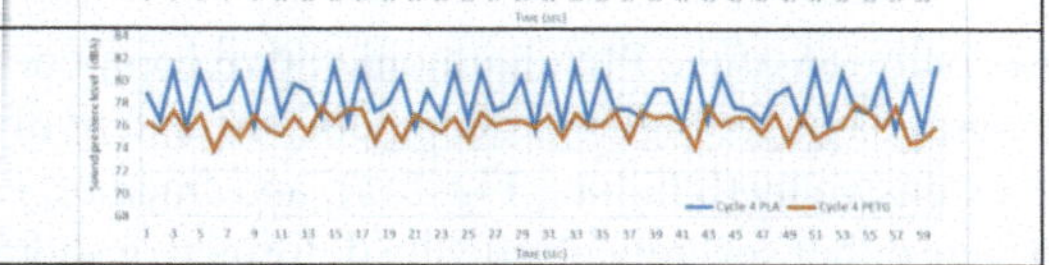
5	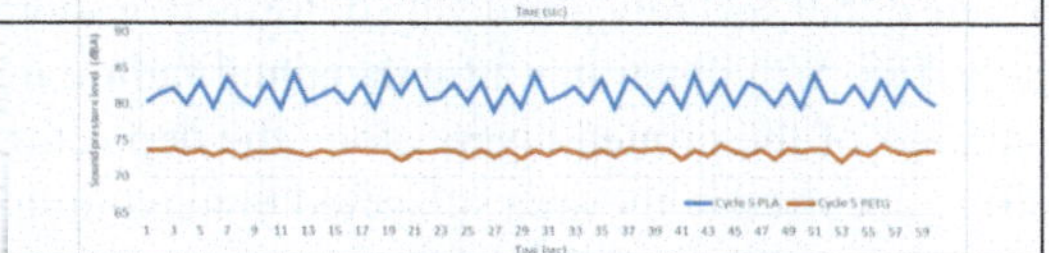
6	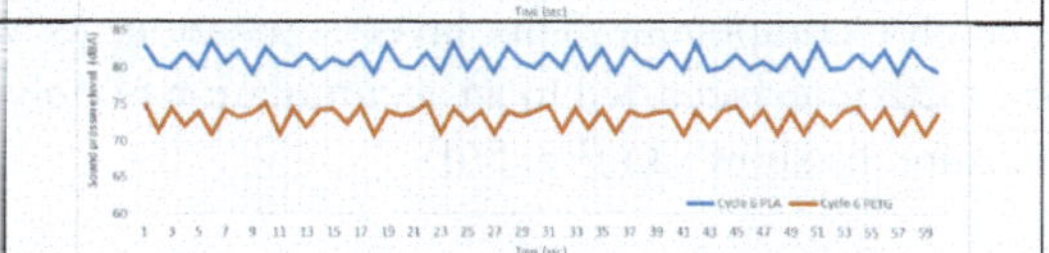
7	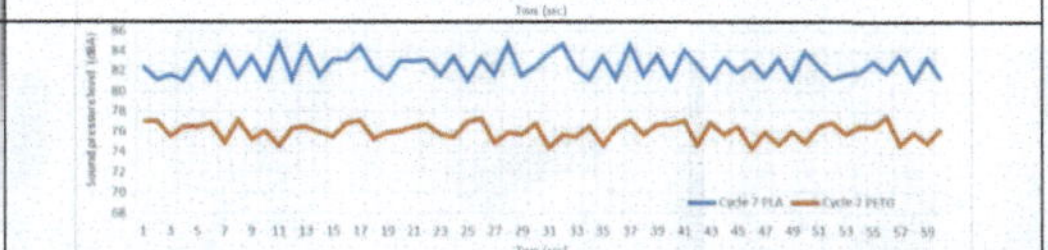
8	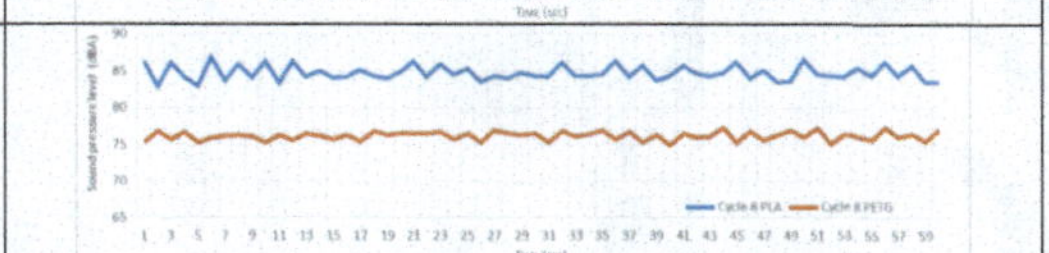
9	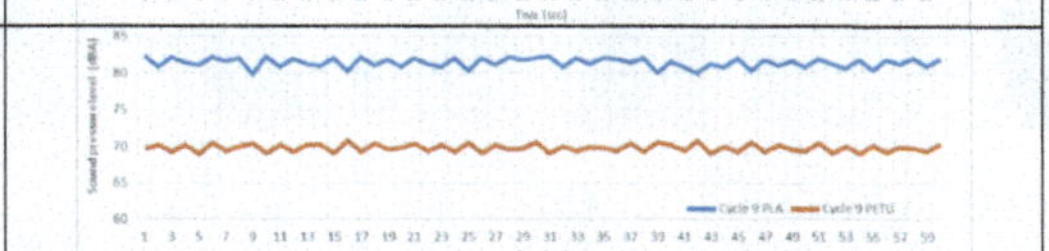
10	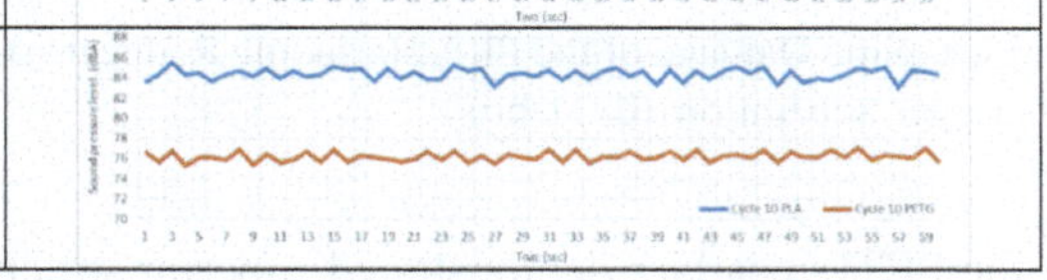

3.3 Mechanical Deformation Results and Analysis

The final phase of the investigation involved a detailed wear examination of the internal components, allowing for a direct correlation between the measured acoustic and thermal instability and the resultant degradation of the polymer structures. This analysis confirmed that the endurance limit of the PLA gearbox was dictated by catastrophic mechanical failure, whereas the PETG gearbox displayed higher dimensional stability.

Figure 4 presents the initial, as-printed condition of the Flexspline and Circular Spline teeth for both materials.

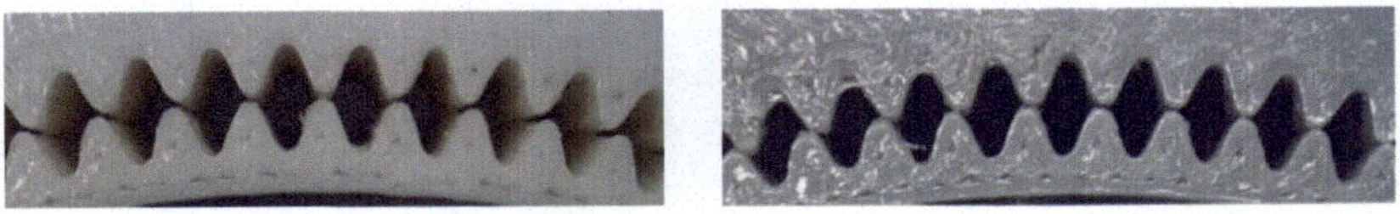

(a) (b)

Fig. 4. Initial Tooth Profile of the Harmonic Reducers - (a) PLA and (b) PETG.

The post-test condition of the PLA harmonic drive components revealed severe and irreversible structural damage. Figure 5 documents the extent of this failure. Within the same figure, the extensive teeth failure, Fig. 5(a), and fracture, Fig. 5(b), are presented. Flexspline teeth were either severely plastically deformed or shorn from the polymer ring structure. Crucially, Fig. 5(c) illustrates a fundamental failure mode specific to the FFF process: the separation of the printed layers along the deposition plane, on the flexible spline. This splitting occurred in the area subjected to maximum cyclic radial deflection by the wave generator component. This layer separation, as well as the mechanical interactions at the teeth interfaces, led to an accumulation of polymer wear debris within the gearbox enclosure, as shown in Fig. 5(d).

(a) (b) (c) (d)

Fig. 5. Post-Test Flexspline Damage of the PLA Harmonic Reducer – (a) Tooth Detachment, (b) Tooth Fracture(c), Layer Separation, (d) Debris.

In contrast, the PETG gearbox demonstrated superior structural integrity. Figure 6 shows the final state of the Flexspline. Figure 6(a) and (b) confirm that the PETG flexible spline did sustain localized tooth failure, but this damage was less catastrophic and less widespread than the multitooth failure and generalized fracturing observed in the PLA gearbox. Figure 6(c) confirms that the PETG component resisted better to the interlayer adhesion failure. Notably, there was no significant quantity of wear debris within the gearbox. The relative integrity of the PETG structure highlights a fundamental difference in the material's ability to maintain its geometric and structural profile.

(a) (b) (c)

Fig. 6. Post-Test Flexspline Damage of the PETG Harmonic Reducer – (a)Flexspline Fracture, (b) Flexspline Delamination, (c) Minimal Layer Separation and absence of debris.

3.4 Multi-modal Performance Correlation

The simultaneous monitoring of acoustic emissions, thermal response, and subsequent mechanical analysis establishes a strong causal relationship between material properties, tribological performance, and the failure modes observed. The observed differences being the results of the different polymeric materials used.

A strong correlation exists between the observed thermal and acoustic instability, particularly evident in the PLA gearbox. The PLA reducer's rapid and sustained elevation in SPL, peaking up to 85 dBA directly correlates with its superior heat generation. This indicates that friction-induced heat immediately pushed the PLA close to its glass transition temperature. This thermal softening initiated rapid plastic deformation and loss of structural integrity, which amplified the kinematic error and increased the noise output, establishing a positive feedback loop of high friction, high heat, and high SPL.

The PETG gearbox exhibited the opposite behavior. Its moderate temperature rise kept the material operating below its critical glass transition temperature. This resilience against thermal softening preserved dimensional accuracy of structural integrity, accounting for the rapid stabilization and sustained low SPL, around 74 to 77 dBA.

The wear analysis links the thermo-acoustic performance to the specific structural failure modes. The catastrophic damage in the PLA gearbox, characterized by multi-tooth fracture and layer delamination is the consequence of its failure to manage the internal heat and cyclic stress. This structural splitting created significant wear and geometric distortion, resulting in extreme acoustic instability in the later cycles. The PETG gearbox, due to its greater toughness and superior interlayer bond, resisted layer delamination. Its failure was limited to localized wear. This structural integrity enabled the PETG unit to maintain a more precise kinematic relationship, thereby preserving its stable and lower acoustic signature, confirming that material selection is the dominant factor governing thermo-acoustic stability in FFF harmonic drive systems.

While the present study employs a no-load protocol to baseline intrinsic behavior, the application of mechanical torque would likely intensify the identified degradation mechanisms. Increased contact pressures at the gear mesh would accelerate frictional heat generation and thermal softening, leading to earlier plastic deformation. Such thermal instability has the potential to induce significantly higher and more irregular acoustic signatures. The introduction of tangential stresses is predicted to intensify the prevalence of specific failure modes, including interlayer delamination and tooth fracture in the Flexspline, further amplifying performance disparity between the gearboxes.

4 Conclusions

The present study conducted a rigorous multi-modal performance analysis to quantitatively assess the acoustic, thermal, and mechanical endurance of two geometrically identical FFF 3D-printed harmonic drive gearboxes fabricated from PLA and PETG under a controlled, no-load cycling protocol. The research established a direct correlation between material intrinsic properties and dynamic degradation mechanisms.

The PLA harmonic gearbox failed rapidly due to thermal and structural instability. The material's specific properties allowed friction-induced heat to cause accelerated thermal softening, leading to geometric distortion of the Flexspline teeth. This degradation was acoustically captured as a persistent SPL instability, peaks reaching 85 dBA. The mechanical failure was confirmed to be catastrophic, including multi-tooth fracture and layer delamination, a critical FFF-specific failure mode under cyclic stress.

Conversely, the PETG gearbox demonstrated significantly greater endurance, driven by its superior toughness and resistance to heat-induced softening. This material successfully mitigated the effects of frictional heating, preserving its dimensional integrity and exhibiting a stable thermal profile. The outcome was a consistent, low acoustic signature, of 74 to 77 dBA, and structural resilience against layer delamination.

In conclusion, for additive manufacturing applications employing harmonic drive reducers, PETG offers superior thermo-acoustic stability and dimensional endurance compared to PLA. The performance disparity underscores that material resistance to heat-induced plastic flow and FFF-specific layer failure are the critical design factors determining component longevity, even in low-torque, high-cycle dynamic systems.

Acknowledgements. The authors wish to express their gratitude for the technical and financial support granted by the Technical University of Cluj-Napoca during the research and development of the paper.

References

1. Nandor, M., Heebner, M., Quinn, R., Triolo, R., Makowski, N.: Transmission comparison for cooperative robotic applications. Actuators. **10**(9), 203 (2021). https://doi.org/10.3390/act10090203
2. Yu, Z., Wang, X., Zheng, Y.: Design study on small transmission ratio harmonic reducer with new profile wave generator. Proc. Inst. Mech. Eng. C J. Mech. Eng. Sci. **237**(15), 3542–3560 (2023). https://doi.org/10.1177/09544062221147799
3. Kim, T., Lee, J., Jeong, W., Jeon, H., Oh, S.: Development of harmonic drive combining four arcs for conventional kinematic application. Machines. **10**(11), 1058 (2022). https://doi.org/10.3390/machines10111058
4. Yu, Z., Ling, S., Wang, X., Zheng, Y.: Modeling harmonic drive with small deformation of flexspline and assessment of tooth wear by sliding coefficient. Proc. Inst. Mech. Eng., Part J: J. Eng. Tribol. **237**(2), 431–450 (2022). https://doi.org/10.1177/13506501221121930
5. Kikuchi, M., Nitta, R., Kiyosawa, Y., Zhang, X.: Stress analysis of cup type strain wave gearing. Key Eng. Mater. **243-244**, 129–134 (2003). https://doi.org/10.4028/www.scientific.net/kem.243-244.129
6. Pacana, J., Pacana, A., Oliwa, R.: Possibilities of using selected additive methods for the production of polymer harmonic drive prototypes. Materials. **16**(11), 4073 (2023). https://doi.org/10.3390/ma16114073
7. Bochnia, J., Kozior, T., Blasiak, M.: The mechanical properties of thin-walled specimens printed from a bronze-filled PLA-based composite filament using fused deposition modelling. Materials. **16**(8), 3241 (2023). https://doi.org/10.3390/ma16083241
8. Zisopol, D.G., Iacob, D.V., Portoacă, A.I.: A theoretical-experimental study of the influence of FDM parameters on FLA spur gear stiffness. Eng., Technol. Appl. Sci. Res. **12**(5), 9329–9335 (2022). https://doi.org/10.48084/etasr.5183
9. Gajjar, T., Yang, R.(C.), Ye, L., Zhang, Y.X.: Effects of key process parameters on tensile properties and interlayer bonding behavior of 3D printed PLA using fused filament fabrication. Prog. Addit. Manuf. **10**, 1261–1280 (2025). https://doi.org/10.1007/s40964-024-00704-y
10. Batista Ponce, M., del Sol, I., Gómez-Parra, A., Vázquez-Martínez, J.M.: Tribological performance of additive manufactured PLA-based parts. Polymers. **16**(17), 2529 (2024). https://doi.org/10.3390/polym16172529
11. Agca, B.V., Tunalioglu, M.S.: Wear and service life of 3-D printed polymeric gears. Polymers. **14**(10), 2064 (2022). https://doi.org/10.3390/polym14102064
12. Song, C., Zhu, F., Li, X., Du, X.: Three-dimensional conjugate tooth surface design and contact analysis of harmonic drive with double-circular-arc tooth profile. Chin. J. Mech. Eng. **36**, 83 (2023). https://doi.org/10.1186/s10033-023-00909-2
13. Alharbi, K.: Tribological analysis and thermal behavior of gear tooth flanks in polymer gears during dry operation. J. Tribol. (2025). https://doi.org/10.1115/1.4067407
14. Ciobanu, R., Arhip, G., Donțu, O., Rizescu, C.I., Grămescu, B.: Additive manufacturing meets gear mechanics: understanding abrasive Wear evolution in FDM-printed gears. Polymers. **17**(13), 1810 (2025). https://doi.org/10.3390/polym17131810
15. Pustan, M., Birleanu, C., Voicu, R., Muller, R.: AFM characterization of temperature effect on the SU-8 mechanical and tribological properties. Polymers. **14**(5), 1009 (2022)
16. Pustan, M., Birleanu, C., Dudescu, C.: Simulation and experimental analysis of thermo-mechanical behavior of microresonators under dynamic loading. Microsyst. Technol. **19**(6), 915–922 (2012)
17. Mao, K.: A new approach for polymer gear design. Proc. Inst. Mech. Eng. C J. Mech. Eng. Sci. **224**(3), 651–658 (2010)

18. Auras, R., Lim, L.-T., Selke, S.E., Tsuji, H. (eds.): Poly(Lactic Acid): Synthesis, Structures, Properties, Processing, and Applications. Wiley (2010)
19. Martins, R.F., Branco, R., Martins, M., et al.: Mechanical properties of additively manufactured polymeric materials—PLA and PETG—for biomechanical applications. Polymers. **16**(13), 2024 (1868). https://doi.org/10.3390/polym16131868
20. Batista, K.L., Tenorio, D., del Sol, I., Vázquez-Martínez, J.M.: Tribological analysis of fused filament fabricated PETG parts coated with IGUS. Appl. Sci. **14**(16), 7161 (2024). https://doi.org/10.3390/app14167161
21. Agca, B.V., Tunalioglu, M.S.: Experimental investigation of the Wear of some polymer gear wheels produced by three-dimensional printer on the variation of fill percentage and Shell thickness. Eurasia Proc. Sci., Technol., Eng. Math. (EPSTEM). **31**, 52–63 (2024). https://doi.org/10.55549/epstem.1593219
22. García-Vilana, S., Adrover-Monserrat, B., Sánchez-Molina, D., et al.: Using acoustic emission to assess stiffness loss in thermoplastics processed through additive manufacturing. Prog. Addit. Manuf. **10**, 8053–8063 (2025). https://doi.org/10.1007/s40964025-01092-7
23. Polanec, B., Glodež, S., Belšak, A.: Noise evaluation of coated polymer gears. Polymers. **15**(3), 783 (2023). https://doi.org/10.3390/polym15030783
24. Nigrin: Nigrin Mehrzweckfett 250 ml Tube, Nigrin Official Product Page. [Online]. Available: https://products.nigrin.com/74145/nigrin-mehrzweckfett-250-mltube/100372?hl=en-US. Accessed 5 Oct 2025
25. Mehrabi, O., Sabri, H., Karamimoghadam, M., Khoran, M., Casalino, G., Moradi, M.: Statistical and experimental study on additive manufacturing of polyethylene terephthalate glycol using fused deposition modeling. Smart Mater. Methods. **1**(2), 119–139 (2024)
26. Moradi, M., et al.: Enhancing 3D printing copper-PLA composite fabrication via fused deposition modeling through statistical process parameter study. Micromachines (Basel). **15**(9), 1082 (2024)
27. ISO 16032:2024 Acoustics - Measurement of sound pressure level from service equipment or activities in buildings - Engineering method. 2nd (ed.) International Organization for Standardization, Geneva, Switzerland (2024). [Online]. Available: https://www.iso.org/standard/83874.html Accessed 23 Oct 2025
28. ICS 17.200.20: Temperature-measuring instruments, International Classification for Standards. International Organization for Standardization, Geneva, Switzerland. [Online]. Available: https://standards.iteh.ai/catalog/ics/17.200.20. Accessed 23 Oct 2025

An Adaptive FEA-Based System for Computationally Efficient Simulation of Cladding in Surface Enhancement Manufacturing Processes

Andrzej Chmielowiec(✉)

Rzeszow University of Technology, Rzeszow 35-959, Poland
achmie@prz.edu.pl

Abstract. This paper presents a computationally efficient finite element analysis (FEA) approach for simulating the cladding process in surface enhancement manufacturing. The proposed method incorporates a transient heat transfer model with temperature-dependent material properties and a moving heat source described by a double Gaussian ellipsoid, following the Goldak model. The formulation accounts for convection and radiation boundary conditions, as well as phase transformation effects represented by a broadened melting interval. A dynamic mesh adaptation strategy was implemented, enabling real-time modification of the computational domain and element size according to the position of the heat source. This approach significantly reduced the computational cost while maintaining numerical stability and accuracy. The simulation results for C45 steel (1.0503) demonstrated excellent convergence, with the relative error not exceeding 0.45%. The entire simulation, performed using open-source software, confirmed that the proposed adaptive FEA framework effectively balances precision and computational efficiency in modeling thermally complex cladding processes.

Keywords: Cladding · Finite Element Analysis · Adaptive Mesh · Heat Transfer · Gaussian Heat Source · Computational Efficiency

1 Introduction

The growing computational power enables precise modeling of increasingly complex processes. Undoubtedly, thermodynamic processes occurring under conditions of varying thermal conductivity and specific heat are among such phenomena. If convection and radiation are also considered in these processes, it turns out that the task becomes very complex from a computational perspective. Numerical simulation of the welding process requires appropriate modeling of the heat source over a wide range of temperatures.

Cladding and related surface enhancement processes involve highly transient heat transfer with strong spatial and temporal gradients, temperature-dependent material properties, and a moving heat source. Capturing these effects is essential for predicting melt-pool geometry, thermal histories, and the downstream thermo-mechanical response

B. Gapiński et al. (Eds.): MANUFACTURING 2026, LNME, pp. 71–81, 2026.
https://doi.org/10.1007/978-3-032-22943-4_6

(distortion and residual stress). Over the past decades, finite element formulations have become the predominant tool for such simulations across welding, directed energy deposition, and additive manufacturing variants, owing to their ability to accommodate nonlinear constitutive behavior, complex boundary conditions (convection and radiation), and evolving thermal fields [1].

A cornerstone of thermal modeling in welding and cladding is the representation of the volumetric heat input. The double ellipsoidal heat source introduced by Goldak et al. [2] provides a flexible three-dimensional Gaussian power density that distinguishes front and rear lobes, thereby spanning shallow arc and deep keyhole regimes and accommodating non-axisymmetric cases. Numerous studies have adopted and further scrutinized this model in different processes and materials. For hardfacing, three-dimensional transient analyses with temperature-dependent properties demonstrated that the double ellipsoid offers accurate temperature fields when contrasted with experiments [3]. In stainless steel GTAW, the double-ellipsoid formulation yields peak temperatures and weld pool sizes in notably better agreement with measurements than classic line or point-source solutions, underscoring the practical fidelity of volumetric heat sources [4].

A central challenge, however, lies in parameter identification for the heat source (ellipsoid semi-axes, front–rear partitioning, absorptivity). Data-driven and optimization-based strategies have emerged to reduce reliance on trial-and-error. Surrogate models such as artificial neural networks can map process inputs to Goldak parameters with competitive accuracy to regression methods [5]. Experimental–numerical calibration frameworks leverage weld-bead or melt-pool metrics to infer parameters systematically [6]. More recently, calibration using both cross-sectional and mid-front melt-pool profiles, coupled with global optimization, has been shown to recover a unique parameter set with very low reconstruction error, improving the robustness and transferability of thermal simulations [7].

Beyond arc welding and hardfacing, volumetric heat-source formulations and transient FE solvers underpin a broad spectrum of thermally driven manufacturing applications. In laser hardening, FE models enable process optimization and reduction of experimental effort by exploiting temperature-dependent $k(T)$, $c_p(T)$, and reflectivity [8]. In powder-bed and beam-based processes, FE frameworks address the distinct effective properties of powders versus solids and their implications for residual stress formation [9]. For wire-arc additive manufacturing and cladding, predictive thermal models inform toolpath design and power modulation, achieving sub-millimetric fidelity in clad geometry prediction after calibration [10]. Coupled thermo–mechanical simulations in wire-DED further demonstrate that physically derived, time-resolved heat-source models are crucial for accurate distortion and residual stress prediction across diverse component scales [11]. At the plant scale, CFD studies of industrial reheating furnaces reiterate the complexity of transient convective–radiative environments that set boundary conditions for reliable solid thermal models [12].

Against this backdrop, the present work contributes a computationally efficient FEA framework for cladding simulations that integrates (i) a Goldak-type moving volumetric heat source, (ii) temperature-dependent conductivity and heat capacity, and (iii) a numerically stabilizing phase-change treatment over a finite melting interval. Emphasis is placed on efficiency-oriented modeling choices that preserve accuracy—most notably, a domain-updating and element-sizing strategy tied to the instantaneous source position, which concentrates resolution where thermal nonlinearities are most pronounced. The resulting approach attains tight error control at substantially reduced degrees of freedom, aligning with the current trend toward calibrated, predictive, and tractable simulations of thermally complex surface enhancement processes.

2 Materials and Methods

The cladding process involves both the delivery of material to the clad layer and highly intensive heat transfer. Consequently, modeling such a process is associated with two fundamental challenges. The first involves the mathematical formulation of heat transfer, while the second concerns the time-dependent evolution of the computational domain. Therefore, the description of the methods employed will begin with the differential equation governing heat transfer, which accounts for convection, radiation, and an internal heat source within the material. In its general form, this equation is expressed as follows:

$$\rho c_p \frac{\partial T}{\partial t} = \nabla \cdot (k \nabla T) + q(x, y, z, t), \quad \text{in } \Omega, \tag{1}$$

$$-k\nabla T \cdot \mathbf{n} = h(T - T_\infty) + \varepsilon\sigma\left(T^4 - T_\infty^4\right), \quad \text{on } \partial\Omega, \tag{2}$$

where: where Ω denotes the computational domain, $\partial\Omega$ its boundary, and $\mathbf{n}$ is the outward unit normal vector, ρ – material density [kg · m $^{-3}$], c_p – material specific heat [J · kg $^{-1}$· K $^{-1}$], T – temperature [K], T_∞ – ambient temparature [K], t – time [s], k – thermal conductivity coefficient [W · m $^{-1}$· K $^{-1}$], q - power density of heat source [K · m $^{-3}$], h – heat transfer coefficient by convection [W · m $^{-2}$· K $^{-1}$], ε – dimensionless emissivity coefficient, σ – Stefan-Boltzmann constant 5.67×10^{-8} [W · m $^{-2}$· K $^{-4}$].

If we assume that the density, specific heat, and thermal conductivity are constant, then Eq. (1) represents a linear differential equation. Unfortunately, in the case of cladding, both the specific heat and the thermal conductivity coefficients depend significantly on temperature. This causes the equation to become nonlinear. As a result, the problem of finding solutions to such an equation becomes substantially more complex and numerically challenging. Figure 1 shows the relationships $k(T)$ and $c_p(T)$ for C45 steel (1.0503), according to the EN 10083–2 specification [3, 13]. It should be emphasized that the c_p plot also accounts for the melting process.

To maintain numerical stability during the solution process, it was assumed that the melting of steel occurs over the temperature range from 1673 K to 1973 K. This is a relatively broad interval, which does not allow for a precise distinction between the solid and liquid phases of the material. However, such an approach avoids the need for excessively small mesh elements, enabling the computations to be completed within a reasonable time frame. Of course, a narrower temperature range could be used, but in that case, the c_p curve within the melting interval becomes much steeper, which in turn requires the use of smaller elements and significantly increases the numerical complexity of the simulation.

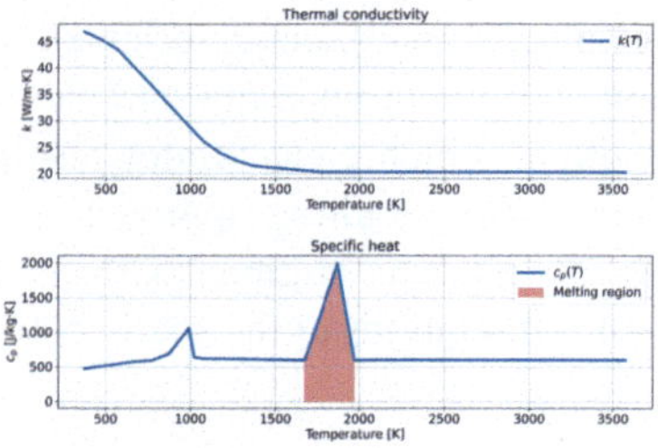

Fig. 1. Plots of thermal conductivity k and specific heat c_p for steel, accounting for the melting process.

The application of FEA requires transforming Eq. (1) into its weak form. If Ω is the domain, $\partial\Omega$ is its boundary, and v denotes the test function, then the weak form of the heat transfer equation in the presented model takes the following form:

$$\begin{aligned} \int_{\Omega}\rho c_p \frac{\partial T}{\partial t} v \, \mathrm{d}\Omega = -\int_{\Omega} k \nabla T \cdot \nabla v \, \mathrm{d}\Omega + \int_{\Omega} q(x, y, z, t) v \, \mathrm{d}\Omega \\ + \int_{\partial\Omega} h(T - T_\infty) v \, \mathrm{d}\Gamma + \int_{\partial\Omega} \varepsilon\sigma \left(T^4 - T_\infty^4\right) v \, \mathrm{d}\Gamma. \end{aligned} \tag{3}$$

The quality of the simulation obtained using the weak form formulated in this way depends largely on defining the heat source density q so that it reproduces the real process as accurately as possible. To achieve this, the double Gaussian ellipsoid model proposed by Goldak [2] was adopted. Let's define two ellipsoids that have the same semi-axes b, c along the y and z axes, respectively, and different semi-axes a_f, a_r along the x axis:

$$\begin{aligned} q_f(x, y, z) = \frac{6\sqrt{3}}{\pi\sqrt{\pi}} \frac{f_f P}{a_f bc} \exp\left(-3\left(\frac{x^2}{a_f^2} + \frac{y^2}{b^2} + \frac{z^2}{c^2}\right)\right), \\ q_r(x, y, z) = \frac{6\sqrt{3}}{\pi\sqrt{\pi}} \frac{f_f P}{a_r bc} \exp\left(-3\left(\frac{x^2}{a_r^2} + \frac{y^2}{b^2} + \frac{z^2}{c^2}\right)\right). \end{aligned} \tag{4}$$

In the above equation, P represents the total power of the source, and the coefficients f_f and f_r satisfy the relations $f_f = \frac{2a_f}{a_f + a_r}$ and $f_r = \frac{2a_r}{a_f + a_r}$.

$$q(x, y, z) = \begin{cases} q_f(x, y, z) & \text{for } x \geq 0, \\ q_r(x, y, z) & \text{for } x < 0. \end{cases} \tag{5}$$

The heat source shown in the Fig. 2 is modeled as a double Gaussian ellipsoid with the following parameters: $P = 1000$ W, $a_f = 3$ mm, $a_r = 5$ mm, $b = 3$ mm, $c = 2$ mm. The figure shows a cross-section through the heat source defined in this way, with surfaces of equal thermal power marked. A characteristic feature of the model is two glued ellipsoids, within which the power density is given by a three-dimensional Gaussian distribution $q(x, y, z)$ that satisfies the following condition.

$$\iiint_{x,y \in \mathbb{R}, z < 0} q(x, y, z) \, \mathrm{d}x \, \mathrm{d}y \, \mathrm{d}z = P. \tag{6}$$

During the experiment, the deposition of a clad layer was simulated at the center of a steel plate with dimensions of $50 \times 50 \times 6$ mm. Due to symmetry, it was assumed that the mesh generation and numerical calculations would be performed only for half

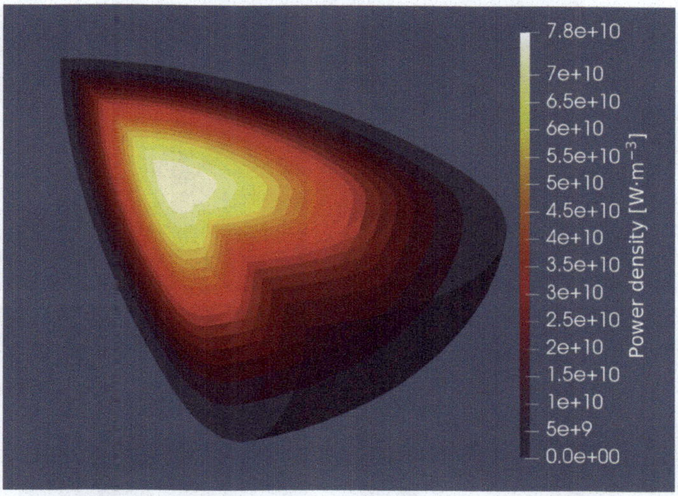

Fig. 2. Cross-section of the heat source modeled using a double Gaussian energy ellipsoid.

of the defined domain. Consequently, a mesh of a rectangular plate with dimensions of 50 × 25 × 6 mm was prepared. The modeled clad was applied along the edge of the mesh and had the shape of a quarter of an elliptical cylinder with semi-axes of 3 mm and 2 mm and a height ranging from 1 to 30 mm. The mesh density and element size were dynamically adjusted and strictly dependent on the current position of the heat source. The element size was determined using the following formula:

$$F(x, y, z) = 0.0001 + 8(x - 0.01)^2 + 8y^2 + 10z^2, \tag{7}$$

where the coordinates x, y, and z indicate the current position of the heat source in SI units (i.e., in meters). Additionally, it was assumed that the maximum element size could not be smaller than 0.002 m.

For simulation purposes, 30 meshes were generated, with the clad length increasing by 0.1 mm in each subsequent case. Figure 3 presents example meshes corresponding to 0.1, 1.0, 2.0, and 3.0 s of the simulation.

Mesh modeling, all numerical computations, and visualizations were performed using open-source software [14, 15]. The meshes were generated with GMSH version 4.13.1 [16]. Numerical computations for the FEA were carried out using FreeFem++ version 4.15 [17] with the built-in UMFPACK64 solver [18]. The visualizations were prepared using ParaView version 5.13.2. All scripts and input files can be provided upon request.

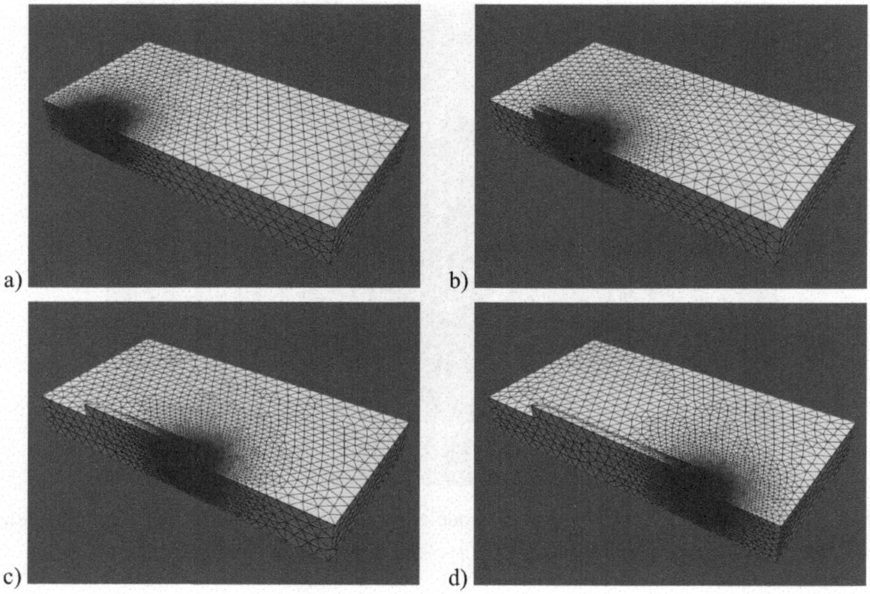

Fig. 3. Mesh with variable geometry and element size for the cladding process after: a) 0.1 s, b) 1.0 s, c) 2.0 s, and d) 3.0 s.

3 Results

All The simulation, conducted using the finite element method, involved a heat source moving at a constant velocity. The heat source traversed the surface of a steel plate, along a 0.030 m segment. The following parameter values were assumed for the simulation: material density $\rho = 7800$ kg $\cdot$ m $^{-3}$, ambient temperature $T_\infty = 300.0$ K, heat transfer coefficient by convection $h = 5.0$ W $\cdot$ m $^{-2}\cdot$ K $^{-1}$, dimensionless emissivity coefficient $\varepsilon = 0.8$. Heat source parameters: velocity $v_s = 0.010$ m $\cdot$ s $^{-1}$, geometry $a_f = 0.004$ m, $a_r = 0.006$ m, $b = 0.003$ m, $c = 0.002$ m, power $P = 2000$ W. On the other hand, the nonlinear dependencies for c_p and k are presented in the Fig. 1.

The entire simulation lasted 5.0 s, with the heat source active only for the first 3.0 s. This means that during the remaining 2.0 s, the material underwent cooling due to heat transfer to the surroundings through convection and radiation. To ensure high simulation accuracy, the computations were performed with a time step of 0.01 s. It should be emphasized that the computational domain defined by the mesh was updated every 0.1 s, corresponding to its modification after every 10 iterations during the active period of the heat source.

Figure 4 presents the simulation results for times of 0.1, 1.0, 2.0, 3.0, 4.0, and 5.0 s. Images (a)–(d) show the cladding process, which accounts for the domain variation resulting from the elongation of the clad layer, while images (e) and (f) illustrate the cooling process after the heat source has been deactivated.

The peak temperatures obtained in the present simulations are consistent with values reported in the literature for welding and cladding processes modeled using Goldak's

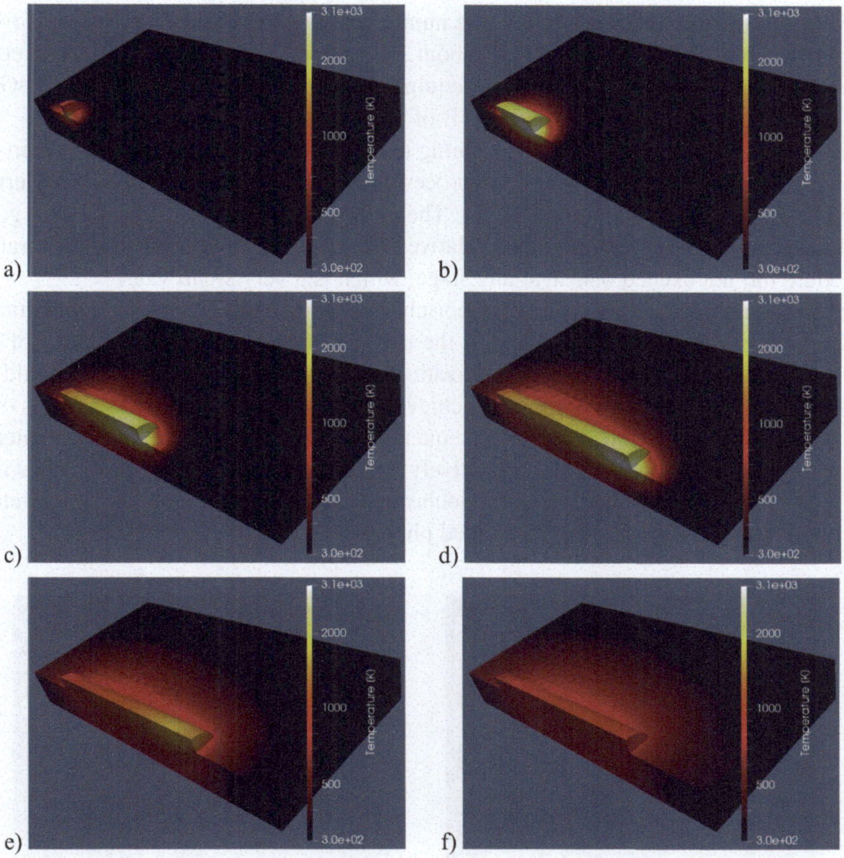

Fig. 4. Simulation results of the cladding process with variable clad geometry after: a) 0.1 s, b) 1.0 s, c) 2.0 s, d) 3.0 s, e) 4.0 s, and f) 5.0 s.

double ellipsoidal heat source. In the current study, the maximum temperature reached approximately 3140 K. For comparison, Chujutalli et al. [6] reported peak temperatures ranging from about 2630 K to 3470 K, depending on the simulation stage and calibrated heat source parameters in an experimental-based welding study. Similarly, Hu et al. [7], in the context of laser-based additive manufacturing and a calibrated Goldak heat source model, reported maximum temperatures on the order of 2960 K. The close agreement between the temperature levels obtained in the present work and those reported in independent experimental–numerical studies supports the physical plausibility of the adopted heat source formulation and parameter selection, despite differences in process type, scale, and calibration strategy.

The phase of clad formation and material heating was simulated using meshes containing approximately 130,000 to 150,000 finite elements (about 23,000 to 25,000 nodes). It should be noted that the use of system symmetry and the application of variable element size depending on the heat source position significantly reduced the computational complexity of the problem. If a uniform element size had been used for the entire trajectory of the moving heat source, the number of elements would have ranged from approximately 1,000,000 to 1,400,000 (about 180,000 to 230,000 nodes). The total computation time was 7.48 h on a computer equipped with an AMD Ryzen 5 7600 3.8GHz processor (6 cores / 12 threads) and 64GB of RAM.

Each computation sequence determining the current temperature distribution in the domain consisted of a five-step iteration process aimed at controlling the maximum error, whose absolute value was limited to 5 K. The applied approach resulted in a very good level of solution convergence, and the relative error obtained after applying the iterative procedure did not exceed 0.45% at any stage, which is a very small value.

Such a small error was achieved precisely due to the applied mesh modifications. A significant refinement of elements in the regions of the highest nonlinearity of the temperature function enabled the stabilization of the numerical process. It should be emphasized that this stability was also achieved by *spreading* the melting process over a temperature range of 300 K. As a consequence of this approach, the layer of material with an undefined phase state has a relatively large thickness. Its geometry is shown in Fig. 5, where the color palette on a monochromatic surface represents the temperature distribution within the region of undefined phase state.

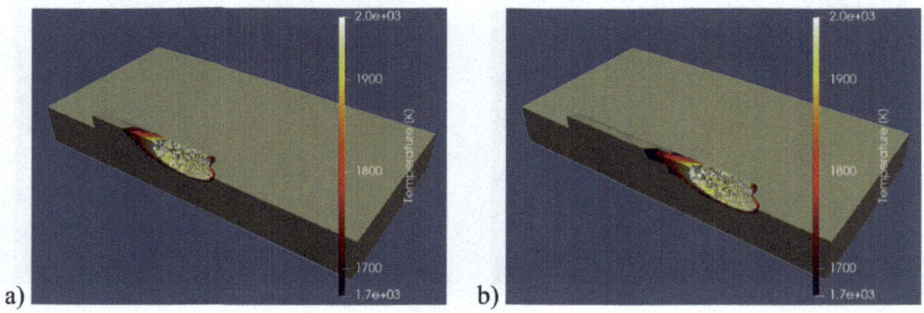

Fig. 5. Material of an undefined phase state on top of a material with a solid phase state at a) 2.0 s and b) 3.0 s of the simulation.

A formal sensitivity assessment of the numerical procedure indicates that the most critical factor affecting convergence and stability is the mesh density within regions where the material temperature enters the melting range. The numerical scheme was found to be highly sensitive to the spatial resolution of the phase-transition zone . A wider assumed melting temperature interval allows for larger finite elements, as it effectively increases the thickness of the transition region. To ensure stable convergence of the temperature field, the melting zone must be resolved by at least 5 to 10 finite elements across its thickness. If this condition is not satisfied, the simulation enters a regime of numerical instability, where convergence cannot be achieved even with multiple nonlinear iterations. These observations guided the selection of the adaptive mesh refinement strategy employed in the present study.

Figure 6 presents the sequence of meshes for the final 0.9 s of the cladding process. The presented images clearly illustrate how the mesh density and the clad model move along with the changing position of the heat source. The applied variable mesh method enables dynamic adaptation of the computational domain model depending on the intensity of temperature changes. This approach increases numerical stability while simultaneously reducing computational complexity.

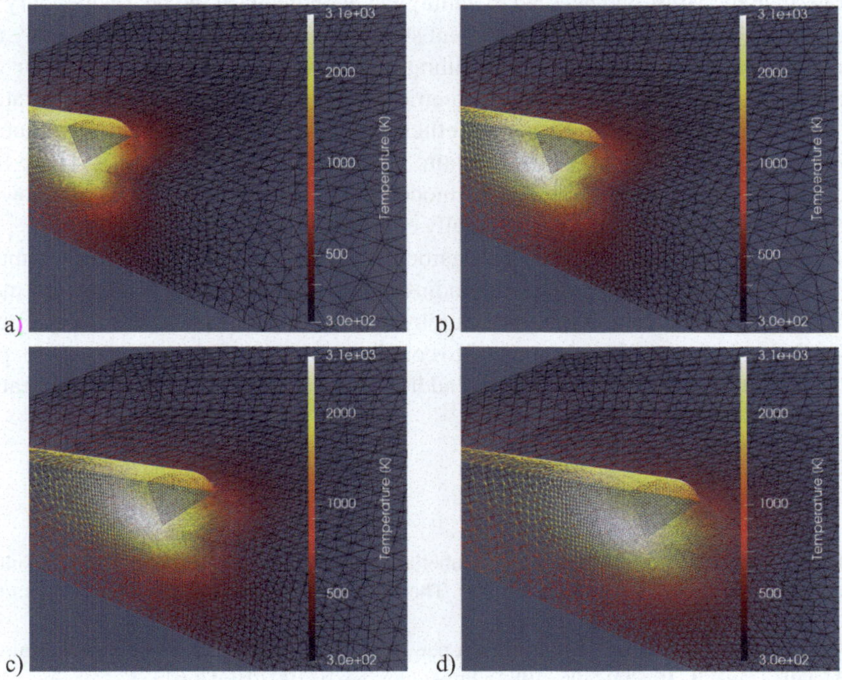

Fig. 6. Simulation results of the cladding process with variable clad geometry and adaptive mesh refinement after: a) 2.1 s, b) 2.4 s, c) 2.7 s, and d) 3.0 s.

4 Conclusions

The conducted research confirmed that the proposed adaptive finite element analysis (FEA)-based system enables computationally efficient simulation of the cladding process, while accurately capturing the variable clad geometry and complex thermal phenomena involved. The developed model combines a precise representation of the heat source, defined as a double Gaussian ellipsoid, with a dynamically modified mesh whose density is directly dependent on the current position of the heat source.

The applied adaptive meshing strategy, together with the extended melting interval of 300 K, ensured high numerical stability without excessive mesh refinement. As a result, the total number of finite elements was reduced by nearly an order of magnitude compared to fixed-density meshes, while maintaining excellent convergence. The relative error did not exceed 0.45%, demonstrating the high accuracy of the proposed approach.

Despite the demonstrated numerical stability and computational efficiency, the proposed approach has several inherent limitations that should be explicitly acknowledged.

First, the phase transformation was modeled using an extended melting temperature interval, which improves numerical robustness but does not allow for a sharp resolution of the solid–liquid interface. As a consequence, the thickness of the partially molten region is larger than in reality, and detailed metallurgical phenomena at the phase boundary are not explicitly captured. Second, the present study focuses on thermal analysis only and does not include direct experimental calibration of the heat source parameters or validation against measured melt-pool geometries. Finally, the adaptive meshing strategy was optimized for thermal accuracy and efficiency; its direct extension to fully coupled thermo-mechanical simulations may require additional refinement criteria. These limitations define the current scope of the model and motivate future extensions toward experimentally calibrated and mechanically coupled simulations.

The developed framework can be regarded as an adaptive and intelligent computational system, capable of automatically adjusting the computational domain and mesh parameters to local thermal conditions, thereby significantly improving efficiency. Future work will focus on extending the method to coupled thermo-mechanical analyses for predicting distortion and residual stresses in additive manufacturing and surface restoration processes.

References

1. Das, A., Kumar, A., Shankhwar, K., Gubeljak, N.: A review of heat source and resulting temperature distribution in arc welding. J. Therm. Anal. Calorim. **147**(23) (2022). https://doi.org/10.1007/s10973-022-11589-w
2. Goldak, J., Chakravarti, A., Bibby, M.: A new finite element model for welding heat sources. Metall. Trans. B. **15**, 299–305 (1984). https://doi.org/10.1007/BF02667333
3. Lazić, V., Ivanović, I., Sedmak, A., Rudolf, R., Lazić, M., Radaković, Z.: Numerical analysis of temperature field during hardfacing process and comparison with experimental results. Therm. Sci. **18**, S113–S120 (2014). https://doi.org/10.2298/TSCI130117177L
4. Jafarzadegan, M., Taghiabadi, R., Mofid, M.: Using double ellipsoid heat source model for prediction of HAZ grain growth in GTAW of stainless steel 304. Materials today. Communications. **31**, 103411 (2022). https://doi.org/10.1016/j.mtcomm.2022.103411
5. Tafarroj, M., Kolahan, F.: A comparative study on the performance of artificial neural networks and regression models in modeling the heat source model parameters in GTA welding. Fusion Eng. Des. **131**, 111–118 (2018). https://doi.org/10.1016/j.fusengdes.2018.04.083
6. Chujutalli, J., Lourenço, M., Estefen, S.: Experimental-based methodology for the double ellipsoidal heat source parameters in welding simulations. Mar. Syst. Ocean Technol. **15**, 110–123 (2020). https://doi.org/10.1007/s40868-020-00074-4
7. Hu, X., Chiu, L., Huang, A., Liu, M., Yan, W.: Application of melt pool profiles for parameter calibration of Goldak's heat source model. Addit. Manuf. **92**, 104379 (2024). https://doi.org/10.1016/j.addma.2024.104379
8. Martinovs, A., Polukoshko, S., Zaicevs, E., Revalds, R.: Laser hardening process optimization using fem. Eng. Rural Dev. Jelgava. **20**, 1500–1508 (2020). https://doi.org/10.22616/ERDev.2020.19.TF372
9. Zafar, M., Wu, C., Zhao, H., Wang, J., Hu, X.: Finite element framework for electron beam melting process simulation. Int. J. Adv. Manuf. Technol. **109**(7), 2095–2112 (2020). https://doi.org/10.1007/s00170-020-05707-x

10. Votruba, V., Fornusek, T., Havlan, T., Kratena, T., Smolik, J.: Modelling the weld cladding process to predict weld clad position and shape error. Int. J. Adv. Manuf. Technol. **132**(5), 2877–2888 (2024). https://doi.org/10.1007/s00170-024-13481-3
11. Nagaraj, P., Gurunathan, S., Amirthalingam, M.: Physically derived instantaneous modelling of complex current-voltage waveform-controlled arc-wire DED process—residual stress and distortion analysis. Int. J. Adv. Manuf. Technol. **138**(2), 687–708 (2025). https://doi.org/10.1007/s00170-025-15547-2
12. Hosain, M., Bel Fdhila, R., Sand, U., Engdahl, J., Dahlquist, E., Li, H.: CFD modeling of real scale slab reheating furnace. In: 12th International Conference on Heat Transfer, Fluid Mechanics and Thermodynamics, HEFAT 2016 (2016)
13. EN 10083-2: Steels for quenching and tempering – Part 2: Technical delivery conditions for non-alloy steels. European Standard (2006). EN 10083–2:2006
14. Bulakh, M., Klich, L., Baranovska, O., Baida, A., Myamlin, S.: Reducing traction energy consumption with a decrease in the weight of an all-metal gondola car. Energies. **16**(18), 6733 (2023) https://www.mdpi.com/1996-1073/16/18/6733
15. Chmielowiec, A., Woś, W., Czyżewski, J.: Numerical analysis of inertia forces in the connecting rod and their impact on stress formation. Materials. **18**(6), 1385 (2025). https://doi.org/10.3390/ma18061385
16. Geuzaine, C., Remacle, J.F.: Gmsh: a 3-D finite element mesh generator with builtin pre-and post-processing facilities. Int. J. Numer. Methods Eng. **79**(11), 1309–1331 (2009). https://doi.org/10.1002/nme.2579
17. Hecht, F.: New development in FreeFem++. J. Numer. Math. **20**(3–4), 251–266 (2012). https://doi.org/10.1515/jnum-2012-0013
18. Davis, T.: Algorithm 832: UMFPACK v4.3 — an unsymmetric-pattern multifrontal method. ACM Trans. Math. Softw. (TOMS). **30**(2), 196–199 (2004). https://doi.org/10.1145/992200.992206

Numerical Modeling of Zirconia: A Synthetic Review of Mechanical Behavior and Manufacturing Processes

Mariam Rajouh[1,2], Szymon Wojciechowski[2,3](✉), El Mehdi Kibbou[1], Maciej Klosak[1], and Marcin Gołaszewski[4]

[1] Laboratoire InterDisciplinaire de Recherches Appliquées, LIDRA, Université Internationale d'Agadir- Universiapolis, Agadir, Morocco

[2] Faculty of Mechanical Engineering, Poznan University of Technology, Piotrowo 3, Poznan 60-965, Poland
szymon.wojciechowski@put.poznan.pl

[3] Faculty of Manufacturing Technologies with a Seat in Presov, Technical University of Kosice, 31 Sturova Street, Presov 080 01, Slovakia

[4] West Pomeranian University of Technology in Szczecin, Piastów 19, 70-310 Szczecin, Poland

Abstract. This paper presents an overview of numerical modeling methods applied to understand and predict the behavior of zirconia (ZrO_2). Zirconia's performance depends on several interacting factors, including its microstructure, phase changes, temperature effects, and mechanical loading. Because these interactions are complex, modeling plays an important role in describing how zirconia behaves during use and during processes such as sintering, machining, and additive manufacturing. This review summarizes the main constitutive models, finite element approaches, and multiscale methods used to simulate elasticity, plasticity, damage, and transformation toughening in zirconia. Although many advances have been made, current models are still incomplete and often lack full multiphysics coupling, realistic process-based material laws, and strong experimental validation. The paper identifies these gaps and suggests future directions, including better multiscale models, the use of machine learning to support physics-based simulations, and virtual manufacturing tools to help optimize zirconia processing and improve final component performance.

Keywords: Zirconia · Numerical Modeling · Finite Element Analysis · Manufacturing · Multiscale Modeling

1 Introduction

Zirconia (ZrO_2) is a high-performance ceramic extensively utilized in dental, biomedical, and structural applications due to its exceptional mechanical strength, fracture toughness, and chemical stability. Its unique stress-induced tetragonal-to-monoclinic phase transformation is a critical feature that provides transformation toughening, significantly enhancing crack resistance and reliability in various applications [1, 2]. However, the

B. Gapiński et al. (Eds.): MANUFACTURING 2026, LNME, pp. 82–97, 2026.
https://doi.org/10.1007/978-3-032-22943-4_7

relationship between zirconia's microstructure, thermal effects, and mechanical loading complicates the prediction of its behavior, as experimental approaches alone may not capture the full spectrum of its performance [3–5]. Finite element (FE) modeling has therefore become an essential tool to investigate zirconia's mechanical response and processing behavior. FE simulations allow the analysis of stress–strain evolution, temperature fields, damage initiation, and phase transformation under controlled virtual conditions, which is essential given the brittleness and small-scale heterogeneities of ceramic materials [6, 7]. By integrating physical constitutive laws with realistic boundary conditions, FE modeling provides a comprehensive study of zirconia from the microstructural to the macroscopic scale as demonstrated in various studies that explore its mechanical behavior and fracture toughness [8, 9]. The processing–structure–property relationship plays a crucial role in determining the final performance of zirconia components. Processing routes such as sintering, micro-milling, or additive manufacturing influence grain size, porosity, and residual stresses, which in turn affect strength and fracture behavior [10, 11]. Accurately capturing these interactions in numerical models requires coupling mechanical, thermal, and phase-transformation phenomena. Despite advancements in modeling these interactions, many studies still rely on simplified constitutive laws and limited experimental validation, leaving gaps in the predictive understanding of zirconia's response under realistic conditions [7, 12]. This paper provides a structured overview of numerical modeling approaches for zirconia, focusing on how constitutive laws, thermo-mechanical effects, and finite element simulations capture its complex mechanical behavior. It first examines models describing elastic, plastic, and transformation-induced deformation, followed by discussions on temperature effects, residual stresses, and phase transformations. Subsequent sections explore FE-based simulations applied to zirconia processing, including sintering, micro-milling, and additive manufacturing. The paper concludes by highlighting current gaps, challenges, and opportunities for developing more predictive, multiscale, and experimentally validated computational frameworks for zirconia-based systems.

2 Characterization of Zirconia

2.1 Crystalline Phases of Zirconia

The characterization of zirconia is fundamental for linking its microstructural features to mechanical and functional performance. Zirconia exhibits polymorphism with three primary crystalline phases: monoclinic, tetragonal, and cubic. Figure 1 shows the temperature ranges of the three polymorphic forms of zirconia: monoclinic (stable up to 1170 °C), tetragonal (1170–2370 °C), and cubic (above 2370 °C), enabling stress-induced transformation toughening that enhances fracture resistance.

This mechanism is particularly critical for applications such as dental implants, where maintaining the tetragonal phase is essential for long-term stability [13, 14]. High-temperature phases can be stabilized at room temperature through doping with oxides such as yttria (Y_2O_3), magnesia (MgO), or ceria (CeO_2), resulting in compositions like 3Y-TZP, 4Y-TZP, or 5Y-PSZ that improve toughness and thermal stability [5, 15–17]. The distribution of phases directly influences toughening behavior, as tetragonal-to-monoclinic transformation absorbs energy and impedes crack propagation, thereby enhancing durability [1, 18].

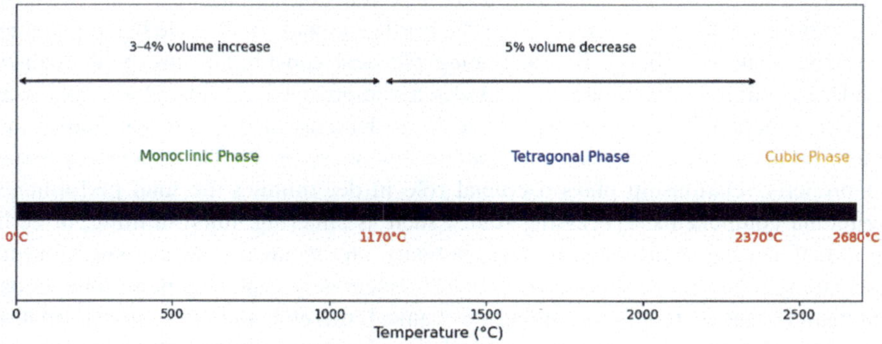

Fig. 1. Temperature-induced phase transitions of zirconia.

2.2 Microstructural Influences on Mechanical Behavior

Zirconia's mechanical performance is strongly governed by its microstructure. At the atomistic scale, defects and phase stability dominate; the mesoscale captures grain structures and boundaries; the microscale reflects porosity, inclusions, and microcracks; and the macroscale describes overall component response, including the effects of machining and sintering [19]. Smaller grain sizes enhance transformation toughening and strength by increasing the volume fraction of retained tetragonal zirconia, promoting microcrack nucleation and compressive surface stresses, which improves fracture toughness and flexural strengths [20]. Porosity plays a critical role in fracture resistance and elastic modulus, with higher porosity leading to decreased compressive strength and a transition from brittle to diffuse damage behavior [21]. Residual stresses, often induced by thermal expansion mismatches in composites, also influence crack propagation and thermal shock resistance. For instance, in zirconia–metal composites, compressive residual stresses in the metal phase enhance overall toughness [22]. In addition to microstructure and thermal properties, surface characteristics and chemical stability play a key role in zirconia's performance. Features such as surface roughness, grain boundary chemistry, and residual stresses can influence crack initiation, wear resistance, and long-term reliability. Advanced characterization techniques provide detailed insights into these aspects, enabling a better understanding of both local and bulk behavior. Considering these factors alongside microstructural and thermal properties helps establish a comprehensive picture of zirconia, supporting more accurate predictive modeling and performance optimization.

2.3 Selected Properties and Fracture of Zirconia

Zirconia exhibits pronounced temperature-dependent behavior in terms of thermal expansion, conductivity, heat capacity, and sintering shrinkage, which are essential for its performance across applications. Phase transitions occur at specific temperature thresholds, with monoclinic-to-tetragonal transformation near 1500 K and tetragonal-to-cubic around 2700 K [23]. The Young's modulus of zirconia ceramics generally decreases with increasing temperature until densification occurs, at which point the modulus rises sharply. Aging studies on dental zirconia indicate negligible effects below 134°C, highlighting stability under physiological conditions [24, 25]. In yttria-stabilized zirconia (YSZ) composites, thermal conductivity remains relatively low (1.5–3.0 $W \cdot m^{-1} \cdot K^{-1}$) across temperature ranges, while heat capacity shows minimal variation with YSZ content [26]. Despite extensive experimental characterization, reported thermo-mechanical properties show substantial variability across commercial zirconia systems. Young's modulus ($\approx$200–400 GPa), Poisson's ratio (0.25–0.35), and Vickers hardness ($\approx$1200–1400 HV) are strongly affected by composition, additives, grain size, and porosity rather than intrinsic material behavior [27–29]. Thermal conductivity is further influenced by stabilizing cations and composites or thin-film architectures, as demonstrated in Sc_2O_3-doped zirconia, $MoSi_2$–YSZ composites, and plasma-enhanced atomic layer deposited YSZ films [30–32]. Fracture behavior varies significantly among zirconia systems, with crack path stability governed by grain boundaries and internal grain interactions, as captured by advanced 3D finite element simulations [8, 33, 34]. However, many numerical models still rely on simplified, averaged, or temperature-independent properties, limiting their ability to accurately predict thermo-mechanical reliability and fracture behavior across processing routes.

3 Modeling of Mechanical Behavior and Processing of Zirconia

3.1 Material Constitutive Models

Constitutive modeling of zirconia aims to capture its complex mechanical response under varying thermo-mechanical conditions. Zirconia exhibits mainly linear elastic behavior with pronounced anisotropy governed by microstructure and processing history. The YSZ coatings show transverse isotropy due to layered defects, with elastic constants strongly dependent on temperature and heat treatment [35]. By contrast, single-crystal zircon displays high incompressibility and weak temperature dependence of elastic module [36]. Thermal effects further influence stability, as reduced thermal expansion in $ZrC(1 - x)Nx$ ceramics indicates improved performance under extreme conditions [37]. At elevated temperatures, 3Y-TZP shows strong coupling between temperature, strain rate, and electric current. Increasing temperature or electric current lowers flow stress, whereas higher strain rates increase it. These effects are critical for modeling high-temperature processing and additive manufacturing [38]. Overall, zirconia's elastic response is highly anisotropic and sensitive to phase stability and microstructure [39, 40]. Zirconia exhibits both rate-dependent and rate-independent plasticity. Rate effects are associated with delayed atomic rearrangements observed in CuZr-based systems [41], while rate-independent frameworks based on internal variables are suitable for quasi-static loading [42, 43]. A key toughening mechanism is the stress-induced tetragonal-to-monoclinic transformation near crack tips, which reduces stress intensity and impedes crack propagation [1, 44]. Transformation depends on applied stress, grain size, and crystallographic orientation [45], and can also be triggered by terahertz excitation [46]. Constitutive models for zirconia have been applied to compression, fracture, machining, and sintering. Elastic and fracture parameters are commonly obtained from experimental–numerical studies [27, 47]. In machining, Johnson–Cook models and brittle-to-ductile transition equations are used under high strain-rate conditions [48, 49], while sintering is modeled using analytical, FE, and ML-based approaches to predict densification and creep [50]. Despite favorable mechanical strength and biocompatibility for dental applications [51–53], limitations related to opacity and long-term performance remain [54], underscoring the need for integrated constitutive models coupling elasticity, plasticity, phase transformation, and effects.

Despite these advances, existing constitutive models for zirconia remain largely fragmented and context-specific (Table 1). Many formulations are calibrated for isolated loading conditions (uniaxial compression, machining, or sintering) and often neglect the simultaneous coupling between phase transformation, anisotropy, temperature, and strain-rate effects. Transformation toughening is frequently modeled using phenomenological approaches that lack direct links to evolving microstructural features such as grain size, orientation, and residual porosity. Moreover, widely used models such as Johnson–Cook, while effective for high-strain-rate machining, were originally developed for metals and may inadequately capture zirconia's brittle–ductile transition and phase-transformation-driven plasticity. Machine-learning-based constitutive laws show strong predictive potential but remain limited by data availability and physical interpretability. These limitations underscore the need for unified, physics-informed constitutive frameworks capable of capturing zirconia's coupled thermo-mechanical, microstructural, and process-dependent behavior across multiple applications.

3.2 Thermo-Mechanical Modeling

Temperature plays a critical role in zirconia's mechanical response, influencing phase transformations, thermal expansion, and micro-cracking. Total strain can be decomposed into elastic, thermal, and transformation components, with elastic strains strongly affected by phase-transition-induced volume changes and thermal strains driven by processing temperature gradients. The martensitic monoclinic–tetragonal transformation involves significant volumetric mismatch, generating elastic strains and micro-cracking, particularly at elevated temperatures. Moreover, the transition may evolve from first- to second-order behavior as temperature decreases, modifying stored elastic energy and mechanical properties in zirconia polycrystals [55]. The resulting anisotropic strain fields, especially at grain boundaries, lead to strain incompatibility and non-uniform stress distributions that remain challenging to estimate accurately [39]. Thermal expansion further complicates thermo-mechanical behavior by altering stress intensity factors and failure likelihood under thermal loading (Fig. 2) [56].

In multi-material systems such as alumina–zirconia composites, residual stresses interact with thermal shock resistance, significantly shifting the critical temperature difference for crack initiation [57]. While thermo-mechanical coupling is essential for

Table 1. Comparative Overview of Constitutive Models for Zirconia

Model	Key Assumptions	Captured Behaviours	Applicable Conditions	Main Limitations
Micromechanics with Shear + Dilatation [39]	Transformed precipitates with dilatational & shear strains; micromechanics interaction theory	Elastic response; dilatational & shear transformation strains; autocatalytic softening; pseudoelastic/shape-memory behaviour	Polycrystalline loading; temperature cycles; mechanical loading with thermal coupling	Homogenization limitations; neglects grain-boundary incompatibilities; limited spatial resolution
Elastic Phase-Field (2D) [44]	2D elastic coupling with order parameters	Elastic behaviour; transformation front evolution; crack shielding/toughening	Stress-induced transformation; crack-tip studies; toughening mechanisms	No explicit plasticity; limited to 2D; requires continuum-scale calibration
Elastoplastic Phase-Field with Defects [57]	Coupled elastoplastic continuum with phase-field; explicit defect modeling	Elastoplastic response; T → M with variant selection; tension–compression asymmetry; defect interactions	Defected microstructures; uniaxial compression/tension; local stress intensification; void nucleation/growth	High computational cost; extensive calibration; limited to local studies; scaling to components difficult
Single-Crystal Micromechanics [36]	Mechanistic single-crystal law; 3D anisotropic elasticity; non-Schmid effects	Anisotropic elasticity; martensite variant evolution; grain-boundary incompatibility; local stress concentration; transformation strains	Micropillar compression; bi-crystal studies; grain-boundary misorientation; small-scale testing	High cost for polycrystals; requires crystallography; single-crystal calibration; limited scalability
Piecewise Linear Elastic-Plastic (Superplastic) [38]	Phenomenological piecewise linear segments from stress-strain curves	Elastic at small strains; staged plastic flow; large deformation; fracture; temperature-dependent	High-temperature superplastic forming; tensile tests; manufacturing simulations	Limited predictive capability; valid only within experimental range; no explicit transformation; temperature-specific
Continuum + Cohesive Zone (Fracture) [7]	Continuum constitutive law with cohesive zone for fracture	Elastic-plastic-transformation response; crack-tip T → M; crack growth/shielding; transformation toughening	Fracture tests; crack propagation; 3D edge cracks; toughening studies	Requires detailed calibration; computationally demanding near cracks; mesh-dependent implementation

predicting crack evolution under thermal shock [57], many models still oversimplify transformation kinetics or assume homogenized material behavior. Phase transformations induce large volumetric and shear strains that contribute to both elastic and inelastic deformation [55, 58, 59], yet transformation-induced plasticity is rarely coupled with temperature-dependent thermo-physical properties in simulations [3].

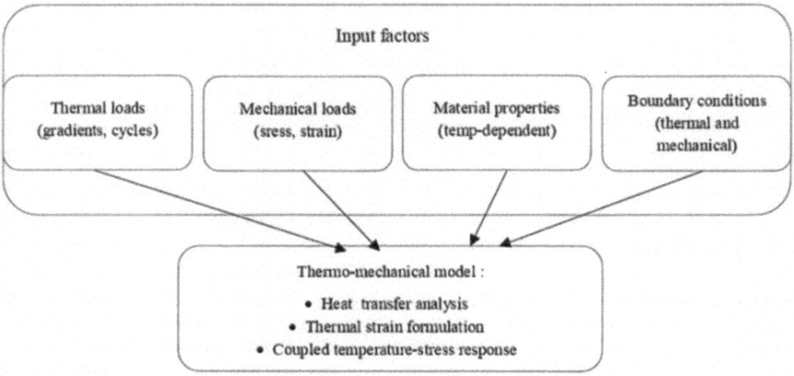

Fig. 2. Process flow of thermo-mechanical material modeling for zirconia.

3.3 Numerical Modeling of Manufacturing Processes for Zirconia

Numerical modeling procedures related to manufacturing of zirconia mainly concern the subtractive manufacturing processes (milling, micromachining) and additive manufacturing (AM) (Table 2). A several distinct modeling approaches, diversified in the scale and the type of model's architecture can be distinguished (Fig. 3). Concerning the scale criterion, the atomistic, meso-, micro- and macroscale models are being applied. In case of modeling type criteria, the molecular dynamics (MD), cluster dynamics (CD) and finite element (FE) methods are usually adapted to simulate a various aspects of zirconia behavior and its processing. Additive manufacturing (AM) enables complex zirconia components for dental, biomedical, and high-precision applications, but performance is highly sensitive to all processing stages, including layer deposition, curing, debinding, and sintering. Numerical modeling is essential to capture multiphysics interactions, residual stresses, and microstructural evolution. In photopolymer processes such as DLP, zirconia-loaded resins are deposited and cured [61–62]. Finite-strain and higher-order FE models predict residual stresses accurately [63, 64]; but these approaches are adapted from polymers or composites and lack full zirconia-specific validation, limiting predictive reliability [65, 66]. Material-jetting (DODMJ) allows highly loaded pastes with near-full density [67], yet rheology, droplet spreading, and shrinkage remain undermodeled, constraining dimensional precision. Sintering governs densification, grain growth, and final properties, and while SLA/DLP parts can reach 337 MPa [61, 62], links between green-body characteristics and sintered performance are weak, especially under ultra-rapid or FFF debinding conditions [68, 69]. Consequently, AM zirconia still shows higher porosity and microstructural heterogeneity than conventional materials, despite improved mechanical properties [70–72]. Alongside AM, zirconia ceramics have evolved from opaque 3Y-TZP to more translucent 4Y- and 5Y-PSZ formulations, improving aesthetic performance while maintaining adequate strength. Digital manufacturing advances, including CAD/CAM milling and rapid sintering have enhanced reproducibility and clinical predictability [73]. Numerical modeling increasingly supports these processes, with machine-learning approaches successfully predicting shrinkage in zirconia block production [74]. Fast-heating techniques such as spark plasma sintering and flash sintering have further expanded manufacturing capabilities, and thermo-electro-mechanical FE models now capture coupled heat transfer and diffusion phenomena [75]. However, most models remain numerical, stage-specific, and fail to provide integrated, process-aware predictions, limiting their application in realistic manufacturing [76]. In zirconia machining, FE simulations predict thermo-mechanical behavior, temperature evolution, and damage initiation with PCD tools [48]. Yet these simulations are largely numerical and rarely include analytical formulations, leaving a gap in fully understanding process–structure–property relationships for zirconia. Thus, the studies related to FE modeling of zirconia machining processes are limited and require further research.

Table 2. Comparison of modeling approaches in zirconia manufacturing.

Ref.	Type of Process	Model Type	Material Model	Simulation Parameters	Main Objectives
[77]	Speed sintering sintering	Thermal-structural FE model	Not specified	Heating rates, sintering cycles	Transient thermal stresses during rapid sintering
[78]	Sintering	Sintering FE model with modified SOVS	Modified SOVS constitutive model	Thermal profiles, densification parameters	Sintering and deformation behavior using data-driven approach
[48]	Milling	Machining FE simulation	Johnson-Cook model	Spindle speed, feed rate, cutting depth	Cutting force, tool wear, and material removal rate
[79]	Pulsed laser polishing	Coupled thermo-mechanical FE model	Thermo-mechanical with fracture criteria	Laser power, scanning speed, pulse frequency, pulse width	Map temperature/stress evolution and microcrack formation
[80]	Ultra-fast sintering	FEM coupled with master sintering curve	Master sintering curve framework	Heating rate, thermal insulation, peak temperature	Predict densification during UHS
[81]	Ultrasonic micromachining (USMM)	3D dynamic FE model	Not specified	Abrasive slurry concentration, feed rate, power rating	Optimize micro-hole quality (MRR, overcut, taper)
[82]	Flash SPS	Multiphysics electro-thermal-l mechanical-microstructura	Not specified	Specimen diameter, current configurations	Examine scale-up effects on thermal homogeneity
[83]	Additive manufacturing (3D printing)	3D fatigue FE analysis	Not specified	Sintering temperature/time, printing direction	Determine sintering parameters and predict fatigue performance

(*continued*)

Table 2. *(continued)*

Ref.	Type of Process	Model Type	Material Model	Simulation Parameters	Main Objectives
[84]	Veneering/ firing process	Viscoelastic FE model	Viscoelastic model for porcelain veneer	Cooling rate, thermal property mismatch	Assess residual thermal stresses during firing/cooling
[85]	Sintered implant components	3D implant/ mandible FE models	Elastic (E, v) from experiments)	Loading intensity, load orientation	Predict mechanical response under masticatory loads
[86]	3D-LCM additive manufacturing	Flexural FE analysis	Not specified	Debinding/ sintering parameters, manufacturing method	Compare flexural properties: 3D-printed vs. milled zirconia

4 Conclusions and Outlook

This review has highlighted the significant progress in numerical modeling of zirconia, including its mechanical behavior, thermo-mechanical responses, and processing. Across elastic, plastic, and transformation-induced deformation, finite element (FE) simulations have provided valuable insights into stress–strain evolution, phase transformations, and damage initiation. Thermo-mechanical modeling has further elucidated the impact of temperature, thermal expansion, and phase transitions on residual stresses, crack propagation, and overall structural reliability. In the context of AM, numerical studies have begun to capture curing, debinding, rheology, and sintering processes, yet modeling efforts remain largely stage-specific and fragmented. The processing–structure–property relationships central to zirconia's performance are therefore only partially described in current computational frameworks. Despite these advances, critical gaps remain. fully integrated multiphysics models linking processing parameters, microstructural evolution, and final mechanical performance are still lacking. Constitutive laws that explicitly connect AM process parameters to microstructural features and properties are scarce, limiting predictive capabilities. Furthermore, comprehensive experimental datasets covering all stages of processing are insufficient, constraining model validation and generalization.

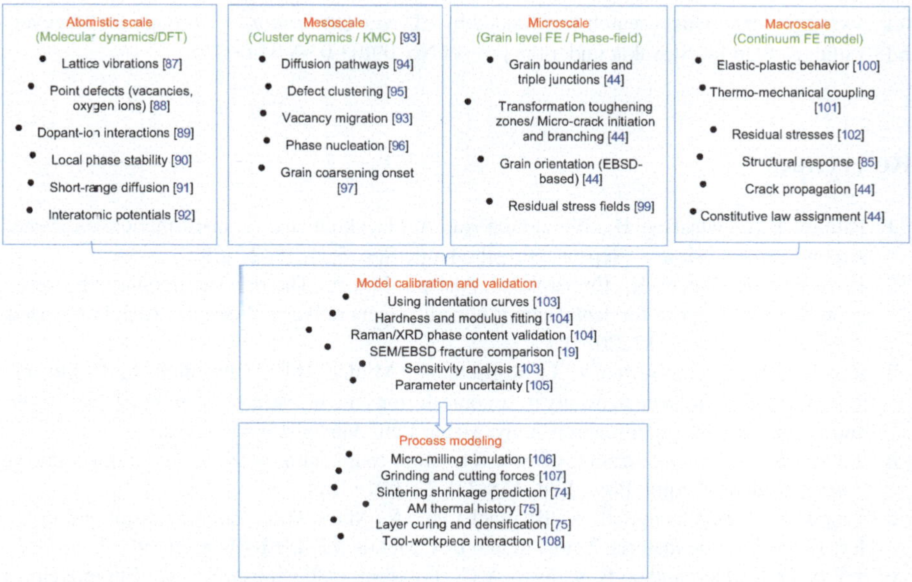

Fig. 3. Scale interactions and modeling approaches devoted to zirconia manufacturing.

Taken together, these findings motivate the development of integrated conceptual frameworks inspired by ICME (Integrated Computational Materials Engineering) and digital-twin approaches, in which additive manufacturing parameters, thermomechanical histories, microstructural evolution (grain size, phase distribution, residual stress), and constitutive response are coupled within a unified multiscale architecture. Such system-level frameworks, already explored in advanced ceramic and metal AM studies, provide a concrete pathway to translate current numerical results into predictive design tools, enabling virtual process optimization and performance-driven material design for zirconia components. Future research should focus on strategies that enhance predictive modeling and process-aware design of zirconia. Key directions include:

- Development of fully integrated multiphysics models that couple processing parameters, microstructural evolution, and mechanical performance.
- Zirconia-specific, process-aware constitutive laws linking AM parameters to microstructure and properties.
- Multi-scale modeling approaches connecting atomistic, mesoscale, and macroscale behaviors for predictive simulations.
- High-fidelity simulations of paste rheology and ultra-fast sintering to improve dimensional accuracy and microstructural uniformity in AM.

In conclusion, while numerical modeling of zirconia has matured considerably, realizing a truly predictive, integrated framework remains an open challenge. Advances that combine multiscale modeling, hybrid physics–ML approaches, and comprehensive experimental validation will be pivotal to enable process-aware design, improve reliability, and expand the applications of high-performance zirconia ceramics.

Acknowledgments. This research was funded by EU *NextGenerationEU* through the Recovery and Resilience Plan for Slovakia under the project No. 09I03-03-V04-00056.

References

1. Pathak, B.K., Maskey, B., Bhochhibhoya, A., Devkota, D.: Transformation-toughened zirconia: an overview. J. Nepalese Prosthodontic Soc. **5**(1), 19–24 (2022)
2. El-gergeni, K., Agela, K., Embarek, B., Alsbani, E.A.A.: The role of zirconia in bioactive ceramics: implications for dental and orthopedic regenerative applications. AlQalam J. Med. Appl. Sci. **8**, 1709–1713 (2025)
3. Reveron, H., Serrano-Munoz, I., Kupsch, A., Müller, B.R., Chevalier, J., Bruno, G.: Transformation-induced plasticity in zirconia during tensile loading: a combined microscopy and synchrotron X-ray refraction study. Mater. Lett. **366**(1), 136445 (2024)
4. Takagi, S. et al.: Visualization of transformation toughening of zirconia ceramics during dynamic fracture. Appl. Phys. Lett. **118**(23), 231901 (2021)
5. Sengupta, P., Bhattacharjee, A., Bhattacharjee, A., Maiti, H.S.: Zirconia: a unique multifunctional ceramic materia. Trans. Indian Inst. Metals. **72**, 1981–1998 (2019)
6. Trillo, J.F.B., Hernandez, B.A., Neto, V.G., Foschini, C.R., Sousa, E.A.C.: Finite element modelling and structural analysis of zirconia and zirconia-cnt composite materials. In: Proc. of the XI Congresso Nacional de Engenharia Mecânica (CONEM 2022), Teresina, Brazil (2022)
7. Liu, B., Liu, N., Luo, J., Xiao, Z.: Study of transformation toughening behavior of an edge through crack in zirconia ceramics with the cohesive zone model. Int. J. Appl. Mech. **10**, 1850066 (2018)
8. Lasko, G., Danilenko, I., Schmauder, S.: 3D finite element simulations of indentation fracture test of ZrO2-ceramics with taking into account its microstructure. In: International Congress on Advanced Materials Sciences and Engineering, pp. 1–5 (2022)
9. Sharanraj, V., Ramesha, C.M., Kumar, V., Sadashiva: Finite element analysis of zirconia ceramic biomaterials used in medical dental implants. J. Adv. Res. Dyn. Control Syst. **68**(3), 24–31 (2019)
10. Olhero, S.M., Torres, P.M.C., Mesquita-Guimarãe, J., Baltazar, J., Pinho-da-Cruz, J., Gouveia, S.: Structural performance of dense alumina-zirconia ceramics: An overview of conventional versus additive manufacturing. Mater. Proc. **8**(1), 5 (2022)
11. Kwon, W., Park, M.-G.: Evaluation of mechanical properties of dental zirconia in different milling conditions and sintering temperatures. J. Prosthet. Dent. **130**, 909 (2022)
12. Barui, S., Lebert, A.W., Jauregui, F., Munjuluri, K.S.A., Kate, K.H.: Microstructure and mechanical properties of 3D-printed zirconia bone scaffold: experimental and computational analysis. J. Am. Ceram. Soc. **107**, 2134 (2024)
13. Frigan, K., Chevalier, J., Zhang, F., Spies, B.C.: Is a zirconia dental implant safe when it is available on the market. Ceramics. **2**(4), 568–577 (2019)
14. Pereira, R., Campos, T.M.B., Bonfante, E.A., Thim, G.P.: A comparative study of mechanical properties of zirconia Y-TZP monolithic and multilayer configuration for dental application. SSRN. 4547913 (2023)
15. Zhu, X. et al.: Effect of Y2O3 doping content on phase composition, mechanical properties and cavitation erosion performance of ZrO2 ceramics. Ceram. Int. **50**, 7831 (2024)
16. Kulyk, V., Duriagina, Z., Kostryzhev, A., Vasyliv, B.D., Vavrukh, V.I., Marenych, O.: The effect of Yttria content on microstructure, strength, and fracture behavior of Yttria-stabilized zirconia. Materials. **15**(15), 5312 (2022)

17. Chen, Y. et al.: Molecular dynamics simulation of Sc2O3-Gd2O3-Y2O3 dopant effect on the crystal structure and thermophysical properties of zirconia-based ceramic materials. Solid State Ionics. **402**, 116382 (2023)
18. You, J., Guo, W., Li, Y., Pan, Y., Cui, T.: Phase evolution and mechanical performance of zirconia ceramics synthesized under high temperature and high pressure. Nano. **15**(16), **1235** (2025)
19. Mafra, M.P.d.A., Júnior, N.S., Santos, C.d., Ferreira, J.L.A., Araújo, J.A., Silva, C.R.M.: Evaluation of the mechanical properties and fatigue resistance of the ZrO2CeYAl2O3 composite. Ceramics. **7**(4), 1600–1615 (2024)
20. Abaza, A. et al.: Fracture properties of porous yttria-stabilized zirconia under micro-compression testing. J. Eur. Ceram. Soc. **41**, 7175 (2021)
21. González Díaz, M., Smirnov, A., Gutiérrez-González, C.F., Estrada, D., Bartolomé, J.F.: Microstructure and mechanical properties of zirconia (3Y-TZP)/Zr composites prepared by wet processing and subsequent spark plasma sintering. Ceram. Int. **46**(3), 53– 64 (2020)
22. Kornienko, O.A., Yushkevych, S.V., Sameljuk, A.V., Andrievskaya, O.R.: Phase equilibrium in ZrO2-CeO2-Eu2O3 system at a temperature of 1500 C. Recent Prog. Mater. **3**(3), 1–27 (2021)
23. Negreiros Ribeiro, F., Ribeiro Ricci Lazar, D., Ussui, V., Batista de Lima, N., Marchi, J., Dalpian, G.M.: Ab initio atomistic description of temperature-induced phase changes: the cases of zirconia and Ti-Y-co-doped zirconia. Phys. Rev. Materials. **5**(2), 023603 (2021)
24. Cao, Y., van der Krogt, C.J.: Temperature dependence in ageing process in commercial zirconia dental ceramics. Adv. Appl. Ceram. **121**(4), 150–153 (2022)
25. Gregorová, E., Nečina, V., Hříbalová, S., Pabst, W.: Temperature dependence of young's modulus and damping of partially sintered and dense zirconia ceramics. J. Eur. Ceram. Soc. **40**(5), 2063–2071 (2020)
26. Parchovianský, M., Parchovianská, I., Hanzel, O., Netriová, Z., Pakseresht, A.H.: Phase evaluation, mechanical properties and thermal behavior of hot-pressed LC-YSZ composites for TBC applications. Materials. **15**(8), 2919 (2022)
27. Abu-Naba'a, L.: A narrative review of recent finite element studies reporting references for elastic properties of zirconia dental ceramics. Ceramics. **6**(2), 898–911 (2023)
28. Alghazzawi, T.F., Janowski, G.M., Eberhardt, A.W.: An experimental study of flexural strength and hardness of zirconia and their relation to crown failure loads. J. Prosthet. Dent. **131**, 320 (2022)
29. Szava, D.T., Szava, I., Vlase, S., Száva, A.: Experimental investigations of the dental filling materials: establishing elastic moduli and Poisson's ratios. Materials. **16**(9), 3332 (2023)
30. Agarkov, D.A. et al.: Thermal conductivity of single crystals zirconia stabilized by scandium, yttrium, gadolinium, and ytterbium oxides. Mod. Electron. Mater. **8**(1), 1–6 (2022)
31. Cernuschi, F., Kulczyk-Malecka, J., Zhang, X., Nozahic, F., Estournès, C., Sloof, W.G.: Thermal conductivity of binary ceramic composites made of insulating and conducting materials comprising full composition range – applied to yttria partially stabilized zirconia and molybdenum disilicide. J. Eur. Ceram. Soc. **43**, 6296 (2023)
32. Park, S.I., et al.: Thermal conductivity of yttria-stabilized zirconia thin films grown by plasma-enhanced atomic layer deposition. J. Am. Ceram. Soc. **106**(9), (2023)
33. Fiege, C.: Vergleichende Untersuchung zur Kaustabilität von monolithischen Seitenzahnkronen aus Zirkoniumdioxid. Justus-Liebig-Universität, Hybridkeramik und Komposit in-vitro (2021)
34. Adachi, K., Kawamata, K., Waki, H.: Elastic constants of plasma-sprayed yttria-stabilized zirconia coatings determined via resonant ultrasound spectroscopy coupled with laser-Doppler interferometry. Surf. Coat. Technol. **424**, 127676 (2021)
35. Luu, A.T., Zhou, W., Zhong, X., Hao, M., Dera, P., Zhang, J.S.: High P-T Single-Crystal Elasticity of Zircon by Brillouin Spectroscopy. American Mineralogist (2025)

36. Alipour, H.R., Hamedani, A.S., Alahyarizadeh, G.: First-principles calculations to investigate the thermal response of the ZrC(1−x)Nx ceramics at extreme conditions. High Temp. Mater. Processes. **42**(1), 20220241 (2023)
37. Wang, K., Zu, Y., Chen, G., Fu, X., Zhou, W.: Effects of an electric current on the superplastic deformation behavior of 3Y-TZP in an oxygen-lean atmosphere. Materials. **16**(20), 6785 (2023)
38. Wang, Z., Lai, A., Schuh, C.A., Radovitzky, R.: Phase transformation and incompatibility at grain boundaries in zirconia-based shape memory ceramics: a micromechanics-based simulation study. J. Mater. Sci. **57**(24), 11132–11150 (2022)
39. Bhattacharya, A., Zaeem, M.A.: A phase-field model for study of ferroelastic deformation behavior in yttria stabilized zirconia. Acta Mater. **276**, 120039 (2024)
40. Amigó, N.: Strain rate effects on the atomic structure and plastic deformation of a CuZr metallic glass. Phys. Scr. **100**, 095944 (2025)
41. Dafalias, Y.F., Manzari, M.T.: Rate dependent but time independent plasticity formulation with application to sand. Mech. Res. Commun. **135**, 104242 (2023)
42. Nielsen, K.L.: Computational rate-independent strain gradient crystal plasticity. J. Mech. Phys. Solids. **148**, 104286 (2021)
43. Zhu, J., Luo, J., Sun, Y.: Study of the fracture behavior of tetragonal zirconia polycrystal with a modified phase field model. Materials. **13**(19), 4430 (2020)
44. Buyakov, A.S., Mirovoy, Y.A., Buyakova, S.P.: Fracture toughness improvement of zirconia based ceramics. In: AIP Conference Proceedings. AIP Publishing LLC (2020)
45. Nagai, M. et al.: Terahertz-induced martensitic transformation in partially stabilized zirconia. Commun. Phys. **6**(1), 88 (2023)
46. Pazhin, A.A., Mirovoi, Y.A.: Mathematical modeling of deformation and fracture of zirconium carbide porous samples subjected to uniaxial compression. AIP Conf. Proc. **2310**, 020246 (2020)
47. Yan, X., Dong, S., Li, X., Zhao, Z., Dong, S., An, L.: Optimization of machining parameters for milling zirconia ceramics by polycrystalline diamond tool. Materials. **15**(1), 208 (2022)
48. Lin, J.Q., et al.: Modelling and experimental study on brittle-to-ductile transition during ultrasonic elliptical vibration-assisted cutting of zirconia ceramics. Adv. Manuf. **14**, 437–451 (2025)
49. Manière, C., Lee, G., McKittrick, J., Chan, S., Olevsky, E.A.: Modeling zirconia sintering trajectory for obtaining translucent submicronic ceramics for dental implant applications. Acta Mater. **188**, 101–107 (2020)
50. Devi, S., Ariga, P., Tulsani, M., Rupawat, D.: Zirconia as a core material-a literature review. J. Contemp. Issues Bus. Gov. **26**(2), 1990–1998 (2020)
51. Qu, Y., Liu, L.: Zirconia materials for dental implants: a literature review. Front. Dent. Med. **2**, 687983 (2021)
52. Singh, P.V., Reche, A., Paul, P., Agarwal, S.: Zirconia facts and perspectives for biomaterials in dental implantology. Cureus. **15**(10), e46828 (2023)
53. Gallardo Rivas, N.V., Luna Domínguez, J.H., Martínez, A.M.: Microstructural and grain size analysis of zirconium materials for dental prostheses restoration. J. Eng. Res. **4**(14), 2–6 (2024)
54. Guinebretière, R. et al.: Coupling between elastic strains and phase transition in dense pure zirconia polycrystals. Phys. Rev. Mater. **6**(1), 013602 (2022)
55. Brahimi, A., Tayeb, K., Benguediab, M., Carpinteri, A.: FEM analysis of thermomechanical effects on stress intensity factors. Stud. Eng. Exact Sci. **5**, 2142 (2024)
56. Papšík, R., Ševeček, O., Schlacher, J., Bermejo, R.: Prediction of thermal shock induced cracking in multi-material ceramics using a stress-energy criterion. Eng. Fract. Mech. **303**, 110121 (2024)

57. Wang, Z., Zhang, S.Y., Shen, Q.: Coupled thermo-mechanical phase-field modeling to simulate the crack evolution of defective ceramic materials under flame thermal shock. Appl. Sci. **13**, 12633 (2023)
58. Verdi, C., Karsai, F., Liu, P., Jinnouchi, R., Kresse, G.: Thermal transport and phase transitions of zirconia by on-the-fly machine-learned interatomic potentials. Comput. Mater. **7**(1), 1–9 (2021)
59. Kelly, J.R., Denry, I. Kelly, J.R., Denry, I.: Stabilized zirconia as a structural ceramic: an overview. Dent. Mater. **24**(3), 289–298 (2008)
60. do Amaral, L.B., Paschoa, J.L.F., Magalhães, D., Foschini, C.R., Suchicital, C., Fortulan, C.A.: Preliminary studies on additive manufacturing of over 95% dense 3Y zirconia parts via digital imaging projection. J. Braz. Soc. Mech. Sci. Eng. **42**(1), 1–8 (2020)
61. Branco, A.C., Colaço, R., Figueiredo-Pina, C.G., Serro, A.P.: Recent advances on 3D-printed zirconia-based dental materials: a review. Materials. **16**(5), 1860 (2023)
62. Levin, V.A., Zingerman, K.M., Vershinin, A.V., Konovalov, D.: Numerical modeling of residual stresses in additive manufacturing products using the theory of repeatedly superimposed finite strains. Math. Mech. Solids. **27**(10), 2099–2109 (2022)
63. Nagaraj, M.H., Maiarù, M.: Micro-Scale Process Modeling and Evaluation of Curing-Induced Residual Stresses in Fiber-Reinforced Polymers Using Higher-Order Fe Models. American Society for Composites (2022)
64. Fonseca, J.C., Paccola, R.R., Zanotto, E.D., Ferreira, E.B.: Evaluating the residual stresses and fracture behavior of glass-ceramics by numerical modeling. Acta Mater. **274**, 120015 (2024)
65. Porojan, L., Topala, F., Porojan, S.: Varied simulation-based stress analyses on zirconia all-ceramic crowns. Mol. Sci. Appl. **2**, 1–4 (2022)
66. Fayazfar, H., Liravi, F., Ali, U., Toyserkani, E.: Additive manufacturing of high loading concentration zirconia using high-speed drop-on-demand material jetting. Int. J. Adv. Manuf. Technol. **109**(9), 2733–2746 (2020)
67. Bhandari, S. et al.: Ultra-rapid debinding and sintering of additively manufactured ceramics by ultrafast high-temperature sintering. J. Eur. Ceram. Soc. **44**(1) (2023)
68. Cano Cano, S. et al.: Additive manufacturing of zirconia parts by fused filament fabrication and solvent debinding: Selection of binder formulation. Addit. Manuf. **26**, 117–128 (2019)
69. Revilla-León, M., Meyer, M.J., Zandinejad, A., Özcan, M.: Additive manufacturing technologies for processing zirconia in dental applications. Int. J. Comput. Dent. **23**(1), 27–37 (2020)
70. Zhang, X., Wu, X., Shi, J.: Additive manufacturing of zirconia ceramics: a state-of-the-art review. J. Mater. Res. Technol. **9**(4), 9029–9048 (2020)
71. Nasiry, K.L., Salazar Rios, A., Tahmaseb, A., Zandinejad, A.: Additive manufacturing of zirconia ceramic and its application in clinical dentistry: a review. Dent. J. **9**(9), 104 (2021)
72. Braz, Í.A.R., et al.: Avanços do óxido de zircônia para próteses odontológicas: revisão integrativa. Caderno Pedagógico. **22**(11), e19669 (2025)
73. Kim, H., Chi, S., Oh, H.W.: Predicting and optimizing shrinkage rates in zirconia block production through machine learning. In: 5th International Conference on Information and Communication Technology Convergence (ICTC), pp. 221–222 (2024)
74. Semenov, A.S., Trapp, J., Nöthe, M., Eberhardt, O., Kieback, B., Wallmersperger, T.: Thermo-electro-mechanical modeling of spark plasma sintering processes accounting for grain boundary diffusion and surface diffusion. Comput. Mech. **67**, 1395 (2021)
75. Kuang, N., Xiao, M., Qi, H., Zhao, W., Wu, J.: Optimization of resin composition for zirconia ceramic digital light processing additive manufacturing. Polymers. **17**(6), 797 (2025)
76. Alshahrani, A.M., Lim, C.H., Kim, J., Zhang, Y.: Transient thermal stresses developed during speed sintering of 3 Mol% yttria-stabilized tetragonal zirconia polycrystals. Dent. Mater. **39**(5), 522–528 (2023)

77. Song, W., Liu, X., Chen, M., Bai, C., Feng, Z., Fang, L.: Data-driven prediction of sintering and deformation behavior of ZrO2 ceramics. Ceram. Int. **51**, 27379 (2025)
78. Zhou, Z., Zhao, Z., He, J., Shi, R.: Pulsed laser polishing of zirconia ceramic microcrack generation mechanism and size characterization study. Crystals. **14**(9), 810 (2024)
79. Dong, J., et al.: Thermally-insulated ultra-fast high temperature sintering (UHS) of zirconia: a master sintering curve analysis. Scr. Mater. **203**, 114076 (2021)
80. Banerjee, B., Pradhan, S., Das, S., Dhupal, D.: Surface topography characterization of USMM during machining of zirconia ceramic using silicon carbide abrasives: an experimental and simulation approach. CIRP J. Manuf. Sci. Technol. **51**, 1–19 (2024)
81. Manière, C., Harnois, C., Riquet, G., Lecourt, J., Bilot, C., Marinel, S.: Flash spark plasma sintering of zirconia nanoparticles: electro-thermal-mechanical-microstructural simulation and scalability solutions. J. Eur. Ceram. Soc. **42**(1), 216–226 (2022)
82. Zhao, Y., Li, P., Dong, P., Zeng, Y., Chen, J.: Investigation on 3D printing ZrO2 implant abutment and its fatigue performance simulation. Ceram. Int. **47**(1), 1053–1062 (2021)
83. Shrestha, R., Reddy, N., Fredeen, S., Zhang, Y., Kim, J.: Functionally graded bi-material interface for porcelain veneered zirconia dental crowns: a study using viscoelastic finite element analysis. Dent. Mater. **40**(8), 1267–1281 (2024)
84. Ward, P.A., et al.: A strength behavior approach for 3Y-TZP ceramics dental implants based on finite elementsimulations. Mater. Res. **26**, e20220110 (2023)
85. Alshamrani, A., Fahimipour, A., Alsarani, M., Ellakwa, A.: Flexural properties and finite element analysis of novel 3D-printed ceramic materials via 3D-LCM technology. Aust. Dent. J. **70**(1), 58–69 (2025)
86. Salikhodzha, Z.M. et al.: DFT study of oxygen ion migration in Mg-doped cubic zirconia. Solids. **64**(4), 55 (2025)
87. Muser, M.H., Sukhomlinov, S.V., Pastewka, L.: Interatomic potentials: achievements and challenges. Adv. Phys.: X. **8**(1), 2093129 (2022)
88. Hutama, A.S., Marlina, L.A., Chou, C.P., Irle, S., Hofer, T.S.: Development of density-functional tight-binding parameters for the molecular dynamics simulation of zirconia, yttria, and yttria-stabilized zirconia. ACS Omega. **6**(31), 20530 (2021)
89. Pan, R. et al.: Unraveling the characteristics of adsorption and diffusion of oxygen atoms on the surface of Zr and Zro2 crystals with point defects from molecular dynamic simulations. J. Nucl. Mater. **597**, 155143 (2024)
90. Rozhkov, A.V., Ignatov, S.K., Suleimanov, E.V.: Potential parameters for evaluating the activation characteristics of ion diffusion in solid electrolytes based on stabilized zirconium oxide. Vestn. Mosk. Univ. Ser. 2: Khim. **79**, 279–287(2024)
91. Liu, J., Song, X., An, Q.: Defect formation and migration in zirconium carbide under charge variation: a first-principles study. J. Am. Ceram. Soc. **108**(9), e20689 (2025)
92. Marinopoulos, A.G., Vilão, R.C., Alberto, H.V., Gil, J.M., Vieira, R.B.L., Lord, J.S.: Local environment and migration paths of the proton defect in yttria-stabilized zirconia studied by ab initio calculations and muon-spin spectroscopy. Hydrogen. **5**(3), 374–386 (2024)
93. Schmidt, E.M., Neder, R.B., Martin, J.D., Minelli, A., Lenée, M.H., Goodwin, A.L.: Direct interpretation of the X-ray and neutron of yttria stabilized zirconia. arXiv:2212.10907 (2022)
94. Kirilkin, N. et al.: Exploring metastable phase formation: swift heavy ion effects on partially stabilized zirconia. J. Nucl. Mater. **602**, 155369 (2024)
95. Vikrant, K.S.N., Wang, H., Jana, A., Wang, H., García, R.E.: Flash sintering incubation kinetics. npj Comput. Mater. **6**, 1–8 (2020)
96. Li, C., Li, J.B.: Phase field study of ferroelastic domain switching behavior related to grain size in polycrystalline tetragonal zirconia. J. Am. Ceram. Soc. **108**, e20495 (2025)
97. Fang, Z., Luo, J., Sun, Y., Huang, H., Zhong, J.: Phase field study of crack growth in t' yttria stabilized zirconia with initial domain structures. Mater. Today Commun. **37**, 107489 (2023)

98. Ebrahimi, S., Rabczuk, T., Anitescu, C.: Plastic crack-tip fields under thermoplastic loads. Eur. J. Mech.-A/Solids. **97**, 104805 (2023)
99. Chen, H., Natarajan, S., Ooi, E.T., Song, C.: Modelling of coupled thermo-mechanical crack propagation in brittle solids using adaptive phase field method with scaled boundary finite element method. Theor. Appl. Fract. Mech. **129**, 104158 (2023)
100. Clayton, J., Leavy, R.B., Knap, J.: Phase field mechanics of residually stressed ceramic composites. Philos. Mag. **102**(19), 1891–1944 (2022)
101. Kulyk, V.V.: The Effect of Sintering Temperature on Crack Growth Resistance Characteristics of Fine-Grained Partially Stabilized Zirconia Determined by Various Test Methods Springer Proceedings in Physics, pp. 331–345 (2023)
102. Giniyatova, S.G. et al.: Investigation of the phase composition, structural, mechanical, and dielectric properties of (1 − x)•ZrO2-x•CeO2 ceramics synthesized by the solid-state method. Appl. Sci. **14**(6), 2663 (2024)
103. Lamnini, S., Pugliese, D., Baino, F.: Zirconia-based ceramics reinforced by carbon nanotubes: a review with emphasis on mechanical properties. Ceramics. **6**, 1705 (2023)
104. Chevalier, J., Gremillard, L., Virkar, A. V., Clarke, D.R. The tetragonal-monoclinic transformation in zirconia: lessons learned and future trends. J. am. ceram. soc. **92**(9), 1901–1920 (2009)
105. Manea, H., Lu, H., Liu, Q., Xiao, J., Yang, K.: Micro-grinding parameter control of hard and brittle materials based on kinematic analysis of material removal. Mathematics. **12**, 1589 (2024)
106. Pang, J., Ji, X., Niu, Y., Chen, S.: Experimental investigation of grinding force and material removal mechanism of laser-structured zirconia ceramics. Micromachines. **13**(5), 710 (2022)
107. He, R., Ghantasala, V., Polak, P., Saleem, B., Pan, J.: Physics-based neural network as constitutive law for finite element analysis of sintering. Ceram. Int. **50**(19), 37291 (2024)
108. Lopata, O., Smirnov, I., Zinkovskii, A., Lopata, L.: Dependence of the Elastic Modulus of Powder Coatings on their Porosity in Electrical Contact Hardening, vol. 26, p. 28. Problems of Tribology (2021)

Adaptive Process Control in Injection Molding Using Cavity Pressure Sensors: Recent Advances and Perspectives

Tomasz Olszewski(✉) and Danuta Matykiewicz

Faculty of Mechanical Engineering, Poznan University of Technology, Poznań, Poland
tomasz.olszewski@doctorate.put.poznan.pl

Abstract. Cavity pressure transducers have become one of the most powerful diagnostic and control tools in modern injection molding, providing direct insight into the conditions inside the mold cavity. Their integration into adaptive and self-tuning control systems – ranging from rule-based iterative learning to adaptive model predictive control (MPC) and data-driven machine-learning controllers – has enabled shot-to-shot compensation for material and environmental disturbances. This paper presents recent empirical and methodological contributions, highlighting representative experimental validations and industrial implementations (including ENGEL's iQ Weight Control). The review emphasizes both process gains (reduced weight variation, improved defect detection, increased process capability) and practical constraints (sensor integration, data handling, control stability), and it presents a focused discussion on hot-runner balancing via cavity pressure measurement and its operationalization in production. The discussion addresses pressure-based V/P switchover detection, the use of the pressure integral as an indicator of part mass, gate freeze-off identification, and the effects of temperature gradients and geometric tolerances on flow balance. Reported cases show up to 85% reduction in shot weight variation, about 50% shorter setup, and 20–30% scrap reduction in multi cavity production. The biggest gains have been seen in single cavity tools and valve gated multi cavity molds; for cold runner molds, comparable benefits depend on balancing capability and targeted sensing.

Keywords: Cavity Pressure Transducers · Adaptive Process Control · Injection Molding

1 Introduction

Injection molding remains one of the most widely employed manufacturing methods in the plastic processing industry due to its high productivity, repeatability, and ability to produce complex geometries [1]. Despite being a well-established and mature technology, the growing expectations of Industry 4.0 have introduced new challenges regarding process efficiency, data integration, and product quality [2]. The use of multi-cavity molds is considered one of the most effective strategies for improving manufacturing

B. Gapiński et al. (Eds.): MANUFACTURING 2026, LNME, pp. 98–110, 2026.
https://doi.org/10.1007/978-3-032-22943-4_8

throughput and cost efficiency. However, ensuring consistent part quality across all cavities remains difficult, as the injection molding process involves numerous interrelated parameters that collectively influence final product characteristics. Product quality can be evaluated based on dimensional and weight stability [3], surface quality (e.g., roughness, sink marks, weld lines) [4], and physical properties, including mechanical, optical, and electrical performance [5–7].

A global trend in modern manufacturing emphasizes establishing quantitative relationships between process parameters and product quality [8]. Nevertheless, because injection molding involves nonlinear interactions among thermal, mechanical, and rheological variables, identifying these relationships is not straightforward. Consequently, hybrid approaches that integrate data analytics with physical understanding have become increasingly necessary. Statistical modeling techniques [9, 10] and artificial intelligence (AI) methods – including machine learning (ML) – are now widely applied for process optimization and quality prediction [11–14]. However, many of these data-driven approaches neglect the underlying physics of polymer flow and solidification. Therefore, developing methodologies that incorporate both physical measurements and data-driven interpretation remains essential.

Part weight is one of the most reliable indicators of process stability and product quality because it reflects variations in melt behavior and cavity filling [15–18]. In multi-cavity molds, deviations in part weight are often exacerbated by subtle geometric tolerances and temperature differences between cavities, leading to melt flow imbalance in the runner system [19, 20]. While direct weight measurement can be used for offline quality inspection, it cannot support real-time control. Hence, in-mold sensing – particularly pressure and temperature monitoring – has gained attention as a promising solution for online process evaluation [17, 21, 22]. Monitoring these parameters provides valuable insights into melt behavior and enables the prediction of part weight through relationships such as the pressure integral [17]. Maderthaner et al. [23] also demonstrated that a combination of in-cavity pressure and temperature measurements can effectively estimate part mass, confirming the potential of in-mold sensing for closed-loop control.

The injection pressure curve typically includes filling and holding stages, separated by the switchover point (V/P). Accurate detection of the V/P is critical for process stability and part consistency [24]. Traditional methods based on mass measurement are time-consuming and inadequate for multi-cavity molds. To overcome these limitations, researchers have proposed data-based methods, such as grey prediction models [25] and least-squares regression [26], to define optimal V/P conditions. Furthermore, optimizing the holding phase is equally important, as holding pressure and time strongly influence shrinkage and warpage [27–31]. While increasing holding pressure can reduce shrinkage [29, 30], its effect on warpage is not universally agreed upon [32, 33]. Gate freeze-off time marks the end of effective pressure transmission into the cavity and can be identified using either cavity pressure profiles [34, 35] or the saturation of part weight with increasing holding time [27, 36].

Despite ongoing efforts in the field, a fully integrated, real-time, sensor-based control approach for multi-cavity injection molding systems remains lacking. This work explores the use of the in-mold pressure sensors to monitor and manage critical process stages within a mold. The goal is to support the development of data-driven, sensor-integrated control strategies for more innovative polymer manufacturing. Among available process signals, cavity pressure provides the most direct link to part mass, density, and internal stresses, making it an ideal feedback variable for control. For this reason, pressure transducers have become essential for advanced process monitoring, automatic compensation, and intelligent optimization [4, 17].

Commercial systems (iQ Weight Control, ComoNeo) are in use, but fully integrated, real-time control that synchronizes all cavities, especially in complex multi-cavity tools, has not yet been achieved.

2 Pressure Transducers: Technology, Mounting, and Measurement Characteristics

2.1 Principle and Function

Cavity pressure transducers convert the physical pressure exerted by molten polymer within the cavity into an electrical signal, typically through piezoelectric or piezoresistive mechanisms. This signal provides a continuous measurement of in-mold conditions throughout the entire injection cycle. The resulting pressure curve can be divided into three phases – filling, packing, and cooling – each providing diagnostic and control information [37, 38].

For example, the peak cavity pressure correlates with the amount of polymer packed into the mold. In contrast, the integral of the pressure curve (area under the curve) correlates strongly with shot weight. Meanwhile, the pressure at the velocity-to-pressure (V/P) switchover point is an indicator of proper filling balance. As such, the pressure curve serves both as a process fingerprint and as a real-time quality predictor [39].

2.2 Mounting Strategies and Geometric Considerations

Mounting approaches are (Fig. 1): (A) flush cavity mounting (preferred for most precision applications), (B) recessed mounting with pressure-transmitting media like the ejector pin (used to avoid marking Class A surfaces), and (C) contact-free (less cavity specificity). Cavity mounting requires careful pocket design, protective sealing, and routing channels for wiring or, where available, use of slip-ring or quick-connect harnesses for automated mold handling. For multi-cavity tools, ensure consistent sensor depth and orientation among cavities to preserve signal comparability [40].

Advantages of Pressure Transducers [41, 42]

- **Real-time quality monitoring**.

  High-speed piezoelectric or strain-gauge pressure transducers yield detailed cavity pressure curves that correlate strongly with part quality (weight, dimensional accuracy, weld lines, sink marks, etc.). By analyzing features of the pressure profile (peak, integral, fill-to-pack transition, etc.), the process can detect defects early. For example, distinct inflections in the pressure curve indicate gate freeze time, and the integrated pressure under the curve has been used to predict part weight accurately. Real-time pressure monitoring thus allows the machine controller or operator to identify short shots, flash, or other deviations immediately.

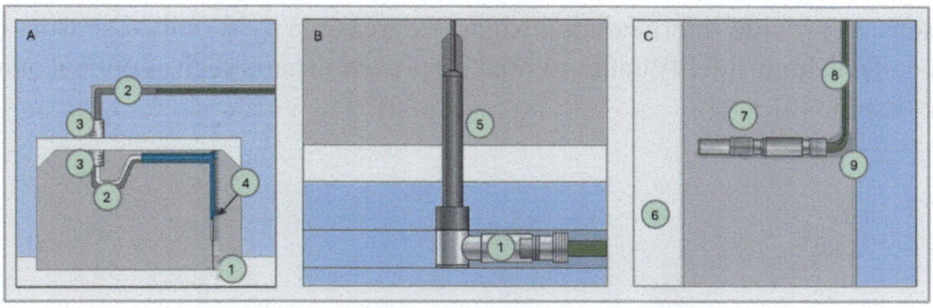

Fig. 1. Cavity Pressure Transducer Placement: A - Direct cavity pressure measurement; B - Indirect cavity pressure measurement; C - Contact-free cavity pressure measurement. Where: 1 – Sensor, 2 – Protective Tube, 3 – Contact Element, 4 – Spacer Sleeve, 5 – Ejector Pin, 6 – Cavity, 7 – Contact-free Cavity Pressure Sensor, 8 – Cable (Single-wire or coaxial cable), 9 – Mounting Hole

- **Process optimization and consistency**.
  Transducer feedback can be used to optimize and stabilize the molding cycle. For instance, Párizs et al. demonstrated that in-cavity pressure monitoring enables determination of the optimal clamping force and precise control of the velocity-to-pressure (V/P) switchover [43]. With suitable algorithms, the pressure signal can trigger the switch from constant velocity to holding pressure mode at the ideal time, reducing overpacking or incomplete fills. Advanced controllers use this data to tighten control over shot-to-shot variability: experiments have shown that adaptive adjustment of switchover and holding time based on pressure can dramatically improve repeatability.
- **Adaptive process control** (closed-loop).
  With pressure feedback, the injection process can be run in a closed loop. Specialized control software such as ENGEL's iQ Weight Control uses the measured pressure (e.g. at the injection screw) and compares it to a reference profile. Deviations due to material viscosity or injection quantity are detected almost instantaneously, and the controller may automatically adjust the injection speed profile and the V/P switchover to maintain constant shot weight [44]. Studies and industrial case reports indicate that such adaptive systems significantly reduce weight variance and compensate for material batch changes or ambient fluctuations [41].
- **Energy and material savings**.
  Because pressure transducers enable precise control, operators can avoid over-clamping and excessive holding pressure. An optimally set clamp prevents flash without wasting energy. In addition, early detection of abnormal pressure increases can prevent mold damage and reduce cycle time. In multi-cavity molds, identifying filling imbalance via pressure sensors enables correction (e.g., via sequential valve gating) to minimize scrap. Several works note that leveraging pressure measurements can reduce defective parts and improve process efficiency, fulfilling Industry 4.0 goals [43, 45].

2.3 Sensor Placement and Data Characteristics

Pressure sensors can be positioned in the nozzle, runner, or directly inside the cavity. Nozzle-mounted sensors are easy to integrate but reflect only global flow behavior.

Runner sensors provide intermediate insight, whereas cavity-mounted sensors (Fig. 2) deliver localized, high-fidelity data essential for precision parts such as optical or medical components [43, 45].

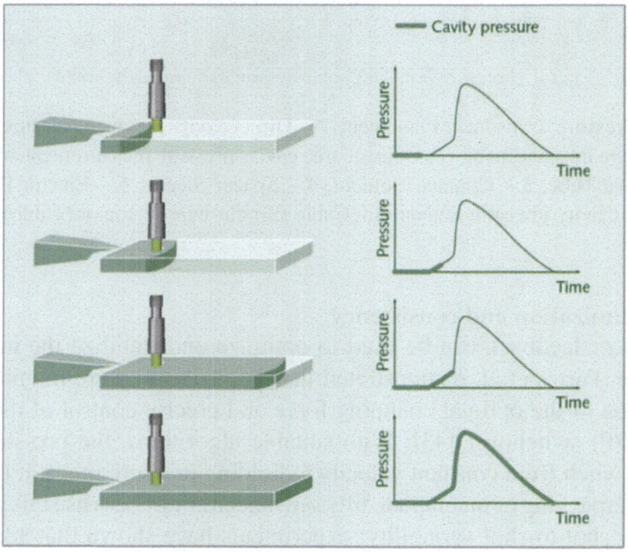

Fig. 2. Correlation Between Mold Filling and Pressure Sensor Response

In high-precision applications, flush-mounted cavity sensors are preferred to prevent distortion of the pressure field. However, these require careful installation and calibration to ensure repeatability. Multi-cavity molds often integrate several sensors, enabling cavity-to-cavity comparison and detection of filling imbalance [43].

Modern data acquisition systems sample pressure at 1–10 kHz, offering millisecond resolution. Combined with digital filtering and signal conditioning, these data streams serve as the basis for real-time control loops and machine-learning models.

2.4 Sensor Lifecycle, Calibration and Diagnostics

Sensors exhibit drift under thermal cycling and mechanical fatigue; therefore, implementing periodic calibration checks (e.g., weekly in high-volume production, monthly in lower volumes) is advisable. Simple health checks include a zero-pressure baseline under cavity-open conditions and a known static-pressure test using calibration fixtures. Embedded diagnostics, such as channel noise statistics and cross-cavity consistency checks, can provide early warnings of sensor degradation [40, 45].

3 Adaptive and Self-Tuning Control Architectures

3.1 Iterative Learning Control (ILC) and Rule-Based Systems

ILC utilizes the repetitive nature of injection molding to refine control parameters cycle-by-cycle. García-Sánchez et al. [37] implemented a pressure-based ILC algorithm that analyzed deviations between measured and reference pressure profiles and adjusted the holding pressure and switchover point for the subsequent cycle. Their approach corrected dimensional errors in molded automotive parts and reduced cycle time by approximately 8%.

Similarly, rule-based adaptive schemes compare key pressure features – such as peak value or integral – with target thresholds and incrementally adjust machine parameters. These algorithms are computationally lightweight and easily integrated into existing controllers but require empirical tuning and stability safeguards [37].

3.2 Adaptive Model Predictive Control (MPC)

More advanced systems employ model predictive control, which forecasts process behavior over a defined horizon and computes optimal control actions under constraints. Vuković et al. [39] proposed an adaptive MPC that used an Unscented Kalman Filter for online parameter estimation. The controller continuously adapted its internal model using cavity pressure feedback and maintained target pressure profiles with steady-state errors below 4 bar. Adaptive MPC provides superior performance but demands accurate system modeling and real-time computational resources.

3.3 Machine-Learning and AI-Based Control

Machine learning extends adaptive control by replacing explicit physical models with data-driven predictors. Tsai et al. [46] developed an adaptive controller using a neural network trained on nozzle pressure and barrel temperature data. The system dynamically adjusted V/P switchover timing and injection speed, reducing shot-weight variation to 0.14%.

Fang-Jiang et al. [38] designed a self-learning control system using a "master" pressure curve to guide cycle-to-cycle corrections, achieving a coefficient of variation (CV) of only 0.05% in part weight. Masato [47] further applied machine learning to recycled polypropylene molding, showing that data-driven control could compensate for material variability and maintain stable output.

Recent industrial research extends these ideas toward artificial intelligence (AI)-driven cognition frameworks. Arslan et al. [48] employed convolutional neural networks (CNNs) to interpret cavity pressure data and predict process reliability zones with 95% accuracy, indicating a feasible pathway to intelligent process autonomy (Table 1).

Table 1. Comparative summary of adaptive control strategies.

Strategy	Advantages	Disadvantages	Implementation complexity
ILC (Iterative Learning Control)	Incremental shot-to-shot improvement; direct use of pressure-profile error; limited computational demand; straightforward integration into existing cycles	Dependence on repeatability; sensitivity to non-repetitive disturbances; learning-gain tuning required; stability safeguards needed	Low–Medium (relatively straightforward algorithms and tuning)
MPC (Model Predictive Control)	Explicit handling of constraints; anticipation of future behavior; good cross-phase tracking; clear optimality criterion	Model identification effort; computation at cycle time; sensitivity to model mismatch; engineering time for commissioning	Medium – High (complex setup: model identification, constraint specification, real-time solvers)
ML (Machine Learning)	Nonlinear relationships captured without explicit physics; adaptation to material variability; quality prediction enabled	Data quantity/quality requirements; validation and explainability burden; lifecycle drift monitoring; IT/edge integration	High (extensive data collection, training, and maintenance effort)

4 Industrial Implementations

4.1 Adaptive Injection Molding Process

ENGEL's iQ Weight Control system is among the most established industrial implementations of adaptive pressure-based control. It compares each shot's measured injection pressure curve with a reference and calculates corrective adjustments to holding pressure and speed within the same cycle. Olszewski et al. [41] demonstrated that iQ Weight Control maintained consistent shot weights in glass-fiber-reinforced polyamide containing 50% regrind, while also improving mechanical strength.

This technology highlights how adaptive control can be directly integrated into commercial injection machines without requiring external sensors or operator input—an essential step toward self-optimizing manufacturing:

- **Deviation detection**. For each shot, the pressure profile is aligned against the trained reference. A pure shift of the curve along the time axis indicates a variation in shot volume (melt quantity), while a change in the curve's slope or amplitude signifies a viscosity change [49]. The software classifies deviations into categories (time shift, slope change, or shape distortion) and quantifies the differences rapidly.

- **Control action**. If a deviation in volume or viscosity is detected early in injection, iQ Weight Control automatically adjusts process parameters within the same cycle. Typically, this involves fine-tuning the V/P switchover point and adjusting the holding pressure. By doing so, the system actively drives the next segment of the shot toward the desired pressure/volume trajectory [49]. For example, a higher-than-expected curve slope (indicating low viscosity) might prompt the controller to shorten the injection profile or advance the switch to avoid overfilling. These corrections occur shot-by-shot, before any defective part is produced [50].
- **Performance**. Case studies report that iQ Weight Control substantially improves part consistency. In one example, an 85% reduction in part-weight variance was achieved for a thin-walled part when iQ was enabled [50]. In medical molding, such adaptive control has enabled compliance with strict validation standards, demonstrating that the process can fully adjust within predefined limits without manual resets [44].

Overall, this adaptive control strategy illustrates how pressure transducers enable smart injection molding: the machine actively learns normal cycles and compensates for material lot changes, moisture, or ambient shifts on the fly, rather than operating in a rigid open-loop mode.

4.2 Closed-Loop Retrofitting

Vasco et al. [42] presented a closed-loop system combining mold pressure sensors with machine-side telemetry. Implemented as a retrofit, the system effectively counteracted external disturbances, such as hydraulic drift and environmental fluctuations. This underscores the viability of adaptive control even in existing production setups.

5 Multi-cavity Scaling and Hot-Runner Balancing

Multi-cavity molds pose additional complexity: each cavity must fill and pack simultaneously to ensure uniform part quality. Traditionally, balancing was achieved by trial-and-error adjustments to hot-runner temperatures or valve timings – a process that is time-consuming and subjective.

Modern systems now use cavity pressure data to perform automatic hot-runner balancing. Pressure traces from individual cavities are compared to identify filling delays or imbalances. Control algorithms then adjust the temperature setpoints or valve timings of each hot-runner zone until cavity pressures synchronize.

Kistler's ComoNeo MULTIFLOW module automates this process by analyzing cavity-specific pressure profiles, calculating corrective setpoints, and transmitting them to the hot-runner controller in real time. The result is synchronized filling across all cavities, reduced start-up scrap, and shorter setup times. This concept demonstrates how cavity pressure transducers not only support quality control but also enable self-balancing mold systems [40] (Fig. 3).

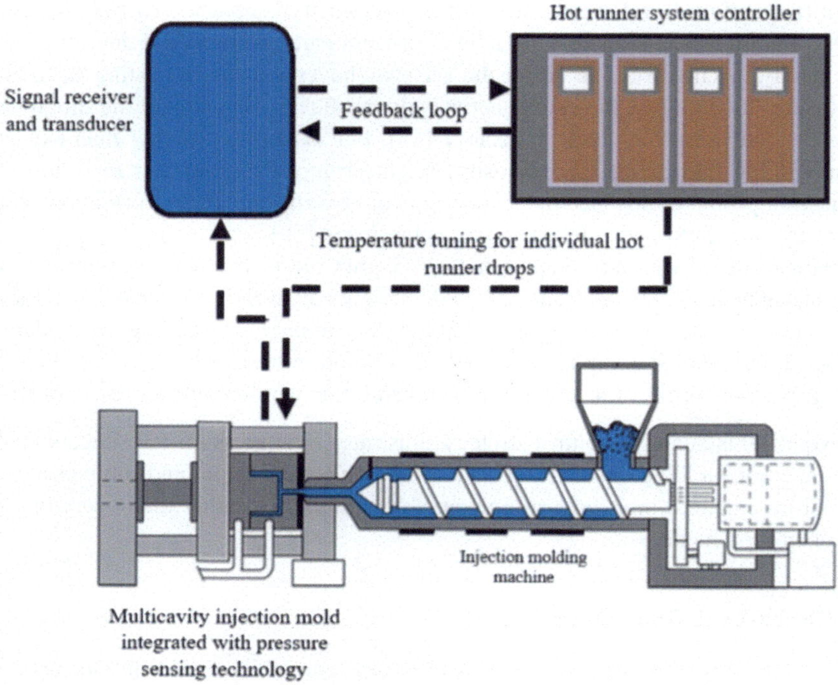

Fig. 3. Multicavity Mold with Integrated Self-balancing System – Schematic Overview.

6 Practical Deployment, Maintenance and Economic Considerations

Sensor integration increases tooling complexity and requires high-temperature cabling, but modular sensor systems now allow easy replacement and calibration. Regular verification of zero-offset and dynamic response is critical for long-term accuracy [40].

Real-time adaptive control generates substantial data. Edge computing solutions are increasingly used to preprocess signals locally, minimizing transmission latency. Commercial controllers achieve shot-to-shot adaptation within milliseconds, a key requirement for stable operation.

The cost of integrating cavity sensors is offset by savings from reduced scrap, improved repeatability, and faster production ramp-up. Automatic hot-runner balancing and adaptive process control reduce setup times by up to 50% and scrap rates by 20–30%, according to industrial case reports [40].

From an economic perspective, the return on investment (ROI) is mainly driven by scrap reduction (20–30%) and shorter setup times (up to 50%) demonstrated in this work, as well as by the avoidance of quality deviations during serial production. Under high-volume production conditions, the investment in sensors and data acquisition is typically recovered through reduced scrap and faster ramp-up, with shorter payback periods when multi-cavity balancing is applied. For transparency, ROI is best evaluated at the mold level by considering sensor costs, expected scrap reduction, machine hourly rate, and ramp-up savings based on the reported case studies.

7 Conclusions and Research Directions

Cavity pressure transducers, combined with adaptive control algorithms, have matured into practical tools that materially improve injection-molding quality and robustness. Across iterative learning, adaptive MPC, and ML approaches, the literature demonstrates consistent reductions in shot-to-shot variability and improved defect detection. Multi-cavity hot-runner balancing using pressure arrays is a proven industrial technique (e.g., Kistler's ComoNeo module) that shortens production ramp-up and reduces scrap [40].

Open research challenges include robust model transfer across machines, lifecycle management of sensors in harsh production environments, and design of hybrid controllers that combine the interpretability of physics-based models with the flexibility of data-driven policies. Addressing these will facilitate broader adoption of self-tuning, pressure-informed injection molding in Industry 4.0 contexts.

As presented in this paper, dimensional variations were corrected and cycle time was reduced by 8% with pressure-based ILC and rule-based schemes. Target pressure profiles were held within $\leq$4 bar by adaptive MPC, and shot-weight CV down to 0.05% was achieved by ML controllers, while up to 85% reductions in weight variance were reported for commercial implementations. In multi-cavity molds, pressure-driven hot-runner balancing synchronized filling, halved setup time, and cut scrap rate by 20–30%, with the largest gains seen in valve-gated and thin-walled applications.

In the context of Industry 4.0, standardized machine–sensor interfaces and traceable data storage were identified as basic requirements for reliable closed-loop control. In addition, integration with plant-level quality systems was recommended to support recipe version control and to track parameter changes across shifts and production lots. When these requirements are met, pressure-based controllers can be implemented as digitally integrated control solutions within the production system, rather than as isolated stand-alone solutions.

Acknowledgments. The research was carried out as part of the Implementation Doctorate Program of the Ministry of Education and Science, Agreement No. DWD/7/0114/2023.

References

1. Zhou, H.: Computer Modeling for Injection Molding: Simulation, Optimization, and Control. Wiley, Hoboken (2013)
2. Aminabadi, S.S., et al.: Industry 4.0 in-line AI quality control of plastic injection molded parts. Polymers. **14**, 3551 (2022). https://doi.org/10.3390/polym14173551
3. Galve, J.E., et al.: Dimensional stability and process capability of an industrial component injected with recycled polypropylene. Polymers. **11**, 1063 (2019). https://doi.org/10.3390/polym11061063
4. Mukras, S.M.S., Korany, H.Z., Omar, H.M.: Achieving optimal injection molding parameters to minimize both shrinkage and surface roughness through a multi-objective optimization approach. Appl. Sci. **15**, 5063 (2025). https://doi.org/10.3390/app15095063

5. Zhao, R., Li, X., Wang, Z., Wu, T., Liang, J.: Effect of injection molding parameters on the tensile strength of short-carbon-fiber-reinforced nylon 6. Polymers. **17**, 1264 (2025). https://doi.org/10.3390/polym17091264
6. Pedroso, L.D., Pontes, A.J., Alves, A., Duarte, F.M., Carneiro, O.S.: From transparent to opaque: a route towards multifunctional parts injected with a single material. Materials. **16**, 6219 (2023). https://doi.org/10.3390/ma16186219
7. Chen, S.-C., Jien, M.-Y., Hsu, C.-C., Hwang, S.-S., Feng, C.-T.: Processing effects on the through-plane electrical conductivities and tensile strengths of microcellular-injection-molded polypropylene composites with carbon fibers. Polymers. **14**, 3251 (2022). https://doi.org/10.3390/polym14163251
8. del Olmo, A., et al.: Tool wear monitoring of high-speed broaching process with carbide tools to reduce production errors. Mech. Syst. Signal Process. **172** (2022). https://doi.org/10.1016/j.ymssp.2022.109003
9. Trotta, G., Cacace, S., Semeraro, Q.: Optimizing process parameters in micro injection moulding considering the part weight and probability of flash formation. J. Manuf. Process. **79**, 250–258 (2022). https://doi.org/10.1016/j.jmapro.2022.04.048
10. Wang, M.-W., Arifin, F., Vu, V.-H.: The study of optimal molding of a led lens with grey relational analysis and molding simulation. Period. Polytech. Mech. Eng. **63**, 278–294 (2019). https://doi.org/10.3311/PPme.13337
11. Cao, Y.L., Fan, X.Y., Guo, Y.H., Li, S., Huang, H.Y.: Multi-objective optimization of injection-molded plastic parts using entropy weight, random forest, and genetic algorithm methods. J. Polym. Eng. **40**, 360–371 (2020). https://doi.org/10.1515/polyeng-2019-0326
12. Lockner, Y., Hopmann, C., Zhao, W.: Transfer learning with artificial neural networks between injection molding processes and different polymer materials. J. Manuf. Process. **73**, 395–408 (2022). https://doi.org/10.1016/j.jmapro.2021.11.014
13. Párizs, R.D., Török, D., Ageyeva, T., Kovács, J.G.: Machine learning in injection molding: an industry 4.0 method of quality prediction. Sensors. **22**, 2704 (2022). https://doi.org/10.3390/s22072704
14. Bustillo, A., Urbikain, G., Perez, J.M., Pereira, O.M., Lopez de Lacalle, L.N.: Smart optimization of a friction-drilling process based on boosting ensembles. J. Manuf. Syst. **48**, 108–121 (2018). https://doi.org/10.1016/j.jmsy.2018.06.004
15. Liou, G.-Y., et al.: Optimize injection-molding process parameters and build an adaptive process control system based on nozzle pressure profile and clamping force. Polymers. **15**, 610 (2023). https://doi.org/10.3390/polym15030610
16. Hassan, H.: An experimental work on the effect of injection molding parameters on the cavity pressure and product weight. Int. J. Adv. Manuf. Technol. **67**, 675–686 (2013). https://doi.org/10.1007/s00170-012-4514-4
17. Cheng, F.-J., Chang, C.-H., Wen, C.-H., Hwang, S.-J., Peng, H.-S., Chu, H.-Y.: Out-of-mold sensor-based process parameter optimization and adaptive process quality control for hot runner thin-walled injection-molded parts. Polymers. **16**, 1057 (2024). https://doi.org/10.3390/polym16081057
18. Yang, Y., Gao, F.: Injection molding product weight: Online prediction and control based on a nonlinear principal component regression model. Polym. Eng. Sci. **46**, 540–548 (2006). https://doi.org/10.1002/pen.20522
19. Chan, I.W.M., Pinfold, M., Kwong, C.K., et al.: A review of research, commercial software packages and patents on family mould layout design automation and optimisation. Int. J. Adv. Manuf. Technol. **57**, 23–47 (2011). https://doi.org/10.1007/s00170-011-3268-8
20. Lee, B.H., Kim, B.H.: Automated design for the runner system of injection molds based on packing simulation. Polym. Plast. Technol. Eng. **35**, 147–168 (1996). https://doi.org/10.1080/03602559608000086

21. Park, H., Rhee, B., Cha, B.S.: Variable-runner system for family mold filling balance. Solid State Phenom. **116–117**, 96–101 (2006). https://doi.org/10.4028/www.scientific.net/SSP.116-117.96
22. Chen, J.C., Guo, G., Wang, W.N.: Artificial neural network-based online defect detection system with in-mold temperature and pressure sensors for high precision injection molding. Int. J. Adv. Manuf. Technol. **110**, 2023–2033 (2020). https://doi.org/10.1007/s00170-020-06011-4
23. Maderthaner, J., Kugi, A., Kemmetmuller, W.: Part mass estimation strategy for injection molding machines. IFAC Pap. **53**, 10366–10371 (2020). https://doi.org/10.1016/j.ifacol.2020.12.2775
24. Kulkarni, S.: 7 – scientific processing, scientific molding, and molding parameters. In: Robust Process Development and Scientific Molding, 3rd edn, pp. 143–177. Hanser (2024)
25. Huang, M.-S.: Cavity pressure based grey prediction of the filling-to-packing switchover point for injection molding. J. Mater. Process. Technol. **183**, 419–424 (2007). https://doi.org/10.1016/j.jmatprotec.2006.10.037
26. Chen, J.-Y., Hung, P.-H., Huang, M.-S.: Determination of process parameters based on cavity pressure characteristics to enhance quality uniformity in injection molding. Int. J. Heat Mass Transf. **180**, 121788 (2021). https://doi.org/10.1016/j.ijheatmasstransfer.2021.121788
27. De Santis, D., Pantani, R., Speranza, V., Titomanlio, G.: Analysis of shrinkage development of a semicrystalline polymer during injection molding. Ind. Eng. Chem. Res. **49**, 2469–2476 (2010). https://doi.org/10.1021/ie901316p
28. Liao, S.J., et al.: Optimal process conditions of shrinkage and warpage of thin-wall parts. Polym. Eng. Sci. **44**, 917–928 (2004). https://doi.org/10.1002/pen.20083
29. Chang, T.C., Faison, E.: Shrinkage behavior and optimization of injection molded parts studied by the Taguchi method. Polym. Eng. Sci. **41**, 703–710 (2004). https://doi.org/10.1002/pen.10766
30. Mohan, M., Ansari, M.N.M., Shanks, R.A.: Review on the effects of process parameters on strength, shrinkage, and warpage of injection molding plastic component. Polym. Plast. Technol. Eng. **56**, 1–12 (2016). https://doi.org/10.1080/03602559.2015.1132466
31. Wu, H., Zhao, G., Wang, J., Wang, G., Zhang, W.: Effects of process parameters on core-back foam injection molding process. Express Polym Lett. **13**, 390–405 (2019). https://doi.org/10.3144/expresspolymlett.2019.32
32. Li, X., Ouyang, J., Zhou, W.: Simulations of three-dimensional thermal residual stress and warpage in injection molding. Comput. Model. Eng. Sci. **96**, 379–407 (2013). https://doi.org/10.3970/cmes.2013.095.379
33. Barghash, M.A., Alkaabneh, F.A.: Shrinkage and warpage detailed analysis and optimization for the injection molding process using multistage experimental design. Qual. Eng. **26**, 319–334 (2014). https://doi.org/10.1080/08982112.2013.852679
34. Min, B.H.: A study on quality monitoring of injection-molded parts. J. Mater. Process. Technol. **136**, 1–6 (2003). https://doi.org/10.1016/S0924-0136(02)00445-4
35. Leo, V., Cuvelliez, C.: The effect of the packing parameters, gate geometry, and mold elasticity on the final dimensions of a molded part. Polym. Eng. Sci. **36**, 1961–1971 (1996). https://doi.org/10.1002/pen.10592
36. Pantani, R., De Santis, F., Brucato, V., Titomanlio, G.: Analysis of gate freeze-off time in injection molding. Polym. Eng. Sci. **44**, 1–17 (2004). https://doi.org/10.1002/pen.20000
37. García-Sánchez, D.A., et al.: Enhancing injection molding process by implementing cavity pressure sensors and an iterative learning control (ILC) methodology. PRO. **13**(9), 3010 (2025). https://doi.org/10.3390/pr13093010
38. Fang-Jiang, J.-C., et al.: Study of an online monitoring adaptive system for an injection molding process based on a nozzle pressure curve. Polymers. **13**(18), 2986 (2021). https://doi.org/10.3390/polym13040555

39. Vukovic, M., et al.: Adaptive model-based predictive control for cross-phase cavity pressure control in injection molding. J. Manuf. Process. **77**, 730–742 (2022). https://doi.org/10.1016/j.jmapro.2022.02.030
40. Kistler Group: Quality monitoring and process control in injection molding – sensors and Systems (brochure, 2025). Data sheet and software modules (e.g., ComoNeoMULTIFLOW for automatic hot-runner balancing)
41. Olszewski, T., Matykiewicz, D., Jakubowicz, M.: Impact of adaptive process control on mechanical properties of plastic parts and process robustness. Appl. Sci. **15**(16), 8829 (2025). https://doi.org/10.3390/app15168829
42. Vasco, J.C., Martins, J., Oliveira, P., Chaves, P.: Closed-loop control applied to the injection moulding process – an industry 4.0 refurbishment case study. Electronics. **12**(2), 271 (2023). https://doi.org/10.3390/electronics12020271
43. Párizs, R.D., Török, D., Ageyeva, T., Kovács, J.G.: Multiple in-mold sensors for quality and process control in injection molding. Sensors. **23**, 1735 (2023). https://doi.org/10.3390/s23031735
44. https://www.engelglobal.com/en/us/blog/successful-validation-of-iq-weight-control-with-engel-support#:~:text=ENGEL%27s%20iQ%20weight%20control%20assistance,process%20by%20the%20machine%20operator. Last accessed 08.11.2025
45. Ageyeva, T., Horváth, S., Kovács, J.G.: In-mold sensors for injection molding: on the way to industry 4.0. Sensors. **19**, 3551 (2019). https://doi.org/10.3390/s19163551
46. Tsai, M.-H., et al.: Development of an online quality control system for injection molding process. Polymers. **14**(8), 1607 (2022). https://doi.org/10.3390/polym14081607
47. Masato, D.: Machine learning-based process control for injection molding of recycled polypropylene. Polymers. **17**(7), 940 (2025). https://doi.org/10.3390/polym17070940
48. Arslan, E., Konuşkan, Y., Lazoglu, I.: AI-driven cognition for advanced injection molding and industrial implementation. Int. J. Adv. Manuf. Technol. **138**, 2043–2064 (2025). https://doi.org/10.1007/s00170-025-15611-x
49. Hutterer, T.: Data-Driven Fault detection And Identification in Rubber Injection Molding, PhD Thesis, Montanuniversiteat Leoben (2020)
50. https://www.plasticportal.eu/image/staticke/File/iQWeightControl_en.pdf#:~:text=Thin,weight%20control%20iQ%20weight%20control. Last accessed 11.08.2025

Minimally Invasive Bone Cement Removal Using an Ultrasonic Pulsating Water Jet with Anatomical Nozzle

Sergej Hloch[1,2](✉), Jakub Poloprudský[3], Michal Vopálenský[4], and Akash Nag[5]

[1] Faculty of Manufacturing Technologies, Technical University of Košice, with a seat in Prešov, Prešov, Slovakia
hloch.sergej@tuke.sk
[2] Institute of Geonics, Academy of Sciences of the Czech Republic Ostrava, Ostrava, Czech Republic
[3] Institute of Physics of Materials, Czech Academy of Sciences, Brno, Czechia
[4] Institute of Theoretical and Applied Mechanics, Czech Academy of Sciences, Centre Telč, Prague, Czech Republic
[5] Faculty of Mechanical Engineering, VŠB–Technical University of Ostrava, Ostrava, Czech Republic

Abstract. This study investigates the disintegration behavior of bone cements using an ultrasonic pulsating water jet (PWJ) equipped with a custom-designed anatomical nozzle. Two commercially available PMMA-based cements, Palacos and C-ment, were subjected to high-pressure jet treatment at 15 MPa using a fixed nozzle diameter of 0.5 mm. The system operated at an excitation frequency of approximately 20.27 kHz, with acoustic power ranging from 209 to 236 W. Material removal efficiency was evaluated at four traverse speeds (2, 4, 8, and 12 mm/s), with constant standoff distance (2 mm) and jet angle (90°). Key response parameters included erosion depth, eroded volume, and material removal rate (MRR), which were quantified using optical microscopy. Results demonstrated that C-ment exhibited consistently higher MRR and erosion volume compared to Palacos, especially at increased traverse speeds, while Palacos showed more stable performance. The anatomical nozzle facilitated efficient, localized erosion suitable for confined surgical geometries, supporting the integration of PWJ systems in minimally invasive orthopedic cement removal procedures.

Keywords: Ultrasonic Pulsating Water Jet · Bone Cement · Extended Nozzle For Mini-Invasive Surgery

1 Introduction

The increasing number of joint replacement surgeries has led to a corresponding rise in the necessity for revision arthroplasty, a procedure often complicated by the challenge of safely and effectively removing old bone cement. The safe and effective removal

B. Gapiński et al. (Eds.): MANUFACTURING 2026, LNME, pp. 111–127, 2026.
https://doi.org/10.1007/978-3-032-22943-4_9

of bone cement is a critical step during orthopedic revision surgeries, particularly in procedures involving cemented hip and knee prostheses [1]. Traditional removal techniques—including mechanical chiseling [2], rotary drills, and ultrasonic hand tools [3, 4]—pose several limitations such as imprecise control, excessive reaction forces, heat generation, and a risk of damaging [5] or fracturing [6] surrounding healthy bone tissue as well as possible cement residue [7]. These problems can lead to longer operation times, patient discomfort, and undesired surgical outcomes. Water jet technology offers a solution as an advanced cold manufacturing technology based on utilizing stagnation or impact pressure phenomena [8]. Continuous water jet (CWJ) was employed for the interface tissue of both cemented and cementless prostheses [9], while Dunnen et al. [10, 11] proposed a model for the prediction of bone drilling depth and later devised [12] a technique of colliding jets to control the depth of the CWJ bone drill. Unlike CWJ, which relies on sustained high-speed water jet energy transmission into a concentrated material area [13], its modification, Pulsating water jet (PWJ) utilize high-frequency pulsations that produce localized erosion through cyclic high-velocity droplet impacts, inducing the water hammer effect described in [14] followed by cyclical lateral jetting [15]. This results in improved disintegration efficiency, lower applied forces, and minimal thermal damage to adjacent tissues. As the working medium in PWJ is typically water, it has great potential for various bio-applications, such as surface roughening [16] with future potential extension of technology by using sterilized water or saline solution.

Recent studies have demonstrated the effectiveness of PWJ for selectively removing bone cement, with material removal performance strongly influenced by parameters such as nozzle geometry [17], jet pressure [18], traverse speed [19], and environmental conditions [20]. The use of water jet technology for bone cement disintegration was studied by the authors listed in Table 1. Early work by Honl et al. [21, 22] demonstrated that both plain and abrasive water jets (AWJ) can effectively erode PMMA-based cements, with AWJ achieving up to three times deeper cuts compared to plain jets. The addition of abrasive particles caries risk of contamination; alternative use of soluble or biodegradable abrasive including ice, may become relevant in the field [23]. Furthermore, studies [21, 22] also noted a loss of selectivity when bone and cement were exposed together, as both materials were simultaneously eroded at higher pressures. Subsequent efforts published in [18] confirmed the viability of water jet systems in orthopedic surgical environments, emphasizing their ability to operate in narrow anatomical spaces with minimal mechanical force transmission [20]. Recent studies have shifted focus toward the possible clinical applicability of ultrasonic PWJ, reported in [19, 20]. The results show that tested extended and anatomically shaped nozzles with sapphire nozzle inserts at the end of tubular extension have similar or better efficiency compared to standard nozzle inserts. In addition, in the work [20], a submerged environment was also considered, showing that although material removal efficiency drops by approximately 65%, PWJ remains functionally effective under fluid-filled conditions—a relevant factor for in vivo applications. Relatively newly published review [24] suggests the use of water jet for medical applications, highlighting PWJ as a non-thermal, minimally invasive tool that indicates positive features to satisfy the demands of modern surgical practice, particularly for revision arthroplasty.

Despite the research effort mentioned below (Table 1) in using PWJ for bone cement disintegration, current experimental setups often use standard nozzle inserts that aren't suitable for a minimally invasive technique for revision arthroplasty to reach deep sites in the femoral channel. To overcome this engineering-medical gap is important for developing non-thermal, selective PWJ instruments for minimally invasive orthopedic procedures without the need to apply windowing [25, 26]. While PWJ's general feasibility for bone cement disintegration is established (Table 1), significant gaps remain, including limited experimental validation of anatomically shaped nozzles, insufficient comparison between different bone cement formulations under PWJ exposure at various traverse speed, and a lack of performance data correlating traverse speed with material removal rate (MRR) and penetration depth for anatomical PWJ systems.

This study aims to address these gaps by designing and implementing an anatomically shaped nozzle compatible with ultrasonic PWJ systems, experimentally evaluating its disintegration performance on two clinically relevant bone cements (Palacos and C-ment) across a range of traverse speeds, quantifying MRR and penetration depth to assess the anatomical nozzle's efficiency and selectivity, and comparing these findings with existing literature to validate performance and highlight clinical potential.

2 Experimental Setup

2.1 Material

The Palacos and C-ment bone cement samples were prepared and used in the experiments. Each cement type was manually mixed in strict compliance with the respective manufacturer's instructions to ensure consistent composition and curing conditions across all tested specimens. The bone cement mixture was molded in a rectangular preform to facilitate material handling and fixation during the experiments.

2.2 Experimental Method

The experiments were carried out by patented ultrasonically excited pulsating water jet technology [28, 29] present in the Institute of Geonics v.v.i., Czech Republic. The experimental setup consists of a high–pressure pump generating the required pressure $p = 15$ MPa, which flows through high-pressure tubing into a pulsating head, where it enters a high-pressure acoustic chamber. In the acoustic chamber, the high-pressure flow interacts with the vibrating sonotrode excited using an ultrasonic generator with an excitation frequency of $f = 20$ kHz. For the efficient transfer of ultrasonic energy, the optimal setting of the acoustic chamber length is essential, which was set based on the ratio of initial and operating frequency with output power to $lc = 10$ mm under the current experimental conditions. In this investigation, an extended tubular nozzle with a length of 100 mm, fitted to the end of the pulsating head assembly with a diameter of $d = 0.5$ mm, was employed. Moreover, an appropriate standoff distance must be selected to achieve maximum erosion under the specified hydraulic conditions. An optimal standoff distance of $z = 2$ mm was identified following a stair trajectory, as demonstrated in a previous study [30]. Further, to observe the effect of the traverse speed on PMMA-based bone cement, 4 levels of traverse speed, starting from v = 2 mm/s to 12 mm/s, were used. The manipulation of the pulsating head assembly was performed by an ABB robotic manipulator for the precise movement of the jet over the sample surface. The schematic of the experimental setup and the disintegrated sample after the experiment are shown in Fig. 1. All the experimental conditions used in the present study are summarized in Table 2.

Table 1. Summary of recent studies on PWJ for bone cement disintegration – chronological order.

Study	Year	Jet Type	Focus	Key Findings	Nozzle/Tool Geometry
Honl et al. [21]	2000	WJ and AWJ	PMMA vs. Bone cutting at varying pressures	Selective PMMA cutting up to 400 bar; AWJ improved depth but lost selectivity	0.3 mm WJ & AWJ; 15 mm standoff
Honl et al. [22]	2003	AWJ	Prosthesis interface cleaning	AWJ created deeper, smoother kerfs in PMMA; usable in revision surgery	0.3 mm AWJ nozzle, 0.9 mm focus tube
Hloch et al. [18]	2013	WJ and PWJ	General orthopedic use	WJ works in confined spaces with minimal force	Basic orthopedic tool geometry
Nag et al. [27]	2020	Ultrasonic PWJ	Acoustic tuning of chamber	Optimal depth at 22 mm chamber length; up to 615 μm	Tunable chamber, 0.5 mm nozzle
Song et al. [24]	2023	Review	Medical PWJ use	PWJ is promising for selective cement removal; highlights safety	Emphasis on miniaturization
Nag et al. [19]	2024	Ultrasonic PWJ	Standard vs. extension tube	20–25% more removal with extension nozzle at low speed	100 mm extension, 0.3 mm Ø

(continued)

Table 1. (*continued*)

Study	Year	Jet Type	Focus	Key Findings	Nozzle/Tool Geometry
Nag & Hloch [17]	2025	Ultrasonic PWJ	Anatomical vs. standard nozzle	Anatomical nozzle improves localized disintegration and surgical access	Custom anatomical tube
Hloch et al. [20]	2025	PWJ (atmospheric vs. submerged)	Medium comparison	Submerged erosion drops ~65%; still effective in wet surgical fields	100 mm extension tube, 10 mm Ø

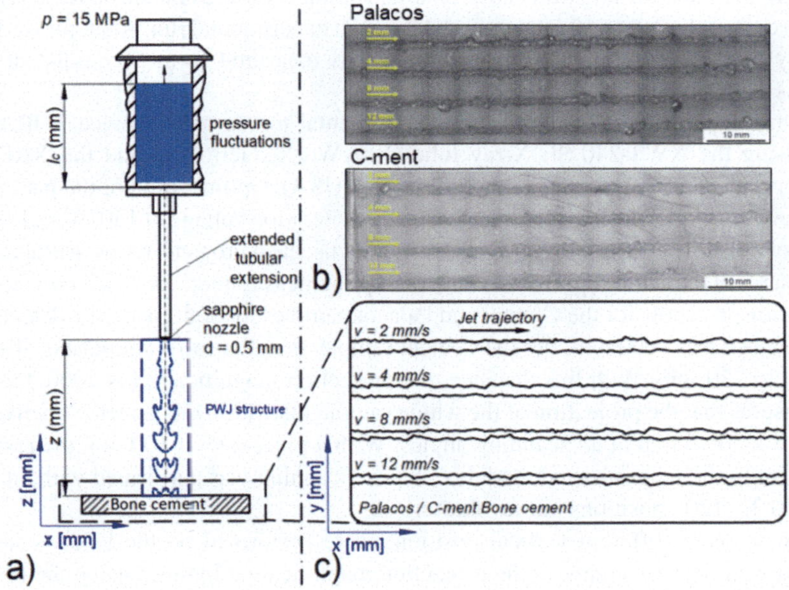

Fig. 1. a) Schematic representation of the PWJ and its interaction with the material, b) optical micrograph of the samples after the experiment, and c) the schematic diagram of the jet trajectory under different experimental conditions.

Table 2. Experimental conditions.

Supply pressure [MPa]	Nozzle diameter [mm]	Nozzle length [mm]	Frequency [kHz]	Chamber length [mm]	Standoff distance [mm]	Traverse speed [mm/s]	Material
15	0.5	100	20	10	2	2,4,8,12	Palacos, C-ment

2.3 Measurements

The erosion performance of the PWJ with varying traverse speeds on different bone cement samples (Palacos and C-ment) was evaluated in terms of the depth, removed material volume, and material removal rate. The entire sample, after exposure to PWJ according to the technological parameters specified in Table 2, was scanned using a VHX-5000 digital microscope with 100x zoom. The 3D topographic data acquired from the scan facilitated the determination of the maximum depth and volume of material removed. For the determination of the material volume removed and the material removal rate, an erosion length of 10 mm was used for each experimental run. Also, for statistical stability of the measurement, erosion depth was measured from 5 equidistant cross sections of the generated grooves.

Both samples were scanned in the experimental tomographic device TORATOM [31], using the XWT-240-SE X-ray tube (XrayWorX, Germany) and the XRD1611 detector with a GOS scintillator (Varex Imaging, USA). Optimization of the parameters to achieve low noise and high contrast led to an acceleration voltage of 150 kV and a target power of 50 W. The exposure time was set to 500 ms; each projection was averaged over 4 exposures at the same angle. Since this was a pilot measurement, we tried two different acquisition protocols for the C-ment and Palacos samples. In both cases, the detector-to-tube distance was 1000 mm. For the C-ment sample, the tube-to-rotational axis distance was set to 250 mm. With this distance, the projection magnification is 1000/250 = 4. This ensures that the projection of the whole sample fits within the detector's active area of 409.6 x 409.6 mm at all scanning angles. With a pixel size of 100 μm, the resulting voxel size in the reconstructed model is 25 μm. Scanning was performed with an angle step of 0.3° (1200 positions per revolution).

Two scans in different vertical positions were performed on the Palacos sample. This allowed for an increase in the projection magnification by decreasing the tube-to-rotational axis distance to 160 mm, leading to a voxel size of 16 μm. The angle step was 0.15° (2400 positions per revolution). Such a scanning protocol should yield a better-quality reconstructed volume at the cost of a quadrupled duration relative to the protocol used for the C-ment sample. Comparison of the two protocols and assessment of the feasibility of prolonged measurements in terms of the expected quality versus the time demand were one of the purposes of this preliminary experiment. The projections were corrected using a standard flat-field correction and reconstructed using the filtered back projection FDK algorithm in VG Studio Max (Volume Graphics, Germany).

SEM observation was carried out using a scanning electron microscope, Tescan Lyra 3 XMH. The accelerating voltage of 10 kV was employed, and micrographs were taken using a secondary electron detector in high vacuum ($< 9 \times 10^{-3}$ Pa). Samples were, prior to SEM observation, coated with a thin carbon layer using a coater Quorum 150 T.

3 Results

Figure 2 illustrate the effect of traverse speed on the erosion performance of two commercially available PMMA-based bone cements—Palacos and C-ment—disintegrated using an ultrasonic pulsating water jet (PWJ) system equipped with an anatomical nozzle. Considered parameters were the depth h (mm), removed volume V (mm^3), and material removal rate (MRR) (mm^3/min).

As shown in Fig. 2a, a clear inverse correlation exists between traverse speed v (mm/s) and erosion depth h (mm). At the lowest traverse speed v = 2 mm/s, both samples exhibit maximum disintegration, which decreases with increasing traverse speed. C-ment consistently shows greater depth of disintegration than Palacos under identical conditions, suggesting material-specific differences in fracture behavior or jet interaction. These findings agree with previous research by [20], which showed that longer PWJ exposure at slower traverse speeds causes significant material removal or disintegration also in submerged conditions. Figure 2b shows that eroded volume is significantly influenced by traverse speed v (mm/s). The highest erosion volumes for both cements were recorded at the lowest speeds, with a steep decline as speed increased. The anatomical nozzle used in this study—specifically designed for access to confined anatomical structures—enabled consistent jet delivery and effective material removal even at variable speeds. Notably, C-ment again exhibited larger eroded volumes across all conditions, which agrees with the extension nozzle findings by Nag et al. [17] where 20–25% higher removal was achieved through better fluid-structure targeting. Figure 2c shows the influence of traverse speed v (mm/s) on MRR (mm^3/min) for two types of PMMA samples Palacos (in red) and C-ment (in blue). The data are presented as vertical bar slopes with dashed trendlines to indicate the erosion evolution within the tested traverse speeds. The results show a clear positive correlation between traverse speed and MRR for both materials, with C-ment consistently achieving higher removal rates, particularly at 12 mm/s. While Palacos exhibits a plateau around 290 mm^3/min at the highest speed, C-ment continues to increase, surpassing 360 mm^3/min, indicating superior responsiveness to higher jet velocities. This behavior suggests that C-ment is more susceptible to fracture under pulsed jet action, making it more efficiently eroded at elevated speeds. The linear trend observed for C-ment highlights the effective use of anatomical nozzle geometry in maintaining disintegration efficiency during rapid scanning.

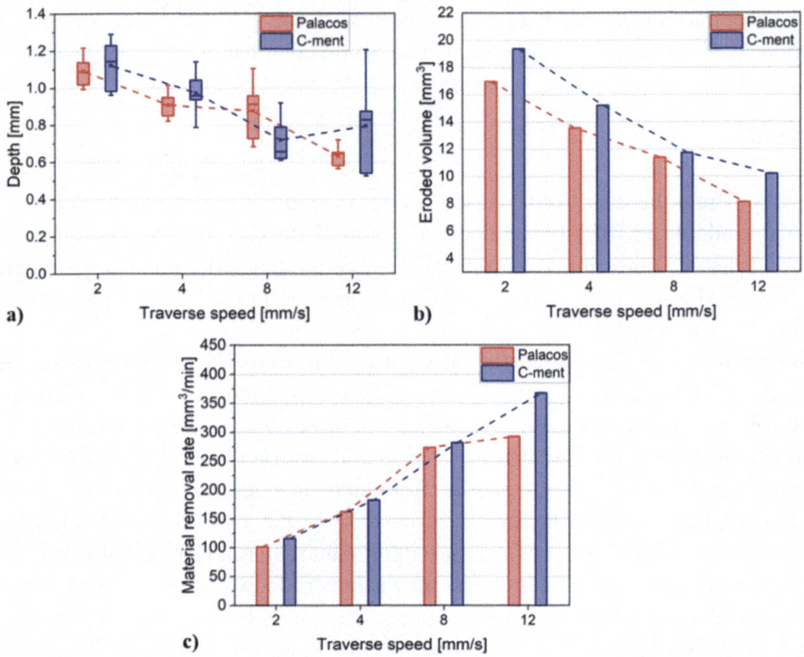

Fig. 2. Influence of the traverse speed of the ultrasonic pulsating water jet on the a) depth penetration, b) removed volume, and c) MRR.

Figure 3 presents SEM micrographs at 50× magnification showing the full width of erosion grooves produced in Palacos (top row) and C-ment (bottom row) PMMA-bone cements at different traverse speeds. The micrograph shows the entire kerf profiles, including central jet impact zones and lateral erosion features visible in detail in Fig. 4. For both materials, lower traverse speeds produce wider, deeper grooves with steeper sidewall geometry and more pronounced material removal, indicating higher erosion intensity due to longer jet-material interaction. With the increase in the traverse speed, the grooves become progressively narrower and shallower, with smoother lateral boundaries. Across all velocities, C-ment grooves appear slightly wider and more irregular than those in Palacos, suggesting greater susceptibility to lateral material loss under identical processing conditions. This is supported by steeper sidewall geometry in Palacos compared to the more gradual sidewall geometry observed in the case of C-ment. These SEM observations correlate with the depth of penetration and eroded volume trends presented in the graphs.

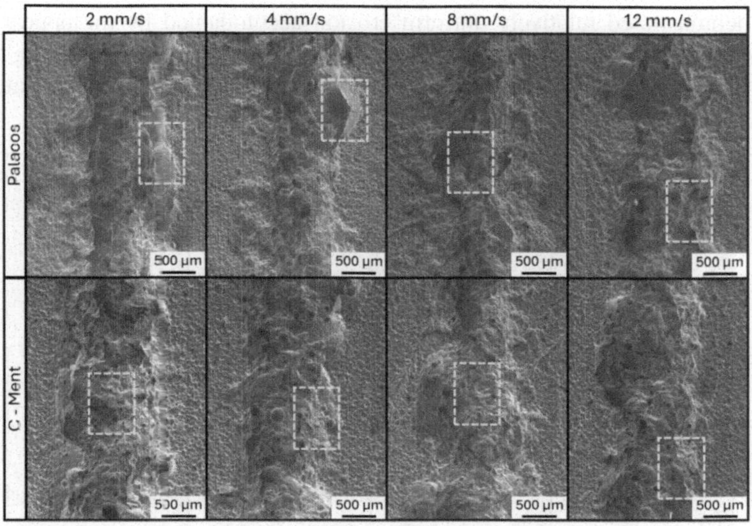

Fig. 3. SEM micrographs showing comparison of the effect of traverse speed on Palacos and C-ment. Print magnification was set to 50 x for observation of the whole width of the erosion grooves. The yellow rectangles show the location of the magnifications shown in the following figure.

Figure 4 of SEM micrographs (200× magnification) shows the effect of traverse speed v (mm/s) on Palacos and C-ment. Both cements exhibit brittle fracture morphologies with planar facets, while C-ment shows smaller dimensions of planar facets and more irregular height steps between the facets compared to Palacos.

At traverse speeds v = 2 mm/s, both materials are eroded and show extensive fracture zones and rough steps between planar facets. Palacos shows larger planar facets with the occasional presence of almost spherical pores. A micrograph of Palacos is taken close to the edge of the eroded groove. The adjacent surface shows typical PMMA particles, with some of them fractured or removed close to the groove edge. Compared to patterns observed in Palacos, C-ment shows a larger number of smaller planar facets with a denser distribution of height steps between them. Inclusions are visible in SE contrast on the fractured surface in the case of C-ment. In the case of v = 4 mm/s, both micrographs of Palacos and C-ment were taken close to the groove edge. The Palacos micrograph shows a large pore with PMMA particles lining its surface. Above the pore brittle fracture is observed. C-ment shows a smaller depth gradient from the edge to the kerf center. The surface shows a higher number of smaller planar facets and a number of semicircular and circular pores. Smaller pores exhibit more circular geometry and a smooth surface, while larger ones exhibit remains of PMMA particles on their surface. At a traverse speed of v = 8 mm/s, the erosion features decrease, with kerfs showing shallower penetration. Palacos surfaces appear smoother and more uniform, while C-ment maintains a rougher morphology with smaller dimensions of planar facets interrupted by pores or height steps. At the traverse speed v = 12 mm/s, the surfaces of both cements were characterized by shallower grooves, with comparatively smoother morphology and shallower height steps between adjacent planar facets reflecting the shorter jet-material interaction time.

Palacos demonstrated relatively uniform erosion accompanied by the appearance of pores, whereas C-ment continued to exhibit microcracking. The underlying fracture mechanism shows morphological similarities, such as hackles between the materials.

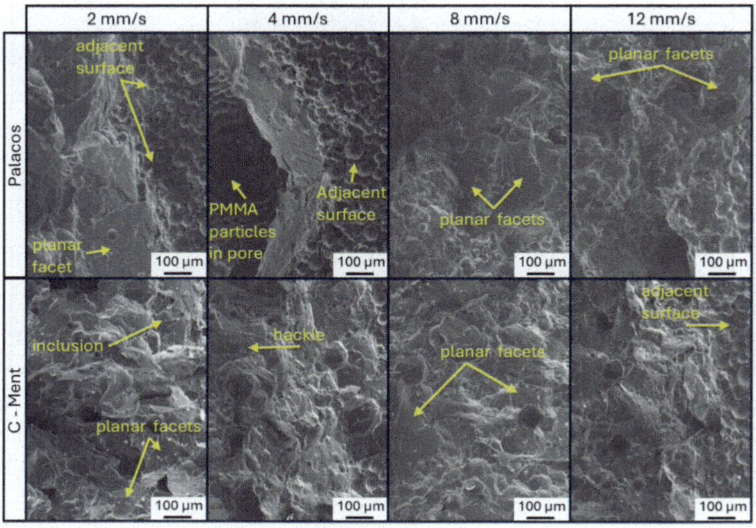

Fig. 4. SEM micrographs showing comparison of the effect of traverse speed on Palacos and C-ment. Print magnification was set to 200 x for observation of fractographic details.

Figure 5 documents a CT analysis of both Palacos and C-ment. While all erosion grooves have been scanned, the boundary traverse speed of 2 mm/s and 12 mm/s are depicted. The erosion grooves were scanned using CT after the SEM analysis to visualize the pores, which were observed frequently on fractured surfaces. The interaction of PWJ impact with pores is critical in erosion development as observed on fractured surfaces. The CT was employed for visualization of cross-sectional planes in both Palacos and C-ment in the case of 2 mm/s. The important role of the interaction of droplet impingement with pre-existing pores has been highlighted in previous research [20]. The results correspond with the crater widening observed in Fig. 3. The crater widening also explains the discrepancy in eroded volume between Palacos and C-ment.

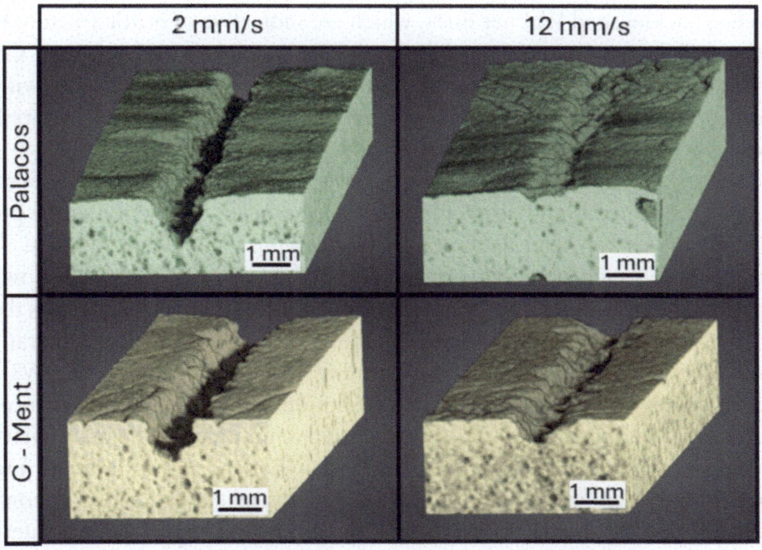

Fig. 5. CT-images showing comparison between Palacos and cement for boundary traverse speeds v = 2 mm/s and v = 12 mm/s.

The quantitative results (Fig. 2) and SEM observations (Figs. 3 and 4) confirm that traverse speed strongly governs the disintegration behavior of bone cements under ultrasonic pulsating water jet with respect to erosion time. While lower speeds enhance penetration depth and eroded volume, higher speeds favor material throughput, with C-ment consistently demonstrating greater susceptibility to erosion than Palacos across all metrics. The combination of anatomical nozzle design and optimized traverse speed of PWJ technology shows a good step for selective, efficient, and minimally invasive removal of bone cement in revision arthroplasty, which is discussed in detail in the following chapter.

4 Discussion

The influence of traverse speed on the disintegration of Palacos and C-ment bone cements under ultrasonic pulsating water jet (PWJ) exposure was quantified in terms of penetration depth, eroded volume, and material removal rate (Fig. 2). A strong dependence on traverse speed was observed: low speeds promoted deeper penetration and larger eroded volumes due to prolonged jet-material interaction, while higher speeds favored material throughout, yielding higher material removal rates (MRR). Across all technological conditions (Table 2), C-ment consistently exhibited greater erosion disintegration than Palacos, indicating differences in microstructural toughness and fracture response. These findings confirm the balance between precision and efficiency in PWJ machining: low speeds optimize selective disintegration, whereas higher speeds support faster material removal.

Fractographic analysis shows details with the following findings. The mechanisms leading to material removal show morphological similarities, including the presence of planar facets, hackles, and Wallner lines, which are indicative of brittle fracture for both bone cements. Palacos displayed smooth planar facets with larger dimensions. Contrary, C-ment showed more fragmented morphology, composed of planar facets with lower dimensions, with numerous interactions with height steps and pores. Both materials show the presence of pores. These can be divided into smaller pores with a smooth surface and larger pores with a surface composed of parts of PMMA particles. The pores were exposed due to repeated liquid droplet impact. The complex interaction of the pulsating water jet with pores can lead to the onset of cracking of adjacent material. C-ment also shows exposed inclusions due to the action of the ultrasonic pulsating water jet. Detachment of the inclusions due to the ultrasonic pulsating water jet action is therefore also possible. The morphological changes correlate directly with the penetration and volume data shown in graphs Fig. 2, demonstrating the interaction between PWJ erosion time and fracture intensity. The stochastic orientations of hackles across surfaces show that ultrasonic pulsating water jet processing leads to multiple local crack initiations on the removed surfaces, which can be linked to the number and size of droplets impacting the surface. These fractographic details complement the macro-scale results, confirming that Palacos, following the described technological history, resists lateral material loss better than C-ment under identical hydraulic conditions.

The experimental findings presented here align well with recent studies on bone cement disintegration using PWJ. Earlier published work [19] shows that lower traverse speeds and higher pressures enhance erosion depth, whereas higher speeds favor efficiency. More recent studies reported by [17] highlighted the benefits of extended and anatomically shaped tubular extension for possible improved bone cement disintegration. Similarly, [20] confirmed that submerged PWJ bone cement disintegration is effective though less efficient than atmospheric conditions, which is important for in vivo applications, as it is expected that the femoral channel will be flooded with body fluid. The following Table 3 shows a comparison between these published results and the present study, highlighting the novelty of using an anatomical nozzle and the direct comparison of Palacos and C-ment under identical PWJ conditions compared to recently gained results.

Comparative overview Table 3 demonstrates the recent progress of PWJ research for bone cement disintegration, from early investigations with standard nozzles to more recent developments employing extended tubes and anatomical nozzle designs. Previous studies have established the influence of traverse speed, pressure, and nozzle extension on erosion efficiency, they were limited either to single-material testing or to standard nozzle geometries, which don't allow for possible controlled erosion disintegration in non accessible places such as the femoral channel. The present work advances this knowledge by being the first to systematically compare two clinically relevant cements (Palacos and C-ment) under identical ultrasonic PWJ conditions using an anatomical nozzle. By combining quantitative metrics (penetration depth, eroded volume, and MRR) with microstructural evidence (SEM analysis at both kerf and fractographic scales), this study not only shows trends reported previously but also provides new insights into the material-specific erosion mechanisms, such as the hammer effect utilised by PWJ, can increase performance in revision arthroplasty. These findings highlight the anatomical nozzle's potential for clinical use, offering improved accessibility in non-accessible surgical spaces while maintaining erosion efficiency.

Table 3. Comparison of the recent works.

Reference	Target Material	Nozzle Type	Environment	Pressure	Traverse speed	Key findings
Nag et al. (2024) [19]	Bone cement (PMMA)	Standard vs. 100 mm extension	Atmospheric	10, 20	0.5–2.5	Extension nozzle improves volume removal by PMMA lower speed and higher pressure enhance efficiency.
Nag & Hloch (2025) [17]	Bone cement (PMMA)	Standard vs. 100 mm extension	Atmospheric	15	1	Extended nozzle achieves smoother surfaces and deeper erosion.
Hloch et al. (2025) [20]	Bone cement (PMMA)	100 mm extension	Atmospheric & Submerged	20	Fixed (variable exposure time)	Submerged cutting reduces efficiency by ~65%; effective in both environments.
Hloch et al. (2013) [18]	Bone cement	Standard nozzle	Atmospheric	10–20	0.5–2.0	PWJ effective for bone cement; depth increases with lower speed and higher pressure.

(*continued*)

Table 3. (*continued*)

Reference	Target Material	Nozzle Type	Environment	Pressure	Traverse speed	Key findings
Present study	Bone cement (Palacos & C-ment)	Anatomical nozzle	Atmospheric	15	2–12	C-ment consistently shows higher erosion depth, volume, and MRR than Palacos. Lower speeds enhance penetration; higher speeds maximize MRR. SEM confirms more fragmented fracture morphology in C-ment.

5 Conclusion

This study presented an in-depth experimental evaluation of two commercially used bone cements—Palacos and C-ment—subjected to disintegration by ultrasonic pulsating water jet (PWJ) using a custom-designed anatomical nozzle. Both cement samples showed the highest eroded volume at a traverse speed of 2 mm/s, with a steep decline at faster traverse speeds, whereas the material removal rate showed a reverse tendency. C-ment exhibits a larger eroded volume across all conditions compared to Palacos. These trends were further analysed by SEM using multiple magnifications, which revealed deeper and more fragmented morphology, including planar facets of smaller dimensions in C-ment due to the content and mechanical properties of the bone cement.

The study's limitations include its laboratory setting, where PWJ with an anatomical nozzle was controlled by a robotic manipulator. Key issues are: impact was only tested perpendicularly, whereas real-world applications often involve angled impacts; the experiment was in atmospheric conditions, but practical use involves submersion due to jet flow, requiring consideration of submerged conditions and excess water pathways; and while the manipulator's position and speed are precisely controlled, further evaluation of using this method as a hand-operated tool is needed.

To extend the present findings and move toward clinical application, future work should focus on:

- Real-time PWJ bone cement disintegration process monitoring (e.g., acoustic emission or visual) for adaptive control of erosion depth and material selectivity.

- Optimization of nozzle geometry and actuation frequency, including flexible or multi-axis configurations tailored for specific surgical approaches.
- Biomechanical and thermal safety studies to verify that surrounding bone and soft tissues remain structurally and functionally intact after jet exposure.

The anatomical nozzle used in this study provided precise, directional jet control suitable for non-accessible areas like femoral canals without the need to compromise the femoral bone. The combined quantitative and morphological findings demonstrate the potential of PWJ as a minimally invasive, nonthermal, and selective tool for bone cement disintegration, offering a better alternative to mechanical or thermal methods in revision arthroplasty.

Acknowledgments. This study was financially supported by the Slovak Research and Development Agency under Contract No APVV-22-0391. The presented work was also supported by a project for the long-term conceptual development of the research institution RVO: 68145535. This work is part of the project Advanced X-ray imaging methods for industrial applications (AXIMEIA), which is co-financed by the European Union within the Interreg Austria – Czechia 2021–2027 programme.

References

1. Kavanagh, B.F., Ilstrup, D.M., Fitzgerald, J.R.H.: Revision total hip arthroplasty. J. Bone Joint Surg. Am. **67**, 517–526 (1985)
2. Porsch, M., Schmidt, J.: Cement removal with an endoscopically controlled ballistically driven chiselling system. A new device for cement removal and preliminary clinical results. Arch. Orthop. Trauma Surg. **121**, 274–277 (2001). https://doi.org/10.1007/s004020000233
3. Gardiner, R., Hozack, W.J., Nelson, C., Keating, E.M.: Revision total hip arthroplasty using ultrasonically driven tools. J. Arthroplast. **8**, 517–521 (1993)
4. Goldberg, S.H., Studders, E.M., Cohen, M.S.: Ultrasonic cement removal in revision arthroplasty. Orthopedics. **30**, 632–635 (2007)
5. Goldberg, S.H., Cohen, M.S., Young, M., Bradnock, B.: Thermal tissue damage caused by ultrasonic cement removal from the humerus. J. Bone Joint Sur. Am. **87**, 583–591 (2005). https://doi.org/10.2106/JBJS.D.01966
6. Haertlé, M., Tücking, L R., Derksen, A., Reulbach, M., Jakubowitz, E., Windhagen, H.: The use of an ultrasonic cement removal device in revision hip and knee arthroplasty – a matched case-control study. J Exp Orthop. **12** (2025). https://doi.org/10.1002/jeo2.70171
7. Roitzsch, C., et al.: Ultrasonic bone cement removal efficiency in total joint arthroplasty revision: a computer tomographic-based cadaver study. J. Orthop. Res. **41**, 1365–1375 (2023). https://doi.org/10.1002/jor.25465
8. Foldyna, J., Svehla, B.: Method of generation of pressure pulsations and apparatus for implementation of this method (2010)
9. Kraaij, G., Tuijthof, G.J.M., Dankelman, J., Nelissen, R.G.H.H., Valstar, E.R.: Waterjet cutting of periprosthetic interface tissue in loosened hip prostheses: An in vitro feasibility study. Med. Eng. Phys. **37**, 245–250 (2015). https://doi.org/10.1016/j.medengphy.2014.12.009
10. den Dunnen, S., Mulder, L., Kerkhoffs, G.M.M.J., Dankelman, J., Tuijthof, G.J.M.: Waterjet drilling in porcine bone: the effect of the nozzle diameter and bone architecture on the hole dimensions. J. Mech. Behav. Biomed. Mater. **27**, 84–93 (2013). https://doi.org/10.1016/j.jmbbm.2013.06.012

11. den Dunnen, S., Dankelman, J., Kerkhoffs, G.M.M.J., Tuijthof, G.J.M.: How do jet time, pressure and bone volume fraction influence the drilling depth when waterjet drilling in porcine bone? J. Mech. Behav. Biomed. Mater. **62**, 495–503 (2016). https://doi.org/10.1016/j.jmbbm.2016.05.030
12. den Dunnen, S., Dankelman, J., Kerkhoffs, G.M., Tuijthof, G.: Colliding jets provide depth control for water jetting in bone tissue. J. Mech. Behav. Biomed. Mater. **72**, 219–228 (2017). https://doi.org/10.1016/j.jmbbm.2017.05.009
13. Zeleňák, M., Říha, Z., Jandačka, P.: Visualization and velocity analysis of a high-speed modulated water jet generated by a hydrodynamic nozzle. Measurement. **159**, 107753 (2020). https://doi.org/10.1016/j.measurement.2020.107753
14. Cook, S.S.: Erosion by water-hammer. Proc. R. Soc. Lond. Ser. A Contain. Pap. Math. Phys. Character. **119**, 481–488 (1928)
15. Burson-Thomas, C.B., Wellman, R., Harvey, T.J., Wood, R.J.K.: Water droplet erosion of aeroengine fan blades: the importance of form. Wear. **426–427**, 507–517 (2019). https://doi.org/10.1016/j.wear.2018.12.030
16. Steeves, A.J., et al.: In Vitro Response of Osteoblastic and Fibroblastic Cells to Water-Jet Treated Titanium Surfaces
17. Nag, A., Hloch, S.: Disintegration of bone cement using pulsating water jet: a comparative study of standard and extended nozzles. Spectr. Mech. Eng. Oper. Res. **2**, 93–103 (2025). https://doi.org/10.31181/smeor21202525
18. Hloch, S., et al.: Disintegration of bone cement by continuous and pulsating water jet | Dezintegracija koštanog cementa pomoću kontinuiranog i pulsirajućeg mlaza vode. Tehnicki Vjesnik. **20** (2013)
19. Nag, A., Stolarik, G., Hloch, S.: Comparison of the effects of standard nozzles and extension tubes on the erosion of PMMA using pulsating water jet technology. MM Sci. J. **2024**, 10.17973/MMSJ.2024_11_2024099 (2024)
20. Hloch, S., et al.: Atmospheric and submerged bone cement erosion by a pulsating water jet with a straight tubular nozzle extension. Sci. Rep. **15**, 6224 (2025). https://doi.org/10.1038/s41598-025-91213-7
21. Honl, M., et al.: The use of water-jetting technology in prostheses revision surgery – first results of parameter studies on bone and bone cement. J. Biomed. Mater. Res. **53**, 781–790 (2000). https://doi.org/10.1002/1097-4636(2000)53:6<781::AID-JBM20>3.0.CO;2-G
22. Honl, M., et al.: The water jet as a new tool for endoprosthesis revision surgery–an in vitro study on human bone and bone cement. Biomed. Mater. Eng. **13**, 317–325 (2003)
23. McGeough, J.A.: Cutting of food products by ice-particles in a water-jet. In: Procedia CIRP, pp. 863–865. Elsevier (2016). https://doi.org/10.1016/j.procir.2016.03.009
24. Song, X.F., Zhao, J., Yan, H., Yu, W., Yin, L.: Waterjet Machining of Biological Tissues in Medical Surgeries: From Soft Tissue Dissection to Bone Cutting (2023). https://doi.org/10.1016/j.jmapro.2023.10.067
25. Klein, A.H., Rubash, H.E.: Femoral Windows in Revision Total Hip Arthroplasty (1993)
26. Buehler, K.O., Walker, R.H.: Polymethylmethacrylate removal from the femur using a crescentic window technique. Orthopedics. **21**, 697–700 (1998)
27. Nag, A., Hloch, S., Dixit, A.R., Pude, F.: Utilization of ultrasonically forced pulsating water jet decaying for bone cement removal. Int. J. Adv. Manuf. Technol. (2020). https://doi.org/10.1007/s00170-020-05892-9
28. Foldyna, J.: Use of acoustic waves for pulsating water jet generation. In: Acoustic Waves – From Microdevices to Helioseismology, pp. 323–342 (2011). https://doi.org/10.5772/18862
29. Foldyna, J.: Use of acoustic waves for pulsating water jet generation. In: Acoustic Waves-From Microdevices to Helioseismology. InTech (2011)

30. Poloprudsky, J., Klichova, D., Stolarik, G., Chlupova, A.: Effect of technological parameters of the pulsating water jet on erosion effectivity on a linear trajectory. MM Sci. J. **2024**, 7526–7533 (2024). https://doi.org/10.17973/MMSJ.2024_10_2024073
31. Fíla, T., Vavřík, D.: Multi-axial device for carrying out X-ray measurements, especially computed tomography. European Patent Application No. EP3254095A1 (2017)

Interactive Computer-Aided Balancing of Machining Lines – A Declarative Modeling Approach

Grzegorz Radzki[1], Grzegorz Bocewicz[1], Mariusz Piechowski[2], Zbigniew Banaszak[1], and Małgorzata Jasiulewicz-Kaczmarek[3(✉)]

[1] Faculty of Electronics and Computer Science, Koszalin University of Technology, Koszalin, Poland
[2] WSB Merito University Poznan, Powstańców Wielkopolskich 5, 61-895 Poznań, Poland
[3] Faculty of Engineering Management, Poznan University of Technology, Poznań, Poland
malgorzata.jasiulewicz-kaczmarek@put.poznan.pl

Abstract. This paper presents a decision-support approach for balancing machining lines that is designed for on-to-fly use by shop-floor planners. The method relies on a declarative representation of the line-balancing problem and is solved with constraint programming, which makes it possible to quickly regenerate feasible production plans when priorities, orders, or operating conditions change. The proposed framework consolidates heterogeneous production information into a single, consistent data model and can be integrated with computer-aided planning and manufacturing supervision systems. To illustrate industrial relevance, the approach is evaluated in a real setting involving machining tasks for components such as rolling bearing parts and cast-iron hubs. The computational study highlights the benefits of an open, extensible architecture that allows company-specific rules and technological limitations to be added without redesigning the core model. Overall, the results indicate that the proposed solution improves responsiveness and transparency in dispatching decisions and can deliver operational advantages compared with more rigid balancing procedures.

Keywords: Machining Line Balancing · Declarative Modeling

1 Introduction

The growing demand for regular revision of manufacturing plans and workflows - including system reconfiguration and schedule revision, rebalancing, the reallocation of personnel, and the changing of order priorities creates substantial difficulties for shop-floor and production control. These challenges, in turn, require the design of interactive, online decision aids adjustment of shop-floor plans and schedules, capable of flexibly responding to unplanned disturbances that impact their execution. In practice, such corrective actions are often undertaken concurrently or within feedback loops, further increasing the complexity of effective decision-making [1]. Most existing research in this area is narrowly tailored to the requirements of schedule revision [2], line workstation-load

B. Gapiński et al. (Eds.): MANUFACTURING 2026, LNME, pp. 128–142, 2026.
https://doi.org/10.1007/978-3-032-22943-4_10

balancing [3] and reconfiguring machine-line layouts [4] as well as staffing structures [5].

The absence of methods that can evaluate how decisions across these re-planning domains interact with one another served as the main motivation for this work. Accordingly, the central goal is to develop a prototype approach that allows planners and supervisors to react dynamically and interactively to unexpected disruptions across multiple decision layers. Such disruptions arise from gaps between planned and observed performance, as well as from performance shifts that either open opportunities for improvement or demand risk reduction. More specifically, the study proposes a method that, by restricting manufacturing re-planning to line balancing and configuration changes constrained to equipment rearrangement [1]), makes it possible to tailor the machining-line baseline model [6] to the distinctive features of a chosen process - here, manufacturing of iron castings that require precision machining operations. Consequently, the method should help dispatching personnel define online scenarios for selected cases of machining-line rebalancing and/or equipment reconfiguration. An open reference model structure was adopted, which allows for easy addition and integration of data from various sources, decision models or scenarios, as well as natural integration and aggregation of newly entered data with structures that determine real-time decision-making. This solution enables dynamic updates that take into account continuously aggregated input variables (e.g., changes in orders, machine failures, tool wear), resulting in timely decisions that integrate the recorded changes [7–9].

For validation, the framework was examined through a practical case example. A set of computational runs demonstrated that it supports fast and well-grounded decisions in highly customized, demand-responsive manufacturing settings. The paper is organized as follows: Sect. 2 surveys related work; Sect. 3 describes the methodological basis, including the model formulation and the role of interactive follow-up control in the decision-support workflow; Sect. 4 reports the computational results. Section 5 concludes with a summary and directions for further study.

2 Literature Review

The implementation of online balancing mechanisms and production re-planning strategies is widely recognized as a necessary condition for sustaining manufacturing operations. Within this context, Machining Line Balancing (MLB) is understood as a class of methods concerned with the distribution of workloads among workstations in machining production lines [10]. Typical performance criteria addressed in MLB include idle time reduction, bottleneck mitigation, and the preservation of continuous material flow [11]. These effects are commonly achieved through coordinated adjustments of cycle times, tool paths, and machine allocations, with the objective of improving resource utilization under fixed production targets.

Existing solution approaches to MLB problems are usually categorized into experimental, exact, and heuristic methods. Experimental approaches are predominantly associated with Design of Experiments (DOE) techniques [12–14], which provide a structured framework for systematic data collection and exploratory analysis. Although general methodological principles for DOE are well established, the literature does not provide a unified or prescriptive procedure for identifying an optimal experimental design within the large space of admissible configurations. Exact methods are most frequently represented by Pareto-based approaches, which formalize MLB as a multi-objective optimization problem and focus on identifying non-dominated solutions, i.e., solutions for which improvement in one objective necessarily entails degradation in at least one other objective [15]. The Pareto frontier serves as a canonical representation of such trade-offs and is commonly used as a decision support artifact for selecting compromise solutions consistent with predefined preference structures [16]. Heuristic and metaheuristic approaches constitute a third major class and have increasingly incorporated machine learning techniques, including reinforcement learning [10, 17], to enable adaptive responses to real-time disturbances, thereby limiting delays and uneven workload distribution. Due to their reliance on data-driven adaptation and online learning, these approaches are frequently discussed within the conceptual framework of Industry 4.0 [18].

In parallel to model-based balancing methods, production re-planning is strongly influenced by line reconfiguration strategies [19]. Reconfiguration refers to the adaptation of existing machining systems to accommodate new products, modified specifications, or evolving demand patterns. Such adaptations are typically triggered when an existing production line is required to manufacture a different product or when changes in product characteristics impose modifications on the operational structure [20, 21]. Reconfiguration activities generally involve the introduction of new operations into the existing equipment layout, accompanied by the removal of obsolete ones. Current research in this area predominantly addresses reconfiguration under disruptive conditions, including product variability, machine breakdowns, or revised performance constraints. These studies often concentrate on the selection of alternative machines or the modification of processing routes. Nevertheless, a significant proportion of the proposed solutions focus on incremental adjustments, such as machine relocation or buffer capacity modification, particularly in scenarios characterized by relatively minor disturbances [12].

Decision support for multi-criteria problems arising in machining line balancing and, more broadly, in production re-planning has been investigated extensively over several decades due to the intrinsic structural and computational complexity of these problems [22, 23]. The persistence of research interest in this area can be attributed primarily to two factors. First, problem formulations exhibit substantial variability with respect to the expected form of the decision outcome (e.g., selection of a single variant versus ranking of alternatives), the structure of the admissible decision space, and the interaction mechanisms employed for preference elicitation. Second, these problems are typically associated with high computational complexity, which motivates the development of dedicated heuristic approaches capable of delivering approximate solutions within the time constraints imposed by online decision-making environments. The relatively limited number of reported industrial implementations is often attributed to intellectual property constraints and to the early developmental stage of AI-driven, domain-specific decision support systems [22, 24].

From the perspective of computational complexity, production re-planning tasks are widely viewed as difficult to solve (often NP-hard) [6] and may involve nonlinearities as well as nondeterministic effects. Therefore, decision-support techniques for such settings are usually customized to the particular process structure and its constraints. In practice, method design must balance the quality of the obtained solution against the time and resources needed to compute it. Against this background, declarative modeling is often considered attractive because it enables quick development and testing of alternative solution concepts within model structures that are open and easy to extend. One of the defining properties of declarative models [6, 25–27] is their ability to incorporate additional dependencies among decision variables without a substantial loss of computational efficiency. Furthermore, these models support both analysis-oriented reasoning (e.g., counterfactual or sensitivity analysis) and synthesis-oriented problem formulations aimed at identifying parameter configurations that induce desired system behavior. When embedded in a feedback-based control loop, these features help keep shop-floor execution consistent with the planned schedule and coordinated with overarching supervision (e.g., dispatcher-level oversight). As a result, the declarative paradigm supports human-guided, real-time steering of recurring production processes, especially where demand changes often due to either internal scheduling updates or customer-side variation. Finally, the considered re-planning tasks remain computationally difficult-typically NP-hard and they are commonly associated with nonlinear and nondeterministic effects.

3 Methodology

3.1 Problem Formulation

The problem addressed here is formulated for a serial machining system L shown in Fig. 1. The system comprises $n$machines m_i, collected in the set $M = \{m_1, \ldots, m_i, \ldots, m_n\}$. Line L is assumed to produce a fixed quantity TN. Each unit corresponds to a task J that follows an ordered operation chain $o_1, \ldots, o_i, \ldots, o_n$, executed according to the prescribed processing order as the part progresses through machines in M. Each o_i is described by three attributes:

- the processing time t_i (i.e., the cycle time on machine m_i),
- the tool lifetime z_i consumed/required during the execution of o_i,
- the energy consumption e_i.

The values t_i, z_i, e_i depend on the operational state of the corresponding machine. Let $x_i \in SX_i \subseteq \mathbb{N}$ denote the assumed state of machine m_i, where SX_i is the set of admissible states of m_i. The state x_i determines the vector of production parameters for m_i, denoted by: $P_i = (t_i, z_i, e_i) : SX_i \rightarrow \mathbb{P}_i$, where $\mathbb{P}_i$ denotes the set of admissible production-parameter tuples for machine m_i.

In this setting, the line state is captured by the machine-state vector $X = (x_1, \ldots, x_i, \ldots, x_n)$; all feasible such vectors constitute the set $\mathbb{X}$. Modifying X changes the effective line cycle time, $T = \max \{t_1, \ldots, t_i, \ldots, t_n\}$, which in turn impacts the Total Operating Cost (TOC). In particular:

- TOC aggregates multiple contributors related to each machine m_i, including energy consumption e_i, tool wear/lifetime z_i, and the individual cycle time t_i. Together, these elements determine the labor cost cw_i associated with machine m_i.

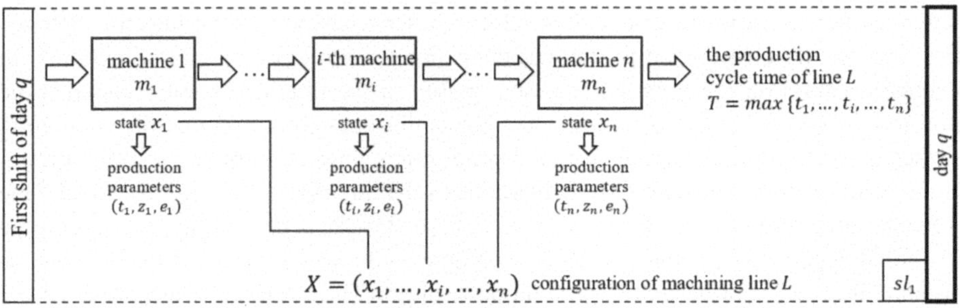

Fig. 1. Schematic of the nominal operating configuration for line *L*.

- The value of T influences the Production Completion Date (PC): PC scales with the amount of available working time WT (e.g., shifts, overtime, weekend work) and decreases as the line cycle time T becomes shorter.

The evaluated measures are further conditioned by the selected operating structure $SL \in \mathbb{SL}$. This structure captures the disturbance context under which the line runs, such as equipment failures, scheduled maintenance downtime, or slowdowns resulting from personnel unavailability. Because such conditions may vary during order execution, the structure is subject to frequent adjustments, which in turn influences both TOC and PC. In this study, the considered structures are $\mathbb{SL} = \{sl_1, sl_2, sl_3, sl_4\}$, where sl_1 represents uninterrupted production, and sl_2, sl_3, sl_4correspond to scenarios in which a single machine fails. Specifically, the analyzed structures reflect the following operating modes:

- **Nominal operating mode:** In the nominal configuration, the full set of machines M remains available, and production is performed within a single-shift schedule (configuration sl_1, Fig. 1).
- **Machine failure – two-shift operating mode:** Upon the unavailability of machine m_i, the operations assigned to this machine are reallocated to machine m_{i-1} and executed within a second shift (configuration sl_2, Fig. 2a). The resulting intermediate products are processed by machines m_{i+1}, m_{i+2}, ..., m_n during the first shift of the subsequent production day. The admissibility of this mode is conditioned by the substitutability of the consecutive machines $m_{i-1}$and m_i.
- **Machine failure – single-shift operating mode with operation aggregation:** If machine m_i is unavailable, its operations are transferred to machine m_{i-1} and carried out as a single aggregated operation within one shift (configuration sl_3, Fig. 2b). This mode is feasible only under the assumption of interchangeability between machines m_{i-1} and m_i.
- **Machine failure – single-shift operation mode with the use of an alternative machine:** When machine m_i becomes unavailable, its assigned operations are executed on the alternative resource m_i' (structure sl_4, Fig. 2c). This variant can be applied only if m_i' is explicitly defined as the replacement for m_i and its involvement does not create extra penalties - for example, by causing downstream delays in the production line in which the substitute machine operates.

In real production environments, only selected line-structure layouts can be considered admissible. The cases discussed above were derived from practical solutions applied in bearing manufacturing. Within these premises, the problem is to choose an admissible pair (*SL*, *X*), where *SL*specifies the operating structure of line *L* and $X = (x_1, \ldots, x_i, \ldots, x_n)$ defines the corresponding machine-state configuration, while meeting the following requirements:

- the Total Operating Cost (TOC) does not exceed a prescribed limit *TR*, i.e., $TOC \leq TR$ and/or $TOC \rightarrow \min$,
- the Production Completion Date (PC) does not exceed a prescribed limit *PD*, i.e., $PC \leq PD$ or $PD \rightarrow \min$.

The feasible decision space is represented by the family of admissible production scenarios $AS = \{(SL, X) \mid SL \in \$\mathbb{L}, X \in \mathbb{X}\}$, where $\mathbb{X}$ is the collection of allowable configurations of machining line *L* and $\$\mathbb{L}$ is the collection of allowable structures for *L* . The task is then to select a scenario (*SL*, *X*) that improves one of the performance measures (TOC or PC) subject to the constraints specified below:

- For one unit of the product (task *J*), the operation chain $o_1, \ldots, o_i, \ldots, o_n$I s carried out in the prescribed order using machines $m_1, \ldots, m_i, \ldots, m_n$ that constitute line *L* under the chosen structure *SL*.
- A machine may execute only one operation at a time.
- After completing operation o_i on machine m_i, the product leaves m_i and proceeds to operation o_{i+1} on machine m_{i+1} (or on a machine substituting for it), provided that the next resource is available.
- If the next machine is busy, the product waits until it becomes available. The intermediate buffer (where tasks queue while waiting for machine availability) is assumed to have unlimited capacity.
- To prevent excessive queue accumulation, a synchronization requirement is imposed on the machines. In particular, the pairwise deviation of their cycle times is bounded by *TD*, i.e., $|t_i - t_j| \leq TD, \forall\{m_i, m_j\} \subset M$.

In the considered problem, the machining line is assumed to operate in a dynamic environment where it must continuously adapt to external factors such as volatile energy prices or unexpected machine breakdowns. As a result, line balancing transforms into an ongoing, feedback-driven decision-making process, centered on addressing several critical operational questions: (1) Which *production scenario* (*SL*, *X*) of the machining line *L* guarantees a predefined Total Operating Cost (*TOC*)?; (2) Which *production scenario* (*SL*, *X*) ensures a desired *Production Completion Time (PC)?;* (3) Is it possible to adjust the current *production scenario* (*SL*, *X*) to maintain cost objectives if production parameters such as labor or energy expenses shift?; (4) Can the setup be reconfigured to uphold *PC* targets in the face of emergencies or disruptions that compromise production flow?

The above considerations indicate that line balancing calls for a method that can evolve with changing shop-floor conditions. Beyond planning production routes in advance, the solution should allow fast adjustments when disturbances and variability occur during execution. This leads to the main research question: Does the software realization of the model developed in this study provide sufficient decision support to

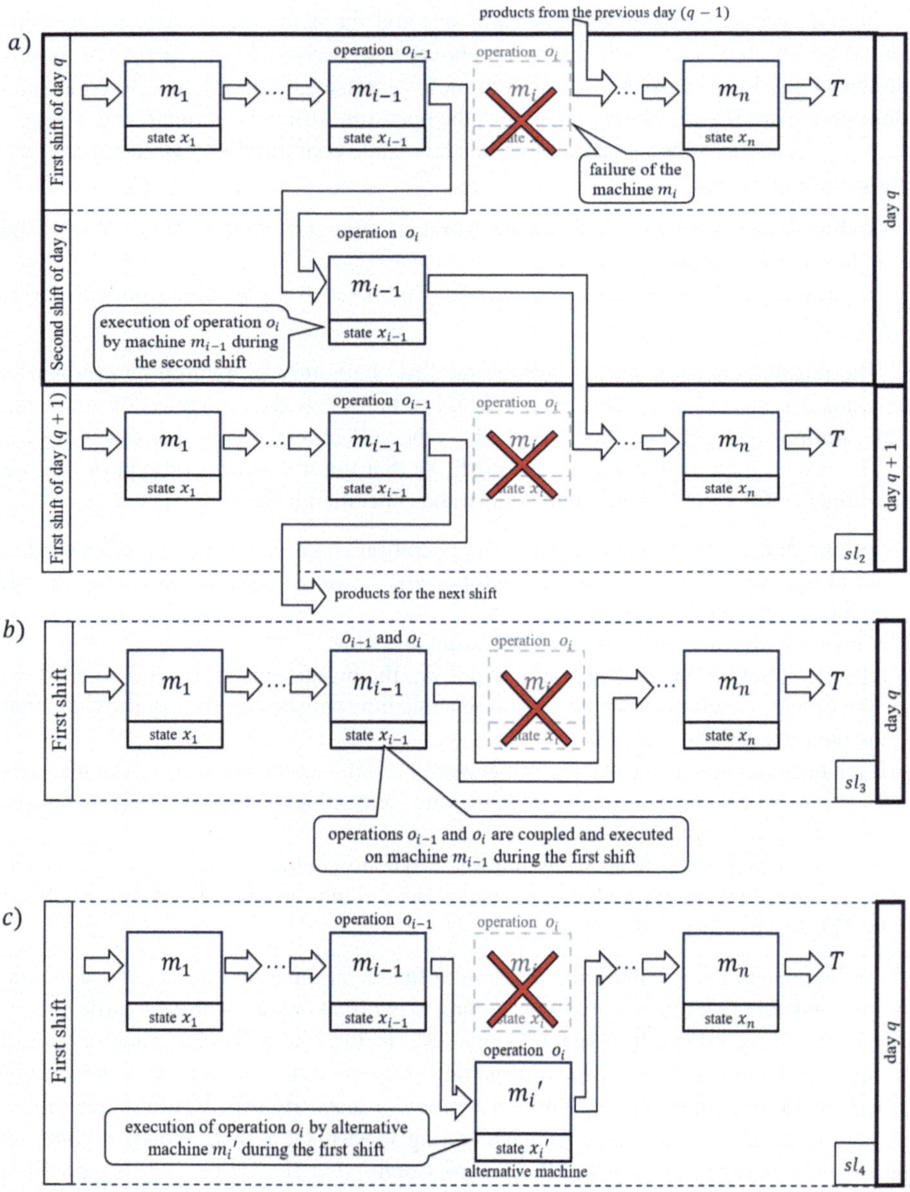

Fig. 2. Machining line L structures SL under failure of machine m_i: sl_2 - two-shift operation mode (a); sl_3 - one-shift operation mode in which the operations of the failed machine m_i are reassigned to machine m_{i-1} (b); sl_4 - one-shift operation mode in which the operations of the failed machine $m_i$are executed by the substitute m_i' (c).

restructure and rebalance the line on the fly, so that the system can react online to unexpected events?

The following section outlines a framework that addresses this challenge through a declarative, constraint-driven method.

3.2 Human-in-the-Loop Follow-Up Control of Production Flow

Figure 3 illustrates the key elements and their relationships within the presented method. Because production line planning is inherently interactive, the method relies on a declarative problem representation. The open architecture allows new, user-specific requirements to be incorporated as they arise, without changes to the core solution algorithm.

The approach assumes that machining line configurations must be adjusted in response to operational dynamics - such as order variations, unexpected interruptions, or demand shifts - and proceeds through the following steps:

- Detection of external variables currently affecting operations (e.g., tooling costs, labor rates, energy charges, and fixed operating overhead).
- Evaluation of the current production state and determination of the minimum daily output required to maintain throughput.
- Specification of user-defined goals and parameters, including objective functions (e.g., minimizing *TOC* or *PC*), operational constraints (e.g., cost upper bounds), and the allowable set of machining line structures S_L.
- Reconfiguration is first checked for necessity. If an adjustment is needed, the decision task is formulated within the COP paradigm [6] and handled by a dedicated solver such as Gurobi. The computation yields multiple admissible options, each described by a selected configuration X together with a corresponding line structure SL.
- Iteration of the procedure to maintain real-time responsiveness as conditions change.

This method supports proactive monitoring of both internal process execution and external influences that may affect cost or scheduling targets. A practical case study involving a two-machine system, presented in the next section, demonstrates the framework's performance under real-world conditions.

4 Application Example

Consider line L consisting of the set $M = \{m_1, m_2, m_3, m_4\}$, used to manufacture bearing parts via typical machining steps (e.g., cutting and milling). For each m_i, a catalog of 100 admissible operating conditions is assumed ($lk_1 = \cdots = lk_4 = 100$). Each condition defines a specific tuple of machine characteristics - namely the processing time t_i[s], the wear/life measure z_i, and the electricity demand e_i[kWh] - as reported in Table 1. Synchronization is enforced by limiting timing differences to at most $TD = 10s$.

The total cost of production accounts for both labor and machine operation expenses (see Table 2). The manufacturing goal is to complete $TN = 10{,}000$ units within a 23-day production window.

A workday WD is set to 8 h, with a minimum required throughput of $DP = 435$ units per day. The analysis also enforces a per-unit cost bound of $TR = 5$ [$m.\ u.$]. Under normal conditions, the machining line adheres to a configuration labeled $SL = sl_1$, shown in

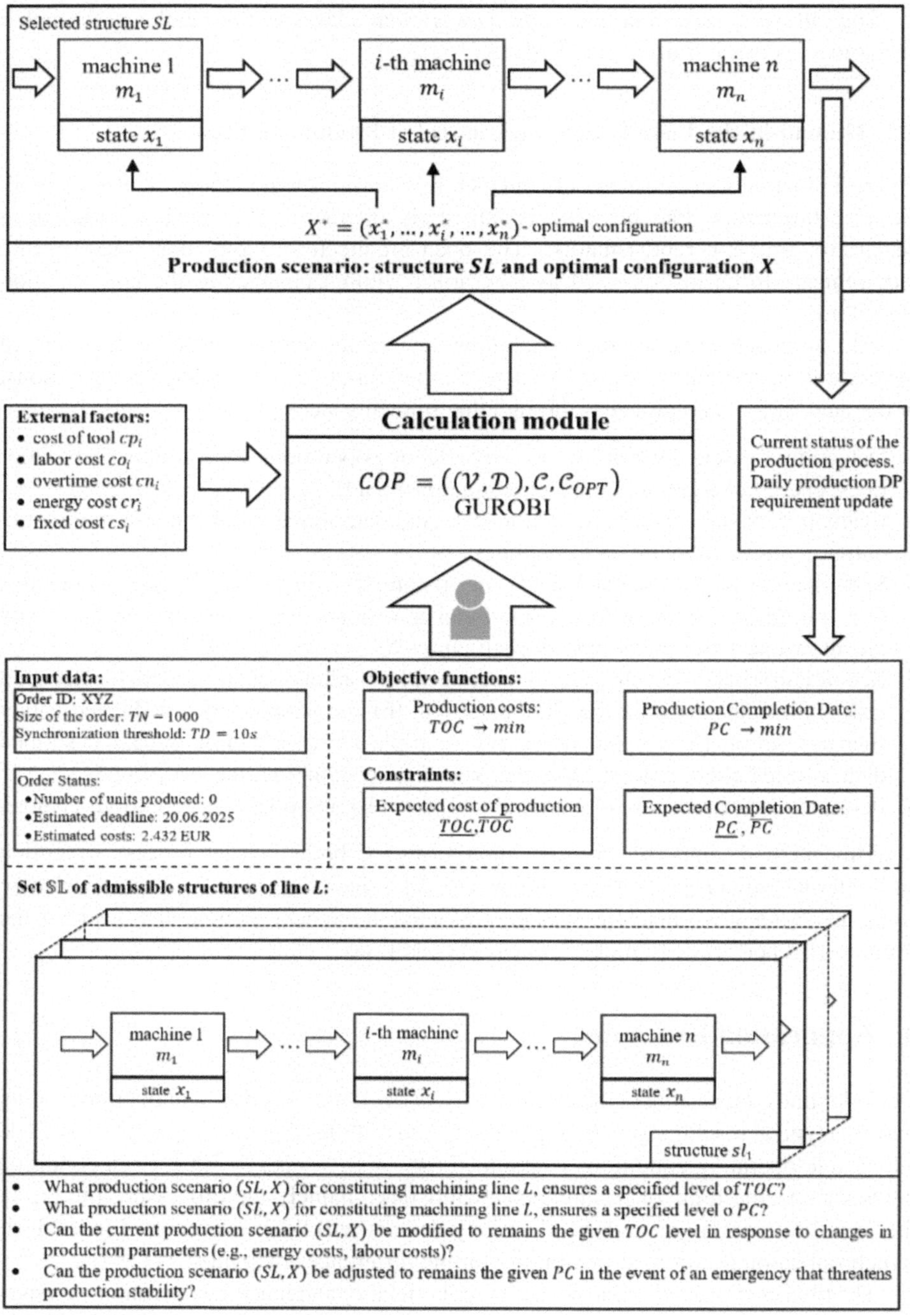

Fig. 3. Adopted interactive follow-up decision-making procedure.

Fig. 4. The main optimization goal is to find a valid production scenario (*SL*, *X*) that

Table 1. Machines distinct states

m_1				m_2				m_3				m_4			
x_1	t_1	z_1	e_1	x_2	t_2	z_2	e_2	x_3	t_3	z_3	e_3	x_4	t_4	z_4	e_4
1	120	52	5.5	1	289	101	4.6	1	183	87	4.6	1	328	107	4.6
2	123	48	5.6	2	296	97	4.5	2	170	85	4.5	2	194	65	4.5
3	126	62	5.7	3	304	97	4.5	3	335	120	4.5	3	246	117	4.5
...	...	...	...	...	...	...	...	...	...	...	...	...	...	...	...
100	43	69	5.1	100	43	69	4.2	100	80	43	4.2	100	56	69	4.2

Table 2. Manufacturing cost elements

Costs	m_1	m_2	m_3	m_4
cost of purchasing a tool cp_i [m.u./unit]	36	45	25	34
labor cost co_i [m.u./h]	38	41	38	41
overtime labor cost cn_i [m.u./h]	47	51	47	51
energy cost cr_i [m.u./kWh]	1.11	1.11	1.11	1.11
fixed cost cs_i [m.u.]	16	16	16	16

minimizes the Total Operating Cost (*TOC*) while satisfying the fixed delivery deadline of 01.08.2025, as per the optimization framework presented in Fig. 3.

Formally, the task is stated as minimizing TOC under a prescribed runtime limit. The optimization routine follows the declarative formulation reported in [6] and was carried out in Gurobi (v11.0.3) accessed via its Python interface [28]. All computations were performed on a workstation with an Intel Core i7-M4800MQ CPU (2.7–4.82 GHz) and 32 GB RAM. Solution process took less than 10 s. Out of the evaluated configurations, 166 were found to satisfy the *TOC* constraint ($TOC_i \leq PR$). The optimal solution was $X^* = (98,98,91,96)$, which resulted in a total cost of $TOC = 4.79$ [$m.\ u.$]. The corresponding machine cycle times were: t_1=53 [s], t_2=43 [s], $t_3 = 47$ [s], $t_4 = 51$ [s], yielding a production cycle time $T = 53$ [$m.\ u.$].

Let's now examine a scenario in which, after producing 7,395 units (on the 17th day of operation), machine m_3 fails and is taken out of service. In a reconfigured line type $SL = sl_3$ that represents a single-shift layout with merged operations, the tasks previously handled by m_3 are reassigned to m_2. This reallocation extends the operation time for m_2, increasing from $t_2 = 43$ [s] $\rightarrow t_2^{'} = 98$ [s], and raises the cost due to intensified tool wear.

In such a disrupted setting, a new question arises: Can the original delivery deadline still be met, or if not, what is the shortest achievable production completion time (*PC*) under the constraint that $TOC \leq 5$ [$m.\ u.$]? According to the optimization flow in Fig. 3, this constitutes a cost-constrained time minimization problem: *Minimize PC subject to* $TOC \leq 5$ [$m.\ u.$] (refer to Fig. 5 for a visual overview).

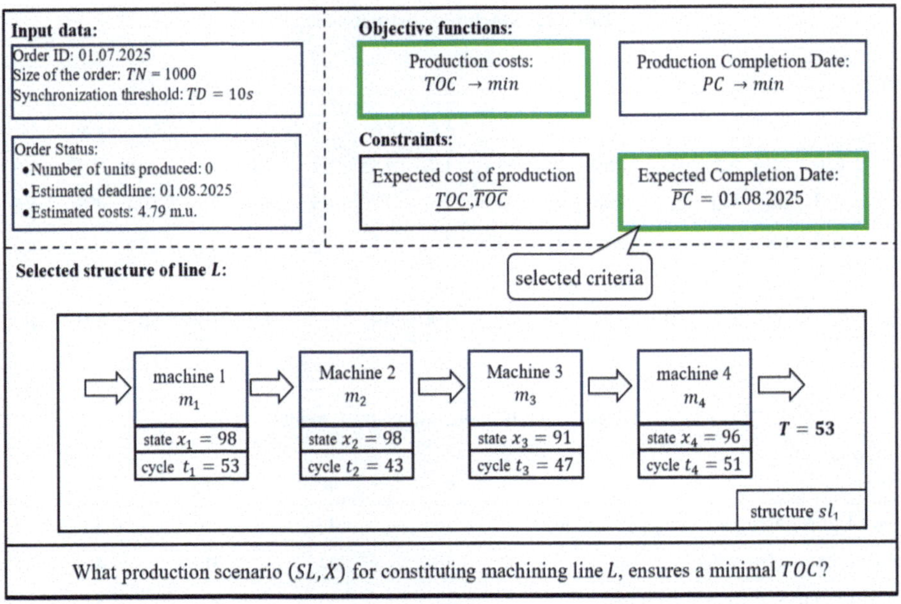

Fig. 4. Nominal machining-line structure $SL = sl_1$ parameterized with Fig. 3 data.

To reflect the updated dataset, the Gurobi solver was reinitialized, delivering a solution in under 2 s. Table 3 presents a comparison of production data prior to and following the disruption.

Specifically, executing the operation on the machine leads to an additional unit cost of 0.10 monetary units, increasing the overall value to a level that remains within the permitted limit. This observation underscores that the proposed model can serve as an element of a decision-support framework for production-line management. Nevertheless, computational performance depends strongly on the size of the configuration search space and on the complexity of scenarios that meet synchronization requirements. A summary of the outcomes is provided in Fig. 6.

The findings show that solving the problem at hand (that is, finding a feasible solution in under 10 min) can be achieved only for the following cases: ($n = 8$) machines with at most 100 states; ($n = 6$) machines with at most 150 states; ($n = 4$) machines with at most 200 states; and ($n = 3$) machines with at most 250 states. These settings are meant to represent a typical standalone production line that in practice rarely contains more than three machines.

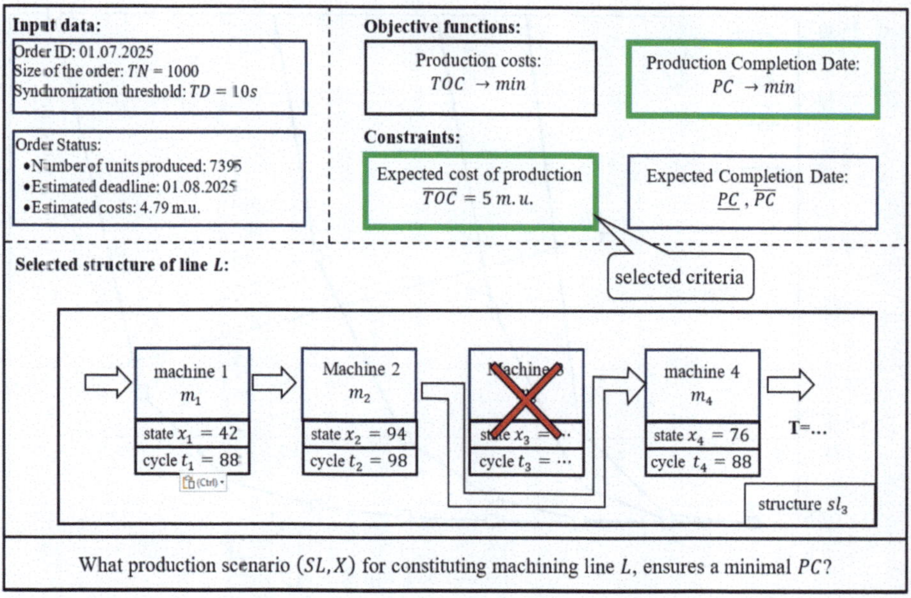

Fig. 5. Line-structure transition from $SL = sl_1$ to $SL = sl_3$ using the dataset shown in Fig. 3.

Table 3. Before–after comparison of line scenarios following failure of m_3.

Parameters	x	t	x'	t'
m_1	98	53	42	88
m_2	98	43	94	98
m_3	91	47	x	x
m_4	96	51	76	88
TOC	**4.79**		**4.89**	

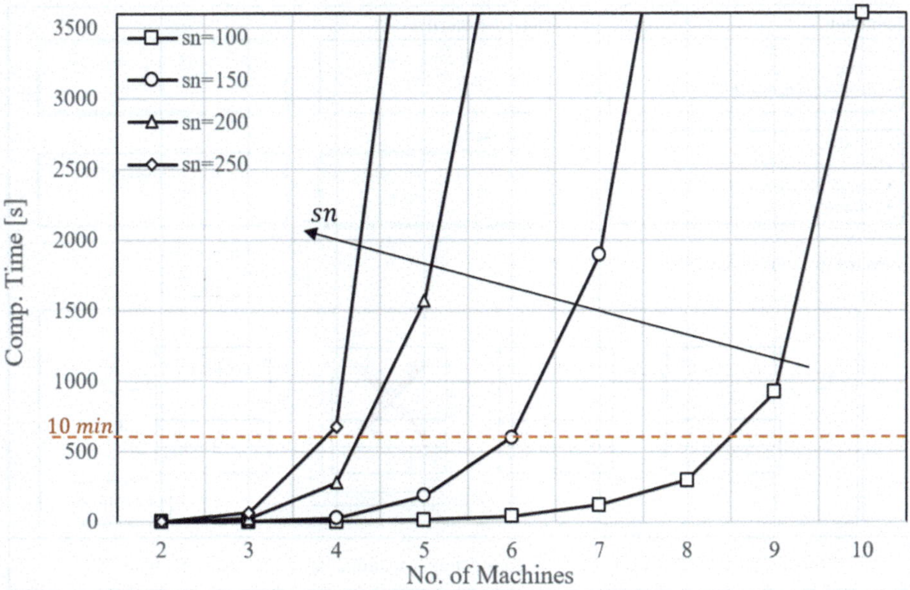

Fig. 6. Scalability analysis: runtime versus problem size (*n*, *sn*).

5 Conclusion

The present work introduces a declarative, constraint-based model intended for dynamic balancing of serial machining transfer lines under changing shop-floor conditions. The model is implemented in a constraint-programming setting and arranged as an open, extensible structure, allowing rapid plan updates in response to disruptions such as equipment malfunctions or shifts in demand, yet remaining consistent with production goals (e.g., TOC and PC). The proposed method formalizes planner goals, live shop-floor information, plus operational limits as a Constraint Optimization Problem (COP) [6]. As a result, dispatchers can generate and evaluate feasible production plans on demand, supporting timely, well-grounded decisions. Computational experiments together with an industrial case example show that the model handles complex scenarios and performs better than conventional balancing procedures, contributing to agile and sustainable planning in machining operations. Future work considers linking the method to predictive-maintenance platforms and shop-floor sensing/connected equipment to broaden applicability across diverse manufacturing settings.

References

1. Wang, J., Liu, H., Lin, T.: Optimal rearrangement and preventive maintenance policies for heterogeneous balanced systems with three failure modes. Reliab. Eng. Syst. Saf. **238**, 109429 (2023). https://doi.org/10.1016/J.RESS.2023.109429
2. Salido, M.A., Escamilla, J., Barber, F., Giret, A.: Rescheduling in job-shop problems for sustainable manufacturing systems. J. Clean. Prod. **162**, S121–S132 (2017). https://doi.org/10.1016/J.JCLEPRO.2016.11.002
3. Battaïa, O., Dolgui, A., Guschinsky, N.: An exact method for machining lines design with equipment selection and line balancing. Int. J. Prod. Res. **62**, 71–91 (2024). https://doi.org/10.1080/00207543.2023.2289182
4. Makssoud, F., Battaïa, O., Dolgui, A.: Reconfiguration of machining transfer lines. Stud. Comput. Intell. **472**, 339–353 (2013). https://doi.org/10.1007/978-3-642-35852-4_22
5. Hashemi-Petroodi, S.E., Dolgui, A., Kovalev, S., Kovalyov, M.Y., Thevenin, S.: Workforce reconfiguration strategies in manufacturing systems: a state of the art. Int. J. Prod. Res. **59**, 6721–6744 (2021). https://doi.org/10.1080/00207543.2020.1823028
6. Bocewicz, G., Piechowski, M., Radzki, G., Banaszak, Z.: Toward follow-up driven machining line balancing. In: Mehmood, R., et al. (eds.) Distributed Computing and Artificial Intelligence, Special Sessions I, DCAI 2025 Lecture Notes in Networks and Systems, vol. 1631. Springer, Cham (2026). https://doi.org/10.1007/978-3-032-05745-7_1
7. Cai, S., Gallina, B., Nyström, D., Seceleanu, C.: Data aggregation processes: a survey, a taxonomy, and design guidelines. Computing. **101**, 1397–1429 (2019). https://doi.org/10.1007/S00607-018-0679-5
8. Schallehn, E., Sattler, K.U., Saake, G.: Advanced grouping and aggregation for data integration. In: Proceedings of the ACM CIKM, pp. 547–549 (2001). https://doi.org/10.1145/502585.502685
9. Schmied, S., Grossmann, D., Mathias, S.G., Mueller, R.K.: Integration of manufacturing information via dynamic information model aggregation. Vietnam J. Compute. Sci. **8**, 245–262 (2020). https://doi.org/10.1142/S219688882150010X
10. Vaisi, B.: A review of optimization models and applications in robotic manufacturing systems: industry 4.0 and beyond. Decis. Analyt. J. **2**, 100031 (2022). https://doi.org/10.1016/J.DAJOUR.2022.100031
11. Beldar, P., Fathi, M., Nourmohammadi, A., Delorme, X., Battaïa, O., Dolgui, A.: Transfer line balancing problem: a comprehensive review, classification, and research avenues. Comput. Ind. Eng. **201**, 110913 (2025). https://doi.org/10.1016/J.CIE.2025.110913
12. Barad, M.: Design of Experiments (DOE) – a valuable multi-purpose methodology. Appl. Math. (Irvine). **5**, 2120–2129 (2014). https://doi.org/10.4236/AM.2014.514206
13. Jankovic, A., Chaudhary, G., Goia, F.: Designing the design of experiments (DOE) – an investigation on the influence of different factorial designs on the characterization of complex systems. Energ. Buildings. **250**, 111298 (2021). https://doi.org/10.1016/J.ENBUILD.2021.111298
14. Spišák, E., Majernikova, J.: Increasing of durability of cutting tools. Adv. Sci. Technol. Res. J. **11**(141–146), 10.12913/22998624/78168 (2017)
15. Zhang, T., Owodunni, O., Gao, J.: Scenarios in multi-objective optimisation of process parameters for sustainable machining. Procedia CIRP. **26**, 373–378 (2015). https://doi.org/10.1016/J.PROCIR.2014.07.186
16. Heidari, A., Rojas Gonzalez, S., Dhaene, T., Couckuyt, I.: Lower confidence bound for preference selection in interactive multi-objective optimization. In: GECCO 2024 Companion, pp. 339–342 (2024). https://doi.org/10.1145/3638530.3654317

17. Liu, C., Li, Y., Huang, C., Zhao, Y., Zhao, Z.: A meta-reinforcement learning method by incorporating simulation and real data for machining deformation control of finishing process. Int. J. Prod. Res. **61**, 1114–1128 (2023). https://doi.org/10.1080/00207543.2022.2027041
18. Serrano-Ruiz, J.C., Mula, J., Poler, R.: Smart manufacturing scheduling: a literature review. J. Manuf. Syst. **61**, 265–287 (2021). https://doi.org/10.1016/J.JMSY.2021.09.011
19. Yu, H., et al.: An effective multi-part dedicated flow-line reconfiguration model considering the optimal selection of machining process path and machines. Proc. Inst. Mech. Eng. B J. Eng. Manuf. **237**, 154–170 (2023). https://doi.org/10.1177/09544054221100873
20. Borisovsky, P.A., Delorme, X., Dolgui, A.: Balancing reconfigurable machining lines via a set partitioning model. Int. J. Prod. Res. **52**, 4026–4036 (2014). https://doi.org/10.1080/00207543.2013.849857
21. Makssoud, F., Battaïa, O., Dolgui, A.: A mathematical model for a reconfiguration problem of transfer machining lines. IFAC Proc. **45**, 438–443 (2012). https://doi.org/10.3182/20120523-3-RO-2023.00419
22. Eom, S.B.: Decision support systems. In: Oxford Research Encyclopedia of Politics (2020). https://doi.org/10.1093/acrefore/9780190228637.013.1008
23. Karlsson, I., Ng, A.H.C., Syberfeldt, A., Bandaru, S.: An interactive decision support system using simulation-based optimization and data mining. In: Proceedings of the Winter Simulation Conference, pp. 2112–2123 (2016). https://doi.org/10.1109/WSC.2015.7408325
24. Dastres, R., Soori, M.: Advances in Web-Based Decision Support Systems (2022)
25. Radzki, G., Bocewicz, G., Piechowski, M., Banaszak, Z.: Declarative modeling approach to interactive computer-aided balancing of machining lines. In: Antosz, K., Trojanowska, J., Machado, J., Stadnicka, D. (eds.) Advances in Lean Manufacturing ELEC 2025. Lecture Notes in Mechanical Engineering, vol. 1. Springer, Cham (2026). https://doi.org/10.1007/978-3-032-09806-1_10
26. Vlk, M., Hanzálek, Z., Tang, S.: Constraint programming approaches to joint routing and scheduling in time-sensitive networks. Comput. Ind. Eng. **157**, 107317 (2021). https://doi.org/10.1016/J.CIE.2021.107317
27. Wikarek, J., Sitek, P., Jagodziński, M.: A declarative approach to shop orders optimization. Appl. Comput. Sci. **15**, 5–15 (2019). https://doi.org/10.35784/Acs-2019-25
28. Gurobi Optimization, LLC: Gurobi Optimizer Reference Manual. https://docs.gurobi.com/projects/optimizer/en/current/. Accessed 25 Apr 2025

Potential Applications of Bio-Intelligent Systems in Ceramic Manufacturing Industry

Oumayma Drissi Yahyaoui(✉), Matteo De Marchi, and Erwin Rauch

Industrial Engineering and Automation (IEA), Free University of Bozen-Bolzano, Piazzal Università 1, 39100 Bolzano, Italy
oumayma.drissiyahyaoui@unibz.it

Abstract. As an important industry in the construction field, the ceramic tile manufacturing sector faces different challenges to achieve enhanced operational performance and environmental sustainability. These challenges include, for instance, its high energy intensity, the quality standards requirements, the materials traceability, and the upcycling considerations at the end-of-life. The current research explores bio-intelligent production concepts to enhance ceramic tile manufacturing through natural systems, which adapt and self-regulate while maintaining the efficient usage of resources. Using a systematic literature review, this research highlights the potential bio-intelligent concepts that show promise for various manufacturing tasks in ceramic production, while determining which production stages require these principles. The research identifies promising applications which include biological homeostasis-inspired adaptive control systems, swarm behavior-based optimization methods, self-organizing production systems, and bio-inspired quality inspection systems. These principles have the potential to be implemented throughout different manufacturing stages, starting from raw material processing until the completion of firing and quality assessment. This study is conceptual and prospective. The identified applications represent theoretical possibilities rather than validated implementations, and it provides vital information which will guide future experimental studies and industrial uses in ceramic tile manufacturing.

Keywords: Bio-intelligent Manufacturing · Ceramic Industry · Sustainable Manufacturing · Industry 5.0

1 Introduction

Ceramic tile manufacturing represents a well-established industry that is actively evolving to address many modern challenges, mainly related to energy consumption, quality consistency, and environmental sustainability [1]. Therefore, this sector requires innovative approaches that can address these challenges while maintaining production efficiency. For instance, bio-intelligent production systems offer potential solutions with the application of principles observed in natural systems to different industrial manufacturing contexts.

B. Gapiński et al. (Eds.): MANUFACTURING 2026, LNME, pp. 143–156, 2026.
https://doi.org/10.1007/978-3-032-22943-4_11

Bio-intelligence in manufacturing refers to the application of biological principles to production systems [2–4]. This includes learning from natural processes such as adaptation, evolution, self-organization, and distributed intelligence. This approach goes beyond simple automation and enables the creation of systems that can respond to changing conditions, optimize themselves, and work cooperatively. In fact, Miehe et al. [2] describe how the biological transformation of manufacturing creates bio-intelligent systems that integrate technical, informational, and biological elements into cohesive production approaches. According to their framework, a system can be termed bio-intelligent if it has been designed following fundamental principles found in nature, uses biological resources, and enables bilateral information exchange between biological and technical systems [3].

Indeed, recent developments in manufacturing technology have expanded the possibilities for implementing bio-intelligent principles [4–6]. The convergence of biological, technical, and informational systems creates opportunities for new approaches to production. Emer et al. [5] identified potential applications of bio-intelligent manufacturing ranging from tissue engineering to waste-to-X production systems, demonstrating the variety of possible implementations. Matt and Rauch [6] provide a comprehensive overview of biological transformation in manufacturing, proposing key areas for practical application in organizations. In particular, their work emphasizes how bio-intelligent manufacturing systems combine technological innovations with principles derived from nature and biotechnology, which represents a significant shift in how production systems can be designed and operated.

Additionally, the transition toward bio-intelligent manufacturing represents a logical evolution beyond Industry 4.0 concepts [7, 8]. While digital technologies enable data collection and connectivity, bio-intelligent principles provide frameworks for using that information intelligently. According to Rauch et al. [7], one of the greatest challenges ahead is transferring Industry 4.0 concepts and technologies such as cyber-physical systems, IoT, and smart manufacturing into SMEs. They also highlight the importance of smart and sustainable manufacturing systems. Natural systems demonstrate remarkable capabilities for adaptation, optimization, and resilience that can inspire manufacturing systems.

This research paper examines how bio-intelligent principles and methods can be applied within the ceramic tile manufacturing sector. In fact, Nazari et al. [9] reviewed advanced manufacturing methods for ceramic and bioinspired ceramic composites, identifying the need for systematic exploration of process applications. The study identifies the specific bio-intelligent approaches from the literature and analyzes their potential application to different stages of ceramic tile production. The aim is to provide a foundation understanding and detect where and how bio-intelligent systems might be integrated into ceramic tile manufacturing processes.

The research addresses two main questions: Which bio-intelligent principles and methods identified in literature could potentially be applied to ceramic tile manufacturing? And in which specific manufacturing stages could these principles be integrated?

The research paper is structured as follows: in Sect. 2, a systematic literature review is conducted, Sect. 3 presents the potential applications of bio-intelligent principles in different ceramic tiles manufacturing phases, including raw material, forming, drying, glazing and decorating, firing, and quality inspection. Section 4 is a discussion of the principles mentioned and applications, and finally Sect. 5 presents a conclusion and future research work.

2 Research Methodology

2.1 Literature Review Approach

The systematic literature review conducted follows the Preferred Reporting Items for Systematic Reviews and Meta-Analyses (PRISMA) framework to ensure methodological rigor, transparency, and reproducibility [10].

The literature search was conducted in the Scopus database. The choice of Scopus as a database for conducting the SLR is explained by its high relevance and usage as a scientific source for publications in the engineering and manufacturing areas. The search string used Boolean operators to link all terms from the bio-intelligent and ceramic manufacturing fields into a single string. The search strategy included specific industrial terms for ceramic manufacturing, such as ceramic, pottery, porcelain, kiln, firing, and glazing, as well as terms to represent bio-intelligent concepts, including bio-intelligence, bio-inspired, and bio-integrated. Moreover, the document type was limited to articles, reviews, and conference papers published in journals and conference proceedings, to ensure peer-reviewed academic quality.

Search string: TITLE-ABS-KEY: (biointelligen* OR bio-intelligen* OR bio intelligen* OR bioinspir* OR bio-inspir* OR bio inspir* OR biointegrat* OR bio-integrat* OR bio integrat*) AND (ceramic* OR pottery OR porcelain OR kiln OR firing OR glaz*).

The search yielded 137 records during the initial phase of the investigation. Following the PRISMA 2020 guidelines, a systematic screening process was implemented in two stages (see Fig. 1). In the first stage (identification), no duplicate records were identified as the search was conducted using a single database. The second stage (screening) required researchers to assess titles and abstracts by using established criteria, which determined which studies to include or exclude. This step has resulted in the exclusion of 116 papers, which are mainly on ceramic applications in biomedical and dental fields, utilizing implants, bone tissue engineering, medical scaffold development for bioenergy, biofuel production, water and wastewater treatment, marine biology, agriculture, and food processing.

The research team established the following criteria for study inclusion: The research investigated application field of biointelligent, bio-inspired, and bio-integrated systems. Moreover, the research focused on ceramic processing and production, and manufacturing system development.

In the second step of screening, a check of the full paper was conducted for the remaining 21 articles, and they were assessed according to whether they are related to ceramic processes and mention bio-intelligent concepts 5 papers were left; however, they don't provide a specific framework or potential applications of bio-intelligence concepts. Indeed, a major remark made is related to the gap in the literature on the different applications of bio-intelligence concepts in the ceramic tile manufacturing processes.

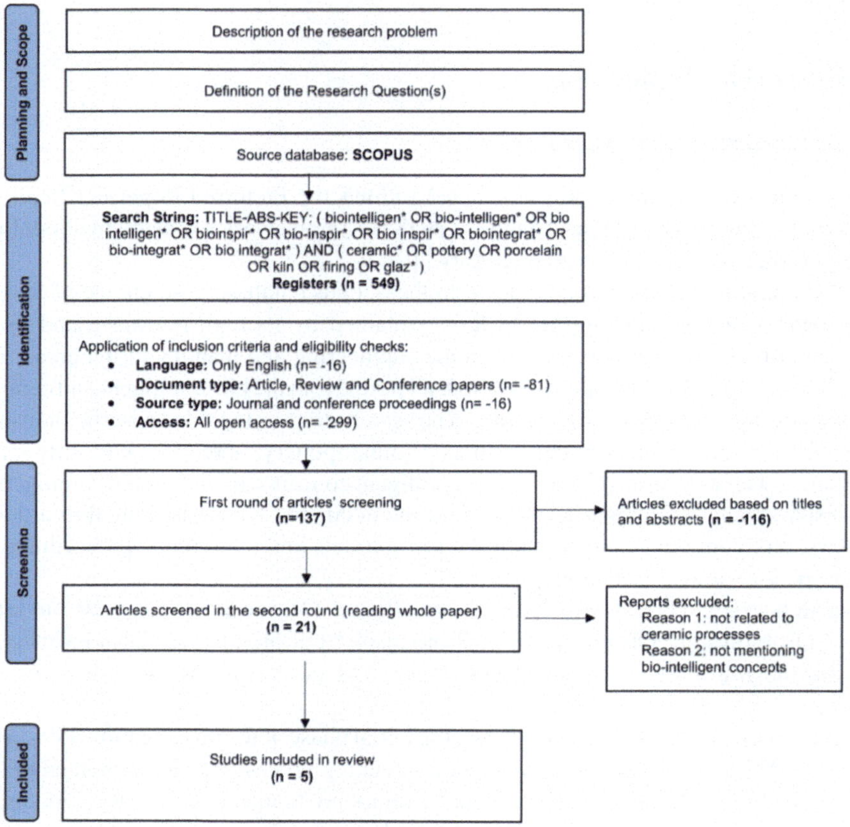

Fig. 1. Search approach based on PRISMA 2020 methodology.

2.2 Framework Analysis

In their research, Matt and Rauch [6] propose a bidirectional approach combining technology pull (top-down problem solving) and biology push (bottom-up opportunity identification). This integrative process of the two perspectives aims mainly to systematically leverage biological insights for manufacturing innovations, to foster a more comprehensive exploration of bioinspired solutions. By iteratively applying both approaches, researchers can better identify relevant biological principles and translate them into practical manufacturing strategies, thereby enhancing the potential for biological transformation in engineering contexts [6].

The ceramic tile manufacturing stages were mapped in Fig. 2, to create a framework for assessing where different bio-intelligent principles might be applied. The framework examines six main manufacturing stages: raw material preparation, forming, drying, glazing and decoration, firing, and quality inspection. For Farzaneh et al. [11], such systematic mapping helps identify which manufacturing systems can perform best as bio-inspired, bio-integrated, or bio-intelligent systems, and which interfaces exist between biological and technical elements.

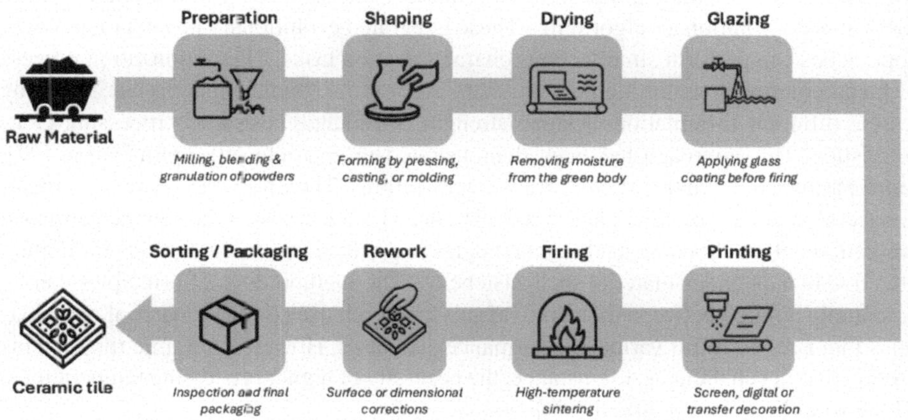

Fig. 2. Ceramic tiles manufacturing processes.

2.3 Classification of Bio-Intelligent Principles

Bio-intelligent principles can be classified into various categories based on their underlying biological inspiration and functional characteristics [2–4]. Matt and Rauch [6] describe mainly three levels of biological transformation, which are:

- Bio-inspired manufacturing (inspiration level): Systems that mimic biological forms, structures, or strategies without incorporating living components (biomimetics).
- Bio-integrated manufacturing (integration level): Systems that incorporate biological materials or processes alongside technical elements.
- Bio-intelligent manufacturing (evolution level): Systems that enable bidirectional information exchange between biological and technical components, exhibiting adaptive, self-organizing processes.

These categories include, for instance, adaptive control systems, evolutionary and genetic algorithms, swarm intelligence, self-organization and morphogenesis, hierarchical structures and modularity, and distributed sensing and information processing. Each category represents a distinct biological concept that has been translated into manufacturing applications. Indeed, adaptive control draws from homeostasis and neural networks [12, 13], evolutionary algorithms mimic natural selection [14, 15], swarm intelligence emerges from collective behaviour [16–18], self-organization reflects biological development processes [19, 20], hierarchical structures mirror natural organizational patterns [21–24], and finally distributed sensing parallels biological nervous systems [25, 26].

3 Potential Applications of Bio-Intelligent Principles

3.1 Raw Material Stage

Bio-intelligent systems can optimize raw material blends through different concepts, for instance, evolutionary algorithms. These integrated evolutionary and neural network approaches can perform simultaneous characterization tasks. The optimization process of blend compositions through evolutionary algorithms handles material variations by testing different formulations against strength, shrinkage, cost, and processing characteristics. This approach has been demonstrated in concrete mix optimization [27], representing a domain transfer to ceramic applications. Therefore, the system can maintain a collection of potential blend formulations, which it assesses through performance prediction before choosing the best candidates to produce new combinations for finding effective blends under material inconsistencies. The methodological principles can be potentially applied between the two fields, as they both use complex material combinations that need to fulfill various performance standards. However, ceramic tile systems present distinct challenges, for instance, the necessity of higher processing temperatures, which would require empirical validation.

Another potential application is related to neural network-based systems. In fact, the system can employ neural networks to improve raw material characterization through advanced testing methods, which analyze multiple material characteristics for processing behavior prediction, using deep learning methods that can discover complex patterns in data, and can automatically learn layered features, which allows systems to analyze multiple characteristics simultaneously and make predictions about material behavior [28]. Hence, the system can use already learned patterns from accumulated material data to predict material behavior through rapid measurements, which improves its accuracy with each new material experience.

3.2 Forming Stage

This stage, also called shaping, is the step where the prepared ceramic body (clay mixture usually made from clay, feldspar, sand, and additives) is shaped into tiles before drying and firing. Current dry pressing systems use fixed parameters (pressure, dwell time, speed) based on predetermined recipes, and they react to problems after they occur [29]. Therefore, self-organizing control systems could coordinate multiple press parameters to achieve uniform density distribution [30].

Unlike Reconfigurable Manufacturing Systems RMS [31], which emphasize structural flexibility, the self-organizing approach adjusts pressure profiles based on powder characteristics and target specifications. Thus, multiple control zones could coordinate their actions to address variations across the press area. The potential result would be more uniform tiles even when powder properties vary across the press bed.

The development of alternative forming techniques should use biological processes that create structures as their foundation. Bargardi et al. [32] show how bio-inspired self-shaping ceramics can be created. Chen et al. [33] developed large toughening effects in biomimetic geopolymer composites through their ability to control structure development. The methods show that materials can organize themselves to produce structures that match desired forms without needing mechanical fabrication. These principles show promise for developing new tile designs which combine improved performance with traditional tile manufacturing methods. The self-organization process allows manufacturers to create structures with precise porosity levels and aligned reinforcement and gradient properties through a simple production method. This represents longer-term potential as manufacturing technologies evolve.

3.3 Drying Stage

This stage is one of the critical phases in ceramic tile manufacturing where improper moisture removal can lead to warping, cracking, and other defects in the tiles. The current traditional drying systems operate based on predetermined and defined temperature and humidity levels, which fail to respond to changes in tile properties and environmental factors [34, 35].

A suggested bio-intelligence application for this stage would be homeostatic moisture control with distributed sensing. In fact, bio-intelligent drying systems could use homeostatic control methods, which biological organisms use to stabilize their internal states against external changes to achieve automatic optimal conditions [36]. The system would monitor tile moisture levels through distributed sensors to create adaptive temperature and humidity and airflow settings, which optimize drying performance. The system uses biological homeostasis principles to monitor tile moisture levels through distributed sensors, which send data to control systems [37]. The drying chamber would achieve complete monitoring of moisture levels through its distributed sensor network, which can detect moisture changes before defects form [38]. The system uses multiple feedback loops, which operate together to handle changes in tile dimensions, moisture levels, and environmental factors. As a result, this application reduces the drying defects and minimizes energy consumption through real-time feedback optimization of drying parameters.

3.4 Glazing and Decoration Stage

Although digital inkjet printing of modern decorative ceramic tiles is largely automated, there is still a need for real-time quality assessments and fine-tuning of quality control parameters [39, 40]. These challenges could be solved through bio-intelligent systems with embedded quality control systems and real-time adaptive control.

A potential bio-intelligence application would be specifically in quality assessment using sensors. Indeed, deep learning systems for machine vision defect detection have surface defect detection accuracy rates for ceramic tiles ranging from 85% to over 95% [41, 42]. However, these figures are context-dependent: performance varies with defect type (e.g., crack, pin-hole, hole), imaging conditions, dataset diversity, and validation protocols. Currently, these systems function in isolation as unintegrated inspection systems. A bio-intelligent approach would assess quality through multiple, cooperative, and integrated inspection system modalities, including visual, thermal, and spectroscopic sensors. A similar approach would integrate multiple sources and sensors to assess and build a coherent environmental awareness, helping to overcome environmental control challenges in defect detection systems [37].

Another application would be to employ swarm intelligence behavior to enhance the integration of several inspection systems in quality assessment [43]. Additionally, the work of Rubenstein et al. [44] illustrates the efficacy of simple agents in the collaborative functioning of systems to achieve problem-solving. The combination of several inspection units could enable the system to analyze the tiles with different sensors at different positions, maintaining active collaboration to achieve overall quality evaluation. Each of the inspection units would be specialized to one attribute of the quality while communicating other results to the other units. The swarm approach would also enhance redundancy, as the system could continue to function in the event of failure of some of the units. This distributed inspection sufficiently represents how biological systems with several sensors increase the awareness of the environment.

3.5 Firing Stage

Multi-zone kiln systems could incorporate real-time and evolutionary adaptive control of firing operations and real-time control. Each kiln zone could maintain burn control autonomously and in real-time with thermal and atmospheric conditions of adjacent zones. This would be a significant advance in adaptive control of firing schedules. The system would control in real-time based on tile characteristics, production rates, and desired attributes of materials [45–47].

Genetic algorithms could maintain a population of candidate profiles, which vary by time, temperature, and atmospheric composition, and retain those that meet a solution to multi-goal optimization of mechanical strength, water absorption, dimensional accuracy, energy expended, and cycle time [14, 48]. Additionally, ceramic characteristics such as mechanical strength, porosity, and dimensional stability can be predicted during the firing stage through advanced machine learning models that examine temperature profiles, heating rates, and atmospheric conditions and how they interact [49]. Through this ability, the control system can predict the final characteristics of the product and adjust firing parameters proactively, rather than adjust them in reaction to the product characteristics. The system would adapt firing to variations in green body moisture, raw materials, production rates, and, over time, changes in product mix and degradation of equipment. This integrated approach of adaptive multi-zone coordination and continuous learning would result in optimized energy consumption during production transitions while maintaining consistent product quality, accumulating process knowledge that improves with operational experience [50].

3.6 Quality Inspection Stage

The integration of principles related particularly to neural networks and distributed sensor networks in the quality inspection of ceramic tiles can potentially result in improvements and advancement in this field.

For instance, the collective operation of high-resolution cameras and other inspection tools, such as laser profilometers and thermal imagers, stationed across inspection points would allow for a full assessment of quality at the tile's surface, as well as across the other parameters, including colour consistency. The level of automation and sophistication in the quality assessments would be unprecedented and would include surface defect recognition and defect integration [39, 41]. Moreover, the integration of different sensors would provide backup and can mitigate the risks of a singular inspection method, as well as provide additional surface recognition of larger discrepancies (irregularities, surface defects, and colour inconstancies) [42]. The automation of defect recognition and other subsurface patterns would be achieved through the fine layering of a deep learning neural network, in particular convolutional neural network. The refinement process would continue until the system can distinguish defect patterns from normal quality variations well enough to outperform most human inspectors and deliver highly reliable results. The system would also incrementally build its functional capacity to accommodate a different level of defect patterns than it previously dealt with, and will do so without the need for manual specification updates to the functional code [39, 51]. The multi-sensor neural network would become self-adaptive based on its defect classification outputs with an increased prediction accuracy.

4 Discussion

This study builds on the understanding of how bio-intelligent principles can streamline the process of ceramic tile manufacturing through the identification and integration of nature-inspired methodologies at different stages of the process. The study also identifies bio-intelligence principles for operational application in the manufacturing process of the ceramic tiles, in particular adaptive control systems, evolutionary optimization, self-organizing production and bio-inspired quality control, that offer an array of benefits for resolving challenges present in the current operational process. The bio-intelligent systems present different characteristics with respect to the degree of complexity and the current stage of the manufacturing system. Inspired control systems, reflecting biological homeostasis, particularly in the drying stage, are easy to implement and can solve an entry challenge for tile manufacturers. Such systems will require tile manufacturers to solve control issues and determine a process whereby appropriate alterations to the control algorithms can be made without the need for an expensive redesign of the control system. Self-organizing and systems inspired by morphogenetic manufacturing are likely to be left for the more distant future, as they will be more technologically demanding and expensive.

Ceramic tile manufacturers can greatly benefit from the development of bio-inspired, bio-integrated, and bio-intelligent systems. Some of the first deployments can draw from bio-inspired systems to implement adaptive control algorithms into existing production systems. With experience gained and value demonstrated, bio-integrated systems that include, for instance, distributed sensing, swarm inspection systems, can be deployed. This approach allows progression in capability and technical expertise, as well as potential risk mitigation. A key finding from this research is how bio-intelligent applications span across different stages of manufacturing. For instance, the layered neural Network used to characterize the feedstock can improve the adaptive control system to be more effective in the forming and firing stages. Also, the distributed sensing system designed for moisture control in the drying stage can be used for a multi-layered quality inspection system. This indicates that manufacturers should plan for more holistic implementations instead of stage-specific interventions that stand alone.

The distinct quality inspection step assumes massive importance. It stands as its own step in the manufacturing process, yet is also incorporated into other manufacturing stages, as in the case of the in-process real-time detection of defects during glazing. This dual-character system should give the bio-inspired quality assessment systems throughput potential value as early implementable systems. The evolutionary optimization of firing profiles addresses one of the top priorities as tile production energy status is among the highest. Homeostatic drying control is also an energy sink and should provide a decrease in manufactured defects. Additionally, manufacturers must balance the energy and material requirements of implementing sophisticated sensor networks and computational systems against the efficiency gains these systems can enable. Potentially, the usage of life cycle assessments would be valuable for quantifying the net sustainability benefits of different bio-intelligent implementations.

The systematic literature review identified a marked discrepancy between the theoretical promise of bio-intelligent manufacturing and actual documented practices in the production of ceramic tiles. While both literatures on bio-intelligent manufacturing and ceramic manufacturing processes are significant separately, literature discussing both is very scarce. Due to this gap in research and the absence of documented applications, the proposed applications of the present research are conceptual and present potential applications, with the need for validation through experimentation and pilot studies.

Although this research delves into specialized bio-intelligent systems, achieving the research objective is a systems design problem, mainly due to the need to integrate the systems with humans. Operators in manufacturing have important skills and knowledge pertaining to process dynamics, material properties, and quality metrics that are not likely to be captured by automated systems. It is more likely that successful implementations of bio-intelligent systems will be those that augment, rather than displace, the human thinking model, such as those that adopt human-in-the-loop frameworks where operators inform the system and supervise the system's automation. Moreover, the transformation to bio-intelligent manufacturing will necessitate the development of training and education programs aimed at workforce skills in data-driven and supervisory control, and in iterative improvement of algorithms. These training programs will have to be supported with technological investments to fulfil the promise of bio-intelligent systems.

Controlled pilot testing in manufacturing settings would provide the best form of experimental validation, which should be prioritized in future studies and executed for the more straightforward applications, such as adaptive drying control or neural network quality inspection. Moreover, studies contrasting swarm intelligence and hierarchical control for multi-zone kiln optimization would be helpful in understanding the performance trade-offs and implementation complexities. Economic analysis of the implementation costs of bio-intelligent systems to specific industries and the potential profits would further enhance the manufacturing implementation of the bio-intelligent systems.

5 Conclusion and Future Research Work

This research paper provides a conceptual framework for understanding how biointelligent principles might be applied to ceramic tile manufacturing. It distinguishes the intersection of principles of bio-intelligent manufacturing and the production of ceramic tiles, demonstrating the value of biological re-engineering in this sector. The systematic literature review highlights a paradox. While much has been published on bio-intelligent manufacturing and ceramic production, the intersection of the two remains virtually unstudied. This paradox highlights the originality of this research and the further gap in knowledge that could be addressed through the implementation of a pilot study.

The proposed integration framework illustrates how the recurring, nature-inspired methodologies could alleviate some of the unresolved challenges of the manufacturing sector, such as energy consumption, stability of output quality, and control of the processes when the production environment changes. This research should assist technology developers in making informed decisions on which production phase they should concentrate on to map biological processes.

Additionally, digital infrastructure, biological insights, and manufacturing expertise are converging, and the ceramic tile industry is positioned to capture the upcoming transformational innovation opportunity. This research provides the foundational knowledge to connect the theoretical bio-intelligent concepts and the manufacturing realities. Further progress will depend on collaborations among industry players, tech developers, and researchers to implement and test operational approaches for the industry. These collaborations will help the industry become more sustainable, more efficient, and more competitive over time as manufacturing demands increase.

Future research should prioritize: (1) pilot implementations in partnership with ceramic tile manufacturers to test feasibility under real production conditions; (2) economic analyses to assess cost-benefit trade-offs of biointelligent system adoption; (3) comparative studies evaluating biointelligent approaches against conventional control methods; and (4) scalability assessments to determine how these systems perform across different production volumes and facility configurations. Such empirical work will be essential to move from the theoretical framework presented toward practical industrial implementation.

Acknowledgments. This research was co-funded by the Italian Ministry of Enterprises and Made in Italy under the measure "Development Contracts" (DM 31/12/2021), grant number: F/310087/01–05/X56 (START—SusTainable dAta-dRiven manufacTuring).

References

1. Javed, S., et al.: Strategies and pathways to improve circularity in ceramic tile production. J. Clean. Prod. **517**, 145783 (2025)
2. Miehe, R., et al.: The biological transformation of the manufacturing industry – envisioning biointelligent value adding. Proc. CIRP. **72**, 739–743 (2018)
3. Miehe, R., et al.: The biological transformation of industrial manufacturing-future fields of action in bioinspired and bio-based production technologies and organization. Procedia Manuf. **39**, 737–744 (2019)
4. Baumgarten, Y., et al.: Biology-technology interfaces – refining the core principle of biointelligent systems. Proc. CIRP. **126**, 875–880 (2024)
5. Emer, A., et al.: Examples of potential applications of bio-intelligent manufacturing. Proc. Comput. Sci. **253**, 2196–2205 (2025)

6. Matt, D.T., Rauch, E.: Biological transformation in manufacturing: overview and fields of application. IEEE Eng. Manag. Rev. **49**, 115–122 (2021)
7. Rauch, E., Matt, D.T., Brown, C.A., Towner, W., Vickery, A., Santiteerakul, S.: Transfer of industry 4.0 to small and medium sized enterprises. In: Peruzzini, M., Pellicciari, M., Bil, C., Stjepandić, J., Wognum, N. (eds.) Transdisciplinary Engineering Methods for Social Innovation of Industry 4.0. Advances in Transdisciplinary Engineering, vol. 7, pp. 63–71. IOS Press, Amsterdam (2018)
8. Rauch, E., Dallasega, P., Matt, D.T.: The way from lean product development (LPD) to smart product development (SPD). Proc. CIRP. **50**, 26–31 (2016)
9. Nazari, K., Tran, P., Tan, P., Ghazlan, A., Ngo, T.D., Xie, Y.M.: Advanced manufacturing methods for ceramic and bioinspired ceramic composites: a review. Open Ceram. **15**, 100399 (2023)
10. Page, M.J., et al.: The PRISMA 2020 statement: an updated guideline for reporting systematic reviews. BMJ. **71** (2021)
11. Hashemi Farzaneh, H., Borgianni, Y., Forti, D., Rauch, E.: A speculation on the potential support of bio-inspired design to biologicalisation in manufacturing. In: International Conference on Engineering Design (ICED21), pp. 221–230. Cambridge University Press, Cambridge (2021)
12. Suresh, B., et al.: Simulation of integrate-and-fire neuron circuits using HfO2 -based ferroelectric field effect transistors. In: 2019 26th IEEE International Conference on Electronics, Circuits and Systems (ICECS), pp. 229–232. IEEE, Genoa (2019)
13. Nikitin, O., Lukyanova, O.: Homeostatic neural network for adaptive control: examination and comparison. In: Manoonpong, P., Larsen, J.C., Xiong, X., Hallam, J., Triesch, J. (eds.) From Animals to Animats 15, pp. 223–235. Springer International Publishing, Cham (2018)
14. Albadr, M.A., Tiun, S., Ayob, M., AL-Dhief, F.: Genetic algorithm based on natural selection theory for optimization problems. Symmetry. **12**, 1758 (2020)
15. Forrest, S.: Genetic algorithms. ACM Comput. Surv. **28**, 77–80 (1996)
16. Dorigo, M., Birattari, M., Stutzle, T.: Ant colony optimization. IEEE Comput. Intell. Mag. **1**, 28–39 (2006)
17. Kennedy, J., Eberhart, R.: Particle swarm optimization. In: Proceedings of ICNN'95 - International Conference on Neural Networks, pp. 1942–1948. IEEE, Perth (1995)
18. Leitão, P., Queiroz, J., Sakurada, L.: Collective intelligence in self-organized industrial cyber-physical systems. Electronics. **11**, 3213 (2022)
19. Leitão, P.: Self-organization in manufacturing systems: challenges and opportunities. In: 2008 Second IEEE International Conference on Self-Adaptive and Self-Organizing Systems Workshops, pp. 174–179. IEEE, Venice (2008)
20. Estrada-Jimenez, L.A., Pulikottil, T., Nikghadam-Hojjati, S., Barata, J.: Self-Organization in Smart Manufacturing— Background, systematic review, challenges and outlook. IEEE Access. **11**, 10107–10136 (2023)
21. Pan, N.: Exploring the significance of structural hierarchy in material systems—a review. Appl. Phys. Rev. **1**, 021302 (2014)
22. Wei, J., et al.: Bioinspired additive manufacturing of hierarchical materials: from biostructures to functions. Research. **6**, 0164 (2023)
23. Kantaros, A., Ganetsos, T., Pallis, E., Papoutsidakis, M.: Biomimetic additive manufacturing: engineering complexity inspired by nature's simplicity. Biomimetics. **10**, 453 (2025)
24. Behera, R.P., Ng, A.Y.R., Reavley, M.J.-H., Du, Z., Gan, C.L., Le Ferrand, H.: Rational design and fabrication of hierarchical ceramics using bioinspired microstructures for tailorable strength and toughness. Cell Rep. Phys. Sci. **5**, 102140 (2024)
25. Wegener, K., Spierings, A.B., Teti, R., Caggiano, A., Knüttel, D., Staub, A.: A conceptual vision for a bio-intelligent manufacturing cell for selective laser melting. CIRP J. Manuf. Sci. Technol. **34**, 61–83 (2021)

26. Tang, D., Wang, L., Gu, W., Yuan, W., Tang, D.: Modelling of bio-inspired manufacturing system. In: Huang, G.Q., Mak, K.L., Maropoulos, P.G. (eds.) Proceedings of the 6th CIRP-Sponsored International Conference on Digital Enterprise Technology, pp. 1165–1174. Springer, Berlin, Heidelberg (2010)
27. Chen, F., et al.: Advancing concrete mix proportion through hybrid intelligence: a multi-objective optimization approach. Materials. **16**, 6448 (2023)
28. LeCun, Y., Bengio, Y., Hinton, G.: Deep learning. Nature. **521**, 436–444 (2015)
29. Santos-Barbosa, D., Hotza, D., Boix, J., Mallol, G.: Modelling the influence of manufacturing process variables on dimensional changes of porcelain tiles. Adv. Mater. Sci. Eng. **2013**, 1–12 (2013)
30. Ma, H., Wang, D., Ren, J., Qiao, J.: Self-organizing neural intelligent control for nonlinear discrete-time systems with particle swarm optimization. Nonlinear Dyn. **113**, 583–595 (2025)
31. Koren, Y., Gu, X., Guo, W.: Reconfigurable manufacturing systems: principles, design, and future trends. Front. Mech. Eng. **13**, 121–136 (2018)
32. Bargardi, F.L., Le Ferrand, H., Libanori, R., Studart, A.R.: Bio-inspired self-shaping ceramics. Nat. Commun. **7**, 13912 (2016)
33. Chen, X., Lim, S.K.J., Liang, Y.N., Zhang, L., Hu, X.: Large toughening effect in biomimetic Geopolymer composites via Interface engineered 3D skeleton. ACS Sustain. Chem. Eng. **7**, 105–110 (2019)
34. Nait-Ali, B., Alzina, A., Lauro, N., Smith, D.S.: Perspectives in drying of ceramics. Open Ceram. **17**, 100554 (2024)
35. Nandi, V.S., et al.: Fast-drying of red ceramics: a new evaluating approach based on the combined use of optical dilatometry and thermogravimetric analysis. Ceram. Int. **50**, 31198–31210 (2024)
36. Hosseinpour, S., Martynenko, A.: An adaptive fuzzy logic controller for intelligent drying. Dry. Technol. **41**, 1110–1132 (2023)
37. Picot, O.T., et al.: Using graphene networks to build bioinspired self-monitoring ceramics. Nat. Commun. **8**, 14425 (2017)
38. Martynenko, A., Mujumdar, A.: Evolution of control strategies toward intelligent drying. Dry. Technol. **42**, 587–588 (2024)
39. Coskun, H., et al.: Integration of digital quality control for intelligent manufacturing of industrial ceramic tiles. Ceram. Int. **48**, 34210–34233 (2022)
40. Enríquez, E., Reinosa, J.J., Fuertes, V., Fernández, J.F.: Advances and challenges of ceramic pigments for inkjet printing. Ceram. Int. **48**, 31080–31101 (2022)
41. Cumbajin, E., et al.: A real-time automated defect detection system for ceramic pieces manufacturing process based on computer vision with deep learning. Sensors. **24**, 232 (2023)
42. Hanzaei, S.H., Afshar, A., Barazandeh, F.: Automatic detection and classification of the ceramic tiles' surface defects. Pattern Recogn. **66**, 174–189 (2017)
43. Guo, J., Martínez-García, M.: Key technologies towards smart manufacturing based on swarm intelligence and edge computing. Comput. Electr. Eng. **92**, 107119 (2021)
44. Rubenstein, M., Cornejo, A., Nagpal, R.: Programmable self-assembly in a thousand-robot swarm. Science. **345**, 795–799 (2014)
45. Han, G., Zheng, N., Wu, J.: Function optimization of ceramic firing process and reduction of energy consumption based on CFD simulation. J. Phys. Conf. Ser. **2729**, 012003 (2024)
46. Fernández Gassó, R., Sevilla Hurtado, L., Cañero-Nieto, J.M.: Optimizing cement kiln operation with expert systems. Int. J. Adv. Manuf. Technol. (2025)
47. Abd Elrahem, R.F.A., Mohammed, H., Soliman, E.D.A.: Development of a smart ceramic kiln based on effective control of the firing process: an applied study. SVU-Int. J. Eng. Sci. Appl. **6**, 155–159 (2025)
48. Mezquita, A., Boix, J., Monfort, E., Mallol, G.: Energy saving in ceramic tile kilns: cooling gas heat recovery. Appl. Therm. Eng. **65**, 102–110 (2014)

49. Vasić, M.V., Awoyera, P.O., Fadugba, O.G., Barišić, I., Grubeša, I.N.: Advanced machine learning models for the prediction of ceramic tiles' properties during the firing stage. Sci. Rep. **15**, 31397 (2025)
50. Marcineková, K., Janáková Sujová, A.: Multi-objective optimization of manufacturing process using artificial neural networks. Systems. **12**, 569 (2024)
51. Czimmermann, T., et al.: Visual-based defect detection and classification approaches for industrial applications—a SURVEY. Sensors. **20**, 1459 (2020)

Design

AI-Assisted Automation of Grinding Tools CAD Design Focused on 2D Drawings: Two-Way CAD Design Basics

Roman Wdowik(✉)

Rzeszów University of Technology, Al. Powstańców Warszawy 12, 35-029 Rzeszów, Poland
rwdowik@prz.edu.pl

Abstract. The paper examines the automation of CAD 2D (2-dimensional) technical drawings based on CAD 3D (Computer Aided Design 3-dimensional) solid models and also the design of the abovementioned solid models based on 2D technical drawings. In this context an automated two-way design approach is discussed. This approach utilizes selected capabilities of artificial intelligence (AI) tools, such as large language models (LLMs). In particular, it focuses on the design of grinding tools (mainly grinding wheels). The first part of the study focuses on the AI-assisted solid modelling and utilization of created solid models to generate their 2D drawings by using **.dxf* format. In the second part, the author investigates the possibilities of designing 3D solid models using 2D drawings rather than relying exclusively on text-based prompts. The final results demonstrate that the presented procedures can be used by designers, while also taking into account the limitations of the approach. The results indicate that human actions and skills are still required to improve and finalize the project. Moreover, the study demonstrates that selected LLMs are a cost-effective solution for the basic two-way CAD design of grinding tools

Keywords: Technical Drawing · Grinding Wheel · Artificial Intelligence · Two-Way CAD Design

1 Introduction

Mechanical engineers use the design documentation such as 3D models and 2D drawings in order to present the design-related details for further use within production phases such as technological process planning, manufacturing or inspection. Many of the current efforts in the area of design are linked to applications of CAD (Computer Aided Design) software tools that revolutionized the design approaches several decades ago [1]. These tools have been implemented by the industry since their first releases and are indicated by many practitioners as an indispensable solution for current industrial needs.

CAD software tools enable the automation and acceleration of selected design tasks, facilitate the rapid implementation of design changes, and support file exchange within the manufacturing community. Moreover, 3D (e.g. solid models or surface models) or 2D models (e.g. technical drawings) can be utilized within various simulations and product

B. Gapiński et al. (Eds.): MANUFACTURING 2026, LNME, pp. 159–170, 2026.
https://doi.org/10.1007/973-3-032-22943-4_12

development processes in other CAx (Computer-Aided x) environments. For instance, 3D CAD model can be used in FEM-based (Finite Element Method-based) simulations or within the inspection process programming.

Latest developments in the field of neural networks and the expected growing impact of artificial intelligence (AI) on the automation of design tasks, raise questions regarding the benefits and challenges associated with the application of AI in design processes. One of the potential applications is machining sector that uses various tools such as grinding tools (e.g. grinding wheels or grinding pins).

Grinding as a machining process, with necessary machine tools and the abovementioned grinding tools play an important role in manufacturing sector and their applications and improvements help to manufacture high quality parts [2]. Moreover, grinding processes are continuously investigated by research centers and academia, and practical guidelines are presented, such as those included in a recent book by Azarhoushang that focuses on both grinding fundamentals and innovations [3]. In addition, literature studies and research trends in 2025 indicate the importance of CAD developments, *inter alia*, due to the growing number of artificial intelligence–related research studies and the expanding commercial market for AI tools.

For instance, Hunde and Woldeyohannes in their work [4] analyze the prospects of CAD taking into account artificial intelligence and also virtual, augmented and mixed realities and 3D printing. In the case of AI they present selected applications in existing software tools that enrich the design process and support product CAD designers. The next work of Buga et al. [5] discusses an impact of artificial intelligence on the design process and this impact is studied from various important perspectives such as machine learning algorithms, generative design, team collaboration or PLM (Product Lifecycle Management). Another research group of Raffaeli an co-authors [6] investigated automation of drafting execution by using gear motor examples, schemes definition and feature recognition. Wdowik and Bełzo [7] generated factory layouts, for instance as *.*dxf* files, by the use of LLMs (Large Language Models) on the basis of proposed scenario regarding various types of machine tools that were placed in the designed factory area by this technique. The same research team focused on the assembly design issues and noticed that AI-assisted CAD design of assemblies is still a complex issue in terms of satisfactory modelling results [8]. Moreover, they underlined the possibilities of designing simple assemblies. In another work of Wdowik [9] it is demonstrated that grinding wheels and pins modelling using generative AI is a promising design approach and this technique enables creation of not only general tool's geometry but also constructional details such as abrasive grains (for instance diamond grains used in superhard grinding wheels) that may be assembled within 3D CAD models with other tool's detailed elements. Zhou et al. discuss in an extended review paper [10] applications of LLMs in parametric CAD and their future challenges. It should be also stated that LLMs have also other various applications. For instance, Alcaide-Marzal et al. [11] propose open-source tool for an assessment of eco-design by the use of Blender software and LLMs. Moreover, Pizoń et al. [12] discuss the use of LLMs for maintenance tasks. It should be also mentioned that developers of CAD software implement various AI-based solutions in order to increase the efficiency of modern software tools and improve modelling experience of designers.

For example, Autodesk Fusion™ software [13] developers have implemented a functionality based on AI that helps to generate an exemplary drawings from 3D model. It is also possible to choose an appropriate solution from various generated alternatives. Similar functionality is available in the SolidWorks™ software [14] developed by Dassault Systèmes. These examples prove that commercial software tools developers investigate such solutions as they correspond to the current trends.

Taking into account the existing knowledge and tools it is necessary in author's opinion to research also other possibilities, mainly cost-effective techniques such as open access LLMs or low cost LLMs that can lead to further automation of the design. Growing interest in Python language [15] and its powerful libraries is encouraging for research on CAD drawing automation.

Based on the cost-effective LLMs application the study focuses on the analysis of selected approaches towards further automation of two-way automation of CAD modelling. This two-way approach should be explained as automation of 2D technical drawings (e.g. saved as *.*dxf* files) design from the 3D solid models (e.g. saved as *.*step* files) and opposite automation of CAD 3D solid modelling from 2D technical drawings. The study is the next step of author's previous co-authored and authored studies focused on AI-assisted CAD modelling for the manufacturing needs. The study is also aimed at such technique effectiveness in comparison to the traditional design procedures and discusses necessary corrections that are crucial to reach the final design quality for the grinding tools manufacturing and related production tasks.

2 Test Stand Configuration and Presentation of the AI-Assisted CAD Design Idea

The test stand consists of PC connected to the internet network. Prompts (text-based questions or data placed in the *.*xlsx* file) are developed by author. The following LLMs have been used: ChatGPT™ [16], Copilot™ [17], Deepseek™ [18], Mistral AI™ [19]. Their usage requires internet browser to be installed. Verification of developed CAD models of grinding tools was performed by using the FreeCAD software [20]. The expected response is a Python language code that is executed in the FreeCAD software (the Python console) or in the *conda* environment. To execute *.*py* files in *conda* environment the Visual Studio Code™ [21] was utilized similarly to the previous author's studies. Moreover, developed *.*dxf* files were verified or improved by using LibreCAD software [22]. Conda environment is reachable after installation of the Anaconda™ or miniconda [23].

It should be underlined that the Python language-based instructions together with dedicated Python libraries enable to create various grinding tools by the use of simple prompts – even sets of many tools. There are two prompts presented below. They were used to generate Python codes and sets of tools (CAD models of grinding wheels and pins). Moreover, reprompting enables to make corrections to the previously obtained results.

Prompt 1: *Generate a set of 20 CAD models of grinding wheels characterized by various shapes and dimentions. Please create a code in the Python language that will be pasted to the FreeCAD Python console.*

Prompt 2: *Generate a set of 20 CAD models of grinding pins characterized by various shapes and dimentions. Please create a code in the Python language that will be pasted to the FreeCAD Python console.*

Reprompting (Prompt 2): *place the same grinding wheels in two rows*

(Note: Pins were incorrectly named as wheels within reprompting process, however, LLM "understood" it correctly and placed grinding pins in two rows.)

Figure 1 presents the AI-assisted design results. Models (created based on the generated Python code) were opened and studied in the FreeCAD software. The Python code can also be adjusted before pasting it to the software's Python console. It is demonstrated in the current study that various shapes of grinding wheels and grinding pins can be created, however, these tools are represented only by the general geometry and other elements such as diamond grains are not included here. This is here due to the commonly used standards in the technical drawings which do not require to draw separate grains of tools and even in the case of superhard grinding wheels and pins (e.g. diamond or cubic Borone Nitride (cBN) tools) the abrasive layers are marked with dash-dot lines instead of using the drawings of separate grains (Please also refer to another published work [9] in which also the grains were designed by using selected primitives). Moreover, selected shapes of created grinding pins are not typical solutions, nevertheless, they can also be manufactured. This example clearly demonstrates that even complex sets of random grinding tools can be generated by using the AI-assited CAD design approach, and they can be adjusted (if necessary) by using the traditional CAD tools.

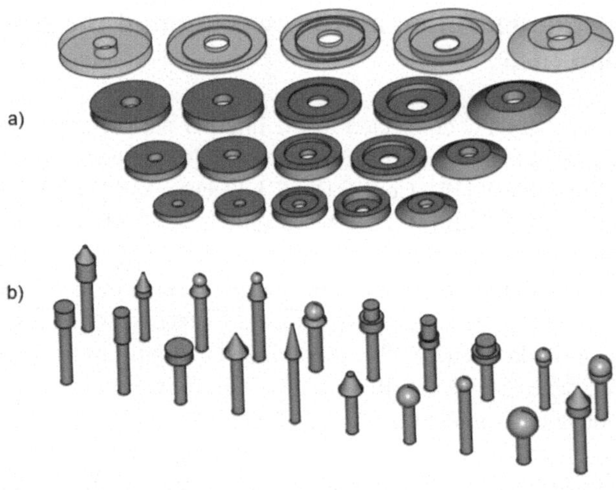

Fig. 1. Grinding tools created in the FreeCAD software on the basis of Python code generated as a response of the ChatGPT™ to the author's prompt: a) set of grinding wheels, b) set of grinding pins

3 Experiments and Their Results

Based on the idea presented in the previous section (i.e. prompting, Python code execution by using the FreeCAD console and model correctness verification in the software by a designer), a solid model of an exemplary grinding wheel was developed in the first stage. The following prompt was used for the abovementioned aim:

*"Generate a Python code that will help to create a grinding wheel CAD solid model in the FreeCAD software. The code will be pasted to the Python console in the software. The details of the grinding wheel are presented in the *.xlsx file".*

The prompt was supported by the data presented in Table 1 transferred to the LLMs by using **.xlsx* file mentioned in the prompt or if not possible (due to LLM limitation) by copying the data from the table. The prompt was executed together with the presented data in order to generate LLM's response. The values of grinding wheel parameters included in the table are presented by the use of units that are typical for metric dimensioning used in 2D drawings. The presented parameters are linked to the general grinding wheel geometry and there is also information regarding geometrical details such as counterbore, central hole, chamfers. Main grinding wheel parameters' values presented in Table 1 were developed on the basis of the offer of Andre abrasive articles [24] in order to test the LLMs based on the data linked to a real grinding wheel data that can be manufactured by a manufacturing firm. Chamfers' parameters were added to test AI-assisted design of such grinding wheels' CAD features.

Table 1. A grinding wheel parameters used in two-way CAD design experiment (main parameters' values (without chamfers) presented in the table were developed on the basis of the offer of Andre abrasive articles company – refer to [24])

Name	Value	Unit
Grinding wheel external diameter	300	mm
Grinding wheel central hole diameter	76	mm
Grinding wheel cylindrical counterbore	200	mm
Cylindrical counterbore height	25	mm
Grinding wheel symmetrical chamfer linked to the edge of central hole	2 x 45	mm x °
Grinding wheel symmetrical chamfer linked to the edge of counterbore	2 x 45	mm x °
Grinding wheel thickness	50	mm

x – *annotation used in 2D drawings*

The obtained code (as a result of LLM response) in the Python language was pasted to the Python console in the FreeCAD software and the results of such modelling experiment, for various LLMs, are presented in Table 2. The table presents obtained solid models and author's comments indicating the results that differ from the expected ones. The main problem concerns the difficulties with general geometry (one LLM) and wheels' chamfers. In the case of the results produced by ChatGPT™ and Deepseek™ there were no chamfers generated. In the case of Copilot™ and Mistral™ chamfers were generated on the edges specified by a designer as well as on additional edges. Moreover, in this specific case, the solid body was missing in the output produced by one LLM (Mistral™). The author also tested Polish large language models such as Bielik™ [25] and PLLuM™ [26], however, the code generated by the models did not enable the creation of a solid model in the FreeCAD software. The generated models were also measured by using FreeCAD *measure* tool in order to verify the dimensions of the features. Similarly to the previous research studies (e.g. refer to [9]), the verification results were positive and the general dimensions such as external or internal diameters, grinding wheel width were correctly generated. It should be also underlined, that in the author's opinion, the results do not clearly indicate a superior solution (better LLM) for the general tool design tasks, however, they indicate that the AI-assisted design process remains susceptible to errors/misunderstanding of prompts created by a human.

Table 2. Comparison of an exemplary grinding wheel modelling results created in the FreeCAD software on the basis of code generated by the following LLMs as response to the author's prompt: a) ChatGPT™, b) Copilot™, c) Deepseek™, d) Mistral™

(a) ChatGPT™	**(b) Copilot™**
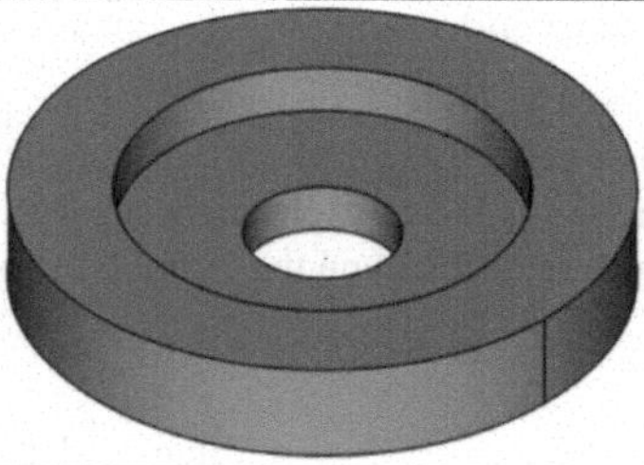	
No chamfers added, the general shape is correct	Additional chamfers added without a designer's clear request but the general shape is correct
(c) Deepseek™	**(d) Mistral™**
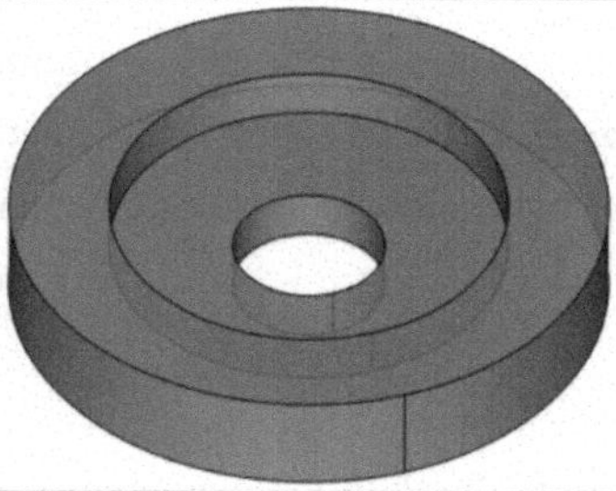	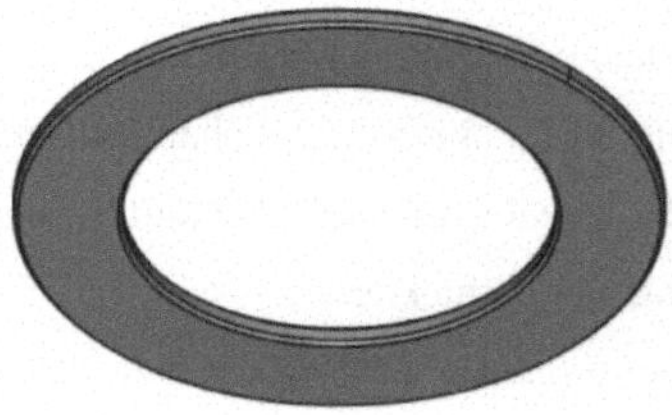
No chamfers added, the general shape is correct	The shape is only partially correct. There is no 76 mm hole, no counterbore and additional chamfers.

The next step of the study concerned the possibility of automated design of 2D technical drawings using Python and its libraries. The tests were performed with the support of ChatGPT™ 5.1 in accordance with the procedure presented in Fig. 2. The figure presents the AI-assisted approach to CAD 2D drawing design from 3D model (top part) and the reverse process used for 3D model design from 2D drawing. As mentioned previously, this indicates that the AI-assisted CAD design process can be performed in both directions (two-way CAD design). Moreover, it is demonstrated which phases of the process require significant human, artificial intelligence or software assistance or reprompting (annotations H, AI, S, H).

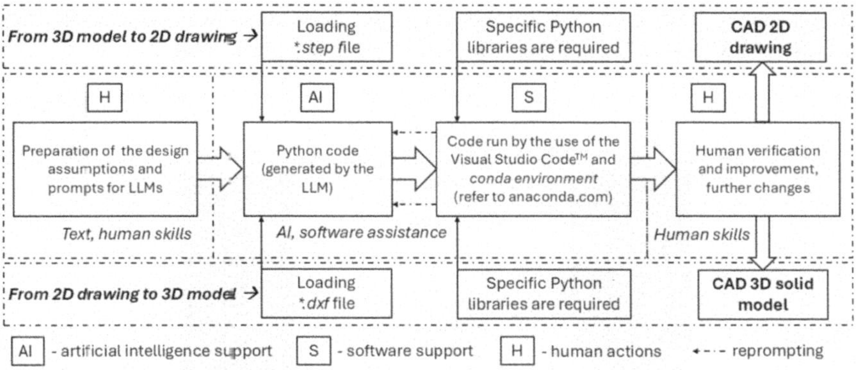

Fig. 2. Procedure leading to a script that automates 2D drawing creation and 3D model creation (two-way CAD design)

In the first stage, the script enabled the creation of a set of 2D views including an isometric view, however, some lines were not generated. The second prompting attempt produced a script that enabled the generation of improved views, yet the dimensioning was still missing. The additional reprompting attempt allowed to generate a script that displays some dimensions, however, there were no dimensions regarding holes and cylindrical features marked with Φ letter. Finally, after the next reprompting attempt, it was also possible for the applied ChatGPT™ to generate such a Python script and the exemplary 2D drawing is presented in Fig. 3. This figure indicates existing problems such as duplicated dimensions, no hatching or incorrect line type, etc. These problems still exist after several reprompting attempts. It was also noticed that generated views represented by the lines contained overlapped lines which should be removed from the drawing in subsequent steps performed by the CAD designer. The problem mentioned above is presented in Fig. 4.

This stage of the AI-assisted design requires the designer's decision about next steps. The prompting attempts can be repeated until the satisfactory results without the guaranteed result or the final drawing can be reached after traditional improvement procedures using existing CAD tools of the CAD software.

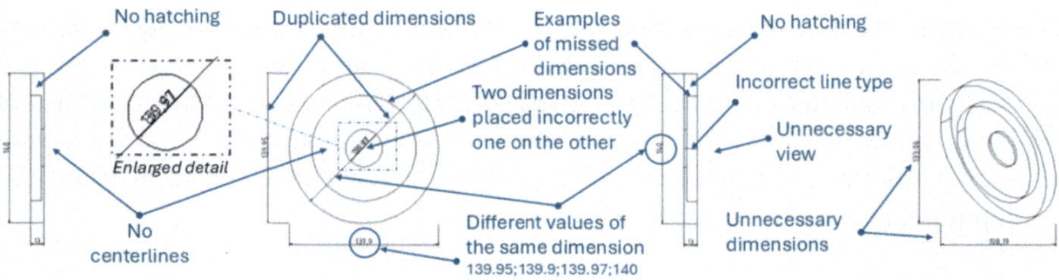

Fig. 3. Exemplary 2D drawing (*.dxf file opened in the LibreCAD software) obtained by the use of the presented AI-assisted design procedure from a 3D model (*.step file)

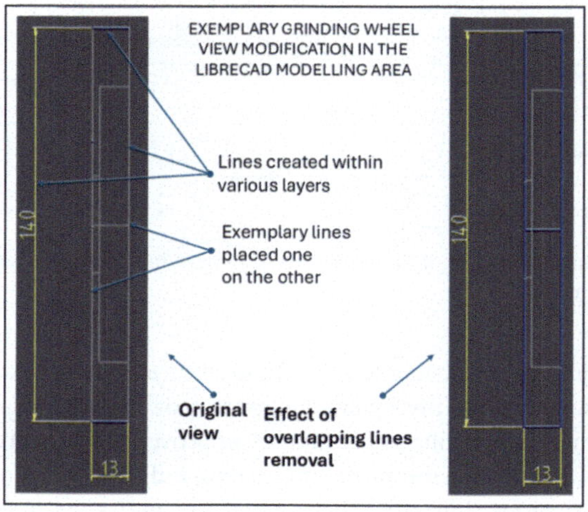

Fig. 4. Grinding wheel view modification in the LibreCAD software: lines created by using various layers and presentation of lines incorrect overlapping

The next step regarding the experiments concerned the opposite action towards creation of 3D solid model from the previously generated 2D technical drawing. Based on the prompts, the new code was created by the ChatGPT™ and verified in the Visual Studio Code™. Finaly it was possible to create the 3D CAD models for specific tool geometries. The procedure and the obtained result are presented in Fig. 5. The figure also illustrates that the 2D drawings can be further improved, as discussed above. Additional lines, incorrect dimensions, missed data may, etc. be added "manually" by a designer using the 2D CAD software toolbox. Similar adjustments may also be applied to 3D models and the 3D CAD software tools characteristic for 3D solid modelling can be applied.

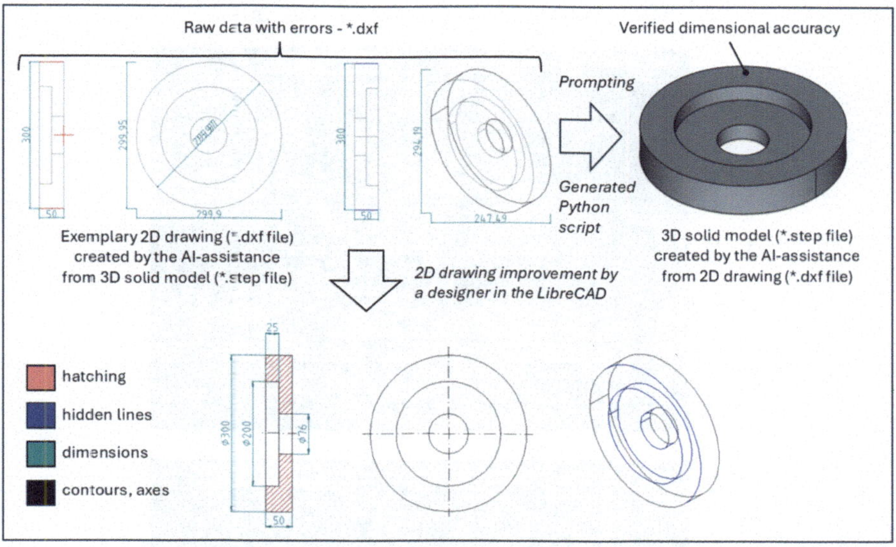

Fig. 5. 2D drawing to 3D solid model AI-assisted design example and presentation of 2D drawing improvements performed by a designer in the LibreCAD software

4 Discussion

The presented research demonstrates that generative artificial intelligence offers various possibilities regarding automation of the CAD design activities. However, the generated data may require additional human changes by using standard/traditional CAD tools and procedures. Nevertheless, in the case of grinding wheels and analyzed examples the obtained results (3D models and 2D drawings) meet very basic technical requirements and the necessary changes are usually linked to adjustments and data quality improvements. General results are usually acceptable for further changes and may be used by the designer for next design steps. Main incorrect results of the AI-assisted CAD design process of grinding wheels concern the presentation of complete set of basic data according to existing standards (in the case of 2D drawings) or incorrect creation of details of 3D solid models. Nevertheless, AI-automation of the design process is feasible for basic and initial design steps and produces encouraging results while indicating that 2D drawings still require improvements. Observations regarding the overall correctness of the CAD models are also similar to the ones that were obtained by the author in the past and presented in the previous authored and co-authored papers. The approach is also worth developing due to its cost-effectiveness because low-cost or free LLMs can help design grinding tools. Positive results were possible because of the LLMs and capabilities of the Python libraries (Fig. 6) utilized within the generated Python codes.

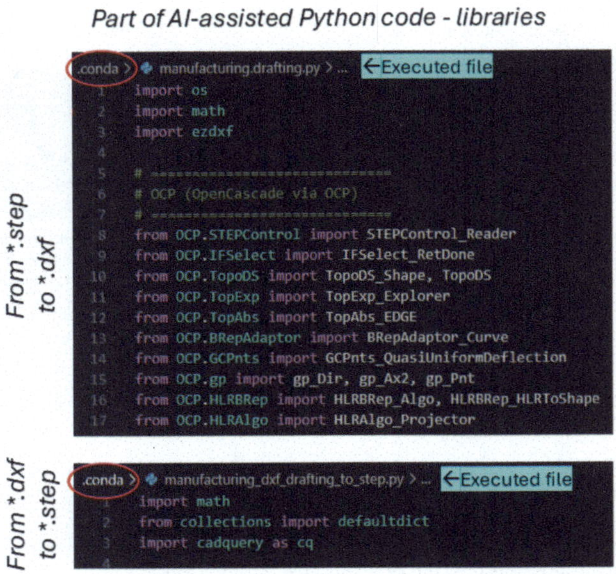

Fig. 6. Python libraries utilized within two-way CAD design procedures displayed in the Visual Studio Code™ text editor

5 Conclusion

The study examined both opportunities and challenges resulting from AI-based automation of technical drawings on the basis of a grinding wheel case study. The following conclusions shall be taken into account to make a final decision regarding the application of the proposed approach for a grinding wheel design:

1. AI-supported data generated based on a designer's prompts or actions should be verified by designers who are responsible for the final project.
2. Most of the generated data requires further improvements in CAD environments (e.g. adding additional dimensions, tolerances, surface characteristics, features etc.).
3. The 2D drawings (**.dxf* files) can be created from 3D solid models (**.step* files) and 3D solid models can be created from 2D drawings. The Python code required adjustments for the specific wheel type.
4. The proposed approach is a great opportunity to improve the efficiency of time-consuming design processes.

Further research can be focused on application of various LLMs and their deeper analysis to verify their usefulness for Python code development.

Acknowledgements. The study was developed within the project entitled *"Teamwork-based Education and Digitalization as an Approach for the Interdisciplinary Engineering Training"* co-funded by the European Union - project reference: 2024-1-PL01-KA220-HED-000257156. Views and opinions expressed are however those of the author only and do not necessarily reflect those of the European Union or the Foundation for the Development of the Education System. Neither the European Union nor the Foundation for the Development of the Education System can be held responsible for them.

Artificial intelligence usage statement: The paper sentences were developed without the artificial intelligence assistance. Python codes were generated by the use of the LLMs (large language models) based on author's prompts to create 3D models presented in Fig. 1 and Table 1 as print screens of the FreeCAD environment. In addition, Figs. 3 and 4 were developed based on the print screens of LibreCAD software environment where **.dfx* files were created by Python scripts that were generated by the LLM on the basis of author's prompts. Figure 5 contains print screens of 3D model and 2D drawing that were obtained on the basis of Python scripts generated by the LMM after author's prompting. Code (libraries) presented in Fig. 6 illustrates Python libraries from the Python code generated by the LLM in the two-way CAD design.

Software usage statement: Author used the FreeCAD software for validation of the *3D* CAD files. The LibreCAD software was used for validation and improvements of the **.dxf* files (2D CAD files). The Visual Studio Code™ software was used to execute the Python scripts generated by the AI based on author's prompts – in the case of a 3D model transformation to a 2D model and similarly in the case of 3D modelling. PowerPoint software (with TM, superscript) was used to support development of figures.

References

1. Schlechtendahl, E.G., Encarnação, J.: Computer Aided Design Fundamentals and System Architectures. Springer, Berlin, Heidelberg (1983)
2. Marinescu, I.D., Hitchiner, M.P., Uhlmann, E., Rowe, W.B., Inasaki, I.: Handbook of Machining with Grinding Wheels, 2nd edn. CRC Press – Taylor & Francis Group (2019)
3. Azarhoushang, B.: Modern Dressing and Grinding Technologies, Volume 1, Principles, Tools, and Process Fluids. Elsevier (2025)
4. Hunde, B.R., Woldeyohannes, A.D.: Future prospects of computer-aided design (CAD) – a review from the perspective of artificial intelligence (AI), extended reality, and 3D printing. Results Eng. **14** (2022). https://doi.org/10.1016/j.rineng.2022.100478
5. Buga, A., Borzan, M., Trif, A.: Artificial intelligence in the CAD process: machine learning models, generative optimisation, and their impact on design. Acad. J. Manuf. Eng. **23**(1) (2025)
6. Raffaeli, R., Cicconi, P., Germani, M.: Automation of drafting execution by schemes definitions and feature recognition. Comput. Aided Design Appl. **13**(4), 459–470 (2016). https://doi.org/10.1080/16864360.2015.1131538
7. Wdowik, R., Bełzo, A.: Production plant layout planning supported by selected CAx tools and artificial intelligence. Adv. Mech. Mater. Eng. **42**(1), 89–100 (2025). https://doi.org/10.7862/rm.2025.8
8. Wdowik, R., Bełzo, A.: Artificial intelligence-based design of assemblies in the FreeCAD software. Technologia i Automatyzacja Montażu (Assembly Techniques and Technologies). **127**(1), 74–82 (2025). https://doi.org/10.7862/tiam.2025.1.6
9. Wdowik, R.: Selected issues regarding AI-assisted CAD design of superhard grinding pins and grinding wheels. J. Mach. Eng. **25**(3), 27–37 (2025). https://doi.org/10.36897/jme/208996
10. Zhou, J., Camba, J.D.: The status, evolution, and future challenges of multimodal large language models (LLMs) in parametric CAD. Expert Syst. Appl. **282**, 127520 (2025). https://doi.org/10.1016/j.eswa.2025.127520
11. Alcaide-Marzal, J., Diego-Mas, J.A.: An open-source application for ecodesign in conceptual 3D modelling with large language models support. J. Clean. Prod. **525**, 146589 (2025). https://doi.org/10.1016/j.jclepro.2025.146589

12. Pizoń, J., Gola, A., Wójcik, Ł., Cieplak, T.: Data-driven maintenance 4.0 with use of large language models. In: Burduk, A., et al. Intelligent Systems in Production Engineering and Maintenance IV. ISPEM 2025. Lecture Notes in Mechanical Engineering. Springer, Cham, 2026, https://doi.org/10.1007/978-3-032-01517-4_20.
13. Autodesk Fusion.: https://www.autodesk.com/pl/products/fusion-360/overview. Last accessed 06 Nov 2025
14. Dessault Systèmes Solid Works.: https://www.solidworks.com/lp/evolve-your-design-workflows-ai. Last accessed 18 Nov 2025
15. Python.: https://www.python.org. Last accessed 18 Nov 2025
16. ChatGPT.: https://chatgpt.com. Last accessed 17 Nov 2025
17. Copilot.: https://www.copilot.com. Last accessed 17 Nov 2025
18. Deepseek.: https://www.deepseek.com. Last accessed 17 Nov 2025
19. Mistral AI.: https://mistral.ai. Last accessed 17 Nov 2025
20. FreeCAD.: https://www.freecad.org. Last accessed 18 Nov 2025
21. Visual Studio Code.: https://code.visualstudio.com. Last accessed 18 Nov 2025
22. LibreCAD.: https://librecad.org. Last accessed 17 Nov 2025
23. Anaconda AI Platform.: https://www.anaconda.com. Last accessed 07 Nov 2025
24. Andre Abrasive Articles.: https://andre.com.pl/index.php/oferta-andre/narzedzia-scierne-ze-spoiwem-ceramicznym/sciernice-ceramika/typ-5. Last accessed 10 Jan 2026
25. Bielik.AI.: https://bielik.ai. Last accessed 28 Nov 2025
26. PLLuM.: https://pllum.org.pl. Last accessed 28 Nov 2025

Design Implementation Features of Multi-operation Machine Spindle Head with Integrated Drive and Innovative Belt Transmission

Oleg Krol[1](✉), Vladimir Sokolov[1], Oleksandr Logunov[1], Marin Zhilevski[2], and Mikho Mikhov[2]

[1] Volodymyr Dahl East Ukrainian National University, 17 Ioanna Pavla II str, Kyiv 01042, Ukraine
krolos@i.ua

[2] Technical University of Sofia, 8 Kliment Ohridski Blvd, Sofia 1000, Bulgaria

Abstract. This paper describes the development and 3D modeling of a spindle head (SH) with an integrated drive equipped with an innovative belt transmission. The kinematic and layout features of the head design for a multi-purpose machine tool, taking into account the mounting of interchangeable modular tooling, are analyzed. A preliminary design is completed, and the main components and structure of the drive unit are defined. The spindle clamping mechanism, its disc spring, and clamping amplifier are calculated. 3D modeling of the SH head in the integrated CAD environment of Creo Parametric is performed, utilizing a set of applications for designing various mechanical drive transmissions. The disc spring in Creo Ansys Simulation is modeled. Rendering and drive explode tools were used to represent the designed structure and its competitiveness more realistically. An innovative version of a sectional belt drive is proposed as a synthesis of a V-belt (the elastic element of the section) and metal elements that have the design features of chains. An analytical basis for a calculation method for assessing the durability of sectional V-belts was developed. Durability indicators for the innovative belt drive design were calculated and compared with those of a drive with a standard V-belt. A compromise solution, characterized by a significant reduction in the maximum stress in the contact zone and a slight increase in the belt pretension force, is considered. The proposed drive design for a multi-operation machine tool has the potential to improve operational reliability during the machining of complex machine parts.

Keywords: Machining Center · Main Drive · Sectional belt · Durability

1 Introduction

Multi-purpose machines (MoM) are used as multi-functional metalworking equipment to implement various work processes using a wide range of cutting tools [1, 2]. These machines process workpieces of various configurations using various technological operations, including milling, drilling, boring, and others. The advantages of MoM include:

B. Gapiński et al. (Eds.): MANUFACTURING 2026, LNME, pp. 171–185, 2026.
https://doi.org/10.1007/978-3-032-22943-4_13

a) the ability to machine parts from different sides without re-mounting (the presence of a rotary table and a rotary milling spindle head); b) a wide spindle speed range, which allows for the selection of the optimal cutting mode for various workpiece materials; c) high productivity, etc.

The MoM is equipped with drives that have an extended adjustment range and are configured for a variety of modular spindle heads [3, 4]. All drives use an adjustable electric drive. Based on analyzed main processing modes in such machines, suitable drive systems with DC and AC motors are presented in [5, 6]. Rotation from the electric motor to the spindle head (SH) can be transmitted directly via a clutch or a V-belt drive. The gearbox (GB) can be integrated into the SH or separate from the spindle (Fig. 1, c). The use of various belt drive designs and their improvement is one of the important tasks facing machine designers [7].

Multi-operation machine tools (MoM) with CNC often have a main drive with a two-stage gearbox. In this case, rotation from the electric motor is transmitted to the spindle via a block of gears. Switching this block with a hydraulic cylinder piston allows for two spindle speed ranges (low and high) [8, 9].

One of the problems with SH performance is the reliability of fastening various cutting tools in the spindle taper [10]. In most multi-operation machines, the tool is mounted in the spindle using powerful springs and pressed forward by a pusher (draw-in rod) with a hydraulic or pneumatic drive. In this regard, the use of additional mechanisms to stabilize the tool clamping force is of interest. The spring clamping force is selected from the condition of non-disclosure of the spindle-tool joint, taking into account the cutting force components that arise when machining various types of workpieces.

Analysis of problems in creating an MoM drive with an integrated spindle unit, its layout, and its interface with the tool blocks has demonstrated the importance and relevance of design and modeling work. At the same time, machine tool designers must also be guided by a design paradigm where engineering quality and environmental sustainability are aligned with the sustainable development of the machine tool industry [11]

In the machine tool industry, work related to the design and production of spindle assembly designs and high-performance spindle drive systems is relevant and in demand. One of the most successful is GMN USA [12]. Spindle drives for high-frequency, low-induction motors have been specifically developed and implemented for specific applications. The main advantages of such design and technological innovations are optimized performance for low- and high-speed motors, compatibility with high-frequency spindles, and drive control technologies for any application.

Another technical leader [13] in the development and production of precision spindles and milling heads places great emphasis on drive systems. For production processes requiring increased torque, Setko recommends belt- and gear-driven spindles powered by an external motor and pulley system. A combination of gears and pulley ratios is used to increase or decrease torque as needed.

The design of MoM shaping generates motion mechanisms effectively using a drive based on multiple (2...3) mechanical transmissions designed to increase the range of rotational speeds and torque [14]. This design enables the implementation of increased torque, which enables the implementation of rough machining processes. High-speed machining at speeds up to 10,000 $\min^{-1}$ is also possible. Similar unified drive units for metal-cutting machine tools of sizes 2nd and 3rd using two-stage gearboxes with belt drives from an electric motor [15].

When designing the main drive of a machine, one of the key objectives is to find designs with minimal weight and dimensions. Article [16] examines an approach to determining the minimum weight of a two-stage gearbox, which includes three shaft assemblies on which gear blocks and subsequently connected to the spindle assembly are mounted. A similar problem was considered by the authors of [17]. For this purpose, the Cambrian v.3.2 software release was developed for determining the minimum mass of the gear drive structural components under consideration. One of the problems considered in [17] is related to finding the minimum dimensions of the drive housing.

The problem of finding the optimal drive option using a bevel gear transmission in the first stage and a cylindrical gear transmission in the second is considered in [18]. At the same time, the condition of guaranteed performance indicators for contact strength must be satisfied. One of the results is the obtained analytical dependences of the gear drive length on the geometric characteristics of the gears and the gear ratios in the first and second stages.

At the analytical level, the problem of comparative evaluation of a two-stage bevel-helical drive using the objective function for minimizing its cost was solved [19]. The main influencing parameters, the mass and size characteristics of the main components, are being considered. They are represented in the drive mass determination diagram: shafts, gears, and housing. During the optimization process, technical constraints for a gear ratio range of 1...9 in the first and second stages were formed.

The "Design optimization Gearbox KISSsoft" software, developed by AGMA [15], has proven effective in solving problems with the design of main- and feed drives for machine tools. Minimizing the internal volume of the drive housing while limiting the level of contact and bending stresses in the gear contact zone as an optimization criterion was considered.

A specific feature of such a drive design integrated into the structure of the spindle head is that the design and mounting characteristics of the cutting tool are considered. This is what ultimately influences the stability of the cutting process, productivity, and the required surface quality [19, 20]. Identification and modeling of the contact parameters between the tapered bore of the spindle and the taper of the tool block, taking into account the clamping force, length, and diameter of the clamped tool, is discussed in [20]. This paper analytically defines parameters at the spindle-holder and toolholder-tool interfaces. The following is also indicated here: the importance of determining clamping force.

Determining clamping force is particularly relevant for high-performance cutting (HPC) and high-speed cutting (HSC) applications across various operating speed ranges, spindle speed and high torques at low cutting speeds [21]. Such loads and cutting conditions under HPC conditions are typical for MoM.

Operational reliability and the level of stress-strain state of the main drive elements largely depend on the type and design of the belt transmission [22, 23]. V-belt, poly-V-belt, and toothed belt drives are used in MoM machines. Under the overload conditions typical of these machines, V-belt drives are effective. In addition to the V-belt's ability to transmit power over significant distances and increased traction with the pulleys, they also have the ability to protect the drive from overloads. This capability is lacking with toothed belt drives (including modified ones) due to the lack of slippage [24]. An alternative to belt drives is a chain drive, which is characterized by high strength and compactness, but operational complexity under dynamic loads [25].

In [26], studies of polyurethane heat-welded V-belts were conducted using the FEA in Abaqus program (Dassault Systèmes Simulia Corp.), which includes a whole set of models for calculating the stress-strain state of isotropic and anisotropic materials. Comprehensive research demonstrated the advantage of new belt materials in terms of improving durability and load-carrying capacity characteristics. Trends observed in the application of new belt materials and their manufacturing technology are discussed in [23]. The area of application of these belts and their influence on the design and geometry of V-belt drive pulleys is noted. It is worth noting that there is an increasing interest in exploring new structural and technological solutions for designing belt drives in metal-cutting machines.

2 Development of a 3D Design for a Horizontal Spindle Head with an Integrated Main Drive and an Innovative Belt Drive

2.1 Layout and Sketch Drafting of the SH Project

When machining a wide range of housing parts with widely variable loads and rotational speeds, universal and multi-purpose machines are used [2, 27]. A distinctive feature of the design under study, in terms of its connection to the machine's main drive, is the use of multi-spindle units. Motion in such a drive is transmitted to a series of parallel spindles, represented by interchangeable modular equipment (Fig. 1, a).

The use of spindle equipment, such as a vertical and slotting head, predetermines a number of layout decisions. The location of the horizontal-axis spindle unit is mounted closer to the base of the gearbox (Fig. 1, b), which lowers its center of gravity and makes the gearbox itself more stable. Focusing on the modular principle when equipping a universal machine tool with a set of interchangeable spindle equipment leads to an increase in the length of the pre-spindle shaft. Moreover, the first shafts, as less loaded, must be made shorter. The presented arrangement ensures a reduction in the vibration level (this is especially important for high-frequency methods of processing workpieces).

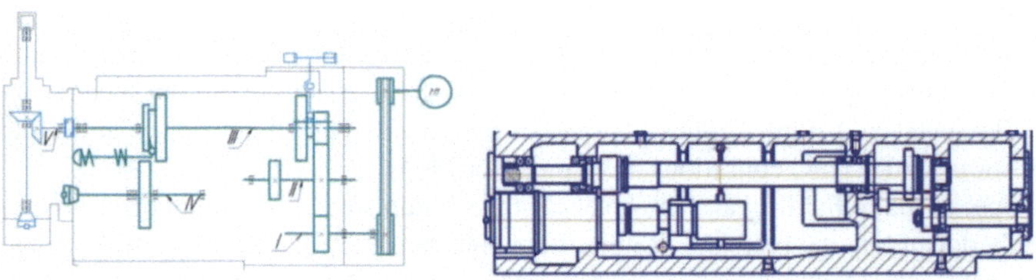

a b

Fig. 1. Layout drafting of the SH project: a) kinematics, b) longitudinal drawing trace

A preliminary design drafting of the SH construction with a horizontal spindle unit is shown in Fig. 2.

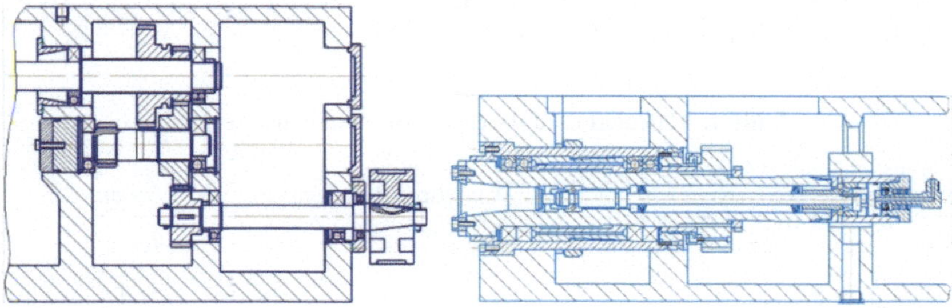

a b

Fig. 2. Preliminary design of the SH project: a) gearbox, b) spindle unit

The SH design (Fig. 2, a) consists of a cast-iron housing, a spindle block with an automatic tool clamping mechanism, and a camshaft transmitting rotation to the horizontal or vertical spindle via an automatic pinion shifting device. The SH layout includes a two-stage gearbox controlled by a hydraulic shifting mechanism [28, 29] and a coolant supply device to the cutting zone, as well as a number of other components and assemblies ensuring its proper operation. Rotation from the electric motor is transmitted via a belt drive to shaft I, then through the two-stage gearbox to shaft III, from which rotation is transmitted to the clutch of the vertical or slotting head, or to the horizontal spindle gear (Fig. 1, a).

For longer gearbox housings (Fig. 2, a), transverse walls between the main walls (which determine the drive dimensions) should be provided to increase rigidity.

The forming spindle assembly (Fig. 2, b) is a hollow shaft mounted on two duplexed angular contact ball bearings 4-46112 bearings mounted in tandem. The tool clamping system in the spindle includes a tool cone, a traction mechanism, and a tool unclamping mechanism. The output characteristics of the forming process largely depend on the strength and rigidity characteristics of the tool-spindle connection (in particular, the clamping force) [30].

2.2 Calculation of the Tool Clamping Force Mounted in a Horizontal Spindle

To fastening the tool in the horizontal spindle of a machine size II, a disc spring No. 111 (Standard 3057-79); corresponds to P/N 170079 DIN 2093, $D_e = 31.5$ mm) with the following dimensions is used: D (D_e) = 30 mm; $D_1 = 15$ mm – spring diameters; $S =$ 2 mm – spring thickness; $f_0 = 0.6$ mm – maximum spring deflection; $h_0 = 2.6$ mm – spring height; $h = 13.8$ mm – spring stroke. Assembly diagram – sequential; force corresponding to maximum spring deformation $P_3 = 7100$ N; force corresponding to tool clamping $P_1 = 2500$ N; force corresponding to collet release $P_2 = 5000$ N;

The initial values for determining spring sizes are forces P_1 and P_2, and the values of preliminary deformation f_1 and working deformation f_2, which are interpolated according to P_1 and P_2 in the intervals $f_1 = 0.4\,f_3 = 0.212$ mm and $f_2 = 0.8{\cdot}f_3 =$

0.424 mm, where the maximum spring deformation is $f_3 = 0.53$ mm. As is well known, the force-deformation characteristic is linear with respect to the ratio $f_3/s < 0.6$ ($f_3/s = 0.265 < 0.6$) and nonlinear $f_3/s > 0.6$.

The calculation of spring characteristics in these intervals for assessing the deformation level is presented in Table 1.

Table 1. Calculation of the deformation of a disc spring.

Parameter Name	Analytical Relationship	Value
Number of springs in a complete set	$n = \frac{h}{f_2 - f_1}$	65.0
Set height in free state	$H_{0k} = h \cdot n$	169.3 mm
Set deformation	$F_1 = f_1 \cdot n$	13.8 mm
	$F_2 = f_2 \cdot n$	27.6 mm
Set height	$H_1 = H_{0k} - F_1$	155.4 mm
	$H_2 = H_{0k} - F_2$	141.6 mm

A special mechanism, a clamping amplifier, is used during tool clamping. To determine the clamping force, consider the force distribution diagram (Fig. 3).

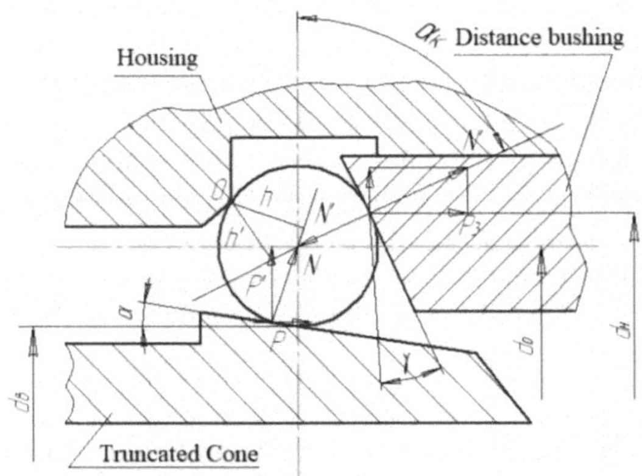

Fig. 3. Schematic diagram of the force distribution in the clamping mechanism

Initial data: component material: ШХ 15 steel; HRC 56...63; $\sigma_B = 200...240$, N/mm^2; contact angle $\alpha_K = 70°$; ball center diameter $d_0 = 30.5$ mm; P_1 –clamping force of disk spring (corresponding to the tool clamp ($P_1 = 2500$ N); truncated cone angle $\alpha = 9^0$; contact force coordinates: $h = 2.7$ mm; $h' = 2.4$ mm; support cone angle $\gamma = 20^0$.

The clamping force using the following algorithm is calculated.

1. Determine the disk spring force $P' = P/tg\alpha = P/tg9^0 = 6.3 \cdot P$. (where P – tangential component of the clamping force, Fig. 3)

2. Calculate the force N (the normal component of the clamping force, Fig. 3) acting on the ball $N = P' \cdot \cos 9^0 = 6.38 \cdot P$.
3. From the condition of the clamping mechanism ball equilibrium (Fig. 3), we find the resulting force acting on the ball N': $N \cdot h = N' \cdot h' \rightarrow N' = N \cdot h/h' = 6.38 \cdot P \cdot 2.7/2.4 = 7.18 \cdot P$.

   In this case, the force acting on one ball is numerically equal to: $(7.18/12) \cdot P = 1500$ N.
4. The clamping force, excluding friction losses, will be equal to: $P_3 = N' \cdot \cos \gamma = 7.18 \cdot P \cdot \cos 20^0 = 6.7 \cdot P$.

2.3 Creating a 3D Design of Horizontal Spindle Head with an Integrated Drive

For developing a set of models of a fairly complex MoM drive mechanism with a shape-generating spindle unit, it is rational to use the integrated CAD Creo Parametric [31]. This system is designed for use in accordance with the Industry 4.0 concept, which is based on advanced digital technologies such as the Internet of Things, artificial intelligence, and 3D printing [32]. The ability to combine parametric and direct modeling using a set of applications for designing various mechanical engineering objects expands the scope of rational use of this CAD system [33, 34].

Three-dimensional models of a horizontal SH in the Creo Parametric environment were constructed (Fig. 4).

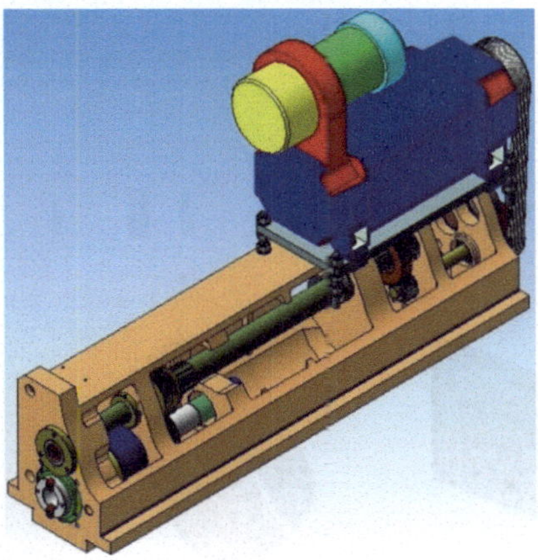

a

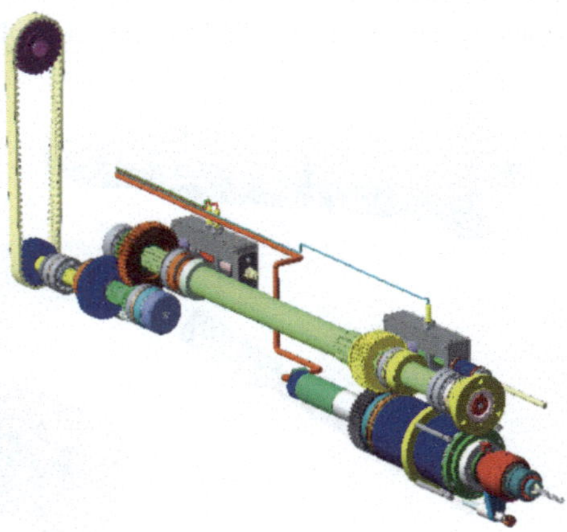

b

Fig. 4. Machine drive: a) assembly; b) kinematics

When working with an assembly in Creo, it's possible to load it in a lightweight format and create a window for designing individual assembly parts and the structure of the assembly units. For these parts, CAD data is loaded for designing mechanical transmissions and other drive components.

When creating 3D models of the parts, the drive housing stands out due to its complexity (Fig. 5).

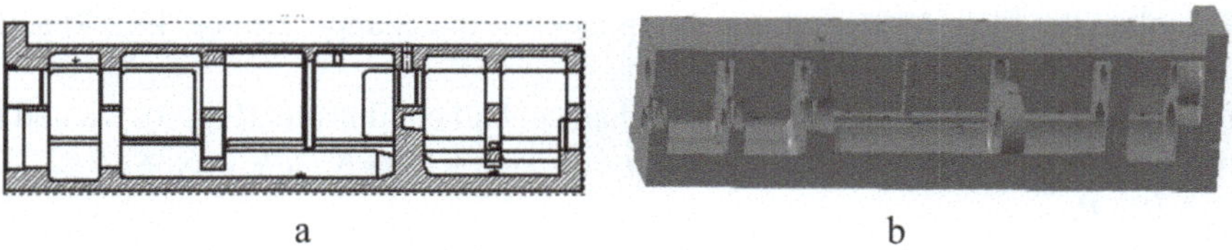

Fig. 5. Housing of drive main motion: a) sketch; b) 3D model (section)

During modeling, the housing as a collection of various geometric bodies, which are processed separately and used as design tools, is represented [31]. These include creating parts from initial bodies, splitting the body, or integrating it with other bodies. CAD Creo widely uses the technology of creating housing structural elements (Fig. 5, b) based on a sketch (Fig. 5, a). This requires creating an internal section with the sketch base structural element tool open. This internal section is independent and not associatively linked to any structural element.

Thus, the sketch structural element can be permanently located in the current model or in another model, providing an effective tooling for multivariate design of the drive housing (Fig. 5).

The main form-generating element – the horizontal spindle unit – is represented as a 3D model built in the Creo Parametric environment (Fig. 6, a; b). The tool clamping mechanism in the spindle cavity is mounted. This mechanism, using a hydraulic drive through a pack of disc springs, is carried (Fig. 6, c; d; e). To simulate the spring pack, we will use the Creo Ansys Simulation application [31]

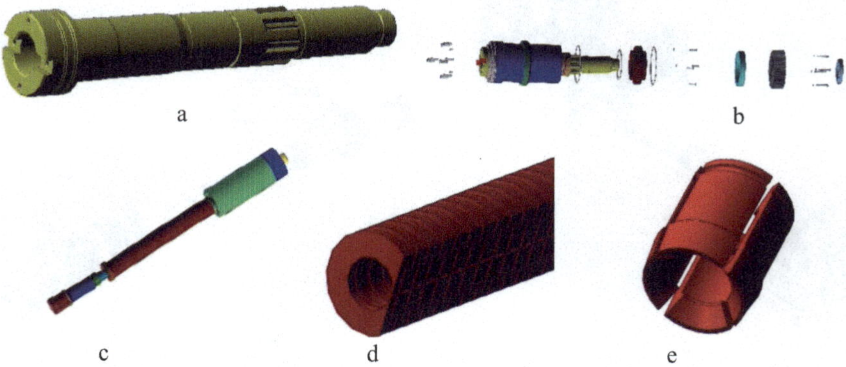

Fig. 6. Spindle assembly: a) spindle; b) exploded; c) clamping mechanism; d) spring; e) collet chuck

The availability of a wide range of cutting tools dictates the choice of the dialog box: Spring Type – option: Torsional (considering the process of twisting and rotation, a 3D spring). In Creo Ansys Simulation [31], the stiffness of the connection between two references in the model is simulated, with one end having a fixed constraint and the other attached to the body of the clamping mechanism.

Spring simulations are performed in the Spring Behavior dialog box according to the following algorithm: 1. Select the Spring Type and define the stiffness along or around the X, Y, and Z axes (initial data is presented in Sect. 3.2). Here, the possible options are to specify a value as a real number, a parameter, or an expression (the value for one or two axes can be zero). 2. Specifying units of measurement for longitudinal or torsional stiffness. 3. Creating the "behavior" of the spring, which is reflected in the modeling tree and is the parent node of the spring, as a basis for further study of its properties. Figure 6, d shows a 3D model of a disc spring for the spindle unit clamping mechanism.

To create a photorealistic rendering, the corresponding Creo resource: Render Studio [31] was used. This resource allows the 3D models developed above to be presented in a more realistic form (Fig. 7), which differentiates the key elements of the SH assembly. In Render Studio, it is possible to select the rendering mode, in particular, using the "auto" option, in which this procedure is performed in the GPU mode. This allows the customer to choose from a wide range of machine tools and evaluate their compliance with the technical specifications. Figure 8 shows a fragment of SH Explode in the Creo environment.

Fig. 7. Rendering of a spindle head with an integrated drive (fragmented)

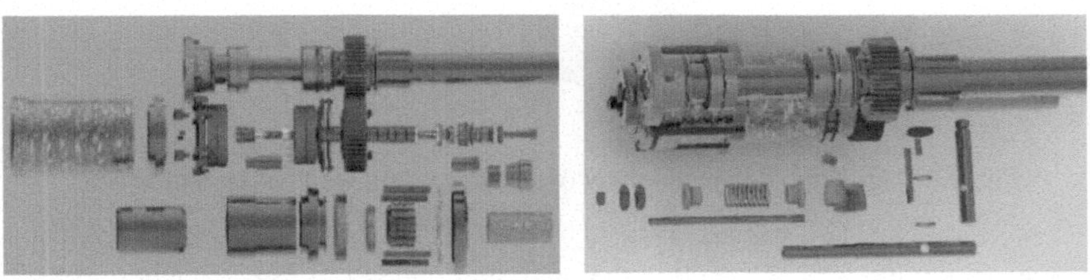

Fig. 8. Explode of SH design elements with integrated drive

3 Innovative Design Solutions for the Spindle Head Drive with Belt Transmission

Various design modifications of V-belt drives [22, 35] are aimed, to varying degrees, at solving one of the significant drawbacks of friction-based power belt transmissions – their short durability.

The conventional V-belt drive has the following disadvantages. Firstly, it has insufficient load-carrying capacity, making it impractical to use in the low-speed stage of a mechanical drive, where torques reach their maximum values. Secondly, the insufficient load-carrying capacity of the conventional V-belt drive results in its considerable dimensions, which limit the range of possible applications for these drives.

This paper aims to improve belt drives by modifying their design, thereby increasing their load-carrying capacity and, consequently, reducing their overall dimensions. To this end, a fundamentally new flexible element – a sectional belt is proposed, supported by a patent [36]. It combines the design features of chains and standard V-belts. The sectional structure of the belt, consisting of individual links, is similar to that of chains. The method of transmitting force from the pulley to the flexible elements of the belt and vice versa is through frictional forces between them, created by pre-tensioning the belt, as in belt drives.

Below are some results from the research of the sectional V-belt drive, covering the main belt parameters and a preliminary assessment of its durability with increased load-carrying capacity. These results can be used for further theoretical and experimental research of the proposed flexible element transmission.

3.1 Design Parameters of a Sectional V-Belt

The innovative design of a V-belt drive with pulleys utilizes the author's patented solution, which combines metallic and non-metallic elements. Figure 9 shows a fragment of a 3D general view of a transmission with a sectional V-belt.

Fig. 9. Fragment of a general view of a transmission with a sectional V-belt

The main parameters of a sectional V-belt are depicted in Fig. 10, where position 1 is the elastic element of the section, and position 2 is the steel plate located within the elastic element.

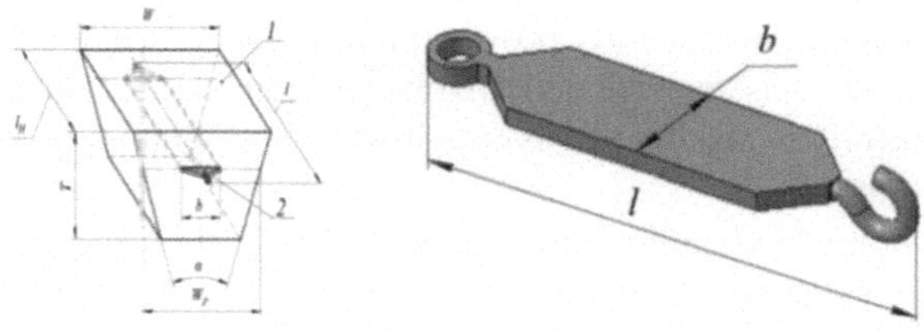

a b

Fig. 10. Main parameters of a belt section: a) dimensional characteristics; b) configuration of the steel rod of the belt section

The non-metallic belt elements are individual V-belt components whose transverse dimensions, shape, and material structure are identical to those of a conventional V-belt drive (Deutsche Normen DIN 2215). The metallic elements of the V-belt drive are joint-connected circular steel rods, which are connected and located within the non-metallic belt components. The steel rods are positioned at the center of gravity of the non-metallic belt components' cross-section.

Their function is to provide the necessary friction force with the pulleys by pre-tensioning the belt. The elastic elements of the sections are connected into a closed, flexible transmission link. This connection is realized by steel plates located within them, width $b \approx W_p/4$, with hooks at the ends, Fig. 10. The length of the elastic element of the sectional belt $l_s = T$, shown in Fig. 10, and the total length of the section (including the plate hooks): $l = 1.2 \cdot l_s = 1.2 \cdot T$.

For sectional [36] and standard [35] V-belts, the arc of contact on the smaller pulley is assumed to have the same minimum angle $\alpha_{\min} = 120^0$. However, the arcs at which the smaller pulley contacts the standard and sectional belts will be different:

$S = 0.5 \cdot d_{p\min} \cdot \alpha_{\min} \approx 1.05 \cdot d_{p\min}$ – for a standard belt; $S^* = z \cdot l_h \approx 0.9 \cdot d_{p\min}$ – for a sectional belt [33], where $d_{p\min}$ – calculated diameter of the smaller pulley; l_h – length of the belt non-metallic parts; $z = S/l \approx 0.9 \cdot d_{p\min}/T$ – number of belt sections on the contact arc of the smaller pulley.

Naturally, the pulley contacts the discontinuous sectional belt along a smaller arc of shorter length than with a standard V-belt: $\Delta S = S^*/S \approx 0.8$, (for the same $\alpha_{\min} = 120^0$), that is S^* approximately 20% less than S. This design change, associated with a reduction in transmission dimensions, necessitates an increase in pre-tensioning force. This leads to a search for a compromise solution to ensure reliable torque transmission without slippage and reduced loads on belt drive components.

3.2 Determining the Durability of a Sectional V-Belt

It is known that the durability L_h of a standard V-belt drive is a nonlinear function of the maximum stress $\sigma_{\max}$ [35]: $L_h = c/\sigma_{\max}^n$, where $c = \text{const}$.

The maximum stress includes four components:

$$\sigma_{\max} = \sigma_0 + \sigma_\nu + \sigma_t/2 + \sigma_b. \quad (1)$$

Currently, there are no sufficiently proven durability test results for standard V-belt drives, so the exponent n has not been definitively determined, other than that $n > 1$. This allows to do the a priori conclusion that even a 20–30% σ_{max} reduction will significantly increase belt durability.

In standard V-belts, the bending stress σ_b is $\approx$ 70–75% of σ_{max}. The fundamental difference between a standard V-belt and a sectional belt is that the latter has $\sigma_b \approx 0$. This is explained by the fact that elastic section belt elements are very short, resulting in virtually no bending deformation on the curved sections of the pulleys. Therefore, it can be considered.

The component σ_v included σ_{max} can also be neglected at this stage, since the stress from centrifugal forces σ_v in general-purpose belt drives is typically less than 10% of σ_{max}.

As for the useful stress σ_t, it is the same in this case for standard and sectional belts, since their durability must be compared for the same initial loads, gear sizes, and pulley speeds.

Thus, for sectional belts, the maximum stress: $\sigma^*_{max} = \sigma_0 + \sigma_t/2$.

The relationship σ_{max} and σ^*_{max} taking into account: 1) a 70…75% reduction due to $\sigma_b \approx 0$; 2) a reduction of approximately 10% due to $\sigma_v \approx 0$. As a result:

$$\sigma^*_{max} \approx (0.15...0.2)\sigma_{max} \tag{2}$$

However, when switching from a standard V-belt to a sectional V-belt, the arc of contact between the sectional V-belt and the pulleys decreases, which will have a negative effect – a decrease in friction forces and, consequently, a drop in the transmission's load-carrying capacity. This consequence can be avoided by increasing the pre-tension force of the sectional V-belt, that is, by using a higher pre-tension value for the standard V-belt σ_0. For standard V-belts, $\sigma_0 = 1.2...1.5$ MPa [35]. Since the arc of contact between a sectional V-belt and a smaller pulley is approximately 20% shorter than with a standard V-belt, for the sectional V-belt, we increase the value σ_0 by 20%, that is, to $\sigma^*_0 \approx 1.5..1.8$ MPa.

Increasing σ_0 to σ^*_0 will not reduce the tensile strength of the elastic element of the sectional V-belt, since this stress is supported by its steel plates. Taking this increase σ^*_0 into account, the final value σ^*_{max} in formula (2) changes to:

$$\sigma^*_{max} \approx (0.3...0.4)\sigma_{max} \tag{3}$$

The increase in the durability L^*_h of a sectional V-belt compared to the durability L_h will be:

$$\Delta L_h = L^*_h/L_h \approx \left(\sigma_{max}/\sigma^*_{max}\right)^n \approx (2.5...3)^n. \tag{4}$$

Based on the results of durability tests of sectional V-belts and the determination of the numerical value of the exponent n, relationship (4) can be used for a preliminary assessment of the durability increase of sectional V-belts.

4 Conclusions

A horizontal spindle head design for a multi-operation milling machine with an integrated gear drive equipped with an innovative belt drive has been developed. A kinematic layout diagram and a preliminary design for the machine's main drive with interchangeable modular spindle equipment have been developed. The proposed design features a basic spindle assembly located closer to the base of the housing, short shafts in the first stage, and an elongated pre-spindle shaft. This spindle head design reduces vibration and provides stable operation of machine tools.

A method and algorithm for calculating the tool clamping force (via a clamping amplifier mechanism) in a horizontal spindle with a 7:24 taper have been implemented. The geometric parameters of the disc spring and the deformation of the spring set have been determined. The diagram of the force distribution in the clamping mechanism, ensuring stable operation of the tool block during shaping of workpieces has been constructed.

A 3D design of a horizontal spindle head with an integrated drive was created using the CAD environment Creo Parametric. For complex SH housing components, Creo utilized its powerful features: the complex design to be represented as a set of various geometric bodies. An original technology for generating housing structural elements based on sketches is realized. A photorealistic representation of the SH design with the "Auto" option in graphics processor mode was created in Creo Render Studio.

A comprehensive set of works to create an innovative design for a sectional belt transmission for the main drive was completed. The choice of an innovative sectional belt drive design as the preferred alternative is based on resolving a tradeoff. This tradeoff lies in increasing the durability of the drive belt drive, on the one hand, and reducing the service life of the drive components as a whole (bearings, shafts, gears), on the other. Analytical studies have demonstrated a significant reduction in maximum stress by 70–75% by minimizing one of the main components—belt bending stress. This factor underlies the achievable (according to the calculations performed in the article) increase in the durability of the innovative belt drive by at least 2.5–3 times. On the other hand, the need to increase the pretension force of the sectional belt during operation leads to increased loads on the bearings, belt drive shafts, and the belt itself. Selecting the optimal pretension force is the key to resolving this tradeoff. This is the subject of future research. The performance improvement, through the durability parameter, is based on an analysis of the nature of the contact between the innovative sectional belt design and the drive pulley. This has been analytically proven and supported by a patent. A reduction in the maximum stress level supported by the belt of at least 2.5 times significantly outweighs the reduction in load-bearing capacity caused by a reduction in contact length of approximately 20%.

Acknowledgements. The authors would like to thank the Research and Development Sector at the Technical University of Sofia for the financial support

References

1. Brecher, C., Weck, M.: Machine Tools Production Systems 3. Springer, Wiesbaden (2022)
2. Lynch, M.: Machining Center. Setup and Operation. CNC Concept, Inc, Illinois (2013)
3. Gomon, V.: Prospects of upgrade of the main drives of CNC machine tools using mechatronic modules. In: III Annual International Conference System Engineering, KnE, pp. 36–42 (2020)

4. Huang, Y.M., Cheng, X., Gao, F., Tong, J.X.: The 3-PRS parallel spindle module with multi-posture vertical and horizontal machining capability. Int. J. Prod. Res. **43**(3), 465–470 (2005). https://doi.org/10.1080/00207540420002689925
5. Mikhov, M., Zhilevski, M.: Study and performance improvement of the drive Systems for a Class of machine tools. In: Proceedings of the 14th International Conference on Modern Technologies in Manufacturing (MTeM 2019), vol. 299, pp. 1–6 (2019). https://doi.org/10.1051/matecconf/201929905003
6. Zhilevski, M., Mikhov, M.: Analysis of mechanical operations in a type of machining centers. In: Proceedings of the 2022 International Congress on Human-Computer Interaction, Optimization and Robotic Applications (HORA), pp. 1240–1244. IEEE Xplore, Ankara (2022). https://doi.org/10.1109/HORA55278.2022.9800103
7. Rodriguez, J., Keribar, R., Wang, J.: A comprehensive and efficient model of belt-drive systems. SAE Technical Paper 2010-01-1058 (2010). https://doi.org/10.4271/2010-01-1058
8. Sokolov, V.: Hydrodynamics of flow in a flat slot with boundary change of viscosity. In: Proceedings of the 6th International Conference on Industrial Engineering (ICIE 2020), LNME, vol. 2, pp. 1172–1181. Springer, Cham (2021). https://doi.org/10.1007/978-3-030-54817-9_136
9. Rogovyi, A., et al.: Using modern mechanical design methods for determining the main characteristics of a cryogenic centrifugal pump. Int. J. Mechatr. Appl. Mech. **13**, 198–208 (2023)
10. Modular tool clamping systems for modern spindle technology: System component specifications. Berg & Co. GmbH Spanntechnik. Bielefeld, Germany (2022)
11. Hamrol, A., Zerbst, S., Bozek, M., Grabowska, M., Weber, M.: Analysis of the conditions for effective use of numerically controlled machine tools. In: Hamrol, A., Ciszak, O., Legutko, S., Jurczyk, M. (eds.) Advances in Manufacturing 2018, LNME, pp. 3–12. Springer, Cham (2018). https://doi.org/10.1007/978-3-319-68619-6_1
12. GMN USA. Precision spindles &spindle repair services. Bristol, USA (2021)
13. Setko Holden Industries.: Top Manufacturer of Spindles, Slides, Milling Heads, and Special Products Global Way. Hebron, KY, USA (2021)
14. Rastvorova, I., Klyucherev, N.: Design and modelling of a universal CN machine. J. Phys. Conf. Ser. **1753**, 012040 (2021). https://doi.org/10.1088/1742-6596/1753/1/012040
15. Patel, N., Gupta, T., Wankhede, A., Wakudkar, V.: Design and optimization of 2-stage reduction gearbox. Int. J. Eng. Dev. Res. **5**(2), 541–552 (2017)
16. Kipcharsky, V.P.: Metalworking machining tools in the beginning. DVNZ PHTU, Mariupol (2018)
17. Buiga, O., Tudose, L.: Optimal design mass minimization design of two stage coaxial helical speed reducer with genetic algorithm. Adv. Eng. Software. **68**, 25–32 (2014). https://doi.org/10.1016/j.advengsoft.2013.11.002
18. Ivanov, K.Y., Galibey, N.I.: Gear two-stage gearboxes. The choice of the optimal scheme. J Bull. Siberian State University Named After M.F. Reshetnov. **1**(27), 45–49 (2010)
19. Hong, T., et al.: Cost optimization study of two-stage bevel helical gearboxes. Technol. Report Kansai University. **62**(07), 3343–3352 (2020)
20. Özşahin, O., Özgüven, H.N., Budak, E.: Estimation of dynamic contact parameters for machine tool spindle-holder-tool assemblies using artificial neural networks. In: Proceedings of the 3rd International Conference on Manufacturing Engineering (ICMEN), pp. 131–134 (2008)
21. Denkena, B., Bergmann, B., Mucke, M., Konigsberg, J., Ponick, B.: Hybrid spindle – an approach for a milling machine tool spindle with extended working range for HSC and HPC. Procedia Manuf. **24**, 159–165 (2018). https://doi.org/10.1016/j.promfg.2018.06.045
22. Krawiec, P., et al.: Testing functional features of V-belt transmissions. IOP Conf. Ser.: Mater. Sci. Eng. **776**, 012008 (2020). https://doi.org/10.1088/1757-899X/776/1/012008

23. Krawiec, P., Warguła, L., Domek, G., Matijošius, J.: Development trends in belt transmissions with V-belt. MATEC Web Conf. **357**, 01003 (2022). https://doi.org/10.1051/matecconf/202235701003
24. Shevchenko, S., Mukhovaty, A., Krol, O.: Research of the modified tooth belt drive for the machining center. In: Radionov, A.A., Gasiyarov, V.R. (eds.) Proceedings of the 7th International Conference on Industrial Engineering (ICIE2021). ICIE2021, LNME, pp. 32–41. Springer, Cham (2022). https://doi.org/10.1007/978-3-030-85233-7_4
25. Wisznovszky, E., Gerocs, A., Culda, L., Komjaty, A., Erdodi, G., Titu, A.: Chain transmission mechanism used to reduce the reaction moment for machinery that performs rotational movement around its own. Acta Technica Napocensis Ser. Appl. Math. Mech. Eng. **67**(II), 211–220 (2024)
26. Gudayu, A., Steuernagel, L., Meiners, D.: The capability of ABAQUS/CAE to predict the tensile properties of sisal fiber reinforced polyethylene terephthalate composites. Compos. Adv. Mater. **7**, 1–13 (2022). https://doi.org/10.1177/26349833221137602
27. Yakovenko, I., Permyakov, A., Dobrotvorskiy, S., Basova, Y., Kotliar, A., Zinchenko, A.: Prospects for the development of process equipment in aggregate-modular design for sustainable mechanical engineering. Int. J. Mechatr. Appl. Mech. **13**, 145–156 (2023). https://doi.org/10.17683/ijomam/issue13.18
28. Shevchenko, S., Mukhovaty, A., Krol, O.: Modification of two-stage coaxial gearbox. In: Radionov, A.A., Gasiyarov, V.R. (eds.) Proceedings of the 6th International Conference on Industrial Engineering (ICIE 2020), pp. 28–35. Springer, Cham (2021). https://doi.org/10.1007/978-3-030-54814-8_4
29. Stoev, L., Bozhikov, S.: 3D-layouts of machine tools for multipurpose machining of rotary components. Tehnicki vjesnik – Technical Gazette. **23**(2), 321–328 (2016)
30. Mahmud, A., Mayer, J.R., Baron, L.: Determining the minimum clamping force by cutting force simulation in aerospace fuselage pocket machining. Int. J. Adv. Manuf. Technol. **80**, 1751–1758 (2015)
31. Wyndorps, P.: 3D-Konstruktion mit Creo Parametric und Windchill: PTC Creo 4.0 und PTC Windchill, Europa-Lehrmittel (in German) (2022)
32. Ivanov, V., Evtuchov, A., Dehtiarov, I., Trojanowska, J.: Fundamentals of Manufacturing Engineering Using Digital Visualization. Springer Tracts in Mechanical Engineering
33. Krol, O.: Modeling of worm gear design with non-clearance engagement. In: Radionov, A.A., Gasiyarov, V.R. (eds.) Proceedings of the 6th International Conference on Industrial Engineering (ICIE 2020), LNME, pp. 36–46. Springer, Cham (2021). https://doi.org/10.1007/978-3-030-54814-8_5
34. Bondarenko, I., Lunys, O., Neduzha, L., Kersys, R.: Dynamic track irregularities modeling when studying rolling stock dynamics. In: Proc. of 23rd Intern. Scientific Conf. "Transport Means. 2019", pp. 1014–1019 (2019)
35. Juvinall, R.C., Marshek, K.M.: Fundamentals of Machine Component Design. Willey, Australia and New Zeeland (2020)
36. Shevchenko, S., Mukhovaty, A., Krol, O.: Belt transmission. UA Patent 142528. Application u201912255, 26 December 2019 (2019)

Applying Reverse Engineering Techniques for the Recreation of a Tractor Cabin Protection Grille

Branislav Katrinič(✉), Monika Töröková, Marek Kočiško, Jozef Török, and Martin Koroľ

Faculty of Manufacturing Technologies of Technical University of Košice, Bayerova 1, 080 01 Prešov, Slovakia
branislav.katrinic@tuke.sk

Abstract. The presented article offers a comprehensive example of the application of reverse engineering methodologies, specifically employing 3D scanning technology with use of 3D laser scanner in conjunction with advanced CAD software, to support the development and optimization of the protective cabin grid of the forestry machine John Deere 1270E. The primary objective of the process was the reconstruction of the existing structural design in order to generate a digital backup CAD model. The secondary objective was the proposal of design modifications aimed at enhancing the operator's visibility conditions. A significant methodological innovation consisted in the design and utilization of specialized 3D printed reference bodies - regular dodecahedrons with integrated magnets which effectively addressed challenges associated with precise scanner positioning on ribbed and dark surfaces. Based on operator expertise, the original design was refined by reducing the number of vertical ribs from 16 to 13 and altering their inclination. These modifications resulted in a substantial improvement of the operator's field of view while maintaining the required protective function. Ergonomic validation was conducted within the Assembly module using a virtual mannequin. To maximize manufacturing efficiency, the vertical rib geometry was standardized, technological locks were integrated, and precise laser marking was implemented. High-strength steel grade S355K2 was selected as the construction material, with laser cutting adopted as the primary fabrication method.

Keywords: 3D Scanning · Reverse Engineering · Creo Parametric · Protection Grille

1 Introduction

This research details the application of reverse engineering techniques, specifically leveraging 3D scanning technology and advanced CAD software, for the development and optimization of a tractor cabin protection grille. This approach integrates the precise geometric data acquired from a ZScanner 700 with the robust modelling capabilities of Creo Parametric to facilitate an iterative design process. This methodology enhances the

B. Gapiński et al. (Eds.): MANUFACTURING 2026, LNME, pp. 186–198, 2026.
https://doi.org/10.1007/978-3-032-22943-4_14

ability to create precise digital models from existing physical components, which is crucial for improving product functionality and safety. Reverse engineering is invaluable across diverse engineering disciplines, facilitating the conversion of physical components into digital representations for subsequent computer aided applications [1]. This process enables engineers to analyse, validate, and fabricate designs with enhanced accuracy and efficiency [2]. This is particularly beneficial for designing protective structures like grilles, where precise fit and integration with complex existing geometries are paramount. The creation of a 3D parametric solid model, which closely mirrors the original object, from raw 3D scanner data is a primary objective of reverse engineering in engineering applications.

This digital reconstruction, often originating from point cloud data, then serves as a foundational dataset for subsequent design iterations and performance analyses [3]. Such detailed models are essential for complex systems, where custom designs and rapid manufacturing are critical for meeting evolving market demands and individual customer needs [4]. Furthermore, the integration of CAD/CAM/CAE tools within this reverse engineering framework enables comprehensive material characterization and mechanical analysis, ensuring the manufactured prototype meets stringent performance requirements [5]. This strategic application of reverse engineering for tractor cabin grilles also offers a significant competitive advantage by enabling the understanding of rival designs and market trends [6]. Moreover, reverse engineering proves instrumental in reconstructing damaged components, thereby reducing production time and costs associated with acquiring new parts or designs.

This method is particularly valuable for re-engineering parts that have lost their original design data, effectively bridging the gap between physical components and their digital counterparts [5]. This approach also significantly accelerates data acquisition for CAD, CAM, and CAE systems, thereby reducing the overall time required for component development, design, and production [7]. The utilization of 3D scanning technologies, such as the ZScanner 700, provides highly accurate point cloud data, which is then processed through specialized software like VXelements to create detailed surface models suitable for direct integration into parametric CAD environments. This digital workflow not only streamlines the conversion of physical objects into editable CAD models but also facilitates detailed dimensional analysis and the development of intricate geometries directly from scan data.

2 Reverse Engineering Applications in the Automotive Industry

This methodology is especially valuable when original design documentation is absent, for intricate parts requiring high precision, or for parts subject to significant wear and tear that necessitate replacement or re-design. This process has found extensive application in industries ranging from automotive to aerospace, where the rapid development of prototypes and customized components is critical for market competitiveness and functional superiority. For instance, reverse engineering combined with additive manufacturing allows for the precise or enhanced replication of complex components, minimizing material usage, energy consumption, and overall time [5, 8]. This approach is particularly effective for components where traditional design methods are inefficient or impractical, offering a streamlined path from physical object to digital model and subsequent fabrication [9]. Moreover, reverse engineering, often coupled with 3D scanning, is a primary method for acquiring the intricate 3D digital geometry of existing physical objects for revision or improvement through digital reconstruction. This process plays a crucial role in modern manufacturing, enabling the creation of editable CAD models from physical objects, which can then be utilized for product development and design modifications [10]. The adoption of such techniques is increasingly prevalent for designing, where personalized fit and intricate geometries necessitate advanced scanning and CAD systems like PTC Creo for modelling and data processing for future production [11]. This integration of advanced manufacturing technologies with reverse engineering allows mass customization, in fields requiring specific solutions [12]. Sheet metal cutting operations, for example, can be optimized through reverse-engineered designs, ensuring precise cuts and minimizing material waste.

2.1 Process of Digitalization

Within the framework of reverse engineering techniques, a reconstruction of the protective structure for the John Deere 1270E forestry machine was carried out. The primary objective of this measurement was the creation of a digital backup CAD model to serve as a safeguard in the event of potential future damage to the existing assembly. The secondary objectives included the proposal of design modifications aimed at improving the operator's visibility conditions during machine operation.

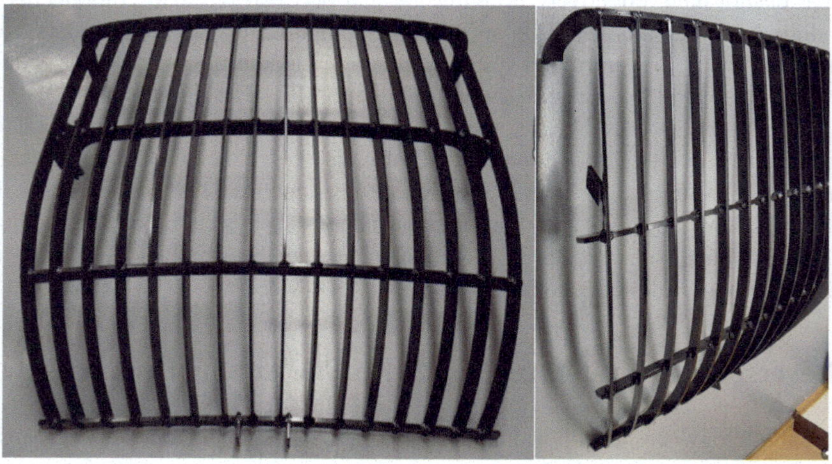

Fig. 1. The original configuration of the protective cabin grid.

As seen in Fig. 1. this assembly is made up of a system of sheet metal sheets that form a grid thanks to their perpendicular installation on the basic or supporting elements. The mechanical connection of all parts is ensured by welds.

The dimensions were measured with high precision using the ZScanner 700 3D scanner. This industrial-grade scanner was selected due to its mobility and sufficient accuracy for the realization of this type of project. Its fundamental parameters are presented in Table 1.

Table 1. ZScanner 700 parameters [13].

Measurement speed	18 000 points/s
Resolution in XY	50 μm
Accuracy in the Z axis	0.1 m
Volumetric accuracy according to ISO	20 μm + 0.2 L/ 1000

The 3D scanning process required the matting of the entire surface of the protective structure due to the scanner's sensitivity to black and glossy surfaces. For spatial orientation and precise point mapping, reference markers were placed on the surface. A key advantage was that the VXelements software can identify these markers and algorithmically reconstructing the surface with geometry corresponding to the surrounding area. This approach prevents distortion and significantly streamlines post-processing, particularly the cleaning of the resulting mesh (Fig. 2).

Fig. 2. Preparing the object for the digitization process.

During the digitization process, it was determined that, in order to ensure the integrity and accuracy of the 3D scanning procedure, the standard planar reference points (adhesive markers) had to be supplemented with specialized spatial reference bodies. This requirement arose due to the limited visibility of the measured pattern when viewed perpendicularly (axially) to the protective grid (ribbed structure).

As a solution, a regular dodecahedron was proposed. This geometry was selected for the following reasons:

- It provided sufficient surface area for secure placement of reference markers, ensuring their visibility from multiple angles.

- Its dimensions allowed positioning within the narrow spaces between adjacent ribs of the grid.
- The installation of multiple reference markers on a single body facilitated the spatial orientation of the scanner (Fig. 3).

a) b)

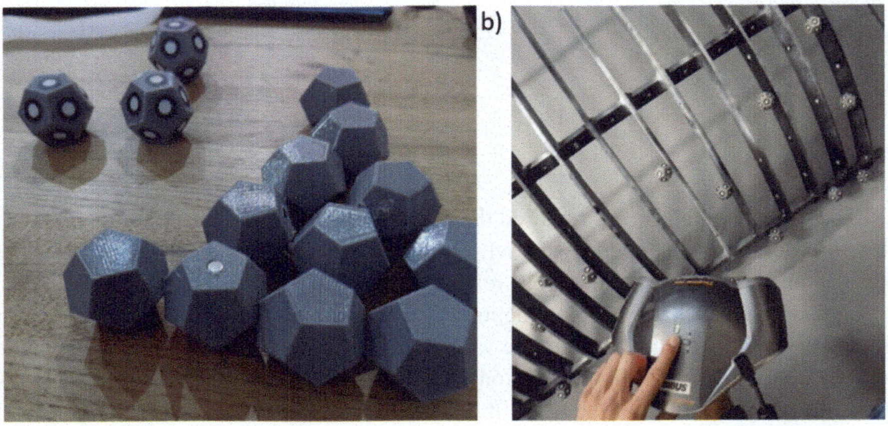

Fig. 3. References bodies: a) Production of reference bodies b) Object ready for the digitization process.

In the CAD design, aimed at enhancing ergonomics and reusability, an opening for embedding a magnet was integrated into one of the twelve faces of the dodecahedron. This system enables rapid and secure attachment as well as repeated reuse of the reference body during successive measurement campaigns. The fabrication of the reference body was carried out using an FFF (Fused Filament Fabrication) 3D printer.

The first step in the VXelements software Version 7 was the creation of a positional map of the reference markers. This approach is particularly suitable when scanning only unrelated or isolated parts of the object rather than the entire surface. Precisely defined points enable the software to reliably position and merge these individually captured segments into a single coherent model. An additional advantage of this method is that by scanning in smaller segments, the risk of losing the entire data set in the event of an error or measurement failure is minimized.

2.2 Data Processing

The subsequent step involves data processing, which includes the removal of excessive noise and small, isolated artifacts from the partial scans. The data are stored in the native format of VXelements, specifically in the CFS file format. If reference objects negatively affect the surface reconstruction, they must be manually identified and eliminated. An example of unwanted data marking is presented in Fig. 4.

In the data merging phase, following the successful cleaning of individual scans, the software submodule Merge Scans offers two methods. The models can be combined either through automatic recognition of overlapping geometric features or by means of

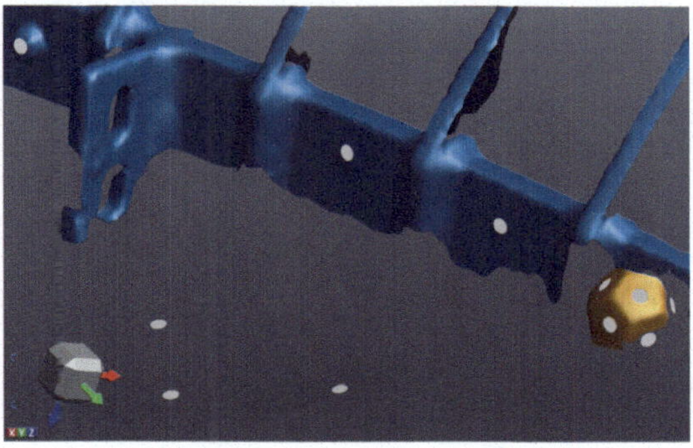

Fig. 4. Scan cleaning process.

a manual procedure, in which the user selects corresponding common points on the polygonal models (Fig. 5).

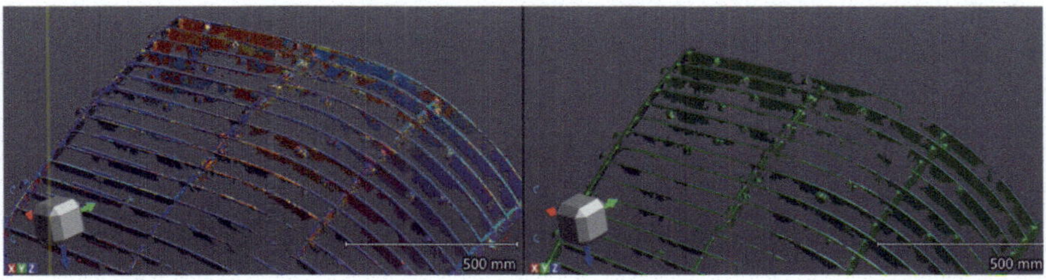

Fig. 5. The process of merging partial scans.

For the continuation of the work, the finalized scan was unified and spatially oriented (rotated) into the position most suitable for modeling within the Creo Parametric environment. Due to the sheet-metal nature of the structure, it was not necessary to generate a fully enclosed model; it was sufficient to ensure the precise spatial placement of surfaces and the rib angles using partial data. Prior to the export of the point cloud (in STL format), reference planes were created on the surface of each rib. These planes were subsequently saved in IGES format for later use within the CAD software.

3 CAD Model Creation Process

Into the Creo Parametric environment, the data were imported using the Get Data/ Insert function, with the polygonal model and the reference planes (IGES format) loaded separately. This approach enables more effective management of the visibility of individual elements according to current requirements. A particular advantage lies in importing each reference plane independently, which simplifies the transparent control of these surfaces throughout the entire design process. The unified data set is presented in the following Fig. 6.

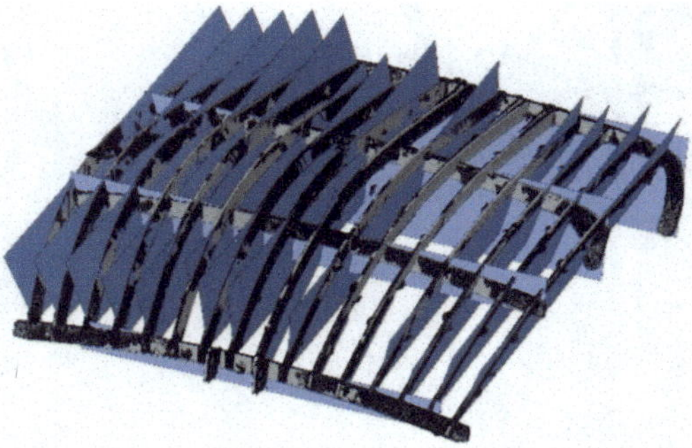

Fig. 6. Imported data in the Creo Parametric software environment.

3.1 Creating a Shape Copy

The process of designing the new structure was guided by the need to ensure its optimal functionality and safety, as well as maximum manufacturing efficiency. To simplify and streamline future production, a universal geometry of the vertical rib was specifically developed, enabling its application across the entire assembly (Fig. 7).

Fig. 7. Design of the initial shape of the vertical rib.

The division of the appropriate sheet material is planned using laser cutting technology. To facilitate the assembly of the structure and the creation of permanent joints through welding, special design gaps and technological interlocking features were integrated into the design to ensure precise alignment of the components. These interlocking features include cut-outs in right-angled corners, which eliminate the issue of "incomplete corner cutting" during laser processing and enable the accurate fitting of parts (Fig. 8).

Fig. 8. Design of technological locks.

3.2 Adaptation of the CAD Design

During the design of the final model, several improvements based on the operator's practical experience were implemented. To enhance visibility, the original design was modified by reducing the number and altering the inclination of the vertical protective slats, thereby effectively improving the field of view from the cabin. A key validation of the correct structural arrangement was carried out in the Assembly module of Creo Parametric. By integrating the CAD model, the precise dimensions of the machine, and a digital mannequin, it was possible to conduct a detailed examination of the virtual operator's field of vision. An example of the assembly, shown from the side view, is presented in the following Fig. 9.

Fig. 9. Position of the dummy in the cabin.

The proposed solution reduces the total number of vertical ribs from the original sixteen to thirteen. This modified grid continues to provide the required level of protection while simultaneously delivering a significant improvement in the operator's field of vision (Fig. 10).

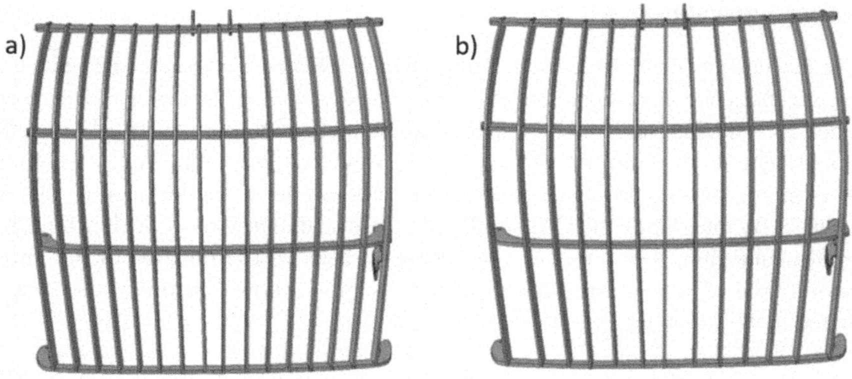

Fig. 10. CAD design of the protective grille: a) original layout b) new layout.

4 Suggestions for Production

By combining reverse engineering with advanced design improvements and precise manufacturing technologies (laser cutting with interlocking features and engraving), a technological procedure was developed that ensures high-quality production, dimensional accuracy, and functional reliability of the safety frame.

For material provision of the critical welded safety structure, high-strength steel S355K2 was selected. This choice is essential, as it guarantees a higher minimum yield strength compared to S235JR and provides assured toughness even at sub-zero temperatures. Such properties are indispensable for year-round operation of the machine in demanding forestry environments (Table 2).

Table 2. Materials comparison table [14]

Property	Classic (S235J)	Selected (S355K2)	Extremely High-Strength (STRENX 960)
Minimum yield strength (fy min.)	235 MPa	355 MPa	960 MPa
Tensile strength (fu max.)	510 MPa	630 MPa	1150–1300 MPa (typical)
Relative strength	Basic	51% higher	308% higher (compared to S355K2)
Toughness (min. impact energy)	Usually test at + 20 °C	Guaranteed at −20 °C	Guaranteed at −40 °C or − 60 °C
Main reason for selection	Low cost, easy processing	Higher load capacity and robustness in cold	Maximization of load capacity and minimization of weight

(continued)

Table 2. (*continued*)

Property	Classic (S235J)	Selected (S355K2)	Extremely High-Strength (STRENX 960)
Typical applications	Non-load-bearing structures, simple frames	Bridges, heavy machinery, dynamically loaded structures	Mobile cranes with long booms, construction machinery, frames for transporting heavy loads
Welding requirements	Very low	Higher caution, possible preheating	Strict technological procedures, high preheating and control of heat input

To produce the safety frame, laser cutting technology was selected, as it is crucial for achieving the required precision and for integrating technological features. The following table compares the key parameters of laser cutting, plasma cutting, and water jet cutting (Table 3).

Table 3. Material separation technology comparison table [15]

Parameter	Laser Cutting	Plasma Cutting	Waterjet Cutting
Primary Use	Thin to medium sheets (up to ≈ 25 mm)	Thick sheets, structures, alloyed steels	Heat-sensitive materials (stainless steel, aluminum, composites)
Accuracy (Tolerance)	Excellent (±0.05 to ±0.1 mm)	Medium (±0.2 to ±0.5 mm)	High (±0.1 mm), precise details
Cutting Speed	Highest (for thin/medium thicknesses)	High (especially for thick materials)	Slow
Heat-Affected Zone (HAZ)	Small (minimal risk of deformation)	Largest (high risk of deformation)	None (so-called "cold cut")
Edge Quality	Excellent (smooth, minimal burr)	Lower (rough edge, burrs, dross)	Excellent (clean, perpendicular)
Marking/Engraving Capability	YES (Laser function, key for positioning locks)	NO (Only perforation/punching)	NO (Only partial, shallow cut)
Operating Costs (Cost per meter of cut)	Low to Medium (high speed compensates costs)	Low (low investment costs, simple maintenance)	High (high consumption of abrasive – garnet sand – and energy)
Need for Additional Processing	Minimal (usually not required)	High (removal of slag and smoothing)	Minimal (usually not required)

The cutting of components is carried out on the TRUMPF LASER L3060 device, based on the prepared CAD model data. In areas where mechanical interlocks could not be applied, the laser engraving function was used to mark precise welding positions. These engraved markers facilitate the welder's work while not compromising the structural integrity. For the welding process itself, the CO_2/MAG welding method (process 135) in a protective atmosphere was selected. To ensure geometric and dimensional accuracy, the parts are positioned on a welding table, and the welding is performed in stages (tack welding followed by full welding), with a specific sequence designed to minimize residual stresses.

4.1 Quality Control and Finalization

Following the finalization of the structure through welding, the next step involves cleaning and shot blasting (e.g., with steel beads) to remove impurities and create an active surface prior to painting. To eliminate hazardous sun reflections that could potentially dazzle the operator, the application of powder coating RAL 9005 (black) with a fine texture was considered. The textured surface diffuses light, thereby reducing the risk of glare-related accidents. Subsequently, a quality inspection is performed before the application of the powder coating. This inspection monitors the geometry of the frame, the integrity of the welds, and the adherence to all laser-marked positions and interlocking features. Only after successful quality control is the coating applied.

The coating process should consist of two layers of RAL 9005 powder paint: a base anti-corrosion layer and a top finishing layer.

5 Discussion

By combining reverse engineering, 3D scanning, and CAD software PTC Creo Parametric, a comprehensive technological procedure was developed for the high-quality production of a safety frame. Reverse engineering was essential for converting the existing physical component into an accurate digital representation.

This process enabled engineers to analyse and design with greater precision.

To overcome challenges in measuring black and ribbed surfaces, the surface was matted, and special reference bodies (regular dodecahedrons) were designed and manufactured. These bodies, equipped with integrated magnets for rapid handling, ensured visibility of reference markers from multiple angles, critical for precise scanner positioning in the complex grid structure. Process efficiency was further enhanced by the VXelements software's ability to automatically identify and replace planar reference markers, saving time during polygonal mesh cleaning.

The design process emphasized functionality, safety, and manufacturing efficiency. The most significant functional improvement was based on operator experience: reducing the number of vertical ribs from 16 to 13 and altering their inclination, which led to a substantial improvement in the operator's field of vision while maintaining necessary protection. Validation of the field of view was performed in the Assembly module of Creo Parametric using a digital mannequin, representing an advanced method of ergonomic verification.

To increase manufacturing efficiency, a universal vertical rib geometry was developed for the entire assembly. Assembly was further simplified by integrating technological interlocks and laser-engraved welding positions, which minimized residual stresses and ensured precise alignment of components.

For the critical safety structure, high-strength steel S355K2 was selected. This material offers a 51% higher minimum yield strength compared to S235JR and guaranteed toughness at −20 °C, which is essential for year-round operation in demanding forestry environments. Laser cutting technology ensures excellent precision (±0.05 to ±0.1 mm) and a minimal heat-affected zone (HAZ).

The proposed workflow concludes with comprehensive quality control after welding and prior to powder coating. The application of two layers of RAL 9005 powder paint with fine texture provides both anti-corrosion protection and eliminates hazardous glare, thereby enhancing operator safety.

In summary, the integration of reverse engineering with advanced design and manufacturing approaches represents an effective strategy for developing safety components that meet strict requirements for functionality, safety, and technological feasibility.

6 Conclusions

This example confirms that the integration of reverse engineering with design and manufacturing approaches is an effective strategy for the development and production of safety components. The application of 3D scanning in combination with VXelements software and the Creo Parametric CAD system enabled the precise reconstruction of the protective structure for the John Deere 1270E forestry machine.

Challenges associated with scanning ribbed and black surfaces were successfully overcome by designing and manufacturing special reference bodies with integrated magnets, which ensured accurate spatial orientation of the scanner. Both primary and secondary objectives were achieved by reducing the number of vertical ribs from 16 to 13 and modifying their inclination, which significantly improved the operator's field of vision from the cabin while maintaining the protective function of the windshield.

For manufacturing optimization, a universal rib geometry was adopted, and the design incorporated technological interlocks and laser-engraved welding positions, minimizing residual stresses and ensuring geometric accuracy during assembly. The selection of high-strength steel S355K2 guarantees the required toughness at low temperatures, thereby fulfilling the requirement for safe year-round operation in demanding forestry environments.

Finally, the application of a two-layer RAL 9005 powder coating was proposed to achieve fine texture which provides not only anti-corrosion protection but also eliminates hazardous glare, further enhancing operator safety during machine operation.

Acknowledgements. The authors thank the Ministry of education of Slovak Republic for supporting this research by the grant KEGA no. 042TUKE-4/2025 and VEGA 1/0344/26 and this work was supported by the Slovak Research and Development Agency under the Contract no. APVV-24-0470. Funded by the EU NextGenerationEU through the Recovery and Resilience Plan for Slovakia under the project No. 09I05-03-V02- 00042.

References

1. Debnath, B., Pourfarash, Z., Ghorpade, B., Raman, S.: Integrating reverse engineering for digital model reconstruction and remanufacturing of mechanical components: a systematic review. Metro. **5**(4), 66 (2025). https://doi.org/10.3390/metrology5040066
2. Kanun, E.: Using photogrammetric modeling in reverse engineering applications: damaged turbocharger example. Mersin Photogram. J. **3**(1), 21–28 (2021). https://doi.org/10.53093/mephoj.901188
3. Marwan, H., Okatan, A., Maftouni, N., Khodami, S.S.: Real-time online education student lecture emotion detection. EURAS J. Eng. Appl. Sci. **3**(1) (2021). https://doi.org/10.17932/ejeas.2021.024/2023.301
4. Algadi, A., et al.: Digital twin-oriented generation of structural data and models with LiDAR scan point clouds. J. Infrastruct. Preserv. Resil. **6**(1) (2025). https://doi.org/10.1186/s43065-025-00140-4
5. Sen, G., Doloi, B.: Rapid product development from an existing product using reverse engineering method. In: Lecture Notes on Multidisciplinary Industrial Engineering, pp. 721–734. Springer, Singapore (2019). https://doi.org/10.1007/978-981-32-9487-5_60
6. Boryga, M., Kołodziej, P.: Reverse engineering in modeling agricultural products. Agric. Eng. **26**(1), 105–117 (2022). https://doi.org/10.2478/agriceng-2022-0009
7. Bushra, J., Rahman, M.H., Shafae, M., Budinoff, H.D.. Distortion minimization in reverse engineering for additive manufacturing: an integrated 3D scanning and simulation framework (Version 2). arXiv. 10.48550/ARXIV.2509.05857 (2025)
8. Ivanov, V., Evtuhov, A., Dehtiarov, I., Trojanowska, J.: Introduction to manufacturing engineering. In: Springer Tracts in Mechanical Engineering, pp. 1–14. Springer Nature, Switzerland (2024). https://doi.org/10.1007/978-3-031-74360-3_1
9. Costa, M., et al.: Optimizing metal AM potential through DfAM: design, performance, and industrial impact. In: Industrial Engineering and Management. IntechOpen (2024). https://doi.org/10.5772/intechopen.1007309Sf
10. Turek, P., Bezłada, W., Cierpisz, K., Dubiel, K., Frydrych, A., Misiura, J.: Analysis of the accuracy of CAD modeling in engineering and medical industries based on measurement data using reverse engineering methods. Designs. **8**(3), 50 (2024). https://doi.org/10.3390/designs8030050
11. Török, J., Vodilka, A., Kaščak, J., Kočiško, M.: Design of Personalized Orthoses with support of PTC Creo and FDM technology. In: Lecture Notes in Mechanical Engineering, pp. 149–162. Springer Nature, Switzerland (2024). https://doi.org/10.1007/978-3-031-56456-7_12
12. Koren, Y., Gu, X., Guo, W.: Reconfigurable manufacturing systems: principles, design, and future trends. Front. Mech. Eng. **13**(2), 121–136 (2017). https://doi.org/10.1007/s11465-018-0483-0
13. Török, J., Dupláková, D.: Integrated practical framework for multidisciplinary prototype design and manufacturing process. PRO. **13**(2), 454 (2025). https://doi.org/10.3390/pr13020454
14. SSAB.: Strenx 960 [Data sheet]. Gama Ocel. https://www.gamaocel.cz/wp-content/uploads/2025/01/strenx-960-data-sheet-2023.pdf (2023)
15. Dillinger, J.: Moderní strojírenství pro školu i praxi. Sobotáles, Praha (2007)

A Review of Fixturing Design Methodology for Machining Thin-Walled Aerospace Components Considering Complex Stress States

Bartłomiej Szczupak[1(✉)], Piotr Paczos[2], and Piotr Szablewski[1,3]

[1] Pratt & Whitney Kalisz, 4a Elektryczna Street, 62-800 Kalisz, Poland
Szczupak2@prattwhitney.com

[2] Faculty of Mechanical Engineering, Poznan University of Technology, 3 Piotrowo Street, 60-965 Poznan, Poland

[3] Institute of Mechanical Engineering, University of Kalisz, 4 Nowy Świat Street, 62-800 Kalisz, Poland

Abstract. Thin-walled aerospace components exhibit low structural stiffness, high aspect ratios and complex residual stress fields originating from forming, heat treatment and previous machining operations. As a consequence, these parts are highly susceptible to elastic deformation, chatter and post-machining distortion. Fixturing design plays a central role in controlling such phenomena and ensuring dimensional accuracy, surface integrity and process stability. This paper presents a structured review of fixturing design methodologies for thin-walled aerospace components, with particular emphasis on approaches that explicitly or implicitly consider complex stress states during machining. Based on 50 publications from 1990 to 2025, the review covers: (I) classical and computer-aided fixture design frameworks; (II) finite-element-based modelling of deformation and residual stresses; (III) fixture layout optimization methods; (IV) adaptive and intelligent fixturing systems; (V) additive-manufactured and topology-optimized fixtures; and (VI) digital-twin and AI-driven design strategies. The analysis indicates a clear evolution from rule-based fixture design towards integrated, model-driven and data-driven methodologies that couple finite element analysis, optimization algorithms, sensor feedback and digital twins. Remaining challenges include high computational cost, limited multi-physics integration, incomplete modelling of evolving residual stresses and the need for robust validation on full-scale aerospace structures. The paper concludes by highlighting research gaps and proposing future directions for holistic, stress-aware fixturing strategies.

Keywords: Thin-Walled Components · Aerospace Machining · Fixture Design

1 Introduction

Thin-walled aerospace components such as ribs, webs, frames, beams and monolithic integral structures, in particular thin-walled aero-engine casings including compressor housings, bearing housings, combustor housings, intermediate housings and turbine housings are extensively used in airframes, aero-engines and space applications.

B. Gapiński et al. (Eds.): MANUFACTURING 2026, LNME, pp. 199–216, 2026.
https://doi.org/10.1007/978-3-032-22943-4_15

Their design typically aims at maximizing stiffness-to-weight ratio, which leads to slender geometries and wall thicknesses that can be as low as 0.8 mm. While beneficial for weight reduction and performance, these characteristics severely reduce structural rigidity during machining operations.

Under cutting and clamping loads, thin-walled components may experience substantial elastic deformation, leading to dimensional errors, poor surface quality and even instability in the form of chatter. Furthermore, residual stresses introduced by prior processes (e.g. rolling, forging, heat treatment or additive manufacturing) can relax during material removal and cause significant post-machining distortion. As a result, geometrical deviations of several times the tolerance limit are frequently reported for thin-walled parts if fixturing is not properly designed.

Fixture design has therefore been recognized as a key enabler for accurate and stable machining. Early reviews of computer-aided fixture design (CAFD) emphasized the importance of systematic fixture planning, support selection, workpiece location and clamping strategies for general components [1–10]. Subsequent works have concentrated specifically on thin-walled parts, highlighting the critical role of fixture layout in controlling deformation and stress redistribution [11–20]. Despite substantial progress, existing fixture design methodologies often treat the workpiece as a rigid body, neglecting the complex interaction between clamping forces, cutting forces, residual stresses and the evolving stiffness of the part as material is removed. For thin-walled aerospace structures, such simplifications are insufficient. There is a growing need for fixturing approaches that are explicitly stress-aware, i.e. capable of modelling, predicting and controlling the combined stress state during machining.

In this context, the present review focuses specifically on fixturing design and layout optimization for machining (primarily milling) of thin-walled metallic aerospace components. Particular attention is given to fixture design methodologies and CAFD frameworks, the finite element environments employed to predict deformation and stress, and the wide range of optimization strategies used to determine effective fixture layouts, including both metaheuristic and mathematically formulated approaches. The review also considers recent progress in adaptive, intelligent and flexible fixturing systems, as well as the emerging role of additive-manufactured and topology-optimized fixtures. Finally, it highlights the growing influence of digital twin and AI-enabled solutions that integrate models, process data and control capabilities. Cutting parameter optimization, toolpath planning and general machining process optimization are discussed only insofar as they relate directly to fixturing, ensuring that the review remains focused on the structural and stress-related aspects of workholding.

1.1 Scope of the Review

The present review focuses on fixturing design and layout optimization for machining (primarily milling) of thin-walled metallic aerospace components. The scope includes:

- Fixture design methodologies and CAFD frameworks [1–10].
- Finite element analysis (FEA) and simulation environments used to predict deformation and stress [11–20].
- Fixture layout optimization methods, including metaheuristics and mathematical programming [21–30].

- Adaptive, intelligent and flexible fixturing systems [31–40].
- Additive-manufactured and topology-optimized fixtures [41–45].
- Digital twin and AI-enabled approaches that integrate models, data and control [46–50].

Cutting parameter optimization, toolpath planning and general machining process optimization are discussed only insofar as they relate directly to fixturing.

2 Research Problem and Scope of the Review

2.1 Characteristics of Thin-Walled Aerospace Components

Aerospace thin-walled parts are characterized by:

- High aspect ratio and low stiffness: long spans with small thickness lead to pronounced bending under relatively low loads.
- Complex 3D geometries: presence of pockets, ribs, stiffeners and varying thickness complicates support and clamping strategies.
- Anisotropic and high-performance materials: aluminium and titanium alloys, nickel-based superalloys and composites, with non-linear and temperature-dependent properties.
- Residual stress fields: introduced by forming, quenching, additive manufacturing or prior machining.

The resulting structural behaviour is strongly path-dependent: the removal of material changes stiffness, natural frequencies and the internal stress state at each stage of the process [13–15]. Fixturing must remain effective throughout this evolution.

2.2 Main Sources of Deformation and Stress

During machining of thin-walled parts, deformation and stress arise from multiple sources:

- Cutting forces: generate bending, torsion and local deflections of walls and ribs. The resulting errors accumulate along the tool path and may exceed tolerance limits if the part is insufficiently supported [11, 12, 14, 27].
- Clamping forces: required to stabilize the workpiece, but excessive clamping can induce local plastic deformation and residual stresses. Upon unclamping, elastic recovery leads to shape distortion [7, 17, 39].
- Residual stresses: pre-existing stresses relax when material layers are removed, causing global warping and twisting. This is particularly critical in large monolithic components [13, 15]
- Dynamic loads and chatter: regenerative chatter and forced vibrations amplify dynamic deflections and surface errors, especially in low-stiffness configurations [14, 31, 38].

The interactions between the major sources of deformation in thin-walled components are illustrated in Fig. 1.

These mechanisms act simultaneously and interact in a non-linear manner. Fixture design must therefore address not only static equilibrium but also multi-source, multi-stage stress states.

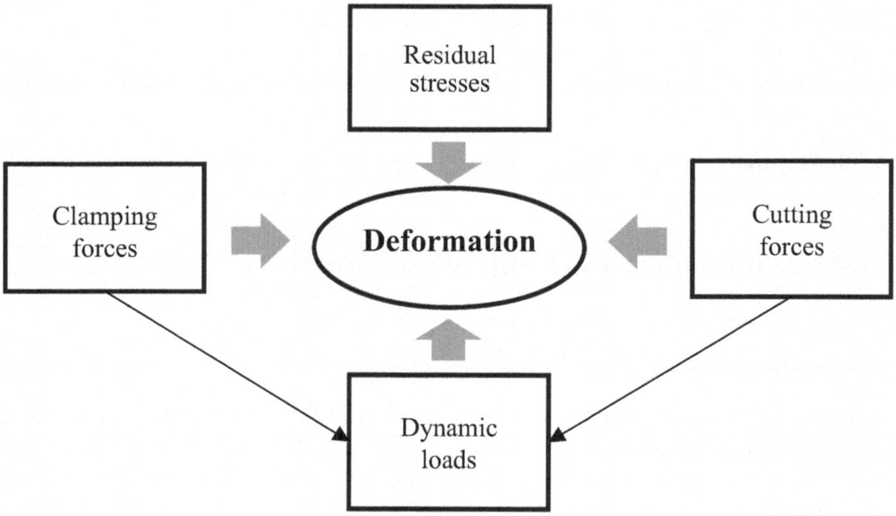

Fig. 1. Main sources of deformation and stress in machining thin-walled aerospace components.

3 Methodological Foundations of Fixture Design

3.1 Classical Computer-Aided Fixture Design

Early work on computer-aided fixture design provided systematic frameworks for locating and supporting rigid workpieces. Boyle et al. [1] and Hargrove and Kusiak [2] reviewed analytical and rule-based CAFD systems, emphasizing locating principle (such as 3–2-1), determination of clamping points and selection of standard fixture elements. Trappey and Liu [3] surveyed fixture design automation with emphasis on geometric reasoning, interference checking and modular fixture configuration.

Subsequent contributions introduced knowledge-based and case-based reasoning (CBR) approaches. Hashemi et al. [4] presented case-based reasoning methods for computer-aided fixture design, enabling reuse of previous fixture solutions for new parts with similar features. Kang and Peng [5] focused on fixture feasibility analysis, formalizing constraints related to accessibility, collision avoidance and support stability.

Wang et al. [8] and Cecil [6] summarized the state of CAFD and identified key requirements for future methods, including integration with CAD/CAM, standardization of fixture components and extension to flexible and intelligent systems. Cai et al. [7] proposed a variational method to design robust fixture configurations that are less sensitive to geometric variation of the workpiece. De Meter [9] developed a min-max load model to optimize clamping forces and improve fixture performance. Nee et al. [10] integrated many of these concepts into a comprehensive monograph on advanced fixture design for flexible manufacturing systems.

While these works laid the foundation for systematic fixture design, they typically treat the workpiece as a rigid or quasi-rigid body and focus on ensuring geometric stability and process feasibility. The effect of complex stress states and large elastic deformation, which is critical for thin-walled aerospace parts, is only addressed indirectly through safety factors and empirical design rules.

3.2 Finite-Element-Based Modelling of Deformation and Residual Stresses

With the advent of higher computational power, finite element analysis became a central tool for modelling thin-walled components under machining and fixturing loads. Rai and Xirouchakis [11] proposed a FEM-based machining simulation environment to analyse part errors generated during milling of thin-walled components. Ratchev et al. [12] combined FEM and compensation techniques to predict and correct milling errors in low-rigidity parts.

Huang et al. [13] investigated residual-stress-related deformation of monolithic components by combining finite element simulation with experimental measurements. Their work demonstrated that residual stress relaxation can be the dominant source of distortion even when clamping forces are moderate. Wu et al. [14] provided a state-of-the-art review of chatter and geometric errors in thin-wall machining, emphasizing the coupling between vibration and elastic deformation.

Wang et al. [15] advanced the state of practice by integrating online measurement and allowance compensation to improve machining accuracy of thin-walled parts. Fei et al. [16] studied the effect of moving fixtures on deformation suppression during milling, while Wang et al. [17] developed a parametric FEA system tailored for fixturing of thin-walled cylindrical components.

These contributions collectively show that fixture design for thin-walled parts cannot rely solely on geometric reasoning; it must instead be model-driven, with FEA providing detailed stress and deformation fields as a function of clamping, cutting and residual stresses. However, the computational expense of fully coupled models remains a major barrier for routine industrial use.

3.3 Residual Stress Characterization and Integration into Fixturing Models

Residual stresses in aerospace thin-walled components originate from multiple manufacturing routes and therefore exhibit fundamentally different spatial patterns and magnitudes [18]. Forming and rolling typically generate near-surface stress gradients and directional anisotropy; forging and quenching can produce large through-thickness gradients driven by thermal non-uniformity; additive manufacturing may introduce heterogeneous, path-dependent stress fields coupled with microstructural anisotropy; and prior machining can super-impose near-surface tensile/compressive layers and re-distribute pre-existing stresses [18]. For fixture design, these distinctions are not merely academic: the distortion response during material removal depends on both the initial stress tensor field and its redistribution under evolving stiffness and boundary conditions [13, 16, 18].

Residual stress measurement and identification commonly rely on diffraction-based techniques (e.g., X-ray diffraction for near-surface stresses), semi-destructive methods (hole-drilling or ring-core for depth profiles), and bulk-sensitive methods (e.g., neutron diffraction) when thick sections or full-field characterization is required [13, 18]. In industrial practice, stress inputs are frequently available only as sparse measurements; there-fore, practical integration into FEM-based fixturing models requires (I) stress field reconstruction/regularization (e.g., fitted profiles or mapped fields), (II) consistent mapping to the FE mesh (including coordinate transformation and smoothing), and (III) staged material removal simulation with stress re-equilibration at each step [11, 13, 17].

Where available, in-process measurement and compensation concepts can partially mitigate uncertainty by updating geometry or allowances during machining, but they do not replace the need for credible initial stress-state characterization when distortion is stress-driven [15]. These considerations motivate the need for "stress-aware" fixture optimization frameworks that treat residual stress evolution as a first-class modeling element rather than a secondary disturbance [18, 19].

4 Fixture Layout Optimization Under Complex stress states

4.1 Optimization Problem Formulation

Fixture layout optimization aims to determine the number, type and positions of locators and clamps that minimize deformation, stress concentration or surface error while satisfying constraints such as accessibility and stability. For thin-walled parts, the objective is often to minimize the maximum deflection or to keep deformation within tolerance throughout machining stages.

The optimization problem is typically formulated as:

- Design variables: positions of locators and clamps, sometimes clamping forces and support stiffness;
- Objective functions: maximum displacement, surface error, stress concentration, or combinations thereof;
- Constraints: fixture-workpiece interference, minimum distances, tool accessibility; geometric constraints and limits on clamping forces;
- Evaluation model: finite element simulation, reduced-order model or analytical approximation.

4.2 Metaheuristic Optimization Approaches

Liu et al. [21] proposed fixture layout optimization for large thin-walled parts based on an improved particle swarm optimization (IPSO) algorithm. The method couples PSO with FEA to determine clamping positions that minimize deformation, showing significant improvement over heuristic layouts.

Other evolutionary approaches have been explored. Liu et al. [24] used evolutionary design strategies to optimize fixture layouts for thin-walled structures, combining genetic algorithms with FEM-based evaluations. Subramanian et al. [25] applied design of experiments (DOE) and FEM to optimize fixture layout for thin-walled components, demonstrating that statistically designed experiments can reduce the number of simulations required.

Calabrese et al. [26] optimized machining fixtures for aeronautical thin-walled components using FEM and optimization techniques. Li et al. [27] modelled surface form error prediction in five-axis flank milling and demonstrated that fixture layout strongly affects error distribution.

Zhong et al. [28] introduced a convex relaxation framework for optimal fixture layout design, providing theoretical guarantees of near-global optimality under certain assumptions but requiring convex approximations of contact mechanics and deformation.

Liu et al. [19] presented a review specifically focused on fixture layout optimization for thin-walled parts, summarizing various optimization algorithms and their strengths and limitations. Together with Wang et al. [20], who studied optimization of fixture number in large thin-walled assemblies based on improved particle swarm optimization, these works provide a comprehensive picture of the evolution of optimization-based fixture design.

4.3 Integration with Complex Stress States

Although many optimization frameworks use FEA to evaluate deformation, explicit modelling of residual stress and dynamic effects is still relatively rare. In most studies, the workpiece is assumed stress-free prior to machining and cutting forces are represented as quasi-static equivalent loads. Only a subset of the literature couples fixture layout with residual stress modelling or chatter prediction.

A key research direction is therefore the integration of:

- Residual stress fields obtained from previous simulations or measurements.
- Time-dependent boundary conditions reflecting changing contact conditions as material is removed.
- Dynamic simulations accounting for chatter and vibration modes.

This integration is essential to achieve truly stress-aware fixture layout optimization.

4.4 Limitations of Metaheuristic Optimization and Scalability

Metaheuristic methods such as PSO and GA are attractive for fixture layout problems because they can handle discrete/continuous variables and non-convex search spaces and can be coupled to FEM for deformation evaluation [21, 24, 25]. Nevertheless, their practical application in stress-aware fixturing is limited by several factors. First, convergence behavior is often sensitive to algorithm hyperparameters and constraint handling, particularly when objective functions become non-smooth due to contact nonlinearity and changing support conditions in low-stiffness workpieces [19, 21]. Second, computational cost scales poorly because each candidate layout typically requires an FEM run; for multi-stage machining and multiple load cases, this can become prohibitive without reduced-order modeling or surrogate assistance [19, 25]. Third, metaheuristics may yield solutions that are difficult to reproduce across runs unless stochasticity and stopping criteria are rigorously controlled, which complicates deployment and auditability in industrial settings [19].

Deterministic approaches (including convex relaxation) can improve repeatability and provide stronger optimization guarantees, but they may require simplifying assumptions to make contact and deformation models tractable [28]. A pragmatic industrial direction is therefore hybrid: use limited global exploration (e.g., PSO/GA) with reduced evaluations, followed by deterministic local refinement and periodic high-fidelity verification, potentially accelerated by surrogate models consistent with digital-twin and data-driven machining analytics [28, 49].

5 Adaptive and Intelligent Fixturing Systems

5.1 Adaptive and Active Fixtures

Adaptive fixtures modify their stiffness or contact conditions in response to changing loads, geometry or process states. Merlo et al. [31] introduced an adaptive fixturing concept for thin-walled aerospace parts, capable of reconfiguring support locations to minimize deformation. Papastathis et al. [32] developed dynamic models of active fixturing systems under moving loads, demonstrating improved vibration control.

Magnetorheological (MR) fluid-based fixtures are particularly attractive for thin-walled parts. Ma et al. [34] proposed an MR fluid flexible fixture to improve stability and suppress vibration in milling. Jiang et al. [35] further investigated vibration suppression for complex thin-walled workpieces using MR fixtures, showing significant reductions in dynamic displacement. Puma-Araujo et al. [38] designed a semi-active MR damper device for chatter mitigation during milling of thin-floor components, integrating sensors and control algorithms.

Low-melting alloy fixtures constitute another approach to adaptively match the part geometry. Wang et al. [39] used low-melting alloys to create conformal supports that provide uniform pressure and improved stability for aeronautical thin-walled components, while allowing relatively easy removal after machining.

5.2 Intelligent and Flexible Fixtures

Intelligent fixtures integrate sensors, actuators and control logic. Gonzalo et al. [33] developed an intelligent fixture concept that uses sensors to detect workpiece deformation and adjust clamping conditions accordingly. Li et al. [36] described design and application of flexible fixtures that can be rapidly reconfigured for different parts, facilitating flexible manufacturing.

Shi et al. [37] presented a flexible adaptive fixture for precision grinding of thin-walled bearing rings, combining mechanical flexibility with sensor feedback to maintain dimensional accuracy.

These systems rely on real-time measurement of forces, displacements or vibrations and on control algorithms that adapt clamping forces, support locations or fixture stiffness. When integrated with predictive models, such fixtures can become key components of smart manufacturing environments and digital twins.

6 Additive-Manufactured and Topology-Optimized Fixtures

6.1 Topology Optimization for Fixture Design

Topology optimization (TO) offers a powerful framework for designing lightweight yet stiff fixtures. Kitagawa et al. [41] applied topology optimization to design fixtures for additively manufactured workpieces, focusing on maximizing stiffness while maintaining accessibility. Zhou et al. [30] developed TO methods for 3D thin-walled structures with adaptive meshing, enabling efficient handling of complex geometries.

Castro et al. [44] used density-based topology optimization to design 3D-printed fixtures for cyclic mechanical testing of lattice structures, demonstrating the ability of TO to concentrate material where it contributes most to stiffness and load transmission.

6.2 Additive Manufacturing and Multifunctional Fixtures

Additive manufacturing (AM) allows for unprecedented geometric freedom, enabling conformal contact surfaces, internal lattice structures and integrated damping features. Plocher and Panesar [42] reviewed design and structural optimization for AM, highlighting the potential for next-generation lightweight structures.

Picavea et al. [43] proposed 3D-printed fixtures that integrate vibration absorption and damping features for thin-wall machining. By embedding tuned damping structures into the fixture body, they demonstrated improved vibration control without increasing overall fixture mass significantly.

Prajapati et al. [45] investigated multi-material additive manufacturing with foam-filled lattice structures, providing inspiration for fixtures that combine stiff external frameworks with energy-absorbing cores.

Such AM-enabled fixture designs can be tailored to match the local stiffness requirements of thin-walled parts, providing high stiffness where needed while reducing mass and inertia. Moreover, conformal surfaces can distribute clamping loads more uniformly, reducing local peak stresses and deformation.

6.3 Practical Constraints of AM Fixtures for Aerospace Machining

While additive manufacturing enables conformal contacts, internal lattices, and integrated damping features, AM-based fixtures face practical constraints that affect stiffness, accuracy, and lifecycle cost in aerospace environments [42, 43]. Dimensional tolerances and surface roughness of printed contact interfaces can degrade locator repeatability and increase local contact compliance unless post-processing (machining, polishing, coating) is applied [41–43]. Material anisotropy and process-dependent defects can reduce fatigue strength and long-term durability under repeated clamping cycles, and these issues become more pronounced when fixtures rely on lightweight lattices or thin features [42, 44, 45].

From a functional perspective, AM may improve damping through lattice architectures or embedded absorbers demonstrated in AM fixture concepts, but the stiffness–damping trade-off must be quantified because excessive compliance can increase deformation even if vibration amplitude decreases [43]. Cost and lead time also remain non-trivial when inspection, post-processing, and traceability are included, especially when the fixture is intended for high-reliability aerospace manufacturing workflows [42, 50]. Therefore, industrial viability requires a validated design-for-manufacturing and verification workflow that explicitly considers tolerances, surface finishing, directionality, inspection plans, and reparability alongside structural optimization objectives [42, 50].

7 Digital Twins, AI and Automated Fixture Design

7.1 Digital Twin Concepts in Machining

Digital twins (DTs) are high-fidelity, continuously updated virtual representations of physical systems. In machining, DTs combine physical models, sensor data and control logic to predict process outcomes and optimize operations. Liu et al. [46] discussed digital-twin-driven rapid individualized design of automated manufacturing systems, laying conceptual foundations for DT-based design workflows.

Grzesik [47] presented case studies of digital twins in removal and additive machining processes, illustrating how twins can support process configuration, monitoring and optimization. Hänel et al. [48] developed digital twins for high-tech machining applications using model-based analytics, highlighting the need for structured data architectures and analytic pipelines.

7.2 AI and Machine Learning for Fixture Design

Jarosz and Özel [49] reviewed machine learning approaches for digital twin development in machining systems. Their work suggests that data-driven models can complement physics-based simulations by providing rapid approximations of complex relationships, such as deformation fields or chatter stability, which are computationally expensive to obtain with FEA alone.

Da Silva et al. [50] provided a comprehensive review of digital twin and Industry 4.0 applications in machining, identifying fixturing as an area with high potential for DT integration but relatively limited current implementation.

In the context of fixture design for thin-walled aerospace components, digital twins and AI can be used to:

- Build surrogate models of stress and deformation as functions of fixture layout, clamping forces and process parameters.
- Enable rapid multi-objective optimization of fixtures.
- Drive adaptive fixtures by providing predictions and control setpoints.
- Accumulate knowledge across multiple parts and production runs, improving predictive accuracy over time.

Combining DTs with adaptive fixtures and AM-enabled designs opens the way towards autonomous, self-optimizing fixturing systems.

8 Discussion

8.1 Evolution of Fixturing Methodology

The literature from 1990 to 2025 shows a clear shift from rule-based CAFD and geometric reasoning [1–10] towards integrated, model-based and data-driven methodologies [11–21, 31–50]. The main stages of this evolution can be summarized as:

- Classical CAFD: automated fixture planning for rigid workpieces using geometric rules and heuristic knowledge.
- FEM-enhanced design: inclusion of elastic deformation and, to some extent, residual stresses to evaluate fixture layouts for thin-walled parts.
- Optimization-driven layout design: systematic search for optimal fixture layouts using metaheuristics and mathematical optimization, often coupled with FEM.
- Adaptive and intelligent fixtures: introduction of sensors and actuators to adjust fixture behaviour in real time.

- AM and topology-optimized fixtures: leveraging additive manufacturing to produce lightweight, conformal and multifunctional fixtures.
- Digital twin and AI integration: towards closed-loop, model- and data-driven fixturing systems embedded in smart manufacturing environments.

8.2 Current Gaps and Limitations

Despite these advances, several important challenges remain:

- Residual stress modelling: only a limited number of studies explicitly integrate residual stress evolution into fixture design and layout optimization [13–15]. Most models start from a stress-free assumption, which is unrealistic for aerospace components.
- Multi-physics coupling: thermal effects, tool wear and material non-linearity are often neglected, yet they can significantly affect stress states and deformation.
- Computational cost: high-fidelity FEM and optimization loops can be computationally expensive, hindering routine industrial application. Surrogate models and reduced-order techniques are still underutilized.
- Validation on industrial parts: many studies use simplified geometries. Systematic validation on full-scale aerospace components with complex geometries and manufacturing histories is relatively rare.
- Integration with process planning: fixture design is often treated separately from toolpath and parameter optimization, although the three are strongly coupled.

Table 1 provides a concise comparison of how different fixturing methodology classes address the main stress sources and physical effects involved in machining thin-walled components. As shown, only a limited number of approaches explicitly incorporate residual stresses, dynamic loads or stiffness evolution, indicating clear opportunities for further research.

Experimental Validation and Reliability
A substantial fraction of the reviewed literature remains simulation-centric, with validation often limited to simplified coupons or laboratory-scale thin-wall artifacts [18, 19]. This is particularly critical for residual-stress-driven distortion, where model credibility depends on the accuracy of initial stress field identification, the representation of stress redistribution during staged material removal and the realism of fixturing contact conditions [13, 16]. Studies combining simulation with experimental measurement provide stronger evidence for deformation-control strategies, for example in residual-stress-related deformation investigations and in approaches using on-machine/online measurement and compensation concepts [13, 15]. Nevertheless, systematic validation on full-scale monolithic aerospace structures remains comparatively rare due to metrology cost, fixture complexity, and limited access to manufacturing-history data, despite the availability of practical fixture concepts demonstrated in applied studies (e.g., intelligent or special-purpose fixturing) [33, 39]. Future work should report validation maturity explicitly and include uncertainty bounds and sensitivity analyses for key assumptions (contact stiffness, friction, clamp compliance, and residual stress inputs) to improve comparability and industrial confidence [18, 19].

Table 1. Coverage of stress sources and physical effects in different fixturing methodology classes.

Fixturing methodology class	Cutting forces	Clamping forces	Residual stresses	Dynamic loads/chatter	Evolving stiffness/material removal sequence
Classical CAFD	NO	NO	NO	NO	NO
FEM-based static fixturing (FEA)	YES	YES	PART	NO/PART	PART
Optimization-based fixture layout (PSO, GA, convex, DOE)	YES	YES	PART	NO/PART	PART
Adaptive/intelligent fixtures (MR, active, sensorized)	PART	YES	NO	YES	PART
AM & topology-optimized fixtures	PART	PART	NO	PART	PART
Digital-twin & AI-enabled fixtures	YES	YES	PART	YES	YES

To clarify the maturity of experimental evidence across the literature, Table 2 proposes a validation-level taxonomy (V0–V4) aligned with the predominantly simulation-driven fixture-layout and thin-wall machining literature [18, 19] while reflecting studies that include experimental components or in-process measurement elements [13, 15, 33, 39].

Table 2. Validation maturity levels for fixturing and distortion studies.

Level	Validation type	Typical artifacts	What it proves	Typical gaps
V0	Simulation only	idealized plates/walls	conceptual feasibility	no credibility for residual stress evolution
V1	Lab coupon/simple thin-wall	small ribs/plates	trend correctness	scale effects, boundary condition realism
V2	Representative subcomponent	pocketed panels/ribs	practical deformation control	limited process history realism
V3	Full-scale aerospace part	monolithic structures	industrial relevance	cost, measurement complexity

8.3 Future Research Directions

Based on the reviewed literature, promising research directions include:

- Integrated stress-aware design frameworks that couple residual stress prediction, dynamic simulation and fixture layout optimization.
- Multi-scale and reduced-order modelling to capture essential physics while keeping computational effort manageable.
- Surrogate and AI models trained on simulation and experimental data to rapidly estimate deformation under varying fixture layouts and process conditions.
- Digital-twin-enabled fixturing where fixtures, machine tools and workpieces are represented in a unified virtual environment, allowing virtual commissioning and real-time adaptation.
- AM-based adaptive fixtures that combine topology optimization, smart materials and embedded sensing.
- Standardized benchmarks and validation cases for thin-walled aerospace components to ensure comparability of methods and to accelerate industrial adoption.

8.4 Comparative Assessment and Industrial Readiness

Across the reviewed studies, physics-based FEM remains the dominant enabler for stress- and deformation-aware fixturing; however, its industrial effectiveness depends strongly on the fidelity of boundary conditions, contact/friction modeling, and the representation of machining stage evolution [11, 12, 17]. Metaheuristic optimization (PSO/GA) can outperform manual layouts in terms of peak deflection and surface error, yet its practical deployment is often constrained by the number of expensive FEM evaluations and by non-deterministic convergence behavior [19, 21, 24, 25]. Deterministic optimization and convex formulations can improve repeatability and computational efficiency, although they typically rely on simplifying assumptions to handle contact and nonlinearity [28]. Adaptive and MR-based fixtures provide a direct pathway to mitigating chatter and time-varying stiffness effects via sensing and controllable fixture dynamics, but they introduce additional hardware integration and calibration burdens [34, 35, 38, 40]. Additive-manufactured and topology-optimized fixtures offer new degrees of freedom in stiffness tailoring and functional integration, yet their industrial value depends on manufacturability and validation of printed structures in realistic loading conditions [41–43]. Digital-twin and AI-enabled approaches are emerging as integrative frameworks; their readiness is currently limited by data availability and validation maturity, but they offer high long-term scalability for multi-part learning and closed-loop fixturing control [46, 48–50].

Table 3 synthesizes a comparative assessment of the major methodology families (FEM [11, 12, 17], PSO/GA-based optimization [21, 24, 25], MR/active fixturing [34, 35, 38], topology optimization and AM-enabled fixtures [41–43], and digital-twin/AI approaches [46, 48–50]) in terms of accuracy, computational cost, scalability, and industrial readiness, highlighting the trade-offs that govern practical adoption.

Table 3. Comparative evaluation of major fixturing methodology families for thin-walled aerospace machining.

Metho family	Typical objective(s)	Accuracy (deformation /stress)	Computational cost	Scalability (part size / stages)	Industrial readiness	Key strengths	Key limitations / risks
FEM-based static fixturing (FEA)	max displacement, surface error	High (if BC/contact correct)	High	Medium	Medium /High	physics-based insight; local stress fields	sensitive to boundary conditions; contact/friction uncertainty; slow for multi-stage
PSO/GA + FEM (layout optimization)	min max deflection / error	High /Medium	Very high (many FEM calls)	Low /Medium	Medium	searches large design space; multi-objective possible	convergence variability; computational burden; constraint handling; reproducibility
Deterministic optimization +reduced models (convex / gradient / ROM)	min compliance / error	Medium /High (model dependent)	Medium /Low	Medium /High	Medium	faster; more repeatable; can give guarantees (convex)	needs simplifications (e.g., contact linearization); risk of model bias
MR-based / active fixtures (adaptive)	vibration suppression, stiffness tuning	Medium /High (dynamic)	Medium (control + sensing)	Medium	Medium	addresses chatter; real-time adaptation	hardware complexity; calibration; reliability; integration cost
Low-melting alloy / conformal supports	distribute pressure; reduce local distortion	Medium	Low /Medium	Medium	Medium /High	good load distribution; practical shop-floor concept	thermal handling; cleanup; repeatability; material compatibility
Topology-optimized fixture bodies (TO)	maximize stiffness /frequencies	Medium /High	Medium /High (TO loop)	Medium	Medium	lightweight +stiff; tailored compliance	manufacturing constraints; validation; cost/time
AM-enabled fixtures (printed)	conformal support + damping	Medium (depends on material/process)	Low /Medium (once designed)	Medium	Medium	geometric freedom; integrated functions	tolerance/roughness; anisotropy; durability; certification
Digital-twin / AI-driven design	fast prediction, closed-loop adaptation	Medium /High (if trained/ validated)	Medium (setup high; inference low)	High	Low /Medium (emerging)	speed, learning across runs, autonomy	data quality; drift; explainability; integration burden

9 Conclusions

This paper has reviewed fixturing design methodologies for the machining of thin-walled aerospace components, with a specific focus on complex stress states arising from cutting forces, clamping forces, residual stresses and dynamic effects.

The main conclusions are as follows:

1. Fixture design is a critical determinant of machining accuracy and stability for thin-walled aerospace components, where small changes in support or clamping conditions can cause large deformation.
2. Classical CAFD approaches, while valuable for systematic planning, are insufficient for thin-walled parts because they neglect elastic deformation and complex stress states.
3. Finite element analysis has become the central tool for stress- and deformation-aware fixture design. However, computational cost and the need for accurate boundary conditions remain key barriers.
4. Optimization-based fixture layout design using metaheuristics and mathematical programming has demonstrated clear benefits in minimizing deformation, but integration with residual stress and dynamic models is still limited.
5. Adaptive and intelligent fixturing systems-including MR-based, low-melting-alloy and sensorized fixtures-offer powerful means of deformation control but add complexity and cost.
6. Additive-manufactured and topology-optimized fixtures enable lightweight, conformal and multifunctional designs tailored to thin-walled geometries, particularly when combined with damping and smart materials.
7. Digital twins and AI are emerging as key enablers for autonomous, stress-aware fixturing, enabling closed-loop design, prediction and control across the entire life cycle of the part.

Overall, the reviewed literature indicates a clear trend towards integrated, multi-physics and data-driven fixturing methodologies. Future research should focus on bridging the gap between high-fidelity modelling and industrial practicality, ultimately enabling robust, efficient and autonomous fixturing solutions for next-generation aerospace manufacturing.

References

1. Boyle, I., Rong, Y., Brown, D.C.: A review and analysis of current computer-aided fixture design approaches. Robot. Comput. Integr. Manuf. **27**(1), 1–12 (2011). https://doi.org/10.1016/j.rcim.2010.05.008
2. Hargrove, S.K., Kusiak, A.: Computer-aided fixture design: a review. Int. J. Prod. Res. **32**(4), 733–753 (1993). https://doi.org/10.1080/00207549408956967
3. Trappey, J.C., Liu, C.R.: A literature survey of fixture-design automation. Int. J. Adv. Manuf. Technol. **5**, 240–255 (1990). https://doi.org/10.1007/BF02601534
4. Hashemi, H., Shaharoun, A.M., Izman, S., Kurniawan, D.: Recent developments on computer aided fixture design: case based reasoning approaches. Adv. Mech. Eng. (2014). https://doi.org/10.1155/2014/484928

5. Kang, X., Peng, Q.: Fixture feasibility: methods and techniques for fixture planning. Comput. Aided Des. Appl. **5**(1–4), 424–433 (2008). https://doi.org/10.3722/cadaps.2008.424-433
6. Cecil, J.: Computer-aided fixture design – a review and future trends. Int. J. Adv. Manuf. Technol. **18**, 790–793 (2001). https://doi.org/10.1007/s001700170004
7. Cai, W., Hu, S.J., Yuan, J.X.: A Variational method of robust fixture configuration design for 3-D workpieces. J. Manuf. Sci. Eng. **119**(4A), 593–602 (1997). https://doi.org/10.1115/1.2831192
8. Wang, H., Rong, Y., Li, H., Price, S.: Computer aided fixture design: recent research and trends. Comput. Aided Des. **42**(12), 1085–1094 (2010). https://doi.org/10.1016/j.cad.2010.07.003
9. De Meter, E.C.: Min-max load model for optimizing machining fixture performance. ASME J. Eng. Ind. **117**(2), 186–193 (1995). https://doi.org/10.1115/1.2803293
10. Nee, A.Y.C., Whybrew, K., Senthil Kumar, A.: Advanced Fixture Design for FMS. Springer, London (1995). https://doi.org/10.1007/978-1-4471-2117-6
11. Rai, J.K., Xirouchakis, P.: Finite element method based machining simulation environment for analyzing part errors induced during milling of thin-walled components. Int. J. Mach. Tools Manuf. **48**(6), 629–643 (2008). https://doi.org/10.1016/j.ijmachtools.2007.11.004
12. Ratchev, S., Liu, S., Huang, W., Becker, A.A.: Milling error prediction and compensation in machining of low-rigidity parts. Int. J. Mach. Tools Manuf. **44**(15), 1629–1641 (2004). https://doi.org/10.1016/j.ijmachtools.2004.06.001
13. Huang, X., Sun, J., Li, J.: Finite element simulation and experimental investigation on the residual stress-related monolithic component deformation. Int. J. Adv. Manuf. Technol. **77**, 1035–1041 (2014). https://doi.org/10.1007/s00170-014-6533-9
14. Wu, G., Li, G., Pan, W., Raja, I., Wang, X., Ding, S.: A state-of-art review on chatter and geometric errors in thin-wall machining processes. J. Manuf. Process. **68**(A), 454–480 (2021). https://doi.org/10.1016/j.jmapro.2021.05.055
15. Wang, G., Li, W., Tong, G., Pang, C.: Improving the machining accuracy of thin-walled parts by online measuring and allowance compensation. Int. J. Adv. Manuf. Technol. **92**, 2755–2763 (2017). https://doi.org/10.1007/s00170-017-0358-2
16. Fei, J., et al.: Investigation of moving fixture on deformation suppression during milling process of thin-walled structures. J. Manuf. Process. **32**, 403–411 (2018). https://doi.org/10.1016/j.jmapro.2018.03.011
17. Wang, Y., Xie, J., Wang, Z., Gindy, N.: A parametric FEA system for fixturing of thin-walled cylindrical components. J. Mater. Process. Technol. **205**(1–3), 338–346 (2008). https://doi.org/10.1016/j.jmatprotec.2007.11.216
18. Liu, H., Wang, C., Li, T., et al.: Fixturing technology and system for thin-walled parts machining: a review. Front. Mech. Eng. **17**, 55 (2022). https://doi.org/10.1007/s11465-022-0711-5
19. Liu, C., Wang, J., Zhou, B., et al.: Development of fixture layout optimization for thin-walled parts: a review. Chin. J. Mech. Eng. **37**, 17 (2024). https://doi.org/10.1186/s10033-024-01004-w
20. Liu, C., Wang, J., Zheng, Y., et al.: Optimization of fixture number in large thin-walled parts assembly based on IPSO. Chin. J. Mech. Eng. **37**, 1 (2024). https://doi.org/10.1186/s10033-023-00972-9
21. Liu, C., Zheng, Y., Wang, J., Jin, K., Yu, J., Liu, J.: Fixture layout optimization for large thin-walled parts based on improved particle swarm optimization algorithm. Int. J. Adv. Manuf. Technol. **132**, 3579–3592 (2024). https://doi.org/10.1007/s00170-024-13567-y
22. Feng, J., et al.: Optimization method of floating fixture layout for distortion control of low-stiffness thin-walled beams. Materials. **17**(17), 4226 (2024). https://doi.org/10.3390/ma17174226

23. Cheng, H., Li, Y., Zhang, K., Luan, C., Xu, Y., Li, M.: Optimization method of fixture layout for aeronautical thin-walled structures with automated riveting. Assem. Autom. **32**(4), 323–332 (2012). https://doi.org/10.1108/01445151211262384
24. Liu, Y., et al.: Evolutionary design of machining fixture layout for thin-walled structure. Math. Probl. Eng. (2022). https://doi.org/10.1155/2022/5216966
25. Subramanian, S., Arunraja, K.M., Praveen, P.: Optimization of machining fixture layout for thin walled component using DOE and FEM. J. Appl. Res. Technol. **21**(4), 552–559 (2023). https://doi.org/10.22201/icat.24486736e.2023.21.4.1398
26. Calabrese, M., Primo, T., Del Prete, A.: Optimization of machining fixture for aeronautical thin-walled components. Procedia CIRP. **60**, 32–37 (2017). https://doi.org/10.1016/j.procir.2017.02.008
27. Li, Z.-L., Tuysuz, O., Zhu, L.-M., Altintas, Y.: Surface form error prediction in five-axis flank milling of thin-walled parts. Int. J. Mach. Tools Manuf. **128**, 21–32 (2018). https://doi.org/10.1016/j.ijmachtools.2018.01.005
28. Zhong, Z., Mou, S., Hunt, J.H., Shi, J.: Convex relaxation for optimal fixture layout design. IISE Trans. **55**(7), 746–754 (2022). https://doi.org/10.1080/24725854.2022.2102272
29. Newaz, A.A.H., Jahan, R.: Design and implementation of fixtures for milling, shaping and drilling operations. Am. J. Mech. Eng. **13**(1), 1–5 (2025). https://doi.org/10.12691/ajme-13-1-1
30. Zhou, Y., Nomura, T., Dede, E.M., Saitou, K.: Topology optimization for 3D thin-walled structures with adaptive meshing. arXiv: Computational Engineering, Finance, and Science (2019). https://doi.org/10.48550/arXiv.1908.10825
31. Merlo, A., et al.: Novel adaptive fixturing for thin walled aerospace parts. IOP Conf. Ser. Mater. Sci. Eng. **26**(1), 012020 (2011). https://doi.org/10.1088/1757-899X/26/1/012020
32. Papastathis, T.N., Ratchev, S.M., Popov, A.A.: Dynamics model of active fixturing systems for thin-walled parts under moving loads. Int. J. Adv. Manuf. Technol. **62**(9–12), 1233–1247 (2012). https://doi.org/10.1007/s00170-011-3868-3
33. Gonzalo, O., et al.: A method to minimize the workpiece deformation using a concept of intelligent fixture. Robot. Comput. Integr. Manuf. **48**, 209–218 (2017). https://doi.org/10.1016/j.rcim.2017.04.005
34. Ma, J., Zhang, D., Wu, B., et al.: Stability improvement and vibration suppression of the thin-walled workpiece in milling process via magnetorheological fluid flexible fixture. Int. J. Adv. Manuf. Technol. **88**, 1231–1242 (2017). https://doi.org/10.1007/s00170-016-8833-8
35. Jiang, X., Zhao, G., Lu, W.: Vibration suppression of complex thin-walled workpiece based on magnetorheological fixture. Int. J. Adv. Manuf. Technol. **106**, 1043–1055 (2020). https://doi.org/10.1007/s00170-019-04612-2
36. Li, H., Chen, W., Shi, S.: Design and application of flexible fixture. Procedia CIRP. **56**, 528–532 (2016). https://doi.org/10.1016/j.procir.2016.10.104
37. Shi, Y., He, Y., Zha, J., Chen, B., Shi, C., Wu, M.: Design and manufacture of a flexible adaptive fixture for precision grinding of thin-walled bearing rings. J. Manuf. Mater. Proc. **9**(5), 139 (2025). https://doi.org/10.3390/jmmp9050139
38. Puma-Araujo, S.D., Olvera-Trejo, D., Martínez-Romero, O., Urbikain, G., Elías-Zúñiga, A., López de Lacalle, L.N.: Semi-active magnetorheological damper device for chatter mitigation during milling of thin-floor components. Appl. Sci. **10**(15), 5313 (2020). https://doi.org/10.3390/app10155313
39. Wang, T., Zha, J., Jia, Q., et al.: Application of low-melting alloy in the fixture for machining aeronautical thin-walled component. Int. J. Adv. Manuf. Technol. **87**, 2797–2807 (2016). https://doi.org/10.1007/s00170-016-8654-9
40. Möhring, H.-C., Wiederkehr, P.: Intelligent fixtures for high performance machining. Procedia CIRP. **46**, 383–390 (2016). https://doi.org/10.1016/j.procir.2016.04.042

41. Kitagawa, H., et al.: Design of a fixture for additively manufactured workpieces using topology optimization. Int. J. Autom. Technol. **19**(1), 4–14 (2025). https://doi.org/10.20965/ijat.2025.p0004
42. Plocher, J., Panesar, A.: Review on design and structural optimisation in additive manufacturing: towards next-generation lightweight structures. Mater. Des. **183**, 108164 (2019). https://doi.org/10.1016/j.matdes.2019.108164
43. Picavea, J., et al.: Synergistic integration of vibration absorption and damping into 3D-printed fixtures for thin-wall machining. CIRP Ann. Manuf. Technol. **72**(1), 357–360 (2023). https://doi.org/10.1016/j.cirp.2023.04.085
44. Castro, J., et al.: Density-based topology-optimized 3d-printed fixtures for cyclic mechanical testing of lattice structures. Polymers. **17**(18), 2468 (2025). https://doi.org/10.3390/polym17182468
45. Prajapati, M.J., Kumar, A., Lin, S.-C., Jeng, J.-Y.: Multi-material additive manufacturing with lightweight closed-cell foam-filled lattice structures for enhanced mechanical and functional properties. Addit. Manuf. **54**, 102766 (2022). https://doi.org/10.1016/j.addma.2022.102766
46. Liu, Q., Zhang, H., Leng, J., Chen, X.: Digital twin-driven rapid individualised designing of automated flow-shop manufacturing system. Int. J. Prod. Res. **57**(12), 3903–3919 (2019). https://doi.org/10.1080/00207543.2018.1471243
47. Grzesik, W.: Digital twin in manufacturing. Part II. Case studies in removal and additive machining processes. Mechanik. **2**, 6–14 (2023). https://doi.org/10.17814/mechanik.2023.2.4
48. Hänel, A., et al.: Digital twins for high-tech machining applications-a model-based analytics-ready approach. J. Manuf. Mater. Proc. **5**(3), 80 (2021). https://doi.org/10.3390/jmmp5030080
49. Jarosz, K., Özel, T.: Machine learning approaches towards digital twin development for machining systems. Int. J. Mechatronics Manuf. Syst. **15**(2–3), 127–148 (2022). https://doi.org/10.1504/IJMMS.2022.124922
50. da Silva, L.R.R., Pimenov, D.Y., da Silva, R.B., Ercetin, A., Giasin, K.: Review of applications of digital twins and industry 4.0 for machining. J. Manuf. Mater. Proc. **9**(7), 211 (2025). https://doi.org/10.3390/jmmp9070211

Robotics

Process of Building Human-Robot Communication Based on the Natural Interfaces Within XR Systems for the Use in Process Automation

Olaf Ciszak[1], Przemysław Zawadzki[1], and Marcin Kasica[2](✉)

[1] Poznan University of Technology, Poznan, Poland
[2] AVR EDU, Poznan, Poland
marcin.kasica@gmail.com

Abstract. The rapid advancement of automation and the evolution of Human-Machine Collaboration (HMC) are particularly significant within modern manufacturing and design processes and fits into the main ideas and development trends of robotics and industry 4.0/5.0. This article presents preliminary tests and a potential pathway for developing a Human-Robotic Collaboration (HRC) system built upon a Digital Twin and integrated within a fully immersive eXtended Reality (XR) environment. The evolution of virtual reality, combined with image processing techniques from augmented reality, has led to term of eXtended Reality (XR). The work objective was to create and make ready to test a low-entry level automation solution by combining a physical collaborative workstations and simulation environment within an XR natural interfaces as unified Cyber-Physical System (CPS). However, further quantitative analysis regarding both usability in both academic laboratories and industrial environments are required.

Keywords: Human-Robotic Collaboration (HRC) · Human-Computer Interaction (HCI) · Automation · Extended Reality (XR) · Natural Interfaces

1 Introduction

1.1 Evolution for Immersive Systems

The evolution of Virtual Reality (VR), together with technologies like Augmented Reality (AR), has led the concept of eXtended Reality (XR). This is aligned with advancements in the interfaces between computer-driven data processing and human communication. The I^3 definition [1] characterises VR through immersion, interactivity and imagination-based engagement. Developments in Artificial Intelligence and enhanced human-feature capture enables the I^4 concept [2] adding Intelligence layer to immersive systems, supporting natural, human-centred interaction (Fig. 1).

Modern Extended Reality (XR) devices concentrate on the multi-sensory aspect to enhance immersion and interaction [3]. Immersion is supported not only by stereoscopic

B. Gapiński et al. (Eds.): MANUFACTURING 2026, LNME, pp. 219–233, 2026.
https://doi.org/10.1007/978-3-032-22943-4_16

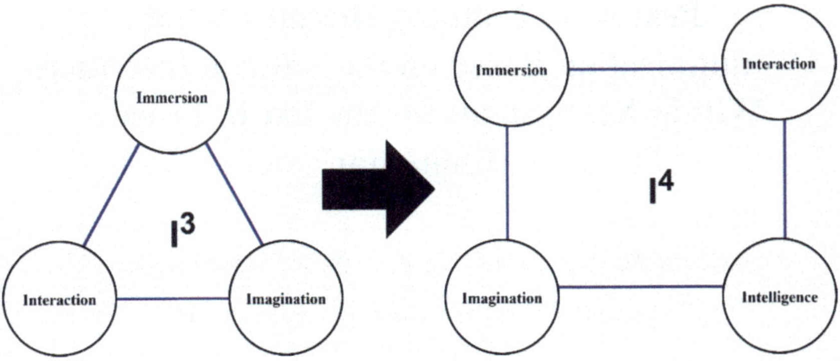

Fig. 1. I^3 definition of the virtual reality by [1] and its evolution to I^4 Intelligence [2].

viewing but also by controller-based interactions and optical tracking systems that provide precise 6DOF and precise user localisation. Many systems offer full-body tracking and gesture recognition.

These devices frequently utilise integrated cameras for depth sensing, RGB colour capture via built-in cameras, and often even LiDAR technology. This sensor suite enables complex spatial awareness, including the capabilities for environment scanning, scene capturing, and digital anchoring. Thanks to Simultaneous Localisation and Mapping (SLAM) an effect of more realistic experience, higher level of immersion and advance possibility of interaction between virtual and real world [4].

Additionally, built-in wireless communication and microphone systems, paired with spatial audio rendering, significantly deepen immersion. These features facilitate interaction via voice and audio, delivering audio cues that complement the visual cues.

Input devices, specifically handheld controllers and other auxiliary equipment, can be used to provide haptic feedback and further interactive modalities.

1.2 Role of Collaborative Robots (Cobots) in Automation

Human-driven programming, which often require significant time and specialised skills, compounded by the unintuitive nature of traditional teaching pendants, collaborative robots (cobots) are increasingly used in both educational settings and as a foundational automation solution for SMEs; this adoption is driven by their inherent benefits like i.e. consistency, flexibility, easy setup. They are utilised for product lines requiring frequent changeovers.

The primary challenge in programming cobots often centres on the traditional Teaching Pendant (TP) interface. While the TP offers the advantage of precise positional data input, essential for industrial accuracy, its brand-specific and non-unified nature necessitates specialised training and experience for each robot manufacturer. Furthermore, this online programming method requires operators to manually handle the physical pendant and, in some cases, the robotic arm itself (via lead-through programming or cobot-mounted buttons) to define waypoints. Although this allows for direct, real-time control, it is inherently cumbersome and limits the use of natural human interfaces (such as voice or advanced gestures) due to the necessity of a dedicated external device for command and control, ultimately increasing the complexity and time required for setup and redeployment, especially in high-mix, low-volume production environments.

1.3 Case-Study Design Process Focus on Potential SMEs Needs

Collaborative robots are strategically suited for SMEs due to their inherent characteristics of consistency, flexibility, and simplified setup. These attributes make them ideal for production lines that demand frequent changeovers and high product quality. The system's design addresses the necessity for flexible, low entry level automation that supports human-centred collaboration. Further adoption is driven by the increasing variety and reducing price of cobots.

Further adoption is driven by the increasing variety of cobots. The strategic suitability of cobots for SMEs is perfectly mirrored in practical, human-centred applications that demand high flexibility. Small laboratories often serve as vital integration and visualization examples for industry-relevant processes:

- Robotic Welding and Assembly (Fig. 2a): One laboratory might focus on demonstrating the process consistency and precision in complex, frequent changeover tasks like robotic welding.
- Assembly line in Smart Factory Laboratory (Fig. 2b): Another laboratory showcases a cobot integrated into a small, versatile production line for light assembly tasks. This example emphasises the cobots simplified setup and flexibility, allowing it to be easily re-tasked for the high variety and low volume production typical of many SMEs, supporting seamless human-robot collaboration and demonstrating a low entry-level automation solution

These laboratory environments, while academic, provide tangible models for how SMEs can implement flexible automation to meet their needs for product variety and quality.

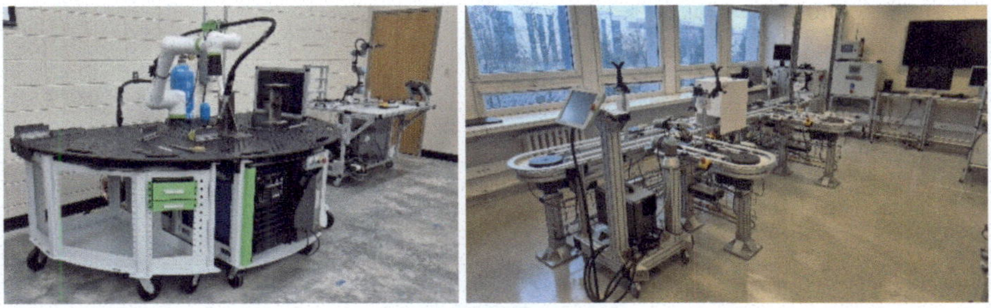

Fig. 2. Reference collaborative robot workstations used in the design. Advanced manufacturing laboratory at MGCCC USA (left - a) and Smart Factory Laboratory [5] at Poznan University of Technology (right - b).

2 System Architecture and Conceptual Design

2.1 Challenges of Traditional Automation

Traditional automation and design workflows present significant barriers to SME integration. These challenges are not only technical but also financial, often making it difficult for smaller enterprises to justify or sustain investment in robotics [6].

- Technical Skills Deficit: Programming requires specialised skills and time.
- Non-Intuitive Programming: The reliance on the teaching pendant is non-intuitive and demands extensive training.
- Simulation Gaps: Design processes, while utilising CAD/CAM, rarely incorporate simulation-based learning.
- High Initial Investment and ROI Constraints: The upfront cost of automation - including robot purchase, integration and training - represents a significant financial barrier.

2.2 Integrative Approach with the Challenge

Modern automation systems are inherently Cyber-Physical Systems (CPS) [7, 25], meaning they are built upon the synergistic combination of several specialised fields [8]. Integration provides the necessary generic view and ability to seamlessly blend knowledge across traditionally separated domains (see Fig. 3):

- Mechanical Engineering: Essential for designing the physical structures, kinematics, and dynamic performance of the robot, machinery, and production line layout.
- Robotics/Automatics: Focuses on the control theory, motion planning, sensor integration, and operational safety protocols for automated devices, including collaborative robots (cobots).
- Informatics/Computer Science: Provides the digital backbone, including real-time data processing, AI algorithms for decision-making and perception, and the software logic for control systems.

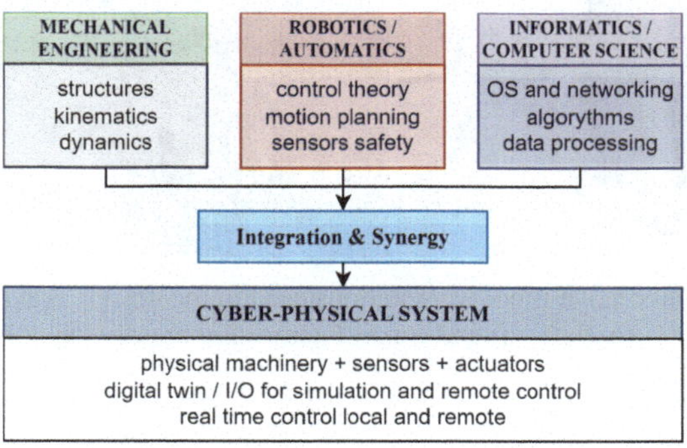

Fig. 3. Integrative approach due to synergetic collaboration various specialities toward design of Cyber-Physical System.

2.3 CAD and DCC, eXtended Reality and Tri-Directional Integration

Implementing robotic simulation and visualisation faces hurdles due to limitations in design workflows and asset management. Education in mechanical engineering and automation shows a deficit in blending industrial informatics with problem-solving [9], as design largely focuses on Computer-Aided Design (CAD) and Computer-Aided Manufacturing (CAM), with minimal simulation-based learning [10]. This is compounded by the high cost, limited accessibility, and technical expertise needed for industrial simulators.

To address this, the system architecture applies a tri-directional integration model (Fig. 4) linking the Physical Robot (C), the Simulation Environment (B), and the XR interactive application (A). As illustrated in Fig. 5, real-world robot data and kinematic parameters are synchronised with the virtual robot instance running in the Virtual Machine-based simulation environment for testing and offline validation using Robotic Operation System ROS2 tools. The major elements of the Human-Computer Collaboration is shifted from the Physical Robot (C) to the immersive level (A).

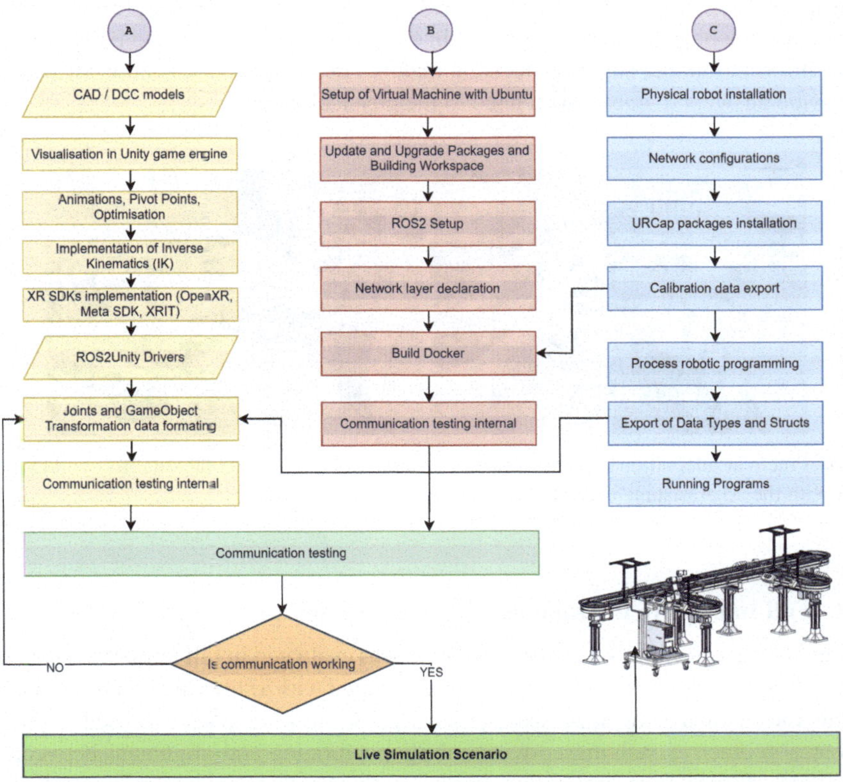

Fig. 4. Analysed process of integration driven in three concurrent paths toward single goal.

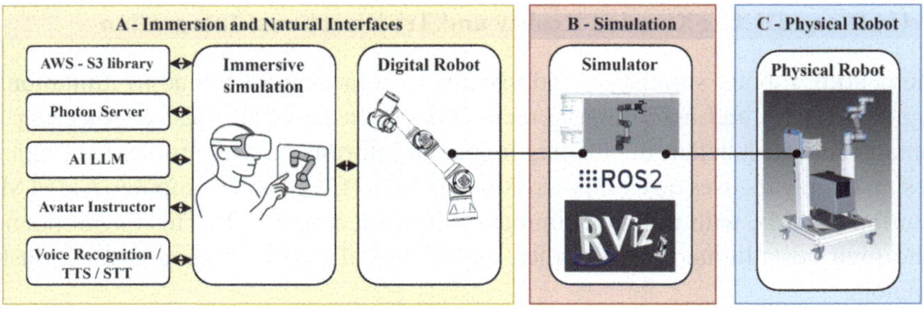

Fig. 5. Concept diagram of the role of eXtended Reality within the HRC.

3 Designing the Solution

3.1 Visualisation of the Collaborative Robot

AR Contextualised – visualisation through the 3D model presentation using augmented reality via mobile devices (Fig. 6) together with the spatial anchor and image processing using EON-XR platform [11]. This approach provides a fast and simple pre-build process to visualise the objects, merge them within environments, annotate and build contextualised information that supports decision making, design processes and is already able to provide answers to questions related to the physical space.

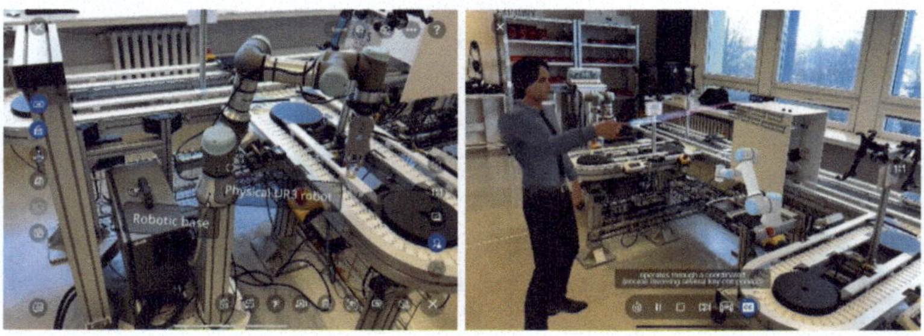

Fig. 6. Practical adaptation of the augmented reality (AR) mobile contextualised collaborative robot with the AI Assistant avatar [11].

3.1.1 Full Immersive Environment/XR Passthrough/Mixed Reality (MR)

During the initial testing phase, the simulation was evaluated in full immersion using the XR headset, where the 3D robotic assets were merged with a simplified representation of the laboratory environment. A basic spatial scan of the workspace was used to provide reference geometry, enabling correct scaling, positioning and alignment between the virtual and physical elements.

At this stage, only fundamental interactions were implemented, serving primarily to validate the feasibility of XR-based manipulation and system synchronisation. Further development is required to refine the environmental 3D assets and increase the visual fidelity of the laboratory representation.

Preliminary observations indicate that operating the robotic arm within XR using handheld controllers appears significantly more natural and intuitive than physical manipulation or programming via a traditional teach pendant. However, this impression requires systematic validation through further studies, following methodologies similar to those presented by Zhang et al. [12].

During testing of the XR passthrough Mixed Reality mode, the system was evaluated within the actual laboratory space (see Fig. 7) to establish a safe and clearly defined collaboration zone. A coarse 3D scan of the environment was used to generate an underlying geometry mesh, which served as the physical layer for AR Raycasting [21] and accurate projection of virtual information onto real-world surfaces. This invisible mesh enabled the system to anchor visual cues, indicators and interaction points directly onto physical objects, improving spatial alignment between the virtual content and the real workspace.

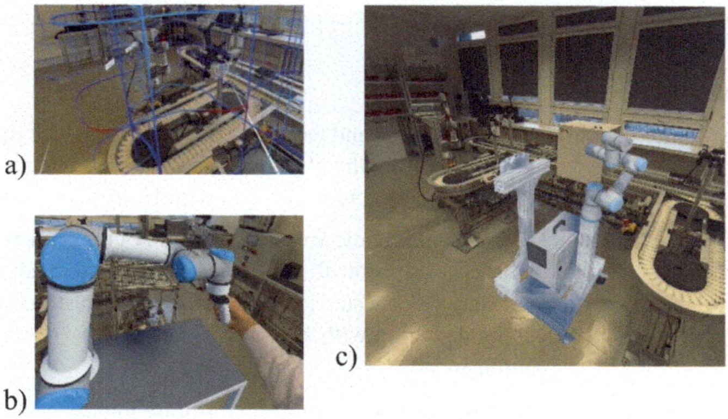

Fig. 7. Adaptation to the mixed reality (MR) solution via physical space mapping (a) 1:1 scale augmentation (c) and virtual robotic arm control (b).

Following this, functional models of the collaborative robots were introduced into the passthrough environment to assess how effectively the virtual robot behaviour could adapt to real operational scenarios. These tests highlighted the need to extend the robot motion library to include more precise joint trajectories, incorporating joint-limit constraints and physical parameters that better reflect the real robot's kinematics. This refinement is necessary to ensure that the virtual and physical systems remain synchronised and that the XR-based workflow accurately represents the operational limits and dynamics of the collaborative robot.

3.2 ROS2 Environment for Integrated Communication

The establishment of a unified communication framework between the extended reality (XR) application and the robotic system require the deployment of the Robot Operating System 2 (ROS2). ROS2 provides a robust middleware architecture based on distributed nodes, topics, services and actions, making it well-suited for cyber-physical systems that demand reliability, modularity and real-time communication [13].

3.2.1 Initial Deployment and Evaluation

The initial implementation was undertaken on a Raspberry Pi 5 running Ubuntu 24.04 with the ROS2 Jazzy (Kilted Kaiju) distribution. Although the system operated correctly in isolation, several constraints emerged [14]. Performance was limited due to the computational demands of ROS2 networking, particularly when visualisation with RViz and messaging loads increased. Additionally, compatibility issues with XR system based on Unity integration restricted seamless data exchange. These factors rendered the Raspberry Pi environment insufficient for the intended real-time XR–robot interaction.

To ensure long-term stability and compatibility, the environment was migrated to Ubuntu 22.04 LTS using the ROS2 Humble Hawksbill distribution [15]. ROS2 Humble remains fully supported until 2027 and provides greater maturity and stability compared to newer releases. It also ensures optimal compatibility with essential development tools, including RViz2 and commonly used robotics libraries (Table 1).

Table 1. Evaluation of ROS2 platform deployment method- Advantages and Disadvantages

ROS2 solution	Advantages	Disadvantages
Microcomputer Kilted Kaiju	Newest version and easy to integrate within the physical robotic infrastructure	Low performance in visualisation and incompatible with Unity
VirtualBox Virtual Machine Humble Hawksbill	Robust and reliable version with network control (Bridging) for faster communication with XR application	Require computer workstation to be running in addition to the other IT infrastructure
Windows Subsystem for Linux (WSL) Humble Hawksbill	Directly operated within Windows.	Limitation on virtualisation requirements from Windows and difficult network communication configurations

3.2.2 Virtual Machine-Based Development Environment

A dedicated development environment was established on a Virtual Machine (VM) configured in Oracle VirtualBox. Allocating enhanced computational resources to the VM improved the performance associated with the earlier Raspberry Pi deployment. The VM installation included also Ubuntu 22.04 and ROS2 Humble distribution, enabling the use of **RViz2** for real-time visualisation of robot kinematics, transforms and sensor data.

This configuration provides a non-problematic testbed for communication between the XR application and the ROS2 ecosystem (Fig. 8).

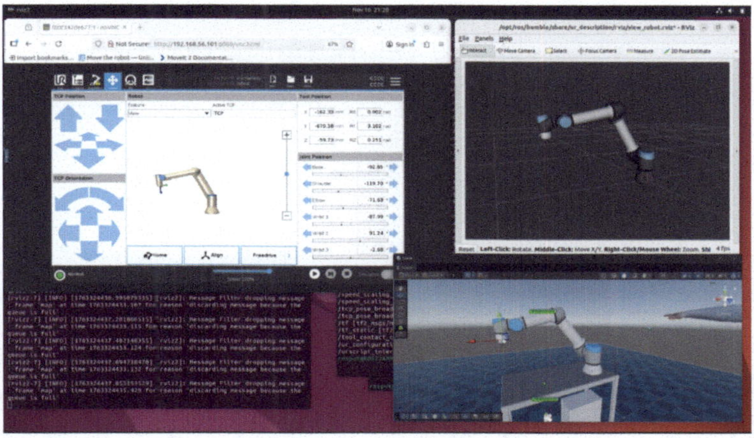

Fig. 8. ROS2 on Ubuntu Virtual Machine with successful signal exchange between Windows Unity instance in both directions.

3.2.3 Evaluation of Unity–ROS2 Integration Solutions

Two primary options were considered for bridging ROS2 with Unity:

Unity ROS–TCP–Connector [16] - Developed by Unity Technologies, this framework is primarily optimised for ROS1 and early ROS2 experimentation. Its limited long-term support and partial compatibility with newer ROS2 message structures restrict its suitability for advanced applications [17].

RobotecAI ros2-for-unity *(chosen solution)* [18] - This actively maintained plugin provides native support for ROS2 Humble and subsequent releases. It utilises standard ROS2 message definitions, facilitates C# message generation within Unity, and supports efficient communication via the Fast DDS middleware. Its architecture is specifically designed to meet real-time simulation and robotic visualisation requirements.

3.3 Common Interface via ROS2Unity Plugin

The developed and presented in this paper technical concept relies on the integration between Unity and ROS2 through the *ros2-for-unity* plugin, which provides a common communication layer for exchanging structured robotic data. While the plugin offers native compatibility with ROS2 Humble, additional refinement of the C# scripts was required to ensure full compliance with ROS2 message definitions [19] and visualisation standards, particularly for use within RViz2.

Initial tests focused on implementing basic publishers (*talkers*) and subscribers (*listeners*) to validate message exchange. These tests employed standard ROS2 data types, including *Bool*, *Float32*, *String*, *Point*, and *KeyValue*. Further development was necessary to define appropriate message handlers and ensure that custom message structures could be correctly interpreted by RViz2 and other ROS2 visualisation tools (Fig. 9).

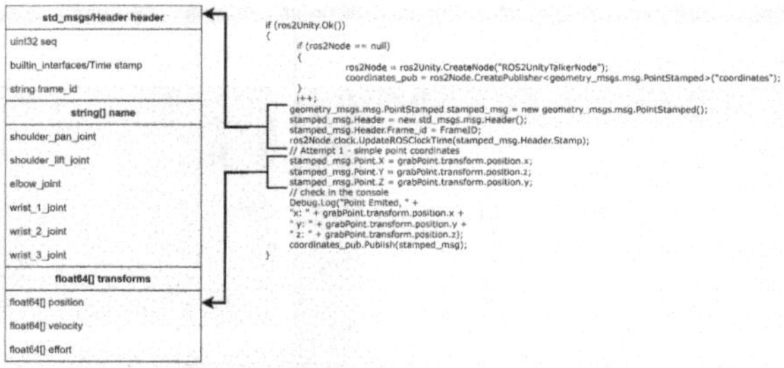

Fig. 9. Structure of the ROS2 Unity plugin and key components to be implemented Own work based on [18, 19].

Key work involved implementing fundamental ROS2 communication concepts (nodes, topics, services, data types) to make Unity a fully compatible, real-time ROS2 participant. As the project advanced, this extended to handling complex data for joint positions, kinematic states, motion commands, and safety parameters. Unlike real robots, the simulation required explicitly programmed physical and collision constraints, based on accurate robot modelling. This real-time, structured communication layer is the foundation for advanced natural, XR, and AI-assisted control interfaces.

3.4 Progressive Integration of Natural Interfaces in XR-Based Robot Control

The development of the XR application initially focused on the visualisation of the workstation, the construction of the robot's kinematic model, and the basic integration of software development kits (SDKs) for both Windows and Android devices. These early steps enabled user engagement within a simulated environment and established the foundation for subsequent interaction mechanisms.

During the preliminary testing phase, interaction with the system began through standard mouse/keyboard inputs and game controllers, providing familiar 2D/3D visualisation control schemes aligned with established open-source robotic interfaces. As development progressed, immersive XR technologies were introduced using OpenXR, the XR Interaction Toolkit and the Meta SDK [20], enabling operation of the Meta Quest 3 via Quest Link software with features such as hand-tracking, gesture recognition, stereo passthrough and 3D environment mapping for a fully interactive XR experience. Further integration incorporated audio input/output through the headset's built-in microphone, allowing Natural Language Processing (NLP) capabilities via external APIs, to support both text-based and voice-based interaction directly within the immersive environment (see Fig. 10).

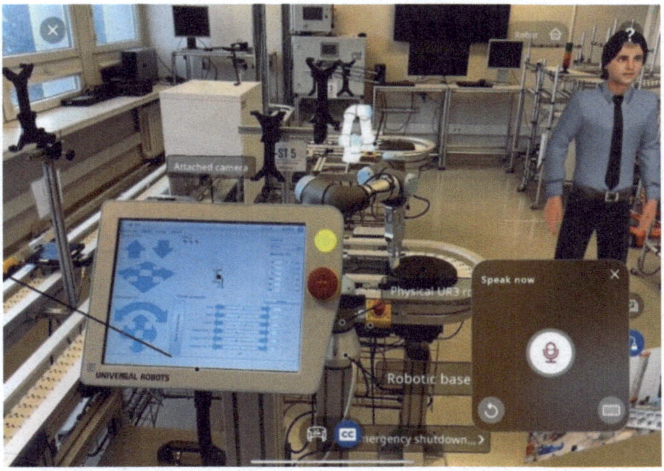

Fig. 10. User in the immersive application providing voice commands and AI agent transform to the reply as text and audio.

3.4.1 Real World Spatial Awareness via Spatial Anchors

To merge the physical and virtual environments more closely, passthrough functionality was combined with the concept of Spatial Anchors. Using depth-scanning capabilities, users are able to place persistent anchors that acted as digital reference points linked to precise physical locations. This approach improved spatial accuracy in terms of scaling, positioning and movement. Mobile-based depth perception technologies such as AR Foundation [21], AR Core [22], ARKit [23] are also employed to support accurate anchor placement. The Meta SDK further enhanced the anchoring process through integrated scene understanding features (Fig. 11).

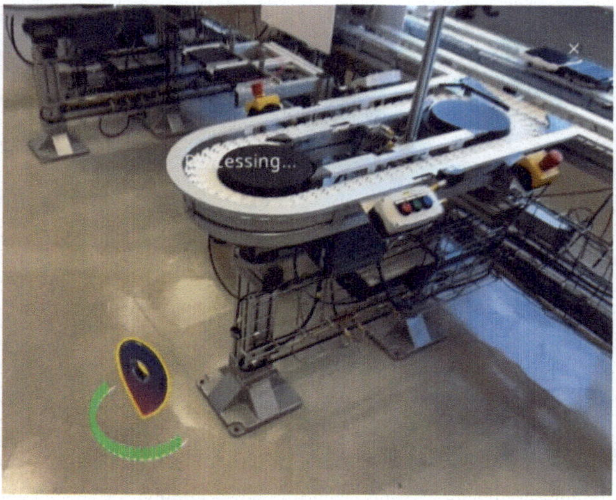

Fig. 11. Spatial awareness via the Spatial Anchors locked within the application and presentation of the AR Raycast to the depth point of location.

Despite these advancements, several challenges remain. Interface elements like buttons used to save positions, can be difficult to manipulate in XR, and navigating code or visual programming interfaces can become cumbersome even with graphical representations. The support in design was provided by ready-made Building blocks [20] within the Meta SDK packages.

Introducing voice control offers an alternative and more natural means of interaction. It enables personalised operator workflows and facilitates hands-free operation. By embedding an AI engine capable of processing contextual information and domain-specific training data, the system can support intelligent, adaptive interaction and assist users throughout the programming process.

3.5 Tests and Implementation

Initial testing of the system was conducted within the Advanced Manufacturing Laboratory at Mississippi Gulf Coast Community College (MGCCC). The early stages focused on validating the visualisation pipeline by creating a 360° virtual tour of the laboratory environment, which served as a contextual reference for later interactive elements. This was followed by the acquisition of three-dimensional scans of the laboratory and the development of preliminary visualisations and animation-based controls for the Fanuc CRX10iA and UR3 collaborative robots. These early demonstrations confirmed the feasibility of XR-based representation of robotic systems and highlighted the importance of integrating real-time communication to improve usability and operational accuracy.

Subsequent testing and development were undertaken at the Poznan University of Technology, where access to a UR3 collaborative robot [5] and an associated basic assembly workstation enabled more rigorous experimentation. This environment made it possible to evaluate the behaviour of the virtual and physical systems in parallel, examining synchronisation, latency, and the fidelity of motion reproduction when transferring data between ROS2, Unity, and the robot controller.

The laboratory at Poznan University of Technology is being prepared as the primary location for future development and evaluation. It will allow students and academic staff to interact directly with both the virtual and physical cobots, providing opportunities to study the correspondence between simulated and real-world robotic behaviour. This approach supports the advancement of robotics education and research without incurring the high costs associated with procuring multiple physical robots. Through the integration of XR simulation, ROS2 communication, and access to industrial-grade hardware, the facility will enable extended experimentation, prototyping and training activities.

4 Conclusions and Implications for Future Research

4.1 XR Devices and Human–Robot Collaboration

The continuing development of XR devices is significantly enhancing the possibilities for natural and intuitive human–robot communication. Improvements in hand-tracking [20], passthrough imaging, spatial mapping and ergonomic design allow operators to interact with collaborative robots through gestures, voice and context-aware interfaces. As these technologies mature, XR will increasingly support accessible, low-barrier robotic programming and more effective use of digital twins for planning and validation.

4.2 Advancements in Artificial Intelligence and Generative AI

Progress in AI - particularly in computer vision and generative models - offers substantial benefits for virtual simulation and human–machine interaction. Real-time analysis of image and video streams enables more accurate scene understanding, enhances virtual environment fidelity and supports natural language interfaces for robotic control. These capabilities improve situational awareness, reduce technical complexity for operators and strengthen the link between virtual and physical robotic systems.

4.3 Importance of Real-Time Communication Protocols

As XR, AI and robotics become increasingly interconnected, efficient real-time communication is essential. Lightweight protocols such as MQTT [6, 24] provide low latency, high scalability and reliable data exchange across diverse devices. Adopting such protocols enables smoother XR interactions, more responsive robot behaviour and safer, more integrated human–robot collaboration.

4.4 Future Research and Practical Validation

While the proposed Cyber-Physical System establishes a robust technical framework for human-robot collaboration, the next critical phase involves rigorous practical testing within both educational and industrial environments. Future work will move beyond initial prototyping to a comprehensive analysis of the system's efficacy in real-world scenarios, specifically focusing on two key aspects: technical precision and operational efficiency.

Consequently, future work will focus on validating the accuracy of virtual motion planning against physical execution. Additionally, a comparative assessment of natural interfaces versus traditional methods will analyse operational efficiency and learning curves to determine the solution's significance in lowering automation barriers for SMEs.

References

1. Burdea, G., Coiffet, P.: Virtual Reality Technology. J. Wiley-Interscience, Hoboken (2003)
2. Bustamante, S., Martinovic, M., Schmitz, M., Saltevo, E.: Innovation in higher education and its role for sustainable development–an introduction to the book. In: Shaping a Sustainable Future, p. 24 (2022)
3. Lyu, K., Brambilla, A., Globa, A., de Dear, R.: An immersive multisensory virtual reality approach to the study of human-built environment interactions. Autom. Constr. **150**(104836), 104836 (2023). https://doi.org/10.1016/j.autcon.2023.104836
4. Wang, Z.-M., Rao, M.-H., Ye, S.-H., Song, W.-T., Lu, F.: Towards spatial computing: recent advances in multimodal natural interaction for extended reality headsets. Front. Comp. Sci. **19**(12) (2025). https://doi.org/10.1007/s11704-025-41123-8
5. Bendowska, K., Zawadzki, P.: Development and verification of a simulation model of an automated assembly line. Appl. Sci. **13**(18), 10142 (2023)
6. Gonzalez-Sanchez, J., Guerra-Silva, R.: Exploring extended reality to optimize human-Cobot interaction. In: 2025 20th ACM/IEEE International Conference on Human-Robot Interaction (HRI), pp. 1314–1318. IEEE (2025). https://doi.org/10.1109/hri61500.2025.10973902
7. Qiu, S., He, Q., Fan, X., Wu, D.: Virtual human hybrid control in virtual assembly and maintenance simulation. Int. J. Prod. Res. **52**(3), 867–887 (2013). https://doi.org/10.1080/00207543.2013.842025
8. Menezes, J.C.: MuModaR: multi-modal framework for human-robot collaboration in cyber-physical systems. In: Companion of the 2024 ACM/IEEE International Conference on Human-Robot Interaction, pp. 755–759. ACM (2024). https://doi.org/10.1145/3610978.3640688
9. Moraes, E.: Impact of new ICT trends for the educational curriculum in the area of industrial automation and engineering. https://doi.org/10.1109/IECON.2015.7392667 (2015)
10. Al-Ahmari, A., Abidi, M., Ahmad, A., Darmoul, S.: Development of a virtual manufacturing assembly simulation system. Adv. Mech. Eng. **8** (2016). https://doi.org/10.1177/1687814016639824
11. EON-XR Platform.: EON-XR - Library. [online] Available at: https://www.eon-xr.com. Accessed 17 Nov 2025 (2025)
12. Zhang, Y., Orthmann, B., Ji, S., Welle, M., Van Haastregt, J., Kragic, D.: Multimodal "Puppeteer": an exploration of robot teleoperation via virtual counterpart with LLM-driven voice and gesture interaction in augmented reality (Version 1). arXiv. 10.48550/ARXIV.2506.13189 (2025)
13. Mahnke, W., Leitner, S.-H., Damm, M.: OPC Unified Architecture. Springer, Berlin Heidelberg (2009). https://doi.org/10.1007/978-3-540-68899-0
14. Flores Gonzalez, J.M., Coronado, E., Yamanobe, N.: ROS-compatible robotics simulators for industry 4.0 and industry 5.0: a systematic review of trends and technologies. Appl. Sci. **15**(15), 8637 (2025). https://doi.org/10.3390/app15158637

15. Open Robotics.: ROS.org | Powering the world's robots. [online] Ros.org. Available at: https://www.ros.org/ (2020)
16. GitHub.: Unity-Technologies/ROS-TCP-Connector. [online] Available at: https://github.com/Unity-Technologies/ROS-TCP-Connector (2024)
17. de Boer, T.A.B., de Winter, J.C.F., Eisma, Y.B.: Augmented reality-based telepresence in a robotic manipulation task: an experimental evaluation. IET Collab. Intell. Manuf., e12085 (2023). https://doi.org/10.1049/cim2.12085
18. GitHub.: RobotecAI/ros2-for-unity. [online] Available at: https://github.com/RobotecAI/ros2-for-unity (2024)
19. Unity.: Advance your robot autonomy with ROS 2 and Unity. [online] Available at: https://unity.com/blog/engine-platform/advance-your-robot-autonomy-with-ros-2-and-unity (2024)
20. Meta.com.: Explore Meta Quest Features with Building Blocks. [online] Available at: https://developers.meta.com/horizon/documentation/unity/bb-overview/ (2025)
21. AR foundation. Unity3d.com. Available at: https://docs.unity3d.com/Packages/com.unity.xr.arfoundation@6.4/manual/index.html. Accessed 15 Nov 2025
22. Overview of ARCore and supported development environments. Google for Developers. Available at: https://developers.google.com/ar/develop?hl=en. Accessed 15 Nov 2025
23. ARKit.: Apple Developer Documentation. Available at: https://developer.apple.com/documentation/arkit. Accessed 15 Nov 2025
24. Medium.: Medium. [online] Available at: https://medium.com/@saumysharma007/ros2-to-unity-mqtt-connection-for-roboticsimulation. Accessed 16 Nov 2025
25. Oks, S.J., et al.: Cyber-physical Systems in the Context of industry 4.0: a review, categorization and outlook. Inf. Syst. Front. **26**, 1731–1772 (2024). https://doi.org/10.1007/s10796-022-10252-x

Spray Painting AMR System

Grzegorz Iskierka(✉), Bartosz Poskart, Kamil Krot, and Paweł Krowicki

Wrocław University of Science and Technology, Wrocław, Poland
grzegorz.iskierka@pwr.edu.pl

Abstract. This paper presents an innovative autonomous mobile spray painting system that integrates a MiR200 mobile platform with a UR10 collaborative robotic arm, offering a flexible, reconfigurable solution for large-scale industrial painting tasks. Designed to address the limitations of traditional stationary systems, the robot combines mobility, precision, and modularity to paint complex components like wind turbine blades and glider wings efficiently and with high quality. By incorporating advanced features such as onboard power supplies, pneumatic systems, and safety mechanisms, the system operates autonomously in dynamic environments while ensuring seamless collaboration with human workers. Its adaptability allows for quick reprogramming and deployment across various production layouts, making it ideal for low-volume, high-cost manufacturing scenarios. The paper highlights the system's transformative potential in industrial automation and explores future directions, including fleet-level coordination, AI-driven navigation, sustainability improvements, and broader applications such as cleaning and inspection. This research demonstrates how autonomous mobile robots can revolutionize industrial processes while advancing smart manufacturing practices.

Keywords: Autonomous Mobile Robots · AMR · Collaborative Robots · Cobots · Spray Painting

1 Introduction

The automation of industrial processes has evolved significantly through the industrial revolutions. Initially, it focused on mechanizing existing tasks, then enabling mass production, and ultimately automating entire workflows, gradually reducing human involvement. The third industrial revolution made automation more accessible, but most systems remained stationary, limiting the scale and flexibility of automated production. Today, in the fourth industrial revolution, manufacturing is becoming more adaptable with reconfigurable workstations, easily movable collaborative robots, and highly reliable autonomous mobile robots. This shift has allowed industries to embrace greater flexibility, accommodating changing demands and enabling more customizable products.

The industry is steadily progressing toward the fifth industrial revolution, marked by the seamless collaboration of diverse fleets of multifunctional autonomous robots. Unlike previous automation waves that focused on isolated, task-specific robots, this new era

B. Gapiński et al. (Eds.): MANUFACTURING 2026, LNME, pp. 234–246, 2026.
https://doi.org/10.1007/978-3-032-22943-4_17

emphasizes interconnected robotic systems that work together dynamically to achieve shared goals. These robots, equipped with advanced AI, real-time communication, and decentralized decision-making, can adapt to changing conditions, optimize workflows, and enhance productivity across various industries.

By integrating different types of autonomous robots, such as mobile robots for material transport, robotic arms for assembly, and drones for inspection, factories and warehouses are becoming more efficient, flexible, and resilient. This level of cooperation not only improves operational efficiency but also paves the way for more intelligent, adaptive production environments capable of handling complex, customized manufacturing needs. The shift toward interconnected automation is transforming industries, bridging the gap between human-robot collaboration and fully autonomous smart factories.

While many processes have already been automated in this way, like warehouse logistics, which has matured through technologies like autonomous mobile robots (AMRs) and AI-driven management systems [1, 2], broader fleet autonomy faces persistent technical and regulatory barriers [3]. Although industries such as logistics and manufacturing are rapidly advancing toward fully autonomous systems, where for example autonomous trucking is already transforming freight movement, manufacturing facilities increasingly adopt autonomous forklifts, loaders, and mobile robots to streamline intralogistics and material handling [4]. These advancements suggest that the shift to autonomous fleets is accelerating across various sectors, driven by breakthroughs in AI, robotics, and connected vehicle technologies.

Other industries are still to adapt to an autonomous fleet approach. For example, the construction sector has yet to fully integrate autonomous machinery for tasks like excavation and demolition on a large scale. Hyung Joo Lee et al. have analysed the construction sector for autonomous robots for deconstruction of concrete walls of buildings, which in many cases are a major stream of waste, and proposed a solution for such deconstruction, utilising robots [5]. Similarly, public transportation and emergency services remain reliant on human operators due to complex operational environments and safety concerns. In some places such public transportation autonomous vehicles are being tested, yet still remote operators need to be present. To adapt the work environment for those remote operators, researchers have compiled a number of scenarios which may occur during emergency situations and gathered feedback in order to create a framework that can be used as a basis for interdisciplinary dialogue between engineers, computer scientists, mobility researchers, human factors specialists, and decision-makers on how to conceptualize and further develop remote operation [6]. In the production industry, while intralogistics automation is gaining traction, it is speculated that by 2027 more than a quarter (26%) of warehouse operations will have some form of automation according to a report by Interact Analysis [7], highlighting the untapped potential for broader adoption of autonomous fleet technologies across diverse industrial contexts, one of which could be industrial spray painting [8].

2 State of the Art

2.1 Automated Spray Painting

Currently, there are multiple solutions in the market that allow to automatically spray paint both small and large objects in a production environment, as well as walls of entire buildings, using different manipulation systems depending on the application. Automated spray painting applications can be found in industries such as automotive [9], aerospace [10], construction [11], and furniture manufacturing [12], delivering consistent, high-quality finishes while reducing material waste and labor costs. Such systems utilize advanced features such as real-time surface analysis, adaptive spray patterns, and predictive analytics to optimize paint application and minimize overspray [13].

Automated spray painting solutions significantly enhance safety by reducing human exposure to hazardous environments and toxic chemicals, such as volatile organic compounds (VOCs) and isocyanates, which are common in industrial paints [14]. Explosion-proof robotic systems eliminate risks associated with flammable substances, while precise automation minimizes overspray and airborne particles, improving air quality in workspaces. By replacing manual labor in confined or high-risk areas, these systems also reduce the likelihood of workplace injuries caused by repetitive strain or accidental contact with machinery. Overall, automated painting technologies create safer working conditions while maintaining high-quality output.

Currently, there are no fully autonomous spray painting robots designed to operate as a coordinated fleet in the same way autonomous vehicles, autonomous mobile robots or drones might. However, there are advanced systems that could be considered precursors to such fleet-based operations. For example, companies like AMBPR or okibo have developed autonomous mobile robots for large-scale applications, such as ship hull maintenance or painting walls at construction sites, where multiple units can work simultaneously but independently on tasks like washing, sandblasting, and painting. Similarly, Dürr's modular painting robots are optimized for automotive production lines and can be deployed in scalable setups, though they function individually rather than as a synchronized fleet. While these systems excel in precision and efficiency, true fleet-level autonomy, where multiple robots coordinate dynamically in real time, is still under development and largely absent from current commercial offerings.

2.2 Multifunctional AMR Fleets

The use of Autonomous Mobile Robots (AMRs) in fleet applications is rapidly expanding across industries, performing tasks such as intralogistics, material handling, and collaborative operations with cobots. However, one of the major challenges lies in the incompatibility of fleet systems provided by different manufacturers. For instance, systems like MiR Fleet enable centralized coordination of AMRs but are limited to their proprietary devices, restricting interoperability with robots from other vendors. This lack of standardization creates inefficiencies and limits the scalability of multi-robot systems in industrial settings.

To address this, an overarching system based on graph theory offers a promising solution. Graph-based frameworks enable the modeling and optimization of multi-robot networks by representing robots as nodes and their interactions or communication links as edges. Such an approach facilitates dynamic task allocation, efficient path planning, and seamless coordination between heterogeneous robots. By leveraging adjacency matrices and graph-theoretic algorithms, a unified system could integrate multidisciplinary mobile robots into a cohesive network, overcoming compatibility issues and enhancing overall functionality. This method not only maximizes efficiency but also provides flexibility for adapting to complex industrial environments as presented in previous work [15].

Algorithms derived from graph theory enable dynamic task allocation, efficient path planning, and real-time coordination between heterogeneous robots. For example, adjacency matrices can analyze communication links between robots, while algorithms like minimum spanning trees or multi-agent pathfinding ensure collision-free navigation and optimized workflows. A centralized graph-based system could unify multidisciplinary mobile robots into a cohesive network, ensuring interoperability and maximizing efficiency across diverse industrial applications. This approach not only addresses the limitations of proprietary fleet systems but also provides a scalable architecture capable of adapting to complex production environments.

3 Research Methodology

The proposed solution leverages commercially available Autonomous Mobile Robots (AMRs) and collaborative robots (cobots) to create a fleet-based system that can be reprogrammed and reconfigured for various tasks and heterogeneous mobile robots. Unlike proprietary systems like MiR Fleet, which are limited to managing robots from a single manufacturer, this approach emphasizes interoperability and flexibility. Current vendor-specific systems often restrict collaboration between robots of different brands, creating inefficiencies in multi-vendor environments. By adopting a centralized fleet management system based on graph theory, the proposed solution overcomes these limitations.

The autonomous mobile painting robot integrates a MiR200 mobile platform with a UR10 collaborative robotic arm, offering a versatile and innovative solution for industrial painting tasks (see Fig. 1). This system is specifically designed to handle large-scale components that exceed the workspace of traditional stationary robots, such as wind turbine blades or glider wings. By repositioning itself along the object, the robot can paint different sections sequentially, eliminating the need for fixed painting zones or specialized facilities like large spray booths. Its workspace, when docked, encompasses a spherical segment with a radius of 1300 mm, enabling precise and repeatable application of paint.

One notable application of this system is the painting of wind turbine blades, which can reach lengths of up to 50 m. Traditional methods require extensive infrastructure, such as transport systems for human painters within specialized booths. The autonomous painting robot simplifies this process by docking at multiple positions along the blade and systematically covering designated sections (see Fig. 2).

Another example includes painting glider wings, showcasing the system's adaptability to various large-scale components, such as glider wings, as presented in Fig. 3.

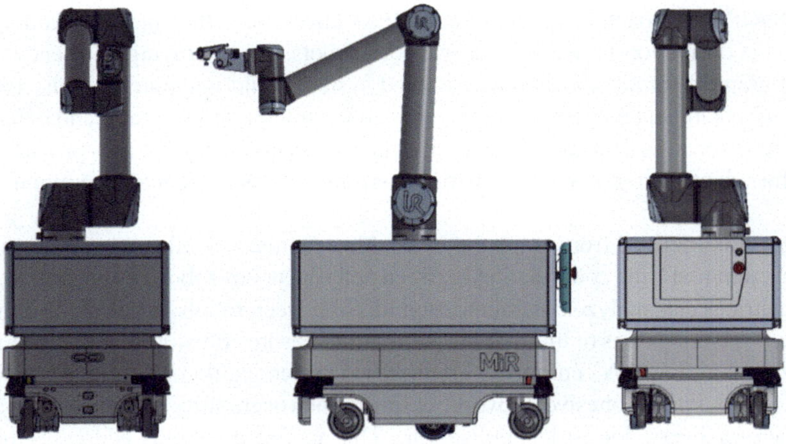

Fig. 1. Autonomous mobile painting robot, consisting of a MiR200 mobile platform and a UR10 collaborative robotic arm.

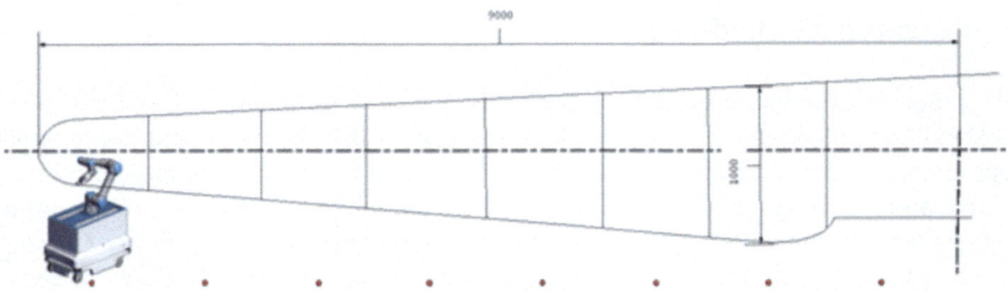

Fig. 2. Diagram showing the painting process for wind turbine blades.

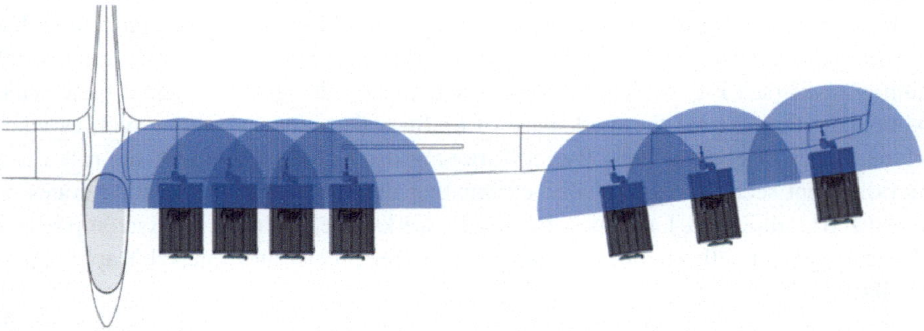

Fig. 3. Painting process for glider wings.

The system's mobility allows it to operate in multiple locations within a single production hall or even move between facilities, providing unmatched operational flexibility. Tasks can be programmed cyclically based on predefined schedules, ensuring consistent output across different environments. Its onboard power supply and integrated compressor further enhance its autonomy, enabling it to function independently of external infrastructure. Designed as a collaborative robot (cobot), the system operates safely in shared environments without requiring physical barriers. It is equipped with safety features that allow it to navigate spaces where workers and other equipment are present, ensuring seamless integration into dynamic production settings.

The air spray painting equipment integrated into the autonomous mobile painting robot is designed to ensure high-quality, consistent, and efficient paint application while maintaining flexibility and safety. The system includes a compressor, which provides the necessary air pressure for paint atomization, ensuring smooth and even coverage. A filter dryer is incorporated to remove moisture from the compressed air, preventing imperfections in the paint finish caused by water contamination. Additionally, a pressure regulator allows precise control of air pressure to optimize paint flow and atomization, ensuring uniform layers and minimal overspray. Finally, an automatic air spray gun is mounted on the collaborative robotic arm (UR10), enabling accurate and repeatable application of paint across complex surfaces (see Fig. 4). The process of spray painting has been deliberately chosen for this application because of the lack of need for too high of a precision in positioning. In this setup, the mobile platform has a positioning accuracy of 50 mm without and 10 mm with a docking prism. Additional hardware can be acquired for precision docking, allowing for positioning accuracy of up to 1 mm. The UR10 collaborative robot does not specify its absolute accuracy but the repeatability is determined as ±0.1 mm. Given these conditions, depending on the task at hand, the system may require different approaches to increase accuracy of the entire system, but it has been foregone for the purpose of robotic spray painting in this research.

Fig. 4. Compartment planned for the pneumatic supply system.

This setup is compactly integrated into the mobile platform (MiR200), with dedicated compartments for each component to maintain balance and mobility. The design ensures that the robot can autonomously perform painting tasks in various locations without relying on external equipment or infrastructure. The modularity of the system also allows for easy reconfiguration to accommodate different types of paints or nozzles, making it suitable for diverse industrial applications like painting wind turbine blades or glider wings. The use of advanced pneumatic components ensures a high level of precision, making this robot an effective solution for large-scale industrial painting tasks. Figure 5 presents the wrist of a UR10 cobot equipped with an automatic paint gun.

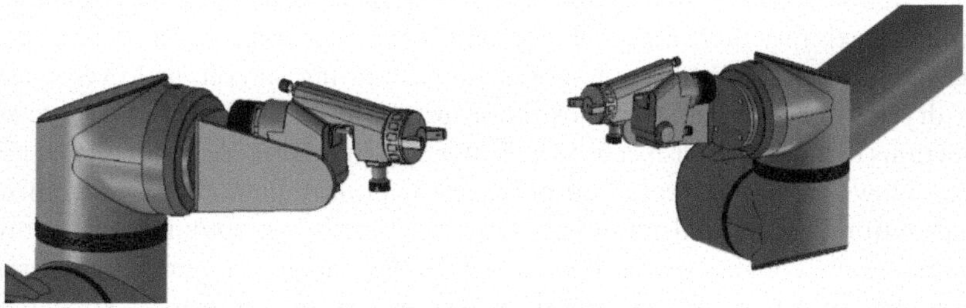

Fig. 5. UR10 robot wrist equipped with an automatic air spray gun.

The autonomous mobile painting robot offers exceptional ease of reconfigurability and reprogramming, making it ideal for flexible production environments, particularly in low-volume, high-cost, and time-sensitive manufacturing. The system integrates a MiR200 mobile platform with a UR10 collaborative robotic arm, enabling it to adapt to various tasks and production layouts. Its modular design allows for quick reconfiguration to accommodate different tools, such as painting equipment or manipulators for other operations. This flexibility is crucial in industries where production requirements frequently change or involve large, complex components like wind turbine blades or glider wings.

Reprogramming the robot is straightforward due to its user-friendly software interfaces. The UR10 arm supports intuitive programming methods, including hand-guided teaching and offline programming, which reduce setup time and minimize production downtime. Similarly, the MiR200 platform can be easily programmed to navigate new environments by creating maps and defining intermediate points for precise positioning. This capability ensures that the system can seamlessly transition between tasks or locations without requiring extensive manual intervention.

The basic operational cycle of the MiR mobile robot program, as illustrated in the diagram shown in Fig. 6, demonstrates how tasks are executed in the robotic painting system. Every cycle starts with the robot docked at a charging station and standing by for the next mission call. Once the object to be painted is positioned correctly, the robot begins its mission by moving to the first intermediate point, where it can position itself at a designated painting spot. Once positioned, it activates the program for the UR10 robotic arm, which performs the planned operation, such as applying paint to a specific section of the object. After completing the task, the robot moves to the next intermediate point and continues its work cycle according to the predefined sequence.

The robot's ability to work collaboratively in human environments further enhances its adaptability. It can operate safely alongside workers and other equipment without physical barriers, thanks to integrated safety systems like LiDAR-based navigation and collision avoidance. This feature makes it particularly suited for dynamic production settings where workflows may need frequent adjustments. Overall, the autonomous mobile painting robot's reconfigurability and ease of programming provide a cost-effective and efficient solution for manufacturers aiming to optimize short production runs or handle diverse product requirements. This autonomous mobile painting system represents a significant advancement in industrial automation by combining mobility, safety, and precision to meet the demands of large-scale and complex painting tasks efficiently.

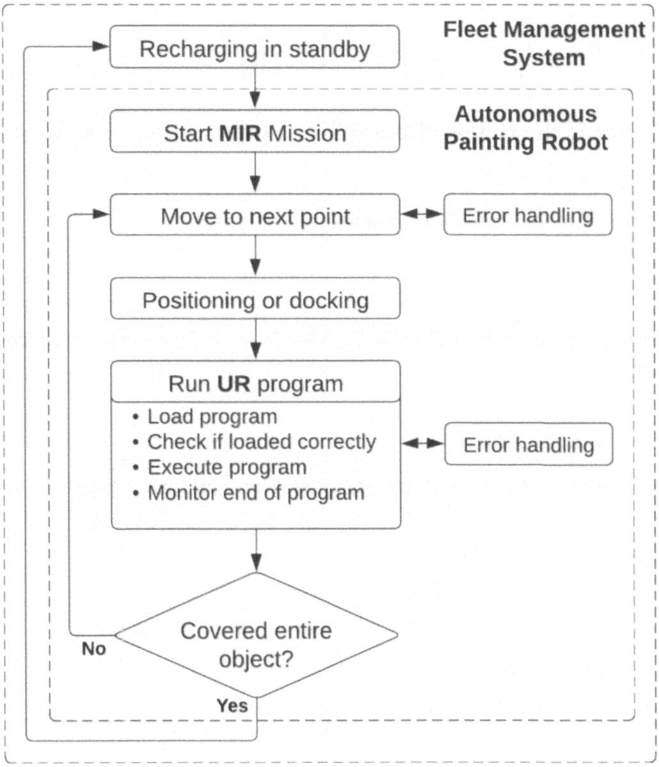

Fig. 6. Flow diagram of the MiR mobile robot program.

4 Results and Discussion

One of the primary objectives of this work was to develop algorithms for calculating the effective painting pass width based on key robot parameters, such as the operational range and the position of the robot's base center relative to the object. In order to determine the width of the painting pass, an algorithm was developed. The pseudocode for this algorithm is presented below.

INPUTS:

H_wall // Wall height [m]
R // Radius of the robot's workspace (sphere) [m]
d // Distance from the robot's base to the wall [m]
z0 // Height of the robot's base (center of the sphere) [m]

OUTPUT:

W_eff // Effective painting pass width [m]

Step 1: Check if the robot can reach the wall

```
IF d ≥ R THEN
W_eff ← 0 // The robot cannot reach the wall
TERMINATE algorithm
```

Step 2: Compute the radius of the sphere's intersection with the wall

```
r_int ← sqrt( R² − d² )
```

Step 3: Define the vertical boundaries of the wall

```
z_bottom ← 0
z_top ← H_wall
```

Step 4: Compute the vertical offsets (absolute differences) with respect to the sphere's center

```
Δz_bottom ← | z0 − z_bottom |
Δz_top ← | z_top − z0 |
```

Step 5: Compute the chord lengths (horizontal cut lengths) at the wall's boundaries

```
IF Δz_bottom ≤ r_int THEN
chord_bottom ← 2 · sqrt( r_int² − (Δz_bottom)² )
ELSE
chord_bottom ← 0
IF Δz_top ≤ r_int THEN
chord_top ← 2 · sqrt( r_int² − (Δz_top)² )
ELSE
chord_top ← 0
```

Step 6: Determine the effective painting pass width as the minimum chord length

```
W_eff ← min( chord_bottom, chord_top )
```

Given the following parameters:

H_wall = 1.8 m (wall height)
R = 1.3 m (radius of the robot's workspace sphere)
d = 0.5 m (distance from the robot's base to the wall)
z0 = 0.7 m (height of the robot's base, i.e., the center of the sphere)

The horizontal chord lengths at the two vertical boundaries of the wall are computed as follows:

chord_top (at the top of the wall, z = 1.8 m) is approximately 0.96 m.
chord_bottom (at the bottom of the wall, z = 0) is approximately 1.95 m.

This means that although the robot can reach a wider area at the lower part of the wall, the overall effective width of the painting pass is constrained by the smaller reachable area at the top of the wall, resulting in an effective width of approximately 0.96 m. Figure 7 illustrates the visualization of the workspace, including the robot's base center position relative to the wall, as well as the calculated lower and upper boundaries of the effective painting pass.

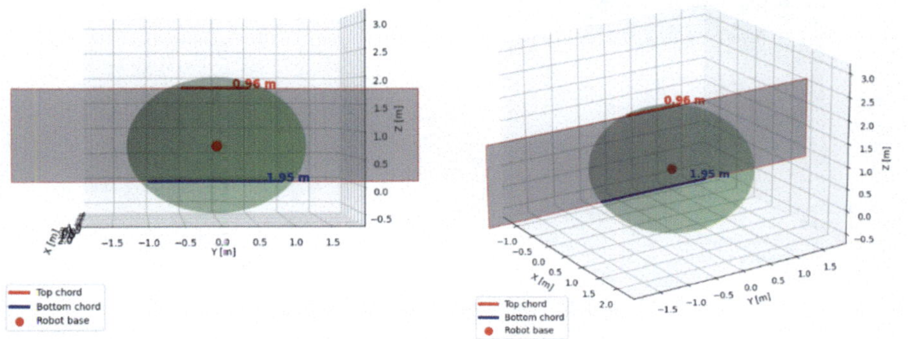

Fig. 7. Visualization of workspace with lower and upper boundaries of the effective painting pass.

This representation provides a clear understanding of how the algorithm interprets the robot's operational parameters, enabling an accurate determination of the pass width. Using the effective painting width, with an assumed 50% overlap of the painted area, and the known coordinates of the start and end points of the painted element, a subsequent algorithm was developed. This algorithm calculates the target coordinates for the mobile robot, which must navigate to these points for the collaborative robot arm with the automatic spray paint gun to initiate the painting program.

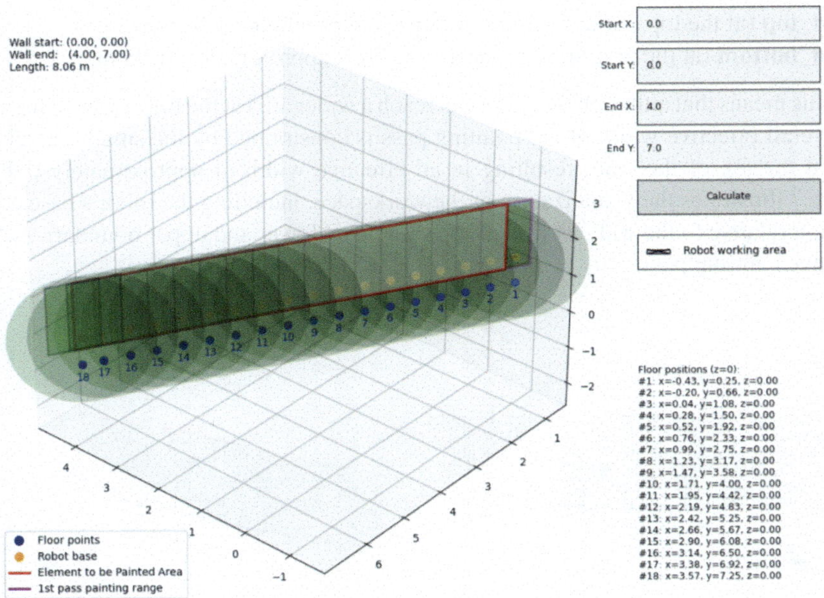

Fig. 8. Workspace visualization for each computed point.

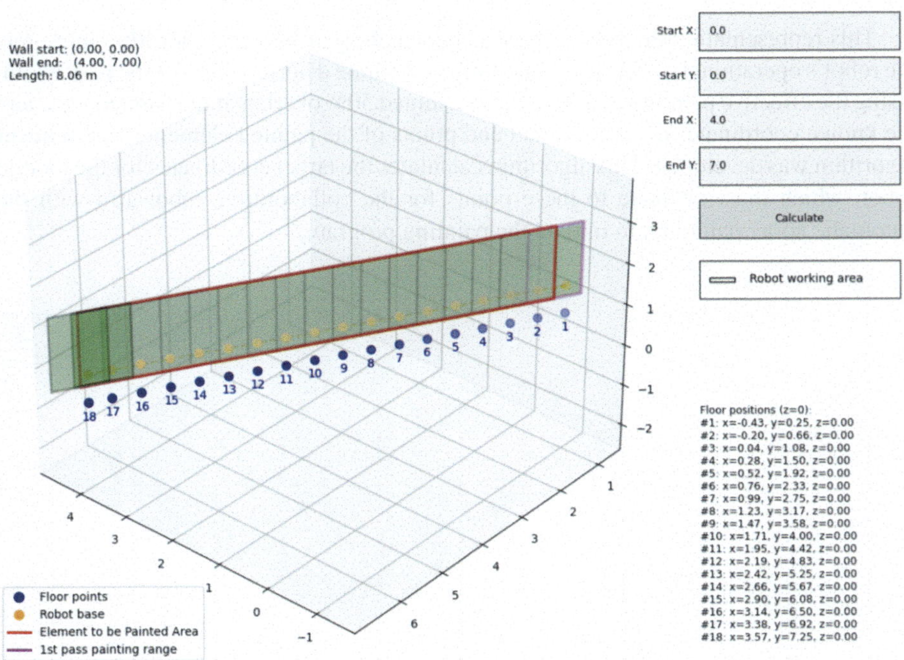

Fig. 9. Paint coverage visualization for each computed point.

Figures 8 and 9 illustrate the workspace and paint coverage visualization and the intermediate points computed for the mobile robot. In this system, the program accepts the coordinates of the beginning and end of the painted element as input and automatically computes all intermediate points in space. This approach enables the automatic preparation for painting of any area, provided its position in space is known.

5 Conclusions and Outlook

In this publication, we have explored the potential of implementing Autonomous Mobile Robot (AMR) platforms within the spray painting industry, particularly for manufacturing or refurbishing large-scale objects such as aeronautic, ship, and railway components. The proposed system combines the mobility of AMRs with the precision of collaborative robotic arms to create a versatile, reconfigurable solution for industrial spray painting tasks. By leveraging advanced features like graph-theoretic fleet management, onboard power supplies, and integrated pneumatic systems, the autonomous mobile painting robot demonstrates significant advantages in terms of flexibility, efficiency, and safety.

The system's ability to adapt to various tasks and production layouts makes it particularly suited for industries where large, complex components need to be painted with high precision. Its modular design and ease of reprogramming allow manufacturers to optimize short production runs or handle diverse product requirements without extensive infrastructure investments. Furthermore, the integration of safety features ensures seamless operation in dynamic environments shared with human workers.

The future of autonomous mobile spray painting systems holds immense potential for innovation and industrial transformation. One key area for development is the coordination of multiple robots within a fleet, enabling real-time collaboration to optimize workflows and reduce downtime. Incorporating advanced AI-driven navigation, obstacle avoidance, and decision-making capabilities allows these systems to operate more effectively in complex environments. Further machine learning algorithms could be employed for real-time surface analysis and adaptive spray patterns, like in the work done by D. Gleeson et al. in [13], further improving efficiency and precision. Furthermore, the modularity of these robots opens opportunities for applications beyond spray painting, such as cleaning, inspection, or surface preparation. Expanding their functionality could make them indispensable tools across various industries. However, as these systems become more prevalent, addressing regulatory challenges related to safety standards and operational guidelines will be crucial to ensure compliance and widespread acceptance.

In conclusion, the outlook for autonomous mobile spray painting systems is highly promising. By addressing current limitations and focusing on innovation in fleet coordination, interoperability, autonomy, sustainability, and versatility, these systems are poised to revolutionize industrial processes and contribute significantly to the advancement of smart manufacturing practices, enabling flexibility in production processes of large-scale industrial objects.

Acknowledgments. The authors declare that the research reported in this paper received no external funding or grant support. The work was conducted within the framework of a team research activity and was supported solely by the internal resources of the University.

References

1. Sodiya, E.O., Umoga, U.J., Amoo, O.O., Atadoga, A.: AI-driven warehouse automation: a comprehensive review of systems. GSC Adv. Res. Rev. **18**(2), 272–282 (2024)
2. Ramachandran, K.K.: Intelligent manufacturing: leveraging autonomous robotics and AI. IJCSITR. **5**(2), 20–32 (2024)
3. Dabic-Miletic, S.: The challenges of integrating AI and robotics in sustainable WMS to improve supply chain economic resilience. J. Ind Intell. **2**(2), 119–131 (2024). https://doi.org/10.56578/jii020205
4. Fragapane, G., de Koster, R., Sgarbossa, F.: Autonomous Mobile robots for material handling in intralogistics. In: Manzini, R., Accorsi, R. (eds.) Warehousing and Material Handling Systems for the Digital Industry. Springer, Cham (2024). https://doi.org/10.1007/978-3-031-50273-6_10
5. Lee, H.J., Heuer, C., Brell-Cokcan, S.: Concept of a robot assisted on-site deconstruction approach for reusing concrete walls. In: Proceedings of the International Symposium on Automation and Robotics in Construction, vol. 39, pp. 422–429. IAARC Publications (2022)
6. Kettwich, C., Schrank, A., Avsar, H., Oehl, M.: A helping human hand: relevant scenarios for the remote operation of highly automated vehicles in public transport. Appl. Sci. **12**(9), 4350 (2022). https://doi.org/10.3390/app12094350
7. Osieczko-Potoczna, K.: Reasons, benefits and challenges on the road to automated internal transportation. Acta Logistica. **11**(4), 605–613 (2024)
8. Okonkwo, C., Sulbaran, T.: A review of commercially available autonomous painting robots in the construction industry. In: Proceedings of 60th Annual Associated Schools, vol. 5, pp. 514–522 (2024). https://doi.org/10.29007/h44l
9. Idris, A.: AM improvement of painting and welding in automotive industry using robots. IOSR J. Mech. Civil Eng. **17**(5), 18–28 (2020). https://doi.org/10.9790/1684-1705011828
10. Becroft, S.A.: Automated painting for aerospace, challenges, newer technologies and lessons learned. SAE Int. J. Aerosp. **5**(1), 22–30 (2012). https://doi.org/10.4271/2012-01-1855
11. Asadi, E., Li, B., Chen, I.-M.: Pictobot: a cooperative painting robot for interior finishing of industrial developments. IEEE Robot. Autom. Magaz. **25**(2), 82–94 (2018). https://doi.org/10.1109/MRA.2018.2816972
12. Karusine, A., Cui, C., Larsey, N., Tanguy-Nolain, K.: The impact of automation in painting furniture parts. Open J. Appl. Sci. **12**, 1995–2003 (2022). https://doi.org/10.4236/ojapps.2022.1212139
13. Gleeson, D., et al.: Generating optimized trajectories for robotic spray painting. IEEE Trans. Autom. Sci. Eng. **19**(3), 1380–1391 (2022). https://doi.org/10.1109/TASE.2022.3156803
14. Moura, P.C., Santos, F., Fujão, C., et al.: Towards the identification of the volatile organic compounds emitted by the coatings used in a car factory painting line. J. Coat. Technol. Res. **21**, 665–682 (2024). https://doi.org/10.1007/s11998-023-00847-7
15. Iskierka, G., Poskart, B., Krot, K., Telesiński, B., Anthony Xavior, M. (2024). Application of graph theory in designing the communication system of a robotic production cell. In: Machado, J., et al. Innovations in Mechatronics Engineering III. Icieng 2024. Lecture Notes in Mechanical Engineering. Springer, Cham. https://doi.org/10.1007/978-3-031-61575-7_7

Integration of Robotic Technologies in Modern Heat Treatment Processes

Ján Duplák, Samuel Mikuláško(✉), Maryna Yeromina, and Róbert Pyteľ

Faculty of Manufacturing Technologies with a Seat in Presov, Technical University of Košice, 080 01 Prešov, Slovakia

samuel.mikulasko@tuke.sk

Abstract. Heat treatment of materials is a technological process that allows control of the microstructure and mechanical properties of metals, but traditional approaches still face limitations in terms of accuracy, repeatability and energy efficiency. The dynamic development of industrial robotics is fundamentally transforming these processes, as robotic systems enable highly accurate, localised and time-controlled application of heat in induction, laser and furnace workplaces. The integration of advanced sensor technology – pyrometry, thermovision, 3D geometry measurement and adaptive power control – enables real-time monitoring of the temperature field and continuous optimisation of process parameters according to the current conditions of the material. At the same time, numerical simulations, the finite element method and digital twins play a key role in predicting thermal, diffusion and microstructural phenomena even before the technological operation is performed, thus shortening development cycles and minimising the costs of experiments. The article provides a comprehensive overview of modern robotic solutions used in the heat treatment of materials, compares traditional heat treatment options with robotic ones, analyses their application possibilities, technical limitations, interaction with artificial intelligence systems, and identifies the main trends associated with autonomy, digitalisation and the Industry 4.0 concept. Modern robotic heat treatment workplaces thus represent a significant step towards more efficient, accurate, safer and sustainable industrial production.

Keywords: Heat Treatment · Robotics · Induction Heating · Laser Hardening · Industry 4.0

1 Overview

1.1 Introduction

Heat treatment is one of the oldest and still indispensable technological processes in manufacturing industries, as it allows targeted modification of the microstructure of metals and alloys in order to obtain the desired mechanical properties, such as hardness, strength, toughness and wear resistance [1]. Traditional technologies such as annealing, hardening and tempering are proven, but often encounter limitations in terms of precise heat field control, energy efficiency and deformation minimisation, especially for complex or large-sized parts [2, 3]. In the last decade, there has been a significant increase in

B. Gapiński et al. (Eds.): MANUFACTURING 2026, LNME, pp. 247–260, 2026.
https://doi.org/10.1007/978-3-032-22943-4_18

interest in combining robotics with heat treatment processes, as robotic arms equipped with laser heads, induction coils or other heat sources allow local heat treatment to be carried out directly on site (in situ) without the need to handle heavy components [4, 5]. Such systems can reduce thermal deformation, shorten processing time and increase the repeatability of industrial processes, while meeting the requirements of modern Industry 4.0. Laser Surface Hardening (LSH) is one of the most important robotic approaches. The technology uses extremely localised heating of the material surface and rapid self-cooling to achieve high surface layer hardness with minimal thermal impact on the core. Studies emphasise that parameters such as laser power, energy density and scanning speed have a significant impact on the resulting microstructure and hardening depth [6, 7]. The finite element method (FEM) is commonly used to simulate temperature and microstructural development during the process, allowing technological modes to be optimised without lengthy experiments [3] In addition to laser technologies, robotic induction is also important, as it enables precise local heat treatment, for example in post-weld heat treatment (PWHT). Research combines electromagnetic and thermal FEM simulations with optimisation algorithms to predict temperature fields and design efficient heating trajectories [5, 8]. Modern solutions implement neural networks that can predict steel sheet deformations and adaptively adjust heating power in real time, thereby increasing processing accuracy [9]. An equally important area is the integration of sensor systems that provide feedback for process parameter control. Advanced monitoring using thermocouples, thermal cameras or FPGA/PLC systems enables precise control of temperature and energy efficiency [10, 11]. Digitisation and the use of digital twins provide the ability to simulate the thermal response of parts, optimise robot trajectories and predict the resulting microstructure even before processing. Research in this area goes beyond traditional metallic materials. Induction-activated composites are emerging in soft robotics, enabling rapid actuation and programmable deformation [12]. Another interesting direction is the development of robotic grippers with controlled heat flow, which classify materials based on their heat transport properties using machine learning methods, in particular LSTM networks [13]. It follows from the above that the integration of robotic technologies into heat treatment represents a significant shift towards precision, digitalisation, energy efficiency and autonomous production systems. Thanks to the combination of advanced sensors, FEM simulations, artificial intelligence and digital twins, thermal processes are becoming more flexible, intelligent and better adapted to the requirements of modern industrial applications.

Foundational studies on conventional heat treatment processes are provided by Ford [3], while recent influential works by Jurči and Novák [1] and Agbedo [2] reflect the transition towards advanced and energy-efficient heat treatment concepts.

1.2 Heat Treatment of Materials

Heat treatment is one of the most important technological processes used in industry to modify the microstructure of metals and alloys. Classic techniques include annealing, hardening and tempering, which are used to improve mechanical properties such as strength, hardness, toughness and ductility. These processes are essential in manufacturing industries, including the automotive, aerospace, energy and engineering industries [3, 14]. In recent decades, advanced heat treatment processes have also emerged, such as vacuum heat treatment (vacuum hardening, tempering, annealing, carburising and nitriding), which allow precise control of temperature and atmosphere. These methods significantly affect the microstructure, mechanical properties and chemical stability of steel components [10]. Surface hardening – especially carburising, nitriding and plasma-assisted techniques – is particularly effective in increasing hardness and wear resistance while maintaining corrosion resistance, especially in the case of stainless steels [2]. The integration of numerical simulation and computational modelling allows for the precise optimisation of heat treatment parameters for specific applications. Modern approaches are complemented by innovative methods such as thermomechanical processing, cryogenic processing and gradient heat treatment, which provide a high degree of control over the resulting properties and are suitable for applications in demanding conditions [2]. These various techniques highlight the key role of heat treatment in the manufacture of components with precisely defined operating requirements [1, 14]. Heat treatment is of fundamental economic importance because it allows the mechanical properties of materials to be improved while reducing energy and operating costs. Current review studies [1] emphasise its role in modern industry, where microstructural control directly affects the functional reliability of components. Energy efficiency is critical in industry. Modern PLC/FPGA-based monitoring systems for induction hardening ensure process stability and reduce energy losses [11]. The design of flexible induction hardening equipment capable of working with different types of steel further increases economic profitability [15] Post-weld heat treatment (PWHT) is another significant benefit — it can restore degraded mechanical properties and extend the service life of expensive components [16]. Furthermore, energy analyses show that combined induction-convection heating strategies can reduce energy consumption by more than 50% [17]. Precise temperature control also plays a key role. Intelligent iPI controllers enable model-free control in furnaces, which is advantageous for non-linear systems without a precise mathematical model [10]. Microstructural evaluations, for example in PWHT processes of aluminium alloys, have demonstrated the importance of calibrated heat treatment on residual stresses and fatigue behaviour [18].

The fundamental principles of heat treatment and their metallurgical background are well established in classical works [3, 14], which form the theoretical foundation for modern approaches. More recent studies focus on advanced techniques such as vacuum heat treatment, surface engineering and energy optimisation [2, 10, 11].

1.3 Types of Heat Treatment

Annealing

Annealing is one of the basic heat treatment processes and offers significant advantages in reducing internal stresses, improving formability and homogenising microstructure. Innovative annealing techniques enable the production of high-strength steels while reducing the need for alloying, thereby improving production efficiency. Its limitations include energy intensity and lower efficiency in increasing strength compared to more advanced techniques such as vacuum heat treatment or cryogenic treatment [2]. Austenitisation temperature: typically above Ac3 for recrystallisation annealing of carbon and low-alloy steels; lower temperatures with long holding times for spheroidisation annealing. Cooling: slow furnace cooling minimises stresses and ensures a uniform microstructure; faster cooling can lead to heterogeneity. Practical application: treatment of sheets before forming, preparation of parts before hardening, improvement of the machinability of tool steels (Fig. 1).

Fig. 1. Annealing of Al-Mg-Si aluminium alloy using a robot [19]

Hardening

Hardening is the process of rapid cooling from the austenitising temperature to form martensite, thereby significantly increasing the hardness and strength of the material. Gas hardening offers increased safety and lower environmental impact compared to liquid hardening [20, 21]. Intensive quenching can also improve fatigue properties and enable the use of more economical steels. Disadvantages include the formation of residual stresses, deformation and the risk of cracking due to high temperature gradients during cooling [22]. Therefore, precise control of parameters is essential — choice of quenchant, controlled cooling (martempering, austempering) and pre-heating [23]. Newer approaches, such as high-pressure gas quenching, represent more environmentally friendly alternatives (Fig. 2).

Tempering

Tempering is the step following hardening, and its purpose is to reduce the brittleness of martensite, optimize the microstructure and adjust the strength-toughness ratio. Rapid tempering offers high energy efficiency, shorter cycle times and better microstructural results [25]. Induction tempering ensures uniform heating conditions and precise control of the resulting properties [23].

Typical temperatures:

- Low tempering (150–250 °C): increase in toughness, minimal change in hardness.
- High tempering (400–650 °C): reduction in hardness, increase in ductility.

Fig. 2. Hardening using a robot [24]

1.4 Application of Robotic Technologies in Heat Treatment Processes

Early foundational research on the integration of robotics into heat treatment focused primarily on automation and repeatability [4, 5]. More recent influential studies demonstrate the use of advanced sensors, numerical simulations and artificial intelligence to enable adaptive and intelligent robotic heat treatment systems [7, 9, 10].

Robotic technologies represent an interdisciplinary field of science and engineering encompassing design, construction, sensors, actuators, control systems and software — including artificial intelligence and machine learning algorithms — that enable machines to perceive their environment, make decisions and perform physical or virtual tasks. Control systems and software — including artificial intelligence and machine learning algorithms — that enable machines to perceive their environment, make decisions and perform physical or virtual tasks with varying degrees of autonomy. This definition covers a wide range of systems, from industrial manipulators and collaborative 'cobots' to autonomous mobile platforms and medical robots to software 'robots' for business process automation; The common denominator is the integration of hardware, communication technologies and control logic for reliable and safe interaction with the physical or information environment. Scientific studies highlight the rapid shift from deterministic programmes to intelligent systems capable of learning, adapting and functioning in unstructured conditions, which opens up new opportunities and new risks in the areas of safety, reliability and ethics of deployment. In terms of applications and research, key topics include perception and fusion of sensory data, motion and manipulation planning, robotic vision, human-robot interaction and integrated networks (IoRT/Industry 4.0), with the literature highlighting the ongoing debate about a precise, universal definition of a „robot "in view of rapidly changing technology and different domain requirements. Robotic technologies increase the accuracy and efficiency of heat treatment processes, leading to improved material properties and reduced operating costs. The integration of robotic technologies increases the accuracy and efficiency of heat treatment processes, ultimately leading to improved material properties and reduced operating costs. In addition, the implementation of robotic systems in heat treatment processes can significantly increase production efficiency and optimise the quality of treated materials. Furthermore, the use of articulated robots in heat treatment processes can streamline operations, ensure consistent quality and reduce the likelihood of human error in production facilities. This approach not only increases efficiency, but also ensures the long-term sustainability of production processes in line with the Industry 4.0 concept [7].

1.5 Types of Heat Treatment Processes Where Robotic Systems Are Used

Induction Hardening: Robotic cells for induction hardening enable precise positioning of induction coils and parts, resulting in a high degree of process repeatability. The robot allows the heating to be localised precisely to critical areas of the part and minimises deformation. Such systems can be fully integrated with quench systems and PLC control, ensuring consistent microstructure and hardness [7].

Laser Hardening: In laser surface hardening (LSH), robotic arms ensure high-precision surface scanning and consistent beam control [26]. The laser allows for highly localised heat input and rapid self-cooling processes, minimising part deformation. Research shows the importance of parameters such as laser power, scanning speed and laser spot size on the hardening depth and microstructure [6] (Fig. 3).

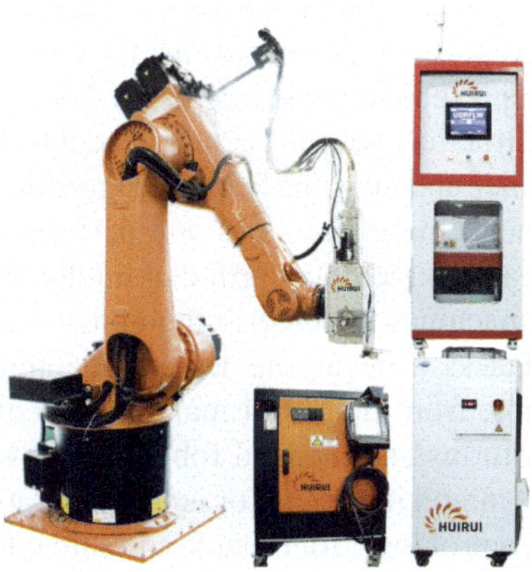

Fig. 3. Laser hardening robot [27]

Automated Hardening and Tempering Furnaces: In automated furnace cells, robots load and unload parts, transferring them between process zones under high-temperature conditions. These cells are often equipped with temperature monitoring, recipes and cycle documentation, enabling full traceability and quality control. An example is the overall growth of automation in the heat treatment industry, which the author [4] refers to as a "paradigm shift".

1.6 Architecture of Robotic Systems

The architecture of robotic systems designed for heat treatment of materials represents a complex integration of mechanical, sensor, energy and software elements, which together create a highly accurate, reliable and robust production workplace. The basis is an industrial robotic arm, the design of which must take into account the specifics of a demanding thermal environment – the presence of high temperatures, radiant heat, hot gases and metal particles. Therefore, robots with increased heat resistance, special protective covers, external and internal liquid cooling, and kinematic assemblies designed so that bearings and servo drives can withstand long-term thermal stress are used in these applications [28]. Heat source technology is mounted on the end of the robot – most often an induction coil, laser head or gripping device designed to handle heated parts. Induction systems require precise shielding of high-frequency cables, intensive cooling of transformers and stable positioning of the coil, as even a minimal deviation can significantly affect the temperature field [29]. Laser systems include collimators, focusing optical elements, and often a galvo scanner, whose dynamic control must be synchronised with the robot's trajectory. A critical element of the entire architecture is the sensor equipment that provides feedback to the control system. Temperature sensors, such as thermocouples and infrared pyrometers, enable point temperature measurement, while thermal imaging cameras provide a spatial distribution of the thermal field – which is essential in laser hardening, where even small deviations can cause undesirable microstructural changes [30]. This data optimize in a closed control loop that continuously compares the current temperature profile with the desired one and automatically adjusts the power of the induction or laser source, the robot's feed rate, and other technological parameters. Modern robotic workstations also integrate distance sensing (LIDAR), 3D scanner-based part geometry measurement, and adaptive algorithms that can compensate for geometry changes occurring during heating. The entire system is controlled by an industrial PLC or MES/SCADA add-on system, which manages technological recipes, archives process data and communicates with other elements of the production environment [5]. The software add-on allows the definition of heating parameters, robot trajectory programming, cycle timing, safety limit setting and equipment diagnostics. Modern solutions increasingly use digital twins of heat treatment processes, which combine physical models with real sensor data. They allow the temperature response of a part to be simulated, the heating trajectory to be optimised, and risk areas to be predicted before production even starts. The architecture also includes advanced safety features – optical barriers, zone scanners and intelligent safety PLCs – which protect workers from hot parts or active heat sources [31]. This architecture – combining robust hardware, advanced sensor technology and adaptive control – not only ensures high precision and repeatability of heat treatment, but also complete process traceability, archiving of technological cycles and the ability to respond autonomously to deviations during heating. The result is a system that significantly improves quality, reduces errors and improves energy and operational efficiency [7].

2 Comparison of Traditional and Robotic Heat Treatment Methods and Their Economic Evaluation

Comparative analyses of traditional and advanced heat treatment methods build on long-term industrial experience documented in classical literature [1, 3], while recent studies provide quantitative evidence of the benefits of robotic and localised heat treatment in terms of energy efficiency, precision and traceability [6, 11, 17].

Traditional heat treatment methods — especially annealing, hardening and tempering — are long-established technological processes whose effectiveness has been proven by decades of industrial practice [1, 3]. However, their main disadvantage is limited control over the temperature field, non-homogeneous heating and high energy consumption, especially in volume heating in furnaces [1]. When processing large or complex parts, deformation and residual stresses often occur, and the repeatability of results depends on the operator. Robotic approaches to heat treatment, including laser hardening, induction hardening and hybrid technologies, enable precisely controlled temperature input with controlled trajectory and timing [5, 11]. The fundamental advantage of robotic systems is their ability to maintain stable temperature profiles in real time based on feedback from sensors such as pyrometers, thermocouples and thermal imaging cameras [10]. Laser hardening achieves high heat localisation, rapid self-cooling and a significant reduction in the thermal impact on the core, which minimises part deformation [6, 7]. Induction robotic hardening is particularly suitable for localised treatments where high repeatability and precise coil positioning are important [8]. A significant difference between traditional and robotic methods lies in quality control. Automated workplaces enable fully digital documentation of cycles, which is crucial for production traceability and for applications in the automotive and aerospace industries [11]. In addition, robotic processes enable the integration of FEM simulations and digital twins, which are used to optimise parameters before production starts [12, 28]. These processes already form the basis of Industry 4.0 today.

2.1 Economic Aspects of Robotic Heat Treatment

The economic benefits of robotisation are one of the most important arguments for its implementation. Modern induction and laser robotic cells consume less energy than traditional furnaces, especially for local modifications [17]. Precise heat targeting reduces energy losses, increases heating efficiency and shortens overall production time. Robotisation also significantly reduces operating costs associated with errors and repairs of parts. Automated processes reduce the number of errors thanks to the high repeatability of thermal cycles and the elimination of variations caused by the human factor [9]. In applications such as post-weld heat treatment, precise temperature control leads to extended component life, which brings additional economic benefits [16]. Significant savings are also evident in labour costs. Robots can operate continuously, in high-temperature environments or in areas that are hazardous to workers. Automated lines with PLC/FPGA control enable the processing of parts with minimal operator intervention and full digital quality control [11]. The resulting reduction in the risk of occupational accidents also has a secondary economic effect.

Table 1. Comparison of traditional and robotic heat treatment.

Parameter	Traditional heat processing	Robotic heat processing
Heating accuracy	Low to medium; depends on the operator and type of furnace	Very high (laser/induction, sensors, adaptive control)
Energy and efficiency	Low (volume heating, long cycles)	High for local modifications; 30–60% energy savings
Repeatability	Medium; variable depending on operator	High, stable, digitally controlled
Part deformation	Often significant for large or thin-walled parts	Minimal with laser/induction hardening
Possibility of local processing	Very limited	Excellent – precise local heating
Labour costs	Higher (manual handling)	Lower (automation, 24/7 operation)
Work safety	Risk of burns, handling in hot environments	Robot takes over risky tasks
Digital documentation	Limited	Comprehensive, full traceability
Possibility of integration with AI/FEM	Low	Very high (digital twins, adaptive control)
Economic efficiency	Medium (especially for large furnaces)	High for medium and small series, local modifications

Advantages of Using Robotics in Heat Treatment

- Increased quality consistency – Robotic positioning and precise heat control reduce microstructure and hardness variability.
- Higher productivity and flexibility – Automated cells enable quick recipe changes and small batch processing without significant downtime.
- Safety – Reduced risk for employees, as hot parts are handled by robots.
- Documentability – Digital records ensure process traceability and support part qualification processes [5].

Technical and Organisational Challenges

- Component resistance: Robotic arms and fixtures must be protected against high temperatures and thermal stress, which requires special materials and cooling structures.

- Accuracy and deformation: In local processes (e.g., induction or laser), optimizme necessary to synchronise the robot‘s movement with the heat source output and real-time measurement to minimise part deformation.
- Integration and validation of recipes: The transition to robotisation requires the validation of new process recipes, staff retraining and adjustments to material logistics. These investments can be a barrier, especially for smaller companies.
- Cyber security: With increasing digitalisation (IoT, cloud), optimize necessary to ensure that recipes, telemetry and control systems are protected against unauthorised access [4].

2.2 Improving Accuracy and Efficiency

The integration of robotics into heat treatment processes can significantly increase accuracy and efficiency, as demonstrated by similar advances in other areas such as CNC machining or heat treatment in the energy sector. For example, in CNC operations, robots ensure greater accuracy and productivity by performing tasks such as material loading, tool changing and quality control, thereby reducing human error and increasing operational efficiency. These automated systems are capable of collecting data in real time and making the necessary adjustments, minimising waste and increasing overall process efficiency [32]. The integration of advanced sensors and automation ensures precise control of parameters such as temperature, pressure and flow, which are essential for maintaining suitable operating conditions. These technologies enable immediate data collection and subsequent adjustment, reducing system failure rates and increasing energy efficiency. Applying these principles to heat treatment in manufacturing can similarly improve accuracy by maintaining consistent process conditions and reducing variability caused by human factors. The use of robotics can lead to more precise heat application, which is crucial in industries where specific heat profile requirements are very important for the properties and performance of materials. The ability of robots to perform complex and repetitive tasks with high precision also contributes significantly to reducing errors and improving the quality of heat treatments. Another significant benefit is the integration of machine learning algorithms with robotic systems, which enables predictive analytics and optimisation of heat treatment processes, ensuring better energy utilisation and reducing environmental impact. As in the power generation sector, the synergistic combination of robotic automation and artificial intelligence in heat treatment brings significant improvements in operational efficiency and sustainability [33].

2.3 The Future of Robotic Technologies in Heat Treatment

The convergence of robotics with Industry 4.0 technologies will also facilitate seamless data integration, predictive maintenance and process monitoring, promoting innovation and sustainability in heat treatment applications across various industries. The implementation of collaborative robots will enable flexible and adaptive interactions with human operators, increasing the efficiency of work processes without compromising safety. Advances in machine learning algorithms will enhance these robotic systems with the ability to learn from historical data, continuously improving process parameters and detecting defects. Integration with digital twins will provide virtual simulations of heat treatment cycles, enabling pre-adjustments and reducing and unplanned downtime. As these technologies mature, the scalability and adaptability of robotic heat treatment solutions for different materials and complex geometries will expand, particularly in the aerospace, automotive and manufacturing industries. The final form of this industry will be characterised by intelligent, interconnected robotic systems that not only improve product quality and performance, but also contribute to environmental sustainability by reducing energy consumption and waste. The deployment of robotic systems in heat treatment, like other advanced robotic systems, faces a number of technological challenges that can be broadly divided into areas such as connectivity, adaptability, software integration, safety and real-world operational complexities [34].

- **Connectivity and integration**: The integration of robotic systems in industrial environments often requires seamless interconnection across different platforms. The use of cloud technologies can improve this area by increasing energy efficiency, real-time performance and reducing costs. Cloud robotics helps coordinate multiple robots and enables higher levels of computing power and storage to process large data sets and support energy-efficient solutions [35].
- **Adaptability and embedded intelligence**: Current robotic technologies often lack the adaptability needed to interact effectively with dynamic environments. The integration of neuromorphic computing and embedded artificial intelligence, which mimics the functioning of the human brain, promises to address these issues through improved adaptability, efficiency, and reliability of robotic systems. Such advances can significantly improve the performance of robotic systems used for heat treatment, where real-time adaptability to changing environmental conditions is critical [36].
- **Software engineering challenges**: The development of maintainable and reusable robotic systems presents a number of software engineering challenges, particularly in terms of ensuring safe and trustworthy self-adaptive capabilities in unpredictable environments. Techniques such as runtime verification and field testing are indispensable for providing safe solutions and verifying the performance of robotic systems deployed for heat treatment [37].
- **Security and privacy concerns:** Security and privacy are significant issues, especially for systems with strong interaction with other devices and networks. The implementation of blockchain and multi-factor authentication can increase the security of robotic systems, making them secure by design and increasing their trust and acceptance [38].
- **Real-world operational testing**: Integrating heterogeneous systems, including various sensors and actuators, into larger frameworks remains a significant challenge. Approaches such as service-oriented systems can facilitate modular design and integration of dynamic software systems, which is essential for multi-component robotic workplaces in heat treatment [39].

While foundational concepts of robotic automation are well documented in earlier studies [35, 36], recent influential research focuses on emerging paradigms such as cloud robotics, neuromorphic computing and blockchain-based security, which are expected to shape the next generation of intelligent heat treatment systems [37–39].

3 Conclusion

The deployment of robotic technologies in the field of heat treatment of materials represents a fundamental qualitative shift towards more precise, efficient and safer production. Thanks to the integration of advanced sensors, artificial intelligence, digital twins and adaptive control, traditional thermal processes are being transformed into highly controlled systems that can monitor and optimise temperature profiles in real time, minimising defects, shortening production cycles and reducing energy costs. The comparison in Table 1 clearly shows that the difference between traditional and robotic approaches is not only technological but also economic – robotic systems exhibit higher repeatability, lower error rates and significantly better energy efficiency, leading to a more sustainable and reliable production process. Robotic systems also improve working conditions by replacing handling in high-temperature environments, reducing risks for operators and enabling safe operation in automated cells. The implementation of these systems also increases production traceability – each cycle is monitored, archived and can be retrospectively analysed, which is particularly important for the automotive and aerospace industries, where the emphasis on quality and repeatability is extremely high. Despite the clear advantages, the integration of robotics in heat treatment also involves a number of challenges. These include the requirements for robust connectivity between control units, the interoperability of diverse sensors and actuators, and ensuring cyber security in a digitalised production environment. However, advanced approaches such as cloud robotics, edge computing, neuromorphic intelligence and blockchain security are paving the way for solutions that will enable the large-scale deployment of these technologies. Another factor for success will be the ability of manufacturing companies to adapt to new process standards, integrate predictive maintenance and develop staff competencies.

The future of robotised heat treatment is moving towards fully intelligent, autonomous and interconnected production systems that will not only be able to perform heat processes, but also predict their outcome, optimise procedures on the fly and communicate within wider production networks. The integration of digital twins will make it possible to simulate the behaviour of materials even before physical heating, which will reduce the number of experiments and accelerate the development of new processes. The combination of robotics, artificial intelligence and physical modelling will create a new generation of technological workplaces that will be able to ensure not only high quality and efficiency, but also the long-term sustainability of production processes.

Acknowledgements. This research was supported by the Cultural and Educational Grant Agency of the Ministry of Education, Research, Development and Youth of the Slovak Republic, grant no. KEGA 024TUKE-4/2023 also by the Scientific Grant Agency of the Ministry of Education, Research, Development and Youth of the Slovak Republic and Slovak Academy of Sciences, grant VEGA 1/0258/24. This research was also supported by project DRP0200194, Moving Plastics and Machine Industry towards Circularity (PLAN-C) under the Interreg Danube Region Programme, co-funded by the European Union.

References

1. Jurči, P., Novák, P.: Heat treatment of metallic materials in modern industry. Materials. **15**(23), 8337 (2022)
2. Agbedo, U.J.: Advanced heat treatment techniques for enhancing strength and toughness in high-performance stainless steels and metallic alloys. Int. J. Res. Publ. Rev. **5**(12), 4917–4930 (2024)
3. Ford, J.A.: Heat treatment processes. In: Metallurgical Technologies. Springer, Berlin (1972)
4. Esteve, V.: Advanced automation in the heat treat industry. In: Proceedings of Heat Treating 2021. ASM International (2021)
5. Gendron, M., Nadot, J., Montmitonnet, P.: Non-linear optimization of a new robotic induction process for in situ local heat treatment. Int. J. Adv. Manuf. Technol. **88**, 1749–1761 (2016)
6. Kluczyński, J., Kirsch, C., Śliwiński, G., Szala, M., Jasiński, J.: Laser surface hardening of carburised steels: key parameters and microstructural effects. Materials. **18**, 3623 (2025)
7. Łach, M.: Recent advances in laser surface hardening techniques, modelling approaches, and industrial applications. Crystals. **14**(8), 726 (2024)
8. Liang, J., et al.: Study on temperature field in the process of induction heat treatment using two-sided coils and auxiliary pulse current. Heliyon. **10** (2024)
9. Xuan Hai, T. et al.: Artificial Intelligence for Predicting Deformation in Steel Plates during Induction Heating (2024)
10. Baciu, A., Lazar, C.: Model-free temperature control of heat treatment process. Energies. **17**(15), 3679 (2024)
11. Esteve, V., Bellido, J.L., Jordán, J.: State of the art and future trends in monitoring for industrial induction heating applications. Electronics. **13**(13), 2591 (2024)
12. Maurin, H., Kuwano, R., Vives, G., Asaka, K., Shintake, J.: Liquid Crystal Elastomer Composites for Induction-Actuated Soft Robotics (2023)
13. Osawa, K., Kase, Y., Domae, Y. et al.: Thermal Flow Driven Material Classification for Robotic Grippers Using LSTM Networks (2022)
14. Zhanna, S.: Steel heat treatment technology. Int. J. Res. Publ. Rev. **5**, 1224–0210 (2020)
15. Stević, Z., et al.: Design of a system for the induction hardening of steels using simulation parameters. Appl. Sci. **13**(20), 11432 (2023)
16. Gergely, R., Hochenauer, C.: Heating strategies for efficient combined inductive and convective heating of profiles. Energies. **16**, 654 (2023)
17. Goo, B.-C.: Effect of post-weld heat treatment on the fatigue behaviour of medium-strength carbon steel weldments. Meta. **11**(11), 1700 (2021)
18. Kosturek, R., Śnieżek, L., Wachowski, M., Torzewski, J.: Influence of post-weld heat treatment on the microstructure and fatigue properties of Sc-modified AA2519 friction stir-welded joint. Materials. **12**(4), 583 (2019)
19. Rübig Aluminium GmbH: Robotic heat treatment cell for aluminium components. https://cdn.rubig.com. Last accessed Aug 2025
20. Goldsteinas, A., Hamid, J.: Redefining quenching technology. In: Proceedings of the ASM Heat Treating Society Conference (2013)
21. Hu, S., Zhu, L., Zhang, M., et al.: Development and prospect of vacuum high-pressure gas quenching technology. Materials. **16**, 7413 (2023)
22. Puga-Patlán, L., López-García, R.D., Medina-Juárez, I., et al.: Study of cooling medium variables during quenching in SAE 4340 steel using statistical and modelling tools. Superalloys. **13**, 1627 (2023)
23. Totten, G.E., Albano, L.L.M.: IFHTSE Global 21: heat treatment and surface engineering in the Twenty-first century. Int. Heat Treat. Surf. Eng. **6**, 5–12 (2012)

24. Ammunition Plant: Hardening shops – hardening workplace Ammunition 155. http://www.msm.sk. Last accessed Aug 2025
25. Vieweg, P., Ressel, G., Raninger, P., et al.: Comparing fast inductive tempering and conventional tempering: effects on microstructure and mechanical properties. Metall. Res. Technol. **115**, 310 (2018)
26. Chabak, Y., Efremenko, V., Barma, Y., et al.: Enhancing dry-sliding wear performance of powder-metallurgy metal matrix–carbide composite via laser surface modification. Eng. **6**(11), 313 (2025)
27. GuoSheng Laser Technology Co., Ltd.: Laser hardening machine in robotic cell. http://sk.guoshenglaser.com. Last accessed Aug 2025
28. Babu, P.D., Buvanashekaran, G., Balasubramanian, K.R.: Experimental investigation of laser transformation hardening of low alloy steel using response surface methodology. Int. J. Adv. Manuf. Technol. **67**, 1883–1897 (2013)
29. Babič, M., Marinković, D., Bonfanti, M., Calì, M.: Complexity modelling of steel-laser-hardened surface microstructures. Appl. Sci. **12**(5), 2458 (2022)
30. Cholkar, A., Chatterjee, S., Kumar, S., et al.: Combining ultrafast laser texturing and laser hardening to enhance surface durability. Adv. Eng. Mater. **26**(20), 2401183 (2024)
31. Babič, M., Kovač, P., Savković, B.: Using genetic programming to model process of robot laser hardening. J. Prod. Eng. **22**(2), 5–8 (2019)
32. Soori, M., Dastres, R., Jough, F.K.G., Arezoo, B.: Robotical automation in CNC machine tools: a review. Acta Mech. Autom. **18**(3), 434–450 (2024)
33. Buratto, W.G., González, G.V., Muniz, R.N., et al.: Review of automation and sensors: parameters control of thermal treatments for electricity generation. Sensors. **24**, 967 (2024)
34. Zhang, X., Sun, J., Li, Y.: Cloud and edge computing for large-scale robotic automation. IEEE Trans. Ind. Inform. **18**(5), 3312–3321 (2022)
35. Wan, J., Yi, M., Zhang, R., Li, D., Wang, S., Imran, M.: Cloud robotics: architecture, challenges and applications. IEEE Netw. **30**(5), 6–14 (2016)
36. Putra, Y.P., Sutomo, S., Hartanto, R.: Neuromorphic computing for robotics: review and future perspectives. IEEE Trans. Neural Netw. Learn. Syst. (2024)
37. Caldas, P., Oliveira, E., Silva, J.: Runtime verification techniques for adaptive robotics: a survey. Robot. Auton. Syst. **158**, 104267 (2024)
38. Marchang, F., Di Nuovo, A.: Blockchain-based security for robotic systems. Comput. Secur. **113**, 102557 (2022)
39. Salcic, Z., Pham, T.T., Gracanin, D.: Service-oriented architectures for robotic system integration: challenges and perspectives. J. Syst. Softw. **155**, 170–188 (2019)

Comparison of Energy Consumption by Robot Due to Velocity and Motion Type

Łukasz Sobaszek[1], Edward Kozłowski[1], Katarzyna Antosz[2](✉), Monika Kulisz[1], Antoni Świć[1], and Sławomir Ordyniec[1]

[1] Lublin University of Technology, Nadbystrzycka 38, Lublin 20-618, Poland
[2] Faculty of Mechanical Engineering and Aeronautics, Rzeszow University of Technology, Powstańców Warszawy 8, Rzeszów 35-959, Poland
katarzyna.antosz@prz.edu.pl

Abstract. This paper investigates the data-driven modelling of electrical power consumption in industrial robots performing different types of motion at various velocity levels. Three representative trajectory classes were analysed: joint motion, linear motion, and circular joint motion. Each class was executed at several commanded speeds. Instantaneous electrical power was measured during experimental runs, pre-processed, and used to identify parametric models relating power demand to selected motion-related variables. The identified model parameters are highly statistically significant, confirming the relevance of the adopted structure. Separate models calibrated for individual motion types achieved very good agreement with the empirical data. For joint and linear motion, correlation coefficients of approximately =0.95 and =0.93 were obtained, with mean absolute percentage errors below 2%. For circular motion, the fit remains satisfactory, with R $\approx$ 0.85 and MAPE below 4%. A unified model, identified using combined data from all motion types, can also reproduce the main trends in power consumption, even though with higher errors (MAPE above 5%), which illustrates the trade-off between accuracy and universality. These results show that relatively simple empirical models can accurately predict instantaneous power consumption. Such models could be employed for cost estimation, energy-oriented optimisation of robot trajectories, and implementing in industrial practice.

Keywords: Industrial Robots · Power Consumption Modelling · Energy Efficiency · Robot Trajectories

1 Introduction

The advent of large-scale robotisation in contemporary manufacturing is now a tangible reality. Industrial robots are no longer limited to major corporations; they are increasingly being integrated into the production processes of small and medium-sized enterprises, reflecting the growing adoption of automation across diverse industrial contexts [1]. The spectrum of robot applications in manufacturing is extensive [2]. As with production systems in general, robotic operations require thorough analysis to ensure high efficiency and measurable benefits [3]. Effective robot utilisation depends on accurately defining

B. Gapiński et al. (Eds.): MANUFACTURING 2026, LNME, pp. 261–276, 2026.
https://doi.org/10.1007/978-3-032-22943-4_19

operating parameters, selecting appropriate end-effectors and auxiliary equipment, and carefully programming trajectories and task execution. Proper planning of robotic operations can yield substantial improvements in real processes [4, 5], yet in practice this step is often underestimated and robots are frequently programmed largely based on the integrator's intuition.

The implementation of robots has been widely discussed in the literature from multiple perspectives, including their impact on overall manufacturing performance [6, 7], task sequencing for specific operations [8, 9], precise motion-path determination [10, 11], and electrical energy consumption [12, 13]. A review of these areas shows a marked increase in publications in recent years, confirming both the relevance of the topic and the timeliness of research on robotics in manufacturing [14].

One of the most prevalent industrial robot applications is palletizing (see Fig. 1), which has therefore attracted extensive study and analysis [15, 16]. Key benefits of robot-based palletizing include higher throughput, reduced risk of worker injuries, increased payload capacity, and consistent task execution times [17]. Since palletizing is typically performed at the end of a production line and is critical for delivering products to customers, it should be examined in detail in terms of motion ranges and trajectories, the balance of productive versus auxiliary movements, and overall workstation organisation [17]. Poorly designed robotic palletizing stations can turn the final production stage into a bottleneck, reducing system efficiency. Electricity consumption is also crucial because it directly affects profitability, and rational energy use remains a major contemporary challenge [18]. To highlight the need for careful planning that accounts for energy-related factors, a model of the analysed workstation was developed in an offline robot programming environment, enabling investigation of the problem using the available tools.

In this context, the research problem addressed in this paper is the quantitative description and prediction of the electrical power consumption of an industrial robot for different motion types and velocity levels. Despite the prevalence of research concentrating on trajectory planning, motion optimisation, and cycle-time reduction, there is a paucity of attention given to simple, empirically validated models that directly relate robot motion parameters (trajectory type, interpolation mode, speed) to instantaneous power demand. This discrepancy hinders the capacity for systematic integration of energy-related criteria into robot programming, production planning and cost estimation within realistic manufacturing environments.

The primary objective of this study is to develop and empirically validate data-driven models for predicting the instantaneous electrical power consumption of an industrial robot for selected motion patterns, namely joint, linear and circular motion. The models are identified on the basis of measured power signals recorded for predefined test trajectories and different commanded velocities. The performance of the models is evaluated using standard goodness-of-fit measures (including , MAE, RMSE and MAPE). Particular emphasis is placed on comparing models calibrated for individual motion types with a unified model identified on combined data. This facilitates the evaluation of the trade-off between model accuracy and universality within the context of practical applications.

The structure of the paper is as follows. The second section of this study outlines the experimental setup, the trajectories that were tested, and the modelling approach that was adopted. This approach includes the structure of the model and the procedure for identifying the parameters. In Sect. 3, the results of the model fitting and prediction for different motion types and velocities are presented and analysed. A detailed comparison of the obtained accuracy indicators is also provided. Finally, Sect. 4 discusses the implications of the findings for energy-efficient robot deployment and the application in manufacturing systems, and summarises the main conclusions together with directions for future research.

2 Materials and Methods

2.1 Research Methodology

The research methodology adopted in this study was designed to quantitatively describe and predict the electrical power consumption of an industrial robot executing different motion patterns. The approach combines experimental measurements with data-driven modelling and statistical validation and consists of four main stages: (1) definition of representative test trajectories and operating conditions, (2) experimental acquisition of electrical power signals, (3) preprocessing and structuring of the measurement data, and (4) identification and evaluation of prediction models.

Three categories of robot motion were considered: joint, linear and circular joint motion. For each type, a dedicated test trajectory was programmed in the robot controller to ensure repeatable execution and sufficient duration for reliable measurement. The trajectories were executed at several commanded velocity levels (expressed as a percentage of the nominal speed), allowing the influence of velocity on instantaneous power demand to be assessed. These trajectories do not represent a specific industrial task but reflect typical motion patterns occurring in real production.

During the experiments, the robot followed the predefined trajectories while instantaneous electrical power was recorded over time. Multiple runs were performed for each motion–velocity combination to reduce random disturbances and obtain statistically meaningful data sets. The recorded signals were then preprocessed, normalised and grouped by motion type and velocity, which enabled both separate and joint model identification. In the final stage, parametric models of instantaneous power consumption were identified using least-squares estimation. The statistical significance of the parameters was verified, and model quality was assessed using the correlation coefficient (R), mean absolute error (MAE), root mean square error (RMSE) and mean absolute percentage error (MAPE). Separate models were first developed for each motion type, followed by a unified model based on combined data, in order to analyse the trade-off between model accuracy and universality.

2.2 Experimental Setup

The workstation conditions were reproduced in a laboratory test setup using a Kawasaki RS003N industrial robot (Fig. 1). The robot was equipped with a suitable end effector—a gripper—which made it possible to verify the developed solution experimentally.

The Kawasaki RS003N is a six-axis industrial robot that can be programmed using a Teach Pendant, the dedicated K-ROSET simulation software, and the AS programming language. The Teach Pendant is an LCD-based handheld controller that enables both programming and operation of the robot, either by executing AS-language commands or via point-to-point teaching. K-ROSET is an advanced programming and simulation environment that supports offline development and validation of robot programs. Robot motion is controlled by a dedicated controller, which also manages communication with the robot through specialized communication protocols. The key specifications of the robot and controller are provided in Table 1.

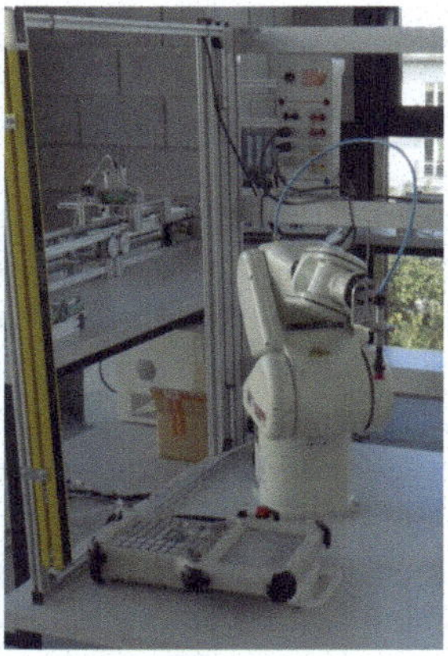

Fig. 1. Laboratory test-set up with Kawasaki RS003N robot

Table 1. Robot and controller specifications [19]

Robot specifications	Controller specifications
Model: RS003N-A Type: articulated robot Degree of freedom: 6 Repeatability: ± 0,05 mm Maximum load: 3 kg Maximum speed: 6000 mm/s Driving motor: brushless AC servo motor	Model: E70/E71 Structure: enclosed Number of controlled axes: 6 Programming: TEACH/REPEAT Teaching: AS language Power requirements: 200–240 V

The study involved measuring the voltage and current in the power supply installation of the robot. For this purpose, the robot was connected to a specially designed extension cable with separated conductors. Voltage probes were connected in parallel to the conductors, while the current-measuring clamp of the multimeter was placed around the phase wire.

The experiment consisted of the robot following paths of similar shape and identical length, depending on the predefined scenario. Each of the four scenarios differed in terms of the combination of path layout and the type of motion with which the robot executed the trajectory (Fig. 2). Additionally, each scenario was divided into four different traversal speeds.

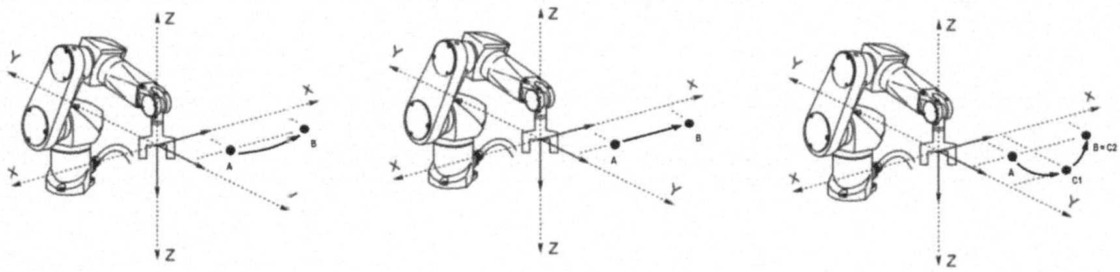

A B C

Fig. 2. Types of robot motions used in the experiment: joint (A), linear (B), and circular (C)

Route 1 had a total length of 1005 mm and consisted of six points arranged at equal distances of 200 mm from one another. The points were connected by straight segments of equal length. The only exception was the final point, SLP06, which was located 205 mm from the preceding point, SLP05. This slight extension was introduced to ensure that Route 1 and Route 2 had the closest possible total lengths. The points, represented as circles with a diameter of 21 mm, were designed to maintain high teaching accuracy, ensuring that the tool's suction cup aligned precisely with each target point during robot programming. Route 1 was executed in two variants: a) the robot traversed Route 1 using joint interpolation (Fig. 3), b) the robot traversed Route 1 using linear interpolation (Fig. 4).

Route 2 had a total length of approximately 1005 mm and consisted of nine points. Most of the segments between the points were connected by arcs to achieve smoother robot motion. If Route 2 ended at point SLK07, it would be more similar to Route 1; however, its total length would differ. Therefore, the arc between SLK07 and SLK09 was added as a complementary segment to obtain a total path length of 1005 mm, matching that of Route.

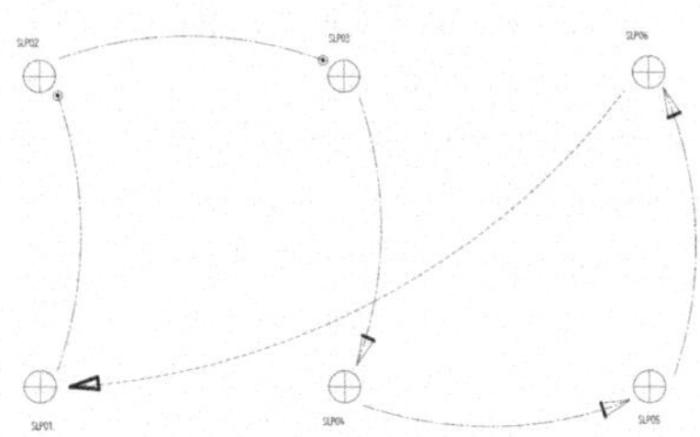

Fig. 3. Route 1 — joint interpolation

Route 2 was executed in two variants: a) the robot traversed Route 2 using circular interpolation with elements of joint interpolation (Fig. 5), b) the robot traversed Route 2 using circular interpolation with elements of linear interpolation (Fig. 6).

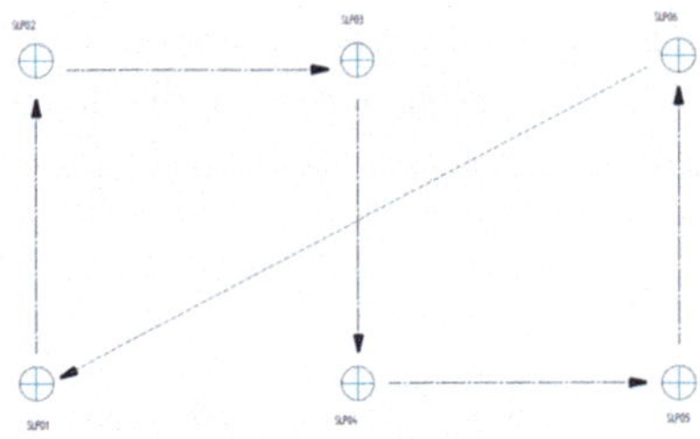

Fig. 4. Route 1 — linear interpolation

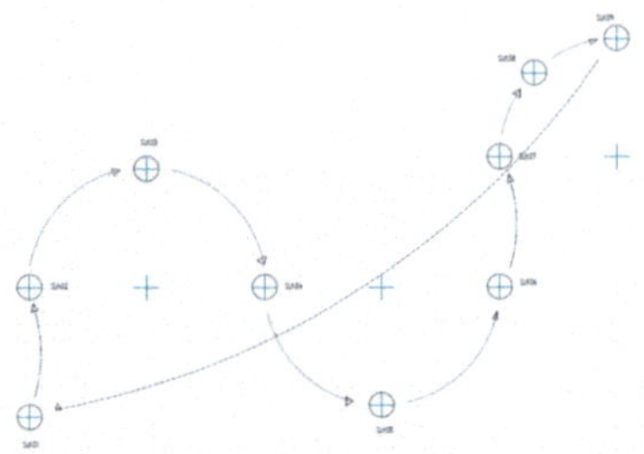

Fig. 5. Route 2 — joint interpolation

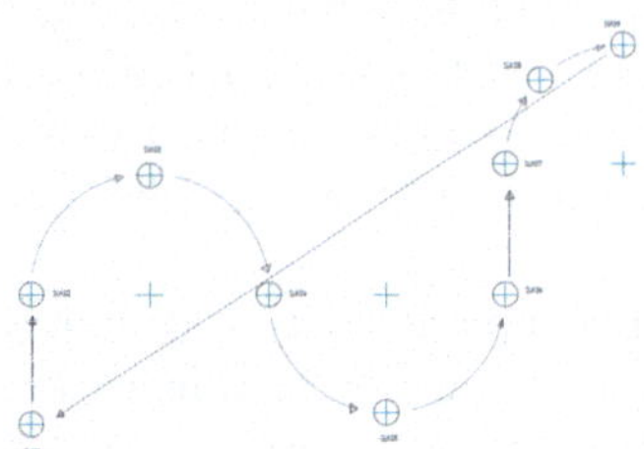

Fig. 6. Route 2 — linear interpolation

In all variants, the robot completed the paths at speeds of 10%, 30%, 60%, and 80%. The robot accuracy parameter was set to 1 for arc points and to 4 for points located on straight segments and those preceding arcs, in order to improve motion smoothness.

The robot was controlled using the KCwinTCP software, which communicated directly with the robot controller. The program issued commands for moving between points and modifying their coordinates. The robot was connected to the computer via an Ethernet cable. To verify the proposed concept and to identify potential inefficiencies during robot programming, an experimental study was conducted to analyze the electrical energy consumption of the Kawasaki RS003N industrial robot.

The basis of the analysis was the determination of the instantaneous power:

$$p(t) = \frac{dW}{dt} = u(t) \bullet i(t) \tag{1}$$

2.3 ARIMAX Model

To identify the influence of predictors (exogenous variables) $X \in \mathbb{R}^m$ into feature $Y \in \mathbb{R}$ we apply the ARIMAX(p,d,q) model

$$(1 - L)^d y_t = \beta_0 + \sum_{i=1}^{m} \beta_i x_{ti} + \sum_{h=1}^{p} \alpha_h (1 - L)^d y_{t-h} + \sum_{j=1}^{q} \theta_j \varepsilon_{t-j} + \varepsilon_t, \tag{2}$$

where $\{\varepsilon_t\}_{t \in \mathbb{N}}$ is independent identically distributed random variables with $N(0, \sigma^2)$ normal distribution and L is lag (backshift) operator ($L^k y_t = y_{t-k}$) and differencing operator $(1 - L)^d = \sum_{j=0}^{d} \binom{d}{j} (-1)^j L^j$. Based on $\{(x_{t1}, \ldots, x_{tm}, y_t)\}_{1 \le t \le n}$ data set we estimate β_0 intercept, $\beta_1, \ldots, \beta_m$ coefficients responsible for impact of exogenous variables, $\alpha_1, \ldots, \alpha_p$ coefficients of autoregressive operator and $\theta_1, \ldots, \theta_q$ coefficients of moving average operator. The model (2) explains both the impact of predictors on the dependent variable and the impact of previous realizations of the dependent variable and external disturbances. The prediction based on (2) model we calculate as follows:

$$\begin{aligned} \hat{y}_t &= (1 - L)^{-d} w_t, \\ w_t &= \hat{\beta}_0 + \sum_{i=1}^{m} \hat{\beta}_i x_{ti} + \sum_{h=1}^{p} \hat{\alpha}_h (1 - L)^d y_{t-h} + \sum_{j=1}^{q} \hat{\theta}_j \varepsilon_{t-j} \end{aligned} \tag{3}$$

where the inverse differencing operator (summation operator) is given as follows;

$$(1 - L)^{-d} = \sum_{j=0}^{\infty} \binom{d + j - 1}{j} L^j \tag{4}$$

and the values $\hat{\beta}_0, \hat{\beta}_1, \ldots, \hat{\beta}_m, \hat{\alpha}_1, \ldots, \hat{\alpha}_p, \hat{\theta}_1, \ldots, \hat{\theta}_q$ denote the estimators of unknown parameters of model (2). By analysing the sequences $\{y_t\}_{1 \le t \le n}$ and $\{\hat{y}_t\}_{1 \le t \le n}$ we assess the fit of the forecast values to empirical data. Then we estimate the indicators:

- correlation coefficient

$$r = \frac{\sum_{i=1}^{n} \left(y_i - \overleftarrow{y}\right)(\hat{y}_i - \tilde{y})}{\sqrt{\sum_{i=1}^{n} \left(y_i - \overleftarrow{y}\right)^2 \sum_{i=1}^{n} (\hat{y}_i - \tilde{y})^2}} \tag{5}$$

where $\overleftarrow{y}$ and $\tilde{y}$ denote the average empirical and forecast values respectively;

- Root Mean Squared Error (RMSE)

$$RMSE = \sqrt{\frac{1}{n} \sum_{i=1}^{n} (y_i - \hat{y}_i)^2} \tag{6}$$

- Mean Absolute Error (MAE)

$$MAE = \frac{1}{n}\sum_{i=1}^{n} \mid y_i - \hat{y}_i \mid \tag{7}$$

- Mean Absolute Percentage Error (MAPE)

$$MAPE = \frac{1}{n}\sum_{i=1}^{n} \frac{|y_i - \hat{y}_i|}{\max(\epsilon, |y_i|)} \tag{8}$$

where $\epsilon > 0$ is an arbitrary small positive value to avoid undefined results when y is zero.

3 Results

We analyse the impact of speed V_t and type of motion: JM_t (joint motion), LM_t (linear motion), CM_t (circular motion) on energy consumption. For each case, the structural parameters of the ARIMAX model and the structural deviations of these parameters were estimated, and the hypotheses concerning the significance of the structural parameters and the values of the test statistics were verified.

3.1 Joint Motion

For joint motion, the model is of the form:

$$y_t = \beta_0 + \beta_1 V_t + \theta_1 \varepsilon_{t-1} + \theta_2 \varepsilon_{t-2} + \varepsilon_t, \tag{9}$$

where the structural parameters are given in Table 2, the standard deviation of these parameters, the test statistic values and the p-value, while the random variable $\varepsilon_t \sim N(0, 71.04)$.

Table 2. Structural parameters of ARIMAX model for joint motion

	Value	Sd. dev	T	p- value
θ_1	0.81980	0.05923	223.59604	1.8865×10^{-296}
θ_2	0.18277	0.05816	50.76938	5.6184×10^{-136}
β_0	232.76981	1.53393	2451.56107	0
β_1	1.04982	0.04497	377.12953	0

From Table 2 we can see that, at the 0.05 significance level, the null hypothesis for each parameter — stating that the parameter does not differ significantly from zero — must be rejected in favour of the alternative hypothesis.

As can be seen from Fig. 7, there is a clear relationship between the robot's traversal velocity and its instantaneous power consumption when it follows the designed path using joint interpolation.

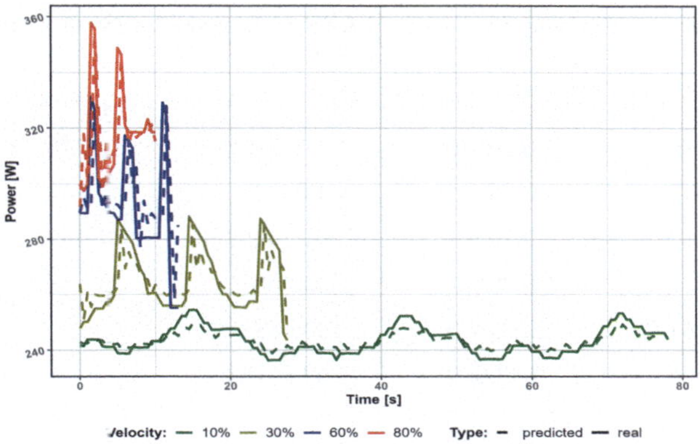

Fig. 7. Energy consumption for joint motion.

As expected, higher velocities lead to significantly higher power peaks. The robot exhibits the highest and most dynamic power fluctuations at 80% of the nominal velocity, with values exceeding 350 W. Reducing the velocity to 60% results in moderately lower peaks (around 320 W), and reducing it further to 30% results in a further reduction, with peaks not exceeding approximately 290 W. The robot operates at a relatively stable level of about 240–250 W with minimal fluctuations at 10% velocity, which is the lowest power consumption observed.

The predicted curves follow the same overall trend as the measured data, demonstrating good agreement across all tested velocities. Although minor deviations are visible, especially at higher speeds where dynamic load variations are more pronounced, the overall correspondence confirms the predictive model's validity for joint-interpolated motion.

3.2 Linear Motion

For linear motion, the model is of the form:

$$y_t = \beta_0 + \beta_1 V_t + \alpha_1 y_{t-1} + \alpha_2 y_{t-2} + \theta_1 \varepsilon_{t-1} + \theta_2 \varepsilon_{t-2} + \varepsilon_t \tag{10}$$

where the structural parameters are given in Table 3, the standard deviation of these parameters, the test statistic values and the p-value, while the random variable $\varepsilon_t \sim N(0, 98.512)$.

As demonstrated in Table 3, each of the structural parameters differs significantly from zero. The individual components of the model were therefore correctly identified, and the selected model enables reliable prediction of energy consumption for the considered case of trajectory execution using linear interpolation. When comparing the obtained results with those for Route 1 executed with joint interpolation, it should be noted that the observed peaks are slightly more pronounced, while the model remains well fitted. This mainly results from the fact that, during linear interpolation, the robot performs movements with higher precision, without sudden increases or decreases in acceleration, leading to more stable motion characteristics.

Table 3. Structural parameters of ARIMAX model for linear motion

	Value	Sd. dev	T	p-value
α_1	−0.48855	0.10640	−75.59135	2.3767×10^{-181}
α_2	0.32817	0.08435	64.04933	2.8512×10^{-163}
θ_1	1.28783	0.10561	200.74067	1.5698×10^{-291}
θ_2	0.44326	0.08401	86.85663	1.0186×10^{-196}
β_0	229.14199	2.08420	1809.88286	0
β_1	0.98666	0.06019	269.86182	0

As illustrated in Fig. 8, the instantaneous power consumption of the robot was measured and predicted as a function of time for four commanded velocities. 10%, 30%, 60% and 80%, respectively. As the velocity increases, the motion cycle becomes shorter and the power peaks become visibly higher. For a velocity of 10%, the robot functions for the entirety of the experiment, exhibiting a comparatively low and steady power demand, typically ranging from 240 to 250 watts, with minor fluctuations. At a velocity of 30%, distinct peaks of approximately 270–280 W manifest during the execution of the most arduous segments of the trajectory. For velocities of 60% and 80%, these peaks become significantly more pronounced, reaching approximately 330 W and above 360 W, respectively. These peaks are concentrated in the initial part of the time interval.

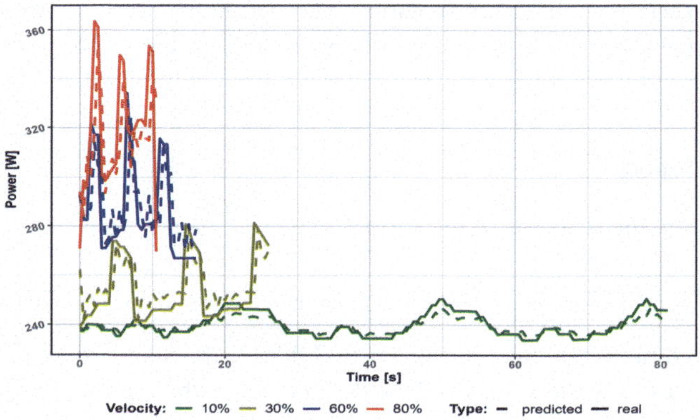

Fig. 8. Energy consumption for linear motion.

The dashed curves, representing model predictions, demonstrate a high degree of correlation with the solid curves, representing measured data, across all velocity levels. The timing and magnitude of the main power peaks are reproduced with a high degree of accuracy, with only minor discrepancies observed, primarily at the highest velocities where dynamic effects are most pronounced. The plot demonstrates that the proposed model reliably captures the dependence of instantaneous power on robot velocity and can be used to predict power consumption over a wide range of operating speeds.

3.3 Joint with Circular Motions

Circular motion is used as the reference case. For a joint following a circular trajectory, the model takes the following form:

$$y_t = \beta_0 + \beta_1 V_t + \beta_2 JM_t + \theta_1 \varepsilon_{t-1} + \varepsilon_t, \quad (11)$$

where the structural parameters are given in Table 4, together with their standard deviations, test statistic values and p-values, and the random variable $\varepsilon_t \sim N(0, 572.984)$.

Table 4. Structural parameters of ARIMAX model for joint with motions

	Value	Sd.dev	T	p.value
α_1	−0.48855	0.10640	−75.59135	2.3767×10^{-181}
α_2	0.32817	0.08435	64.04933	2.8512×10^{-163}
θ_1	1.28783	0.10561	200.74067	1.5698×10^{-291}
θ_2	0.44326	0.08401	86.85663	1.0186×10^{-196}
β_0	229.14199	2.08420	1809.88286	0
β_1	0.98666	0.06019	269.86182	0

Also, as demonstrated in Table 4, each of the structural parameters differs significantly from zero.

The Fig. 9 presents the power consumption characteristics of the robot executing the trajectory using a combination of joint and circular interpolation.

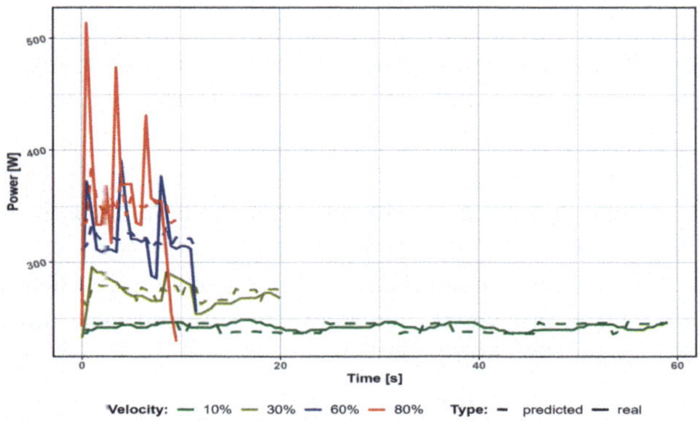

Fig. 9. Energy consumption for joint with circular motions.

The results clearly show that the robot's velocity has a strong impact on instantaneous power demand. At 80% velocity, the motion produces the highest and most dynamic power peaks, exceeding 500 W. These sharp spikes are typical for joint-based motion at high speed, where rapid acceleration and deceleration of individual robot axes generate significant transient loads. At 60% velocity, the peaks remain distinct but their amplitudes are lower and the overall profile becomes more stable. When the speed is reduced to 30%, the fluctuations decrease considerably, and the power curve becomes smoother, although still influenced by the local accelerations occurring during circular segments of the path. The most stable behavior is observed at 10% velocity, where both joint and circular movements generate nearly steady power consumption around 250–260 W with minimal variability. The predicted curves show good agreement with the real measurements for all velocity levels. The model captures both the magnitude and general shape of the power profiles, including the characteristic oscillations caused by transitions between joint and circular motion. The largest discrepancies are visible at the highest speed (80%), which is expected due to the nonlinear dynamic effects and abrupt axis movements typical for fast joint interpolation. A sudden increase in energy consumption is additionally caused by the fact that circular motion requires the simultaneous engagement of multiple robot mechanisms, and in this experiment the robot executed the circular segments with high positional accuracy to faithfully reproduce the intended trajectory. The presented results confirm that combining joint and circular interpolation leads to more dynamic and irregular power patterns at high speeds, whereas at low speeds the robot operates in a stable and energy-efficient manner. The prediction model reflects these behaviors accurately, demonstrating its suitability for forecasting power consumption under mixed-interpolation motion.

3.4 Joint, Linear and Circular Motions

Also in this case the circular motion is used as the reference case. For the joint, linear & circular model trajectory, the model takes the following form:

$$y_t = \beta_0 + \beta_1 V_t + \beta_2 JM_t + \beta_3 LM_t + \alpha_1 y_{t-1} + \alpha_2 y_{t-2} + \theta_1 \varepsilon_{t-1} + \theta_2 \varepsilon_{t-2} + \theta_3 \varepsilon_{t-3} + \varepsilon_t, \tag{12}$$

where the structural parameters are given in Table 5, together with their standard deviations, test statistic values and p-values, and the random variable $\varepsilon_t \sim N(0, 1093.915)$.

As presented in Table 5, the parameters $\alpha_1, \alpha_2, \theta_1, \theta_2, \theta_3, \beta_0, \beta_1, \beta_2$ exhibit a statistically significant deviation from zero. However, for the parameter β_3, at the specified significance level, there is insufficient evidence to reject the working hypothesis (i.e., the structural parameter does not significantly differ from zero, indicating that energy consumption during linear movement is greater than during circular movement, and this discrepancy is not statistically significant).

Figure 10 shows the measured and predicted instantaneous electrical power consumption of the robot during linear motion for four commanded velocities: 10%, 30%, 60% and 80% of the nominal speed.

At 10% velocity, the robot operates with the lowest and most stable power demand, mainly in the range of 240–260 W, and the motion cycle is the longest (about 45 s). The increase in electrical power consumption during linear motion is primarily driven by the required motion precision. To accurately reproduce the predefined linear trajectory, the robot must engage a larger number of mechanical components, which results in higher

Table 5. Structural parameters of ARIMAX model for joint, linear & circular motions

	Value	Sd. dev	T	p- value
α_1	−0.63488	0.25955	−32.07959	2.5147×10^{-72}
α_2	−0.57970	0.13027	−58.36158	3.6747×10^{-111}
θ_1	0.81186	0.24057	44.25915	9.5058×10^{-93}
θ_2	0.90432	0.10208	116.18292	1.1169×10^{-158}
θ_3	0.30385	0.08699	45.80766	5.2926×10^{-95}
β_0	219.72227	5.97519	482.26629	4.8759×10^{-259}
β_1	2.12953	0.14847	188.11126	1.5746×10^{-192}
β_2	6.26929	9.27718	8.86270	1.2954×10^{-15}
β_3	0.87042	6.78460	1.68256	0.09437

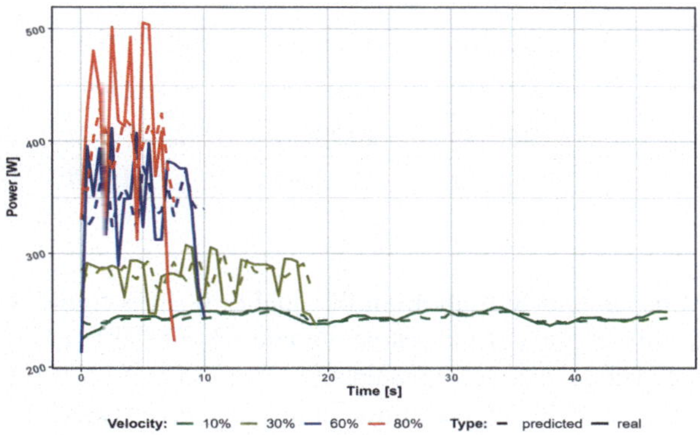

Fig. 10. Energy consumption for linear motion.

energy usage. The observed peaks are caused by changes in the direction of motion and by the robot's accelerations. Increasing the velocity to 30% introduces more noticeable variability, with power oscillating between approximately 270 and 300 W and a visibly shorter process duration of around 20 s. At higher velocities (60% and 80%), the power demand rises significantly, reaching peaks above 400–500 W, and the profiles become much more dynamic, with pronounced fluctuations resulting from rapid motion and higher accelerations.

The dashed lines (predicted values) follow the solid lines (measured data) closely for all velocity levels. The model reproduces both the overall power level and the timing of the main peaks, with only local deviations, particularly in the most dynamic high-speed regimes (60% and 80%). Overall, the comparison demonstrates that the predictive model effectively captures

3.5 Fitting Models

The Table 6 presents a comparison of the model fits to the empirical data. The optimal agreement with the measurements is achieved for the joint motion, for which the correlation coefficient reaches 0.947. This is accompanied by the lowest errors among all cases (mean absolute percentage error [MAE] $\approx$ 4.45 W, root mean square error [RMSE] $\approx$ 8.38 W, mean percentage error [MPE] $\approx$ 1.62%). The linear motion model performs marginally worse, with a high correlation of R = 0.933 and only moderately higher error values (mean absolute percentage error (MAE) $\approx$ 5.21 W, root mean square error (RMSE) $\approx$ 9.83 W, mean absolute percentage error (MAPE) $\approx$ 1.89%). The findings suggest that the proposed approach can predict power consumption with a high degree of accuracy when calibrated for a specific type of motion.

Table 6. Basic parameters for fitting models with different motions

	R	MAE	RMSE	MAPE
joint	0.9471704	4.446318	8.379762	0.0162478
linear	0.9326219	5.205709	9.832999	0.0188897
joint with circular	0.8486130	10.739893	23.759552	0.0358829
joint, linear & circular	0.8488292	17.100626	32.291437	0.0553031

In the context of circular joint motion, the quality of fit is found to deteriorate, as evidenced by a decline in the correlation coefficient (R = 0.849) and an augmentation in the error indices (mean absolute percentage error (MAE) $\approx$ 10.74 W, root mean square error (RMSE) $\approx$ 23.76 W, maximum percentage error (MAPE) $\approx$ 3.59%). The most significant discrepancies are observed when a single model is identified jointly on data from all motion types (joint, linear and circular). In this instance, the correlation remains at a comparable level (R = 0.849), yet the absolute and relative errors increase considerably (MAE $\approx$ 17.10 W, RMSE $\approx$ 32.29 W, MAPE $\approx$ 5.53%). This behaviour confirms the expected trade-off between universality and accuracy: a unified model can describe all trajectories reasonably well, but at the cost of reduced precision compared to models tailored to a specific motion pattern.

Moreover, using models (9)–(12), it is possible to predict the instantaneous energy consumption (i.e., the instantaneous power demand $\hat{y}_t$ in W) as a function of the commanded velocity V_t and the motion type (joint/linear/circular, represented by the corresponding indicators). Consequently, during trajectory design and offline programming, the time history of the predicted power can be generated for each candidate speed profile and interpolation type. This enables estimating not only the instantaneous power demand at every time step, but also the total energy required to execute the entire trajectory, or equivalently as an integral over time. Such an approach supports energy-aware selection of motion type and velocity settings before deployment, while also allowing analysis of peak power levels and transient loads along the trajectory.

4 Conclusions

The studies presented in this paper confirm that effective design and optimisation of robotic processes require detailed, data-driven analysis of robot operation. For all considered motion types, it was possible to develop models capable of satisfactorily reproducing the electrical power consumption of an industrial robot. This demonstrates that relatively simple empirical formulations can capture the essential relationship between motion parameters and power demand. The findings indicate that such models have the capacity to facilitate practical engineering tasks associated with production planning, cost estimation and energy-efficient trajectory design. The capacity to draw parallels between disparate motion strategies is facilitated, encompassing not merely the temporal parameters of cycle time and feasibility, but also the energy consumption dimension, a salient consideration in contemporary manufacturing systems. In this particular context, the proposed approach is consistent with the concepts intended to facilitate the deployment of robots, reduce waste, and enhance the quality and reliability of automated processes.

The work also highlights certain limitations. The experiments were conducted on a single robot, for selected trajectories and operating conditions, which means that the models reflect the behaviour of a specific system. It is therefore recommended that future research endeavours concentrate on validating and extending the approach to different robots, payloads and real industrial tasks, as well as to longer-term, repetitive operation. Nevertheless, the findings provide a solid foundation for the development of practical tools for energy-aware robot programming and the broader integration of energy criteria into the design and implementation of robotic manufacturing systems.

References

1. Knudsen, M., Kaivo-Oja, J.: Collaborative robots: frontiers of current literature. J. Intell. Syst. Theor. Appl. **3**(2), 13–20 (2020)
2. Mohsin, I., He, K., Li, Z., Du, R.: Path planning under force control in robotic polishing of complex curved surfaces. Appl. Sci. **9**, 5489 (2019)
3. Sobaszek, Ł., Świć, A.: Scheduling the process of robot welding of thin-walled steel sheet structures under constraint. Appl. Sci. **11**(12), 5683 (2021)
4. Dawande, M., Geismar, H.N., Sethi, S.P., Sriskandarajah, C.: Sequencing and scheduling in robotic cells: recent developments. J. Sched. **8**, 387–426 (2005)
5. Rudawska, A., et al.: Experimental research and statistical analysis of polymer composite adhesive joints strength. J. Phys. Conf. Ser. **842**, 012074 (2017)
6. Shabtay, D., Arviv, K.: Optimal robot scheduling to minimize the makespan in a three-machine flow-shop environment with job-independent processing times. Appl. Math. Model. **40**, 4231–4247 (2016)
7. Zacharia, P.T., Xidias, E.K., Aspragathos, N.A.: Task scheduling and motion planning for an industrial manipulator. Robot. Comput. Integr. Manuf. **29**, 449–462 (2013)
8. Focant, G., Fontaine, B., Steinicke, L., Joudrier, L.: A robotics task scheduler – TAPAS. In: Tokhi, M.O., Virk, G.S., Hossain, M.A. (eds.) Climbing and Walking Robots, pp. 1031–1036. Springer, Berlin/Heidelberg (2006). https://doi.org/10.1007/3-540-26415-9_123
9. Baizid, K., et al.: Robotized task time scheduling and optimization based on genetic algorithms for non-redundant industrial manipulators. In: 2014 IEEE International Symposium on Robotic and Sensors Environments (ROSE), pp. 112–117. IEEE, Timisoara, Romania (2014)

10. Zhang, J., Meng, Q., Feng, X., Shen, H.: A 6-DOF robot-time optimal trajectory planning based on an improved genetic algorithm. Robot. Biomim. **5**, 3 (2018)
11. Bahani, A., Ech-Chhibat, E.C., Samri, H., Ait El Attar, H., Ait Maalem, L.: Intelligent control of the gripping force of an object by two computer-controlled cooperative robots. Applied computer. Science. **19**(1), 133–151 (2023)
12. Aiming, L., Huan, L., Bitao, Y., Wenjun, X., Ming, Y.: Energy consumption modeling of industrial robot based on simulated power data and parameter identification. Adv. Mech. Eng. **10**, 1687814018773852 (2018)
13. Hou, L., Zhang, L., Kim, J.: Energy modeling and power measurement for mobile robots. Energies. **12**, 27 (2019)
14. Abdelfetah, H., Aouache, M., Abderraouf, M., Akli, I.: Key challenges and open issues of industrial collaborative robotics. In: IEEE International Conference on Robot and Human Interactive Communication (IEEE RO-MAN 2018) Workshop Human-Robot Interaction: From Service to Industry (HRI-SI 2018) (2018)
15. Ning, F., Yun, C., Chen, X.: Dynamic analysis and control of the MJR robot palletizer. In: Xiong, C., Liu, H., Huang, Y., Xiong, Y. (eds.) Intelligent Robotics and Applications. ICIRA 2008 Lecture Notes in Computer Science, vol. 5315, pp. 713–722. Springer, Berlin/Heidelberg (2008). https://doi.org/10.1007/978-3-540-88518-4_77
16. Gou, W., Wang, L.: Design of robots' automatic packaging and palletizing. IOP Conf. Ser.: Earth Environ. Sci. **242**, 032030 (2019)
17. Bernier, C.: Palletizing robots and their impact on production and revenue. https://howtorobot.com/expert-insight/palletizing-robots. Accessed 5 Apr 2023
18. Miroshnyk, O., Kovalyshyn, S., Tomporowski, A., Kruszelnicka, W., Bałdowska-Witos, P.: Research of probability characteristics of current and voltage unbalance based on using graphs of load for the duration at the substation. J. Phys. Conf. Ser. **1426**(1), 012036 (2020)
19. Kawasaki Heavy Industries: Kawasaki Robots User Manual. Kawasaki Heavy Industries, Ltd, Japan (2010)

Systemic Resilience as a Model of Intersectoral Cooperation Through the Prism of the Participation of Unmanned Aerial Vehicles in the Work of Emergency Headquarters During the 2024 Floods

Piotr Gładysz[1], Mateusz Mazur[2](✉), and Izabella Kęsy[3]

[1] Doctoral School, Poznan University of Technology, Poznan, Poland
[2] 3rd School Battalion, Military University of Technology, Warsaw, Poland
mateusz.mazur02@student.wat.edu.pl
[3] Institute of Management, Economics and Logistics, Pomeranian Higher School in Starogard Gdanski, Starogard Gdanski, Poland

Abstract. The purpose of this article is to analyze the participation of unmanned aerial vehicles (UAVs; drones) in the work of emergency headquarters during the floods that hit southwestern Poland in September 2024, in the context of building systemic resilience and intersectoral cooperation. The article attempts to show how the use of UAVs can support the integration of activities of public services, local governments, the private sector and non-governmental organizations in emergency management. A case study method and analysis of industry reports and documents were used to identify key tasks carried out using UAVs during flood control operations. The results of the study indicate that the data acquired with the help of UAVs significantly increases the efficiency of the work of emergency staffs by improving hazard recognition, impact assessment and coordination of activities. The article also highlights the importance of UAVs as an element that strengthens systemic resilience by facilitating information sharing and cross-sectoral interaction in emergency situations.

Keywords: Rescue · Crisis · UAV · drone · Flood · Environment · Emergency

1 Introduction

In recent years, there has been a marked increase in the frequency and intensity of extreme events as a consequence of ongoing climate change. Climate change has increased the frequency and intensity of extreme events, placing growing pressure on emergency management systems. The increase in their frequency is forcing countries to revamp their existing response strategies and invest in more sustainable and crisis-proof infrastructure solutions. Poland is no exception to these challenges. The catastrophic flooding that hit the southwestern part of the country in September 2024 became one of the biggest tests of the functionality of the country's homeland security system since the so-called

B. Gapiński et al. (Eds.): MANUFACTURING 2026, LNME, pp. 277–291, 2026.
https://doi.org/10.1007/978-3-032-22943-4_20

"flood of the millennium" in 1997. Many localities were cut off from energy and water supplies, and thousands of residents had to be evacuated. According to announcements from the Ministry of the Interior and Administration and official government reports, the effects of the 2024 floods in Poland affected around 57,000 people with material losses exceeding 13 billion PLN. The destruction included critical infrastructure, residential buildings, transportation infrastructure and agriculture. It is estimated that a total of more than 2.3 million people lived in the disaster areas, of which more than 11,000 single-family homes and more than 2,000 multi-family buildings were damaged [1, 2]. The disaster highlighted both the strengths of the institutional response of the state and civil society, as well as numerous weaknesses in the coordination of activities, communication and distribution of humanitarian aid, which is emphasized in various emergencies [3, 4]. This indicates the need to strengthen emergency management systems, improve procedures for cooperation between institutions and develop modern IT tools that could support decision-making in emergency situations [5]. Moreover, the results of analyses of post-crisis reports [6] indicate that the key problem of the first 72 h after the floods occurred was information chaos, the lack of a unified operational management system, and insufficient preparation of local structures for rapid resource mobilization. Therefore, the rapid deployment of unmanned aerial vehicles in reconnaissance and monitoring activities proved to be one of the key factors in reducing the effects of these shortcomings. The UAVs have made it possible to acquire data on an ongoing basis, assess the scale of damage more quickly, and coordinate rescue efforts more effectively, thus helping to improve the flow of information and increase the resilience of the disaster management system. This is an important element of systemic resilience, and therefore the main objective of this article is to analyze the role of unmanned aerial vehicles in the activities of emergency headquarters during the 2024 floods, with particular emphasis on their importance in improving intersectoral cooperation and increasing the efficiency of crisis management. The presented scientific considerations were supplemented by the presentation of key concepts in crisis management, analysis of the course of rescue and relief operations, and diagnosis of communication and organizational problems in crisis management during floods. This made it possible to outline the background of the crisis and the framework for intersectoral cooperation, as well as to identify areas and opportunities for the use of UAVs as a tool to enhance situational awareness, improve coordination of activities and fill informational gaps created by the overload or failure of traditional observation systems. The novelty of this article is the interdisciplinary approach to the role of UAVs in the context of the state's systemic resilience, showing their importance not only as a technical tool, but also as a factor supporting cooperation between public administration, emergency services and the private sector. Key to this is the experience of one of the co-authors, who as an officer of the State Fire Service in Poland participated both in this relief operation and in other emergencies in which UAVs played an important role. The operational value of UAVs in crisis management is also documented beyond flooding, including earthquake damage assessment, wildfire perimeter monitoring, and search-and-rescue support in inaccessible terrain. This broader evidence suggests that the mechanisms discussed in this case, such as situational awareness, faster mapping, and shared operational pictures, are relevant across hazard types. This case study can be replicated by compiling the same categories of primary sources (official situation reports, post-incident evaluations, and operational summaries) and by documenting UAV mission types (reconnaissance, patrol, photogrammetry) and the GIS workflow used to generate and annotate orthophotomaps.

2 Literature Review

In order to accurately identify the role of UAVs in the crisis management system, it is necessary to present the key issues in this area. Crisis management in scientific literature has evolved from simple, linear models to complex, adaptive decision-making frameworks. Classical approaches frame crisis management as a cyclical process involving four key phases: prevention, preparedness, response and recovery ("4R") [7, 8]. This model assumes a logical sequence of activities and a hierarchical structure of responsibility, where each link is equally important for reducing the effects of disasters and quickly restoring the functioning of society. On Polish and international grounds, such a division of phases has become the basis of legislation and systems of practice for disaster management. The work of Arjen Boin et al. [9] extends the classic cycle by analyzing the "governance of crises", that is, the complex tensions and dilemmas that arise at the interface of politics, public administration and inter-institutional coordination (multi-level coordination). According to these authors, responding to a crisis often requires transcending rigid procedures and teamwork that goes beyond formal schemes, which is particularly evident during sudden, intense events such as floods or technological disasters. Louise Comfort's [10, 11] concept emphasizes the importance of adaptability and learning from crisis systems in a broad sense (adaptive capacity, learning from crisis). In her view, effective crisis management involves continuous monitoring of the situation, flexible adjustment of procedures and dynamic processing of incoming information. David Alexander [12, 13], on the other hand, points to the growing importance of standards and professionalization and the creation of international and national guidelines (standards, doctrine, professionalization) as the foundation of modern crisis management systems. An important element of a modern crisis management system is also social and infrastructural resilience, defined as the ability of a system - social, technical or institutional to absorb shocks, adapt to changing conditions, and transform in a way that allows it to continue to function [14]. This concept, derived from the classic works of C. S. Hollings [15], was further developed by, among others, S. L. Cutter [16] and colleagues, who proposed the Social Vulnerability Index, and by Linkov and others [17, 18], authors of the so-called Resilience Matrix [17, 18]. Resilience is therefore not a state, but a dynamic process that combines technical, social and organizational aspects, with risk reduction and proper risk management being a key elements [19]. Another key in crisis management is semi-governance and inter-sectoral networks (network governance) [20]. Their importance has been confirmed by both Polish and international experiences [21, 22]. In crisis situations, there is an increasing move away from a centralized command model to network governance, involving public administration, emergency services, NGOs, the private sector and local communities [23]. In Poland, systemic assumptions of intersectoral partnership have been included in the National Crisis Management Plan and promoted by the Government Centre for Security [24]. The perspective and dynamics of information flow as well as the density of relationships between network participants are extremely important in this regard. Research by Kapucu and co-authors [25] shows that the tighter the ties and more frequent the exchange of inter-organizational information, the faster the network responds to changing crisis conditions and more effectively solves operational problems. Experience points to problems associated with the dispersion of communication channels, the lack of a unified warning system and the insufficient use of social media in coordinating information activities. UAVs play a special role in light of these expectations. The literature emphasizes that UAVs are an important tool to support decision-making processes in emergency situations, especially in the context of limited access to disaster areas. Studies indicate that the use of UAVs enables faster acquisition and processing of spatial data, which translates into more precise planning of rescue operations, assessment of losses and monitoring of the dynamics of hazardous phenomena [26]. The authors of the analyses point out that the use of remote sensing, thermal imaging and photogrammetric technologies in UAVs makes it possible to create real-time situational models, facilitating decision-making by emergency headquarters and field services [26, 27]. The concept of using UAVs also fits into the broader framework of contemporary research on systemic resilience and cross-sector cooperation. As numerous studies [28, 29] point out, effective emergency management requires the integration of technological, informational and organizational resources between public administration, emergency services, the private sector and civil society [30]. UAVs can act as a link between these sectors, providing a unified source of reliable data and enabling it to be shared quickly on a common operational platform. As a result, they become not only a technological tool, but also a structural element that strengthens the state's systemic resilience. Significantly, the way emergency management has changed in recent years with the development of information technology. Solution based on real-time data analysis, crowdsourcing of information and artificial intelligence are becoming increasingly important [31]. Studies show that integrating data from variety of sources - including social media, telemetry systems, environmental sensors or mobile applications - enables faster responses and better tailors assistance to the actual needs of the affected. In conclusion, the literature clearly indicates that the effectiveness of emergency management under conditions of violent threats depends on the ability to integrate diverse data sources and quickly convert them into useful operational information. The combination of unmanned aircraft technology with other data acquisition channels has the potential to create a coherent, multi-layered information structure that significantly increases the systemic resilience of the state. UAVs are a key element in this arrangement, enabling rapid and precise recognition of the situation on the ground and supporting cross-sectoral information exchange, so the following section of the article presents the results of an analysis of the use of UAVs during the 2024 floods in Poland as a case study illustrating the practical application of these solutions. The presented experience shows that the effective integration of UAV into the structures of emergency headquarters can be the foundation of a modern, integrated model of intersectoral cooperation, which is one of the key elements of building systemic resilience.

3 The Course and Context of the 2024 Flood Crisis in Poland

Flooding in Poland took place in September 2024. At the beginning of the month, the hydrological situation seemed stable, but on September 11–16 a low pressure system over the southwestern part of the country became active over the Gulf of Genoa, which caused that on September 13–14 there were days with precipitation totals exceeding 100 mm, locally reaching 161–219 mm (including Jarnołtówek - 161.5 mm; Śnieżnik - 218.8 mm; Ustroń-Równica-Wieś - 216.0 mm). On September 15, precipitation decreased, although 60–90 mm was still recorded in places [32]. The highest number of exceedances of alarm levels in the Oder River basin was recorded on September 16 (81 sites), and the culmination of flood waves moved down the Oder River on the following days [32]. The scale of the threat is presented in the Sentinel-2 (Copernicus/EO Browser) satellite image below [33] (Fig. 1).

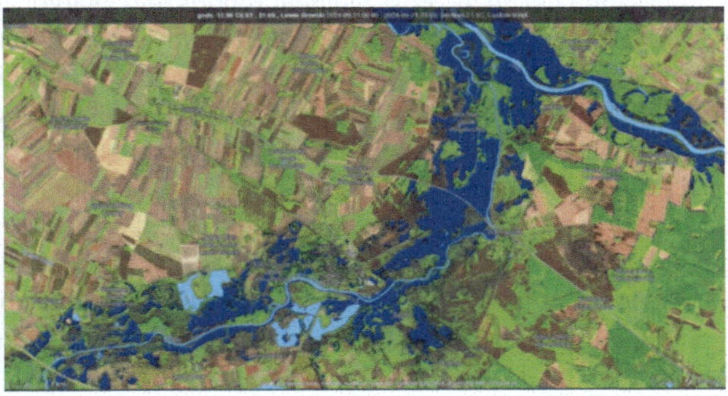

Fig. 1. Flooded areas in the lower section of the Nysa Klodzka River on 21.09.2024 on Sentinel-2 satellite imagery (Copernicus/EO Browser).

On September 17, the government's reconstruction program for flood-affected regions was announced at the operational level. In the following days (September 18–22), briefings were held with governors on the state of reservoirs, roads and railroads, and coordination of services (PSP - State Fire Service, Police, EMS). On September 20, Poland activated assistance under the EU Civil Protection Mechanism, while the EC/JRC and Copernicus EMS conducted rapid mapping of the deluge. The fire department reported that more than 29,000 interventions had been made since September 11, evacuating approx. 4,500 people; nearly 98,000 rescuers and 24,000 vehicles in the affected provinces were involved. Copernicus reported that about 60% of the 56,441 ha of mapped flooded area in the seven countries fell in Poland [34]. According to the Ministry of Interior and Administration and the Chancellery of the Prime Minister (briefing 21–22 IX [35]), around 57,000 people were realistically affected by the threat. As many as 6,544 people were evacuated, and more than 11,000 buildings and more than 6,000 outbuildings were flooded. The area of the state of disaster included 744–749 localities with a population of about 2.39 million people (population in the affected zone). In financial terms, Poland submitted an application to the European Commission for support from the EU Solidarity Fund on November 29, 2024; the total amount of expenses and losses reported by governors and ministries exceeded 13 billion PLN. In parallel, the EC/ECHO mobilized assistance under the UCPM for the Czech Republic and Poland [32, 36]. The dynamically changing situation and such a significant scale of the threat promoted the appearance of information chaos, decision-making delays or barriers to coordination. Technical materials of the Polish Waters and the Institute of Meteorology and Water Management (IMGW) indicate that extreme conditions caused failures and loss of some hydrological telemetry, which limited the current depiction of the situation and forced later "re-estimations" based on water balances, which was a real barrier to warning and operational decisions at the peak of the event [32]. In this situation, the role of unmanned aerial vehicles was significant. They provided the support that made it possible to quickly fill in missing data, continuously observe sensitive sections of embankments and monitor spills in near real time. The use of UAVs meant that despite the failure of telemetry, relatively good situational awareness was maintained, and the services were able to respond more precisely and quickly than would have been possible with traditional observation channels. Post-incident reports and analytical studies (e.g., HumanDOC; Stefan Batory Foundation [6, 37]) additionally pointed to the dispersion of communication channels, ambiguous evacuation messages, overloaded hotlines, and inconsistent information flow between administrative and volunteer levels. The cited reports identified such gaps as the lack of a unified real-time operational information system, poor integration of satellite and field mapping data, and insufficient cross-sector cooperation procedures. In this context, it is important to emphasize the need to ensure that UAV imagery and records are available in a unified form to all actors involved - from emergency management centers to emergency services to volunteer coordinators - which could significantly reduce uncertainty, improve information consistency and enhance real-time operational decisions [38].

4 Operational Efficiency and Intersectoral Cooperation

During the 2024 floods, the emergency services demonstrated a high level of operational efficiency and personnel commitment, which was a key factor in reducing the impact of the disaster. The State Fire Service, supported by Volunteer Fire Departments, carried out more than 29,000 rescue interventions, including more than 4,500 evacuations of people from threatened areas, and secured several thousand public facilities and critical infrastructure. Nearly 98,000 officers and volunteer firefighters, as well as over 24,000 vehicles and units of specialized equipment participated in the operations, making this operation one of the largest mobilizations of emergency services in the history of the Third Polish Republic [32]. In parallel, the Polish Armed Forces, particularly the Territorial Defence Forces (WOT), land army engineering units and the Military Police, played an important role. The Ministry of National Defence has deployed some 6,000 troops to the affected areas, providing support in the construction of temporary crossings, strengthening of flood barriers, logistical transports and medical evacuations. Cooperation of the troops with the PSP and local administration was mostly evaluated positively, and in many places it was the WOT troops that ensured continuity of operations when local structures were overwhelmed. Interdepartmental interaction was largely in line with existing procedures, although local problems with the flow of information and synchronization of decisions were noticeable during the first phase of the response. An advantage of the system was the good integration of rescue structures at the provincial and district levels, including the activation of PSP command centers and points of contact at crisis headquarters [39]. The establishment of crisis headquarters allowed for the creation of a space for the exchange of information on ongoing and planned activities, the coordination of these undertakings, and the ongoing management of the flow of information. This arrangement was a key support for the crisis management groups established by the Plenipotentiary in both municipalities. An important element was the division of the staff team according to the subject areas it was to deal with. PSP officers worked closely with experts and those in charge of the relevant tasks - including civil and military engineers, city and municipal officials, managers overseeing repairs to individual municipal installations, etc. This interaction covered areas such as rescue, security, municipal and road infrastructure. Due to the extensive damage that particularly affected several municipalities in southern Poland, the Minister of the Interior and Administration decided to suspend the powers of the mayors of the four municipalities with regard to flood prevention and recovery activities. Plenipotentiaries drawn from among the officers of the State Fire Service were appointed to carry out these functions, and they headed the crisis staff in the organizational structure and managed the work. Officers who were part of the staffs had experience and competence gained during missions abroad, within search and rescue groups and international rescue modules. These firefighters participated in training courses covering international relief systems and cooperation with local emergency services, administrations and international rescue and civil protection structures in the affected countries. Such preparation, according to the plenipotentiary, guaranteed a broad perspective on operations and provided the greatest opportunity to effectively organize work and meet the demands placed on the State Fire Service. Also important was the extensive professional experience of firefighters, resulting from participation in extensive and complex rescue and firefighting operations carried out throughout the country. A graphic representation of the organizational structure of the emergency headquarters is shown in the figure below (Fig. 2):

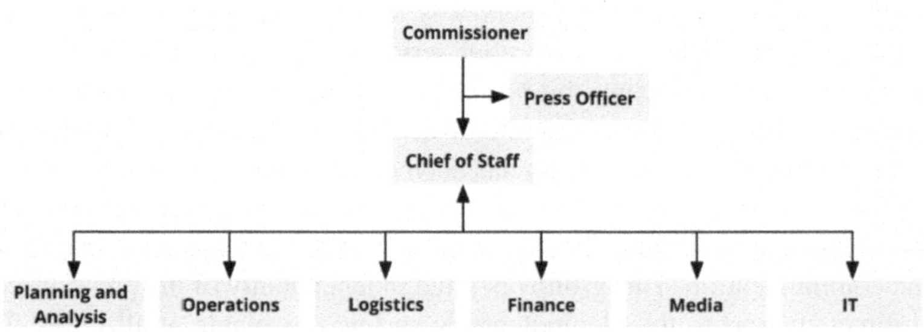

Fig. 2. The organizational structure of the crisis staff.

In its formation, the tasks set before it were guided. Since the catalog of tasks to be carried out was extensive and was changing (mainly enlarging) with each passing day, it was decided to modify the structure adopted in the National Firefighting and

Rescue System (KSGR), using the experience and knowledge of the operation of headquarters under the Incident Command System, used, among others, in the USA, and the internal regulations governing the command system in the fire department in Germany Führung und Leitung im Einsatz - Führungssystem. The first of the headquarters units, responsible for planning and analysis, focused on collecting, processing and analyzing information, as well as communicating conclusions and proposals for task solutions and managing the flow of data both inside and outside the headquarters. From an organizational point of view, the most important role was played by the operations team, which was the most extensive part of the staff. It was the one that was crucial to the smooth implementation of planned activities, dealing with the coordination and implementation of tasks in cooperation with individual chiefs, based on the arrangements made during crisis management team briefings. Within the structure of the operations team there were several separate groups responsible for specific tasks, including rescue (State Fire Service), ensuring public safety (Police), relief operations and cleanup (Territorial Defence Forces), maintenance of road infrastructure (engineering troops), as well as operation of water and sewage, electricity, gas infrastructure or social support. Selected services and entities participating in operational activities used their autonomous staffs, called operational staffs. It was crucial to ensure proper communication and efficient flow of information not only vertically (Fig. 2) between the emergency headquarters and the operational headquarters, but also horizontally (Fig. 3), which was particularly important for the efficiency of operations. The horizontal organization of the headquarters is shown in the figure below (Fig. 3):

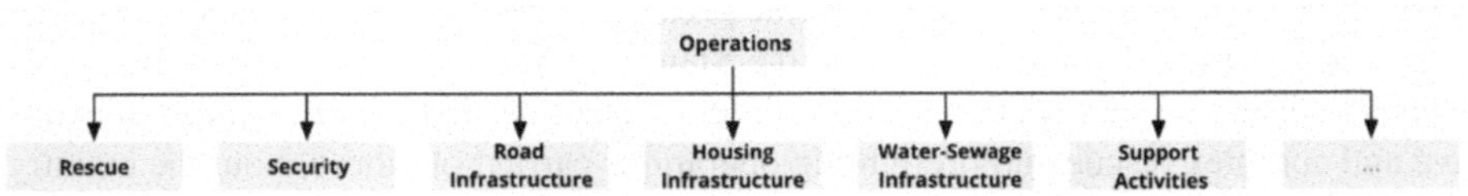

Fig. 3. Structure of the operations team divided into individual sub-teams

Another key part of the headquarters was the logistics team, whose main tasks included taking stock of the needs of relief services (e.g., fuel, food, lodging, equipment), identifying and collecting the needs of local residents, managing warehouses with humanitarian aid, and organizing the distribution of this aid. To increase the efficiency of operations, the logistics team has been divided into specialized sub-teams. In order to ensure high efficiency of the staff's work, a separate competence area related to finance was separated. Additional functional units covered logistics, finance, IT support, and public communication. This group operated independently of the press officer, who provided direct support to the plenipotentiary and was available at all times. The last separate unit was the IT unit, responsible for preparing the ICT environment necessary for the smooth functioning of the emergency headquarters. The complexity of the situation, the multitude of problems of varying magnitude and the enormous dynamics that characterized the situation required the staff team to be highly flexible and able to make decisions quickly, in the absence of sufficient data and thus a high risk of decision-making error. An important support for all staff units, both operational, logistics, as well as planning and analysis, was information obtained from unmanned aircraft. Data collected during aerial operations made it possible to update the picture of the emergency situation on the ground, identify new threats and verify the effectiveness of ongoing operations. Thanks to aerial imaging, it was possible, among other things, to determine the direction of movement of water, the scale of damage to infrastructure, the passability of roads. This information provided not only valuable material for the planning and analysis cell, but also a direct decision-making tool for individual teams, enabling them to plan their operations more precisely and make optimal use of resources [40]. The UAVs also performed an important function in ensuring the proper flow of information between the various entities involved in crisis management. They enabled the rapid transmission of up-to-date data to emergency headquarters. UAV imagery became a reference during crisis staff meetings and briefings, facilitating the synchronization of tasks between sub-teams and minimizing the risk of erroneous decisions resulting from incomplete information. As a result, UAVs have proven to be one of the key elements for enhancing the effectiveness of crisis management, especially in a rapidly changing and complex operational situation. The detailed application and types of tasks performed by UAVs will be discussed using selected case studies in the next chapter.

5 Unmanned Aerial Vehicles as an Essential Element in the Work of Emergency Staff

The potential of unmanned aerial vehicles, especially in terms of rapid collection and transmission of data from the scene of an incident (photos, video recordings, live broadcasts), has proven to be a key tool in the development and implementation of emergency management strategies, as well as in dealing with their consequences through rescue and humanitarian operations. Both UAV manual flights and automatic missions in the form of photogrammetric raids allowed the development of staff materials used by officers of the State Fire Service (PSP) and cooperating other officers from the Police, the Polish Armed Forces, including the Territorial Defence Forces, and other state institutions. During the flood operations, in the first days of the operation, flights were carried out using UAVs from fire protection units, including Volunteer Fire Departments, the Police, the Polish Armed Forces, associations and the private sector. In addition, UAVs belonging to the Polish Air Navigation Services Agency (PANSA), Volunteer Fire Departments and Tatra Volunteer Rescue Service (TOPR) remained at the disposal of rescue services. In view of the number of available UAVs and the intensity of air operations, officers responsible for coordinating the UAV teams of both the PANSA and the other services involved have been designated at the Headquarters of the State Fire Service. In addition, in order to standardize operations, only UAVs belonging to law enforcement, state aviation and military aviation UAVs were allowed to fly. The main type of UAVs used during operations were multi-rotor aircraft belonging to the State Fire Service and the Police. Far less numerous but extremely useful were UAVs of airframe design, which were piloted by soldiers of the Polish Army. In order to ensure safety and effective coordination of the movement of manned aircraft and UAVs involved in operations, the State Fire Service requested that the Polish Air Navigation Services Agency designate restricted geographic zones for unmanned systems at the service's disposal. The Polish

Air Navigation Services Agency, at the request of the State Fire Service, created zones with a ban on UAV flights, managed by the emergency services and without the possibility of obtaining permission, while asking them to refrain from recreational flights. The zones were visible on dronemap.pansa.pl. maps and on the DroneTower app while they were in effect. These messages were also communicated using social media, an example being the following Fig. 4 showing a Facebook post shared by the Pansa Drones profile, where circles just marked the zones where recreational flying was prohibited (Fig. 4).

Fig. 4. A post from Pansa Drones profile published on the social media platform.

The range of use of unmanned aircraft was very wide. One of the main groups of tasks for which unmanned aerial vehicles of the PSP were used for the work of emergency headquarters were reconnaissance missions. The potential of the UAV to safely and quickly reach hard-to-reach places for people and vehicles, and then visualize the actual state of affairs, allowed representatives of state services and institutions to analyze and assess, among other things, the size and extent of spills, the type and scale of damage, as well as the passability of roads, particularly in terms of rescue and engineering vehicles. Missions of a reconnaissance nature were also used to catalog, plan and prioritize work, as well as segregate it according to the range of capabilities and capacities of task forces. Another important type of missions of a reconnaissance nature for which PSP UAVs were used were patrol missions. These were carried out on a regular basis to, among other things, ensure constant control of water levels in rivers and reservoirs, including reading and recording the values of water gauges, monitoring embankments and hydraulic facilities to identify potential damage and take preemptive action. Patrol flights were also tasked with checking for flags on flooded facilities as a form of conventional signs for calling for help. Another key group of tasks for which PSP UAVs were used for the work of emergency headquarters were automatic missions. These were aerial operations carried out according to a mission plan with set parameters, aimed at taking a series of photos. The images in question were the input data for processing and generating orthophotomaps. These maps were processed by representatives of state services and institutions in geographic information systems (GIS) to create, edit, analyze, visualize and publish spatial data depicting the actual state. The next key group of tasks for which PSP UAVs were used were missions providing decision support. These were tasked with real-time situational imagery for emergency headquarters meetings and briefings. Missions were also carried out for the purpose of supervising rescue operations, including monitoring individual combat sections, monitoring dangerous situations that pose a threat to the safety of rescuers, and for faster verification of equipment and material needs. The Polish Armed Forces, including the Polish Army, were also actively involved in the area of floodplain monitoring activities using UAVs. Soldiers carried out reconnaissance and patrol missions using two advanced UAV models from the airframe group, the Bayraktar TB2 and FlyEye. Reconnaissance flights using FlyEye and Bayraktar unmanned aerial vehicles were part of Operation Phoenix, which was announced on 23.09.2024 to help people from flooded areas. In summary, unmanned aerial vehicles were a key tool in the work of emergency headquarters during the floods. Their ability to provide rapid reconnaissance, precise mapping, continuous monitoring, and real-time decision-making and communication support, contributed significantly to the recognition of emergency situations and optimal coordination of rescue efforts. There are also ethical and legal considerations - UAV operations in populated disaster areas require strict compliance with airspace rules, coordination with manned aviation, and clear command authorization (including temporary restricted zones). In parallel, imagery collection raises privacy and data governance issues: recordings may capture identifiable persons or private property, so access controls, retention limits, and purpose limitation (use only for emergency management) should be defined in advance. Establishing standard operating procedures that address both flight safety and responsible data handling is essential for maintaining public trust.

6 Conclusion

The September 2024 flood was one of the biggest tests of the emergency management system since 1997. Government structures and provincial crisis management centers were quickly activated, and services under the Ministry of the Interior Administration - especially the State Fire Service and Police - began rescue operations within the first day. Territorial Defence troops and engineering units supported logistics and infrastructure reconstruction, and the Government Centre for Security ensured the ongoing flow of information, including through SMS alerts. Local governments also acted effectively in many regions, organizing aid stations and coordinating volunteers. One of the key elements that improved the assessment of the situation was data from Unmanned Aerial Vehicles, which supported the identification of threats, monitoring of embankments, assessment of the scale of damage and location of areas requiring urgent intervention.

However, the flood highlighted serious systemic weaknesses, including information chaos, dispersion of competencies and limited digitization of operational processes. The lack of a unified data exchange system between key institutions led to delays and unclear evacuation messages. Low interoperability between central and provincial structures was also problematic, making it difficult to link field reports - including those obtained from the UAV - with hydrological and satellite data. Many local governments also reported staff shortages and insufficient preparation of information system operators. All of this points to the need for system reform toward greater information integration and the development of modern situation centers supported by aerial observation technologies. The analysis confirms that the integration of UAVs with traditional channels of data collection and processing should become a permanent component of the national emergency management system. The use of UAVs during the 2024 floods showed that they are capable of significantly reducing information acquisition time, increasing the accuracy of threat assessment and improving decision-making at every level of command. The results of the study suggest that the development of appropriate procedures, the standardization of data transmission and the integration of UAV operators into permanent staff structures can significantly improve the consistency and efficiency of emergency operations. At the same time, the flood situation revealed that fully realizing the potential of UAV technology requires further investment in system interoperability, personnel training and data integration into unified information platforms. This study is based on a single case (the 2024 floods in Poland) and combines documentary evidence with practitioner-informed interpretation, which may limit generalizability and introduce perspective bias. Operationally, UAV-derived information may be affected by weather, visibility, flight restrictions, battery endurance, and data transmission constraints; mapping products may include georeferencing or stitching errors, and real-time interpretation can be sensitive to operator experience. Finally, information integration gaps (e.g., between UAV imagery, hydrological telemetry, and satellite products) can delay or distort decision support if workflows and responsibilities are not standardized. These findings indicate that the future of emergency management will largely depend on the state's ability to build multi-source, dynamic knowledge bases supported by aerial observation technologies. The proposed integration model is scalable if UAV tasking, data formats, and briefing products are standardized so that multiple teams can contribute to a single operational picture. Transferability to other administrative contexts (e.g., smaller municipalities or cross-border incidents) depends on interoperability (shared GIS layers, common reporting templates) and pre-agreed coordination rules between services and civilian partners. These conditions are applicable beyond floods, including storms, wildfires, industrial incidents, and infrastructure failures. Combining these solutions into a coherent information architecture could become one of the most important factors in enhancing systemic resilience in the face of increasing frequency and intensity of crisis events.

References

1. Ptaszyńska, Z.: Podliczono straty wyrządzone przez powódź. Polska zwróciła się do KE o pieniądze [flood damages have been tallied. Poland has asked the European Commission for funding.]. Radio ZET – Wiadomości (2024). https://wiadomosci.radiozet.pl/polska/podliczono-straty-wyrzadzone-przez-powodz-polska-zwrocila-sie-do-ke-o-pieniadze
2. Bankier.pl. Rząd: 57 tys. Osób jest realnie dotkniętych powodzią. Nikt nie zostanie bez pomocy [Government: 57,000 people are actually affected by the flood. No one will be left without help.]. Bankier.Pl (2024). https://www.bankier.pl/wiadomosc/Rzad-Obecnie-57-tys-osob-jest-realnie-dotknietych-powodzia-Nikt-nie-zostanie-bez-pomocy-8816314.html
3. Borucka, A., Kozłowski, E.: Mathematical evaluation of passenger and freight rail transport as viewed through the COVID-19 pandemic and the war in Ukraine situation. Adv. Sci. Technol. Res. J. **18**(4), 238–249 (2024). https://doi.org/10.12913/22998624/188503
4. Borucka, A., Kozłowski, E.: Selected polynomial identification techniques to evaluate maritime transport trends around Covid-19. Sci. J. Sil. Univ. Technol. Ser. Transp. **120**, 51–68 (2023). https://doi.org/10.20858/sjsutst.2023.120.4
5. Piechowski, M., Meller, A.: Use of KPI indicators to optimize production process performance – case study. In: Mehmood, R., Alves, V., Praça, I., Wikarek, J., Parra-Domínguez,

J., Loukanova, R., De Miguel, I., Pinto, T., Nunes, R., Ricca, M. (eds.) Distributed Computing and Artificial Intelligence, Special Sessions I, 20th International Conference, vol. 741, pp. 3–12. Springer Nature Switzerland (2023). https://doi.org/10.1007/978-3-031-38318-2_1
6. Springer, D.: Powódź: Wnioski, doświadczenia i rekomendacje na przyszłość. Komunikacja, zarządzanie pomocą humanitarną i rola organizacji pozarządowych w trakcie katastrof naturalnych [Flood 2024: Lessons Learned, Lessons Learned, and Recommendations for the Future: Communication, Humanitarian Aid Management, and the Role of NGOs during Natural Disasters Raport. Fundacja HumanDOC 2024 (2025). https://humandoc.pl/wp-content/uploads/2025/03/rekomendacje.pdf
7. Ouyang, L., Zhang, Y., Tang, H.: Research on urban flood disaster response based on the 4R crisis management theory: a case study of the "23·7" torrential rain in Beijing-Tianjin-Hebei. In: Yuen, K.-V., Xiang, P., Ding, D., Zhang, S. (eds.) Proceedings of the 2025 International Conference on Engineering Management and Safety Engineering (EMSE 2025), vol. 267, pp. 302–314. Atlantis Press International B.V (2025). https://doi.org/10.2991/978-94-6463-780-9_29
8. Wang, L.: Construction and application of college student safety management system based on 4R theory. In: 2022 International Conference on Education, Network and Information Technology (ICENIT), pp. 195–199 (2022). https://doi.org/10.1109/ICENIT57306.2022.00050
9. Boin, A.: Public leadership in times of crisis: mission impossible? Public Adm. Rev. **63**, 544–553 (2003)
10. Comfort, L.K., Wang, W.-J.: Managing risk in dynamic conditions: emerging crises, changing technologies, and the collective capacity to learn. Public Adm. Q. **46**(4), 359–383 (2022). https://doi.org/10.37808/paq.46.4.4
11. Comfort, L.K., Sungu, Y., Johnson, D., Dunn, M.: Complex systems in crisis: anticipation and resilience in dynamic environments. J. Contingencies Crisis Manag. **9**(3), 144–158 (2001). https://doi.org/10.1111/1468-5973.00164
12. Alexander, D.: Towards the development of standards in emergency management training and education. Disaster Prev. Manag. **12**(2), 113–123 (2003). https://doi.org/10.1108/09653560310474223
13. Alexander, D.E.: Disaster and Emergency Planning for Preparedness, Response, and Recovery. Oxford University Press (2015). https://doi.org/10.1093/acrefore/9780199389407.013.12
14. Sobczuk, S., Borucka, A.: Recent advances for the development of sustainable transport and their importance in case of global crises: a literature review. Appl. Sci. **14**(22), 10653 (2024). https://doi.org/10.3390/app142210653
15. Holling, C.S.: Resilience and stability of ecological systems. Annu. Rev. Ecol. Syst. **4**(1), 1–23 (1973). https://doi.org/10.1146/annurev.es.04.110173.000245
16. Cutter, S.L., Boruff, B.J., Shirley, W.L.: Social vulnerability to environmental hazards. Soc. Sci. Q. **84**(2), 242–261 (2003). https://doi.org/10.1111/1540-6237.8402002
17. Linkov, I., Eisenberg, D.A., Plourde, K., Seager, T.P., Allen, J., Kott, A.: Resilience metrics for cyber systems. Environ. Syst. Decis. **33**(4), 471–476 (2013). https://doi.org/10.1007/s10669-013-9485-y
18. Linkov, I., Trump, B.D.: The Science and Practice of Resilience. Springer International Publishing (2019). https://doi.org/10.1007/978-3-030-04565-4
19. Jasiulewicz-Kaczmarek, M.: Practical aspects of the application of RCM to select optimal maintenance policy of the production line. In: Safety and Reliability: Methodology and Applications, pp. 1187–1195. Taylor & Francis Group, London (2015)
20. Moynihan, D.P.: The network governance of crisis response: case studies of incident command systems. J. Public Adm. Res. Theory. **19**(4), 895–915 (2009). https://doi.org/10.1093/jopart/mun033

21. Kapucu, N., Hu, Q.: Network Governance: Theories, Frameworks, and Applications. Routledge (2020)
22. Baran, K., (2022). Przywództwo w administracji publicznej: Perspektywa zarządzania kryzysowego [Leadership in Public Administration: A Crisis Management Perspective] (Wydanie pierwsze). Wydawnictwo Naukowe Scholar
23. Provan, K.G., Kenis, P.: Modes of network governance: structure, management, and effectiveness. J. Public Adm. Res. Theory. **18**(2), 229–252 (2007). https://doi.org/10.1093/jopart/mum015
24. Rządowe Centrum Bezpieczeństwa: Krajowy Plan Zarządzania Kryzysowego National Crisis Management Plan. Rządowe Centrum Bezpieczeństwa. (2025) https://www.gov.pl/web/rcb/krajowy-plan-zarzadzania-kryzysowego
25. Kapucu, N., Arslan, T., Collins, M.L.: Examining intergovernmental and interorganizational response to catastrophic disasters: toward a network-centered approach. Adm. Soc. **42**(2), 222–247 (2010). https://doi.org/10.1177/0095399710362517
26. Gładysz, P., Merkisz, J., Borucka, A.: Reliability of unmanned aerial vehicles in the context of selected factors. Eksploatacja i Niezawodność – Maintenance and Reliability. **28**, 2025 (2025). https://doi.org/10.17531/ein/210312
27. Tanda, G., Migliazzi, M.: Infrared thermography monitoring of solar photovoltaic systems: a comparison between UAV and aircraft remote sensing platforms. Therm. Sci. Eng. Prog. **48**, 102379 (2024). https://doi.org/10.1016/j.tsep.2023.102379
28. Sikder, A.S., Harvey, K.: Techno-resilience: unraveling the impact of cutting-edge information technology in crisis management and emergency response for enhanced disaster preparedness and response efficiency. Int. J. Imminent Sci. Technol. **1**(1), 138–169 (2014). https://doi.org/10.70774/ijist.v1i1.17
29. Lee, O.-K., Jiang, S., Hyun, Y., Jeong, H.: The role of information Technologies in Crises: a review and conceptual development of IT-enabled agile crisis management. Commun. Assoc. Inf. Syst. **50**(1), 857–890 (2022). https://doi.org/10.17705/1CAIS.05038
30. Saniuk, A., Jasiulewicz-Kaczmarek, M., Samolejová, A., Saniuk, S., Lenort, R.: Environmental favourable foundries through maintenance activities. Meta. **54**, 725–728 (2015)
31. Rojek, I., Jasiulewicz-Kaczmarek, M., Piszcz, A., Galas, K., Mikołajewski, D.: Review of the 6G-based supply chain management within industry 4.0/5.0 paradigm. Electronics. **13**(13), 2624 (2024). https://doi.org/10.3390/electronics13132624
32. Państwowe Gospodarstwo Wodne Wody Polskie: Raport z przeglądu i aktualizacji wstępnej oceny ryzyka powodziowego w 3 cyklu planistycznym [Report on the review and update of the preliminary flood risk assessment in the 3rd planning cycle]. Państwowe Gospodarstwo Wodne Wody Polskie (2025). https://powodz.gov.pl/www/powodz/aWORP/3W-12_Raport_WORP_20250320_v1.00.pdf
33. Kosieliński, S. (2025). See, Analyse, Help. Micromacro Institute Foundation. https://informacjakryzysowa.pl/uploads/337/see-analyse-help-drones-during-the-flood_1739957407.pdf
34. Komenda Główna Państwowej Straży Pożarnej. Powodziowa statystyka interwencji oraz zaangażowanych sił i środków jednostek ochrony ppoż. [Flood statistics of interventions and the forces and resources of fire protection units involved.]. Gov.Pl – Aktualności (2024). https://www.gov.pl/web/kgpsp/powodziowa-statystyka-interwencji-oraz-zaangazowanych-sil-i-srodkow-jednostek-ochrony-ppoz
35. Ministerstwo Spraw Wewnętrznych i Administracji.. Odprawa z Wojewodami W Temacie Podsumowania Strat Powodziowych [Briefing with Voivodes Regarding the Summary of Flood Losses]. Gov.Pl – Aktualności (2024). https://www.gov.pl/web/mswia/odprawa-z-wojewodami-w-temacie-podsumowania-strat-powodziowych
36. Polska Agencja Prasowa. Polska wystąpiła do KE o pomoc finansową w związku z powodzią [Poland has applied to the European Commission for financial assistance in connection

with the flooding.]. PAP.Pl (2024). https://www.pap.pl/aktualnosci/polska-wystapila-do-ke-o-pomoc-finansowa-w-zwiazku-z-powodzia-1
37. Kotras, M., Kubala, K., Sześciło, D.: Kiedy pękają tamy. O państwie, wspólnocie i zarządzaniu kryzysem [When the Levees Burst: On State, Community, and Crisis Management]. Fundacja im. Stefana Batorego. (2025) https://www.batory.org.pl/wp-content/uploads/2025/10/Raport_Kiedy-pekaja-tamy_wersja-skrocona.pdf
38. Loska, A., Paszkowski, W., Waszkowski, R.: The scenario approach to the concept of maintenance of technical systems of urban engineering. In: Trojanowska, J., Kujawińska, A., Pavlenko, I., Husar, J. (eds.) Advances in Manufacturing IV, pp. 84–97. Springer Nature Switzerland (2024). https://doi.org/10.1007/978-3-031-56444-4_7
39. Meller, A., Piechowski, M., Gola, A.: Indicators of hierarchical structure model of supporting the production management process – a framework. In: Burduk, A., Batako, A., Machado, J., Wyczółkowski, R., Antosz, K., Gola, A. (eds.) Advances in Production, vol. 790, pp. 356–367. Springer Nature Switzerland (2023). https://doi.org/10.1007/978-3-031-45021-1_27
40. Antosz, K., Jasiulewicz Kaczmarek, M., Machado, J., Relich, M.: Application of principle component analysis and logistic regression to support six sigma implementation in maintenance. Eksploatacja i Niezawodność – Maintenance and Reliability. **25**(4) (2023). https://doi.org/10.17531/ein/174603

Green Technology

Influence of Ecological Cooling Methods on Air Pollution Indicators During Turning of Ti6Al4V Titanium Alloy

Weronika Dylewicz[1(✉)], Radosław Wojciech Maruda[2], and Kamila Skoczylas[3]

[1] The Doctoral School of Exact and Technical Sciences, University of Zielona Gora, Institute of Mechanical Engineering, Zielona Góra, Poland
20000941@stud.uz.zgora.pl

[2] University of Zielona Gora, Institute of Mechanical Engineering, Zielona Góra, Poland

[3] Thomas Praefab Międzyrzecz Sp. z o.o, Międzyrzecz, Poland

Abstract. The aim of the study is to determine the impact of selected ecological cooling methods and their application conditions on air pollution emissions during the turning of the titanium alloy Ti6Al4V. In the context of human health, pollutant emissions are associated with the potential risk of respiratory, cardiovascular, and cancer diseases, especially in the case of long-term exposure to PM dust and TVOC. During the experimental tests, pollution levels were monitored for dry machining, minimum quantity lubrication (MQL), minimum quantity cooling and lubrication (MQCL), and minimum quantity lubrication with alumina nanoparticles (MQL+ Al_2O_3). The analysed pollutants included: carbon dioxide (CO_2), total volatile organic compounds in the gas phase (TVOC), and particulate matter PM1.0, PM2.5, PM5.0, and PM10.0. The statistical analysis using the Pareto method showed that the cooling method has the greatest impact on pollutant emissions. The lowest levels of pollution were found during dry machining, while for ecological cooling methods, it is recommended to use the maximum air flow rate and the cooling-lubrication fluid flow rate in the range of 0.7–0.9 g/min. The highest TVOC values were observed for the MQL + Al_2O_3 method, while for PM2.5, the highest emissions were found for MQL. This is due to changes in viscosity, surface tension, and lubrication properties of the cutting fluid (CF). The results obtained from this study make a significant contribution to the search for the most efficient cooling methods in machining processes, considering both health and environmental aspects.

Keywords: Atmospheric Contaminants · PM · Titanium Alloy · Turning · MQCL · MQL · Nanoparticles

1 Introduction

In the context of sustainable industrial development, the analysis of air pollution generated during machining processes becomes a crucial issue. Traditional cooling techniques are associated with the emission of dust and volatile organic compounds, which negatively affect both the environment and human health [1]. Exposure to air pollutants

B. Gapiński et al. (Eds.): MANUFACTURING 2026, LNME, pp. 295–315, 2026.
https://doi.org/10.1007/978-3-032-22943-4_21

created during machining can lead to long-term health consequences, affecting various human organs. These issues may include the development of chronic respiratory diseases, internal organ damage due to the accumulation of harmful substances in the body, as well as an increased risk of cancer development [2]. Figure 1 illustrates the impact of air pollution on human health.

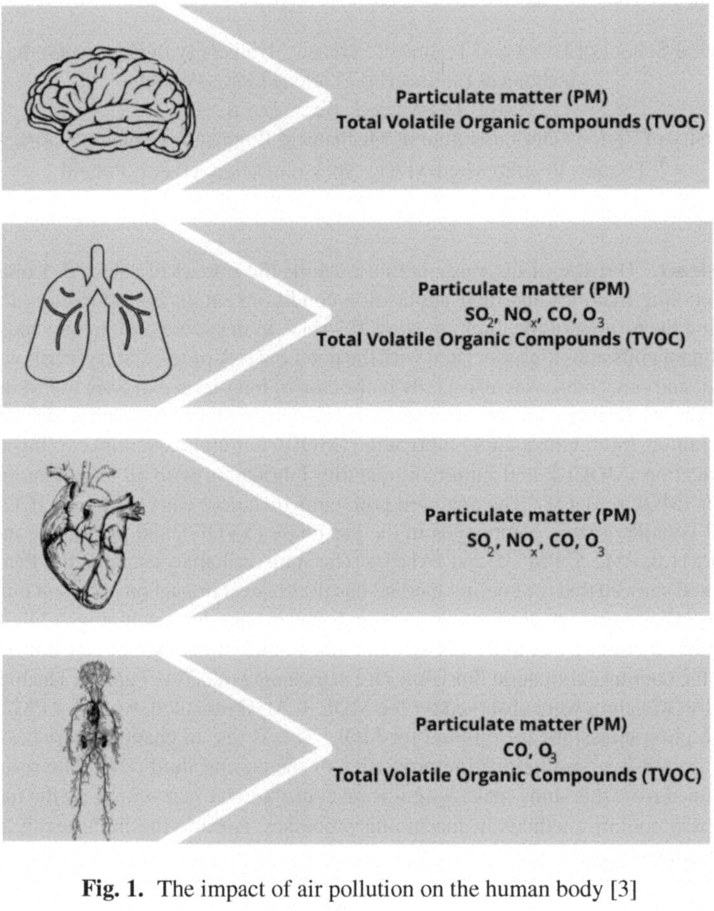

Fig. 1. The impact of air pollution on the human body [3]

These pollutants may appear in the form of fumes, dust, as well as gaseous contaminants. They most commonly originate from the cutting fluid (CF). Other sources of pollution during operation include the contact aera between the workpiece and the cutting tool. During the machining process, various reactions take place between the workpiece surface, the tool, the cutting fluid, and the atmosphere, potentially leading to the formation of harmful substances [4].

In light of increasing environmental requirements and the pursuit of carbon neutrality, the search for innovative cooling solutions becomes crucial. An important aspect of these efforts is the simultaneous reduction of impact on air quality and the improvement of the working environment for machine operators. The integration of these issues, in accordance with the principles of Industry 4.0, forms the foundation for future sustainable development [5]. In the context of green production, the potential benefits of innovations in eco-friendly cooling methods include reduced electricity consumption, improved surface quality, extended tool life, and the reduction of pollutants generated during the machining process.

During machining with cutting fluids (CF), fumes, aerosols, and a tribofilm layer are formed on the surface of both the workpiece and the cutting tool. These byproducts can cause skin and respiratory tract irritation, as well as allergic reactions [2]. Additionally, CF often contain trace amounts of polycyclic aromatic hydrocarbons, with benzo[a]pyrene being a typical representative [6, 7]. Nitrosamines are a group of highly harmful chemical compounds to human health and life. These compounds are formed as a result of reactions between amines and nitrites therefore, and therefore, may be present in some CF that do not contain nitrosamines as part of their composition. Amines are used in coolants as corrosion inhibitors and pH stabilizers, while nitrites, although not currently added to CF, can easily enter them from various sources. These include pollutants from microorganism metabolism, water used to dilute emulsions and concentrates, nitrogen oxides from the atmosphere, and residues from cleaning and washing agents [8]. Formaldehyde, commonly used in coolants as a cleaning agent, has strong biocidal properties. It has been classified as a Group 1 carcinogen, meaning it is a known carcinogen to humans [9]. Research has shown that formaldehyde was detected in all analysed specimens, with concentrations reaching up to 1.3% [10].

Airborne particulate matter also poses significant health risks to operators during machining, consisting of solid particles (PM) with various diameters. These particles contain harmful substances, such as benzo[a]pyrene, furans, dioxins, and heavy metals [4]. The most harmful are considered to be particles in the PM10 and PM2.5 categories. Particles with a diameter $\leq$ 10 μm (PM10) can accumulate in the upper respiratory tract, potentially leading to the development of asthma. PM2.5 particles can penetrate the alveoli and contribute to the development of diseases with cardiological and pulmonological origins [11]. Dust containing solid particles exacerbates asthma, weakens lung function, and increases the risk of lung, throat, and laryngeal cancers, arrhythmias, atherosclerosis, vasculitis, and worsens symptoms of cardiovascular and respiratory diseases [4].

The selection of the appropriate cooling method plays a crucial role in the overall machining process, affecting both its performance and efficiency. However, it is equally important to consider the comfort and health of the operator, as well as the environmental impact. The use of suitable cutting fluids (CF) helps reduce dust emissions [12]. Current trends focus on the complete elimination or significant reduction of CF use. Unfortunately, in some cases, dry machining is not feasible, prompting the search for alternative solutions. As a result, eco-friendly methods such as Minimum Quantity Lubrication (MQL), Minimum Quantity Cooling Lubrication (MQCL), and cryogenic cooling are increasingly being adopted [13].

A holistic approach to assessing the environmental impact of machining is the Life Cycle Assessment (LCA) method. Ginting et al. [14] demonstrated that the use of the

MQL method, compared to wet machining, leads to a reduction in carbon dioxide emissions, thereby decreasing the carbon footprint. The emission of greenhouse gases during the entire process using the MQL method was reduced by 21%. Agrawal et al. [15] conducted a comprehensive analysis of carbon dioxide emissions during turning Ti6Al4V using wet cooling and cryogenic cooling methods. CO2 emissions were reduced by 27% with the use of the cryogenic method. However, replacing wet machining is often associated with an increased risk of air pollution. Additionally, as cutting speed increases, there is also an observed rise in the formation of PM2.5 particles [16].

The literature analysis has shown that the use of alternative cooling methods, compared to wet machining, impacts the emission of air pollutants. However, scientific studies lack data that clearly demonstrate how air pollutant formation varies depending on aerosol creation parameters in minimized cooling and lubrication methods, as well as comparisons between eco-friendly cooling methods. Understanding these relationships is crucial for improving the efficiency of machining processes, especially in the context of minimizing the negative impact on operator health and the environment.

2 Methodology

2.1 Methods

The research was conducted using the Parameter Space Investigation (PSI) method. This is a research-optimization technique that involves systematically searching the parameter space (independent variables) within specified ranges to: identify acceptable areas (satisfying constraints), identify multi-criteria trade-offs (Pareto set), and select the range of independent variable values that most favourably influence the dependent variables [17]. In methods based on minimized cooling and lubrication, the parameters for aerosol formation are critical [18, 19], which is why the study selected values (Table 1) for the variable volumetric airflow rate *P* and the variable mass flow rate of the cutting fluid (CF) *E*.

Variable aerosol formation parameters were applied in the minimum quantity lubrication (MQL) method, minimum quantity cooling and lubrication (MQCL), and minimum quantity lubrication with alumina nanoparticles – MQL + Al_2O_3. Dry machining was also used during the research to compare the values with eco-friendly cooling methods.

During the study, the following air quality indicators were monitored: carbon dioxide (CO_2), total volatile organic compounds in the gas phase (TVOC), and harmful airborne particulate matter (PM – Particulate Matter): PM1.0, PM2.5, PM5.0, PM10.

Table 1. Values of the parameters for the formation of the active medium *E* and *P* corresponding to individual research points according to the PSI method.

Point according to PSI	Mass flow rate of cutting fluid *E* [g/min]	Volumetric flow rate of air *P* [l/min]
1	0.752	25
2	0.46	35

(continued)

Table 1. (*continued*)

Point according to PSI	Mass flow rate of cutting fluid *E* [g/min]	Volumetric flow rate of air *P* [l/min]
3	1.06	15
4	1.182	30
5	0.606	10
6	0.89	20
7	0.388	40

2.2 Turning Process

The research used a CNC lathe CKE6136i. The tool setup consisted of a turning insert holder SVLBR2020K16 and a VBMT160404-F1 insert. The geometry of the cutting edge was as follows: the approach angle of the main cutting edge: $\kappa_r = 95°$, the inclination angle of the main cutting edge: $\alpha = 5°$, corner radius: $\varepsilon_r = 0.4$ mm. The research setup, including the placement of the air quality monitoring device, is shown in Fig. 2.

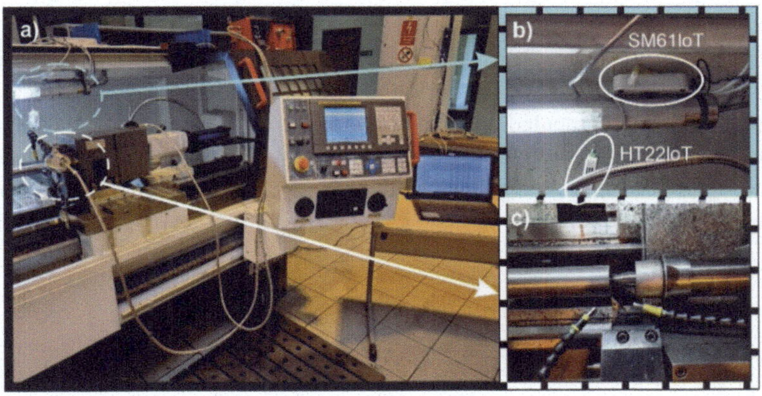

Fig. 2. Research setup: a) CNC lathe; b) air quality sensors (1 - HT22IoT, 2 - SM61IoT); c) nozzles placement in the machining area.

The cutting parameters were selected according to the tool manufacturer's recommendations: cutting speed $v_c = 70$ m/min, feed rate $f = 0.065$ mm/rev, and cutting depth $a_p = 0.6$ mm. The tests were repeated 5 times for each set of changed parameters, and each individual trial lasted 60 s.

The aerosol was delivered to the cutting zone using a Lubrimat L60 device by Steidle (Fig. 3), then transported to the cutting area and sprayed through flexible nozzles (Fig. 2c). The first nozzle was directed at the cutting surface and the main contact surface, while the second nozzle targeted the cutting surface and the auxiliary contact surface of the cutting tool. The aerosol formation occurs by mixing the cutting fluid (CF) with compressed air (at a pressure of 0.4 MPa) in the final design of the nozzle.

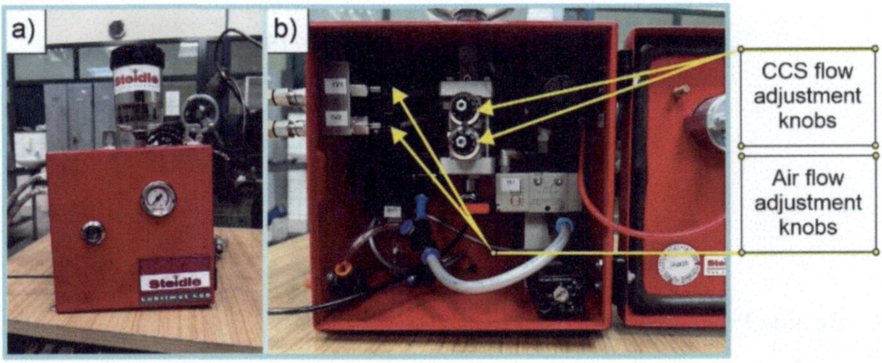

Fig. 3. Lubrimat L60 device by Steidle: a) general view; b) adjustment of variable aerosol formation parameters.

2.3 Workpiece

The workpiece material was the titanium alloy Ti6Al4V (α–β phase, Fig. 4), which is known as hard-to-cut due to its combination of high strength and strain hardening, along with a low elastic modulus (~110 GPa). This combination favors deflections and vibrations, produces segmented chips, and requires low feed rates and cutting speeds. Additionally, its very low thermal conductivity (~6–7 W/m·K) leads to heat accumulation at the cutting edge, while its reactivity with cemented carbides (such as adhesion and diffusion) causes buildup, chipping, and crater wear, ultimately reducing tool life [15, 20].

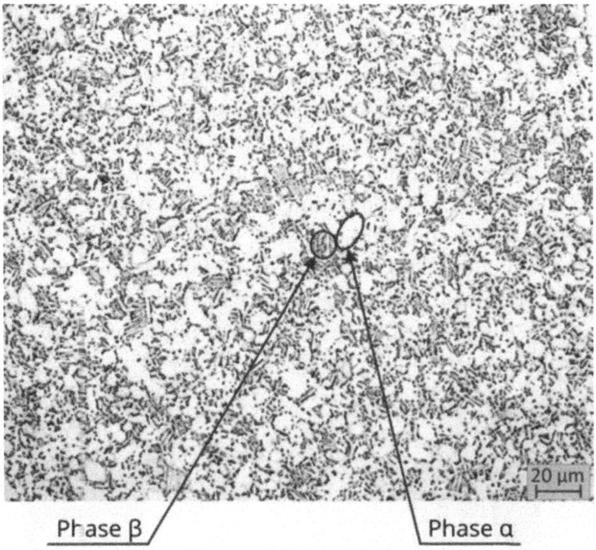

Fig. 4. Structural composition of Ti6Al4V alloy

The mechanical properties of the titanium alloy Ti6Al4V, according to the provided certificate, were: yield strength $Re = 877$ MPa, tensile strength $Rm = 950$ MPa, elongation in 4D $A_{10} = 16\%$, and reduction of area 25 min $Z_{min} = 52\%$. The chemical composition was: Al = 5.96÷6.13%, V = 3.96÷4.03%, $Fe_{max} = 0.15\%$, $C_{max} = 0.014\%$, $N_{max} = 0.008\%$, $H_{max} = 0.0009\%$, $O_{max} = 0.159\%$.

2.4 Cooling-Lubricating Fluids

In the Minimum Quantity Lubrication (MQL) method, Beruform STO 110, a mixture of polyol ester, was used as the cutting fluid (CF). The product's density at 15 °C is 0.915–0.930 g/cm^3, and its kinematic viscosity at 40 °C is 41.4–50.6 mm^2/s [21].

In the Minimum Quantity Cooling Lubrication (MQCL) method, a mixture of the Wedolit K-910 emulsion concentrate (Lubricant Polska Sp. z o.o.) with an 8% concentration mixed with 92% water was used as the CF. This is a semi-synthetic, high-quality emulsion containing 37% mineral oil. At 20 °C, the concentrate has a density of approximately 1.00 g/cm^3, and its viscosity is 270 mm^2/s [22].

In the MQL method with nanoparticles, alumina (Al_2O_3) was used as an additive to the CF, primarily to reduce friction in the cutting zone by forming a tribofilm at the tool-workpiece interface. The creation of such a tribofilm improves surface quality [23], reduces vibrations [24], and also enhances the strength and wear resistance of the cutting tool [25]. In the MQL + nanoparticles method, the active medium used was Beruform STO 110, enriched with alumina nanoparticles (Al_2O_3) at a concentration of 0.5% in the CF, with a particle size of 14 nm [24]. The dispersion of alumina nanoparticles in the CF was carried out using a Hielscher UP200St ultrasonic homogenizer, and the entire nanoparticle stabilization process lasted for 1 h.

2.5 Measuring Equipment

Two sensors were used to measure air pollution parameters, with their placement near the cutting zone shown in Fig. 2b. Data from the sensors were transmitted to a computer with dedicated software. The HT22loT sensor by Lumel was used in the study to measure environmental parameters: temperature, relative humidity, light intensity, CO_2 concentration in ppm and volatile organic compounds (TVOC) in ppb. A data concentrator with a SM61loT server by Lumel, equipped with SENSIRION SPS30 sensors, was used for monitoring particulate matter (PM), focusing on particulate contamination. The sensor provides mass concentrations of PM1.0, PM2.5, PM5.0, PM10 in $\mu g/m^3$.

The graphical representation of the experimental phase of the research is shown in Fig. 5.

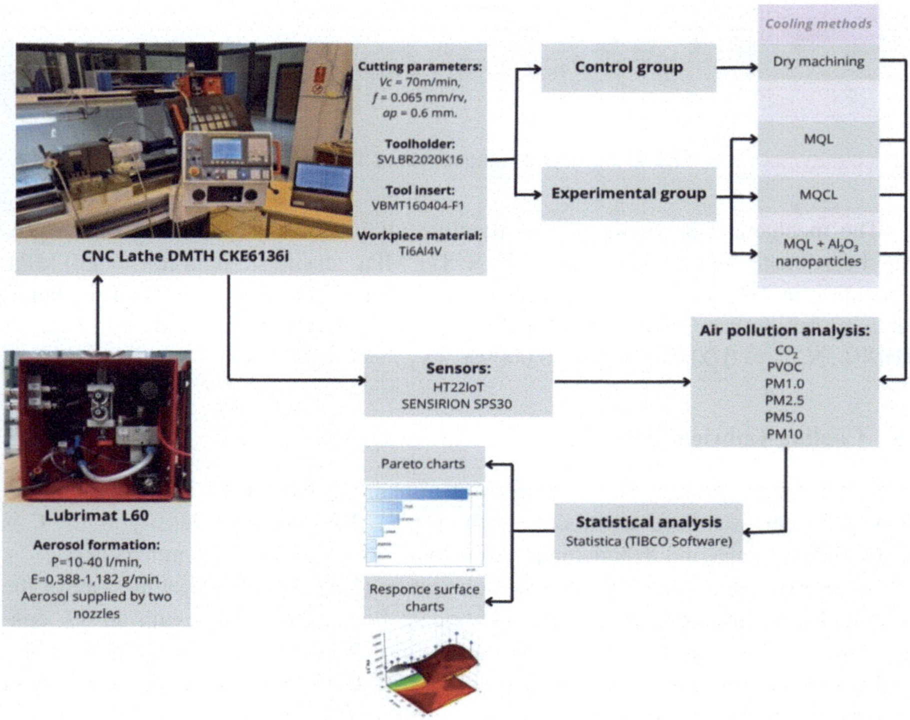

Fig. 5. Diagram of the research workflow showing the workflow of the study and the sequence of the main experimental and analytical steps

3 Results and Discussion

Table 2 presents the measurements results for selected air quality parameters, including carbon dioxide (CO_2), harmful particulate matter (PM1.0, PM2.5, PM5.0, PM10), and total volatile organic compounds in the gas phase (TVOC). For the obtained data, average values and standard deviations were calculated. A percentage change analysis was also conducted between the average value of a given pollutant and the value obtained during dry machining.

All results were subjected to statistical analysis using Pareto charts, which enable the identification of the hierarchy of independent variables affecting the monitored dependent variables. This method allows for the identification of independent variables or groups of independent variables that have the greatest impact on the studied air quality parameters, thus allowing for the mitigation or reduction of the influence of these independent variables, depending on their beneficial or negative effects. The analysis was conducted using Statistica software (TIBCO Software, 2017). The independent variables considered in the Pareto method were: cooling method, volumetric airflow rate P, and mass flow rate of the cutting fluid (CF) E. A surface response chart was then created in Statistica (3D surface/contour) to show how the dependent variable (response) changes as a function of two independent variables while keeping the other parameters constant. This type of model visualization (e.g., polynomial regression/Response Surface) allowed for the evaluation of curvature, interactions, and identification of optimal areas.

Table 2. The results of the pollution measurements taken during the machining

	Cooling method	*E* [g/min]	*P* [l/min]	PM 1.0 [μg/m^3]	PM 2.5 [μg/m^3]	PM 5.0 [μg/m^3]	PM 10 [μg/m^3]	CO_2 [ppm]	TVOC [ppb]
1	Dry	0.388	40	1958	2505	2866	3034	389	142
2	MQL	0.752	25	4782	7717	9924	10950	434	171
3		0.46	35	3744	5741	7219	7906	428	167
4		1.06	15	5063	8601	11295	12548	494	181
5		1.18	30	4011	7622	10560	11908	484	201
6		0.606	10	4017	6288	7870	8633	463	175
7		0.89	20	4423	6733	8438	9230	452	189
8		0.388	40	2281	2999	3487	3713	428	162
Mean:				**4046**	**6529**	**8399**	**9270**	**455**	**178**
Standard deviation:				**906**	**1829**	**2620**	**2991**	**27**	**14**
Difference compared to the control specimen:				**107%**	**160%**	**193%**	**206%**	**17%**	**25%**
9	MQCL	0.752	25	3678	5484	6807	7421	472	208
10		0.46	35	2673	3689	4404	4736	463	195
11		1.06	15	4084	6080	7542	8222	474	216
12		1.18	30	3358	5551	7210	7481	477	225
13		0.606	10	3676	4956	5843	6255	462	212
14		0.89	20	3819	6281	8141	9006	470	206
15		0.388	40	2252	2863	3262	3447	438	185
Mean:				**3363**	**4986**	**6173**	**6653**	**465**	**207**

(*continued*)

Table 2. (*continued*)

	Cooling method	*E* [g/min]	*P* [l/min]	PM 1.0 [µg/m^3]	PM 2.5 [µg/m^3]	PM 5.0 [µg/m^3]	PM 10 [µg/m^3]	CO_2 [ppm]	TVOC [ppb]
Standard deviation:				**663**	**1267**	**1777**	**1974**	**13**	**13**
Difference compared to the control specimen:				**72%**	**99%**	**141%**	**119%**	**20%**	**46%**
16	MQL + Al_2O_3	0.752	25	5769	10060	13344	14871	446	212
17		0.46	35	3371	4376	5050	5363	460	231
18		1.06	15	5596	9760	12947	14429	458	228
19		1.18	30	4767	7087	8784	9573	466	244
20		0.606	10	4033	5142	5869	6208	449	214
21		0.89	20	3267	4040	4525	4750	466	210
22		0.388	40	2499	3316	3874	4133	439	201
Mean:				**4186**	**6254**	**7770**	**8475**	**455**	**220**
Standard deviation:				**1239**	**2762**	**3992**	**4566**	**10**	**15**
Difference compared to the control specimen:				**137%**	**150%**	**171%**	**179%**	**17%**	**55%**

3.1 Total Volatile Organic Compounds in the Gas Phase (TVOC)

The total concentration of volatile organic compounds was measured by the sensor as an equivalent of the isobutylene calibrant. The indicator includes compounds emitted from the cutting fluids (CF) and lubricants. High temperatures and friction during machining promote an increase in the organic compounds content of the CF, which can degenerate its performance properties, reduce machining efficiency, and increase the emission of aerosols and endotoxins – significant health hazards for workers [26, 27].

Figure 6 presents the Pareto chart of standardized effects for TVOC. Factors exceeding the absolute of the standardized rating were considered significant. The greatest deviation was observed for the change in the cooling method – with a value of 9.990 (linear model) – confirming its statistically strongest impact on TVOC emissions during the turning of the Ti6Al4V alloy. The smallest impact was observed for the mass flow rate *E* of the cutting fluid (CF).

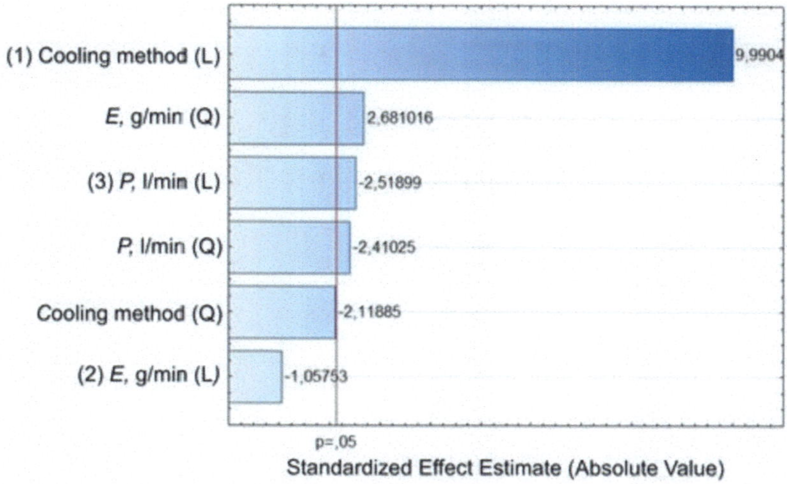

Fig. 6. Pareto charts of the standardized effects for the total concentration of volatile organic compounds (TVOC) in the air

Figure 7 shows the comparison of the total concentration of volatile organic compounds formed during the turning of titanium alloy, depending on the cooling method used along the percentage changes compared to dry machining. The highest TVOC value was observed for MQL method with Al_2O_3 nanoparticles. The implementation of Al_2O_3 nanoparticles into the CF, increasing the viscosity of the nanofluid [28] and also changing its surface tension [29]. Alumina has a large specific surface area and known adsorptive properties for many organic molecules, enabling nanoparticles to "carry" more organic matter and promote the accumulation of TVOC. The change in these properties of the nanofluid increases the Sauter Mean Diameter (SMD), worsening atomization compared to pure oil [30]. Consequently, the lowest TVOC concentration among the selected eco-friendly cooling methods was found with the use of the MQL method, which was only 25% higher compared to dry machining.

Figure 8 presents the surface response for TVOC as a function of the cooling method and the volumetric airflow rate *P*, which, according to the Pareto chart (Fig. 7), statistically have the greatest impact on the considered air quality parameter. For each cooling methods, the lowest concentration was observed at the maximum airflow rate, which corresponded to the values at which the smallest aerosol droplet diameters were delivered to the cutting zone [19]. A smaller SMD improves penetration into the cutting zone and heat dissipation, thereby reducing the thermal processing of organic fractions and compounds [31]. The highest contamination, defined as the total concentration of gaseous organic compounds during the turning of Ti6Al4V alloy, was observed for the MQL + Al_2O_3 method with airflow *P* rates between 15 l/min and 25 l/min. The lowest TVOC emissions in the cutting zone ware observed for MQCL method in the airflow range *P* between 35 and 40 l/min, where concentration of the TVOC was below 120 ppb.

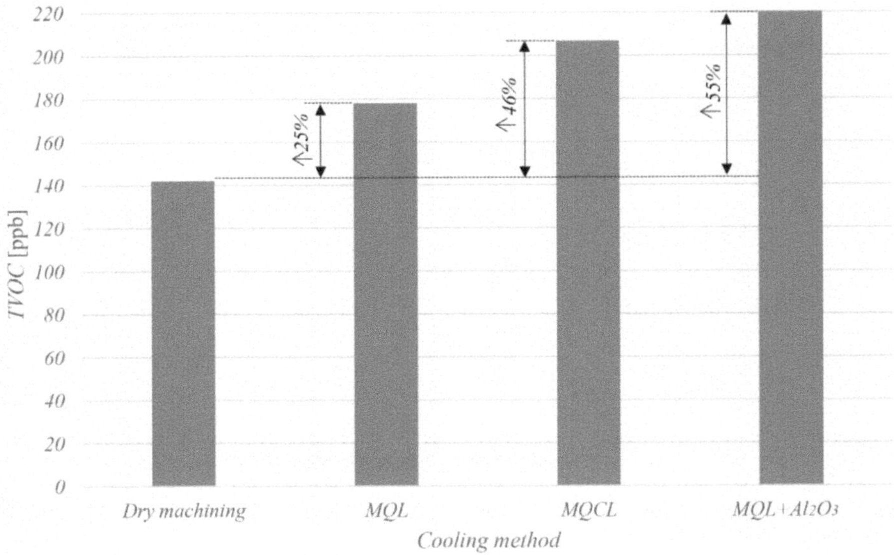

Fig. 7. Influence of the applied cooling method on the total concentration of volatile organic compounds (TVOC) and their percentage comparison to dry machining

Figure 9 presents the surface response chart for TVOC as a function of the cooling method and the changes in the mass flow rate *E* of the CF. The lowest total concentration of gaseous organic compounds occurs during the turning of the Ti6Al4V alloy in the mass flow rate *E* range of 0.7 g/min to 0.9 g/min. When considering the cooling method, the highest TVOC values are observed for the MQL method with nanoparticles Al_2O_3, in the CF mass flow rates from 0.4 g/min to 0.6 g/min and from 1.1 g/min to 1.2 g/min, where the TVOC concentration exceeds 260 ppb.

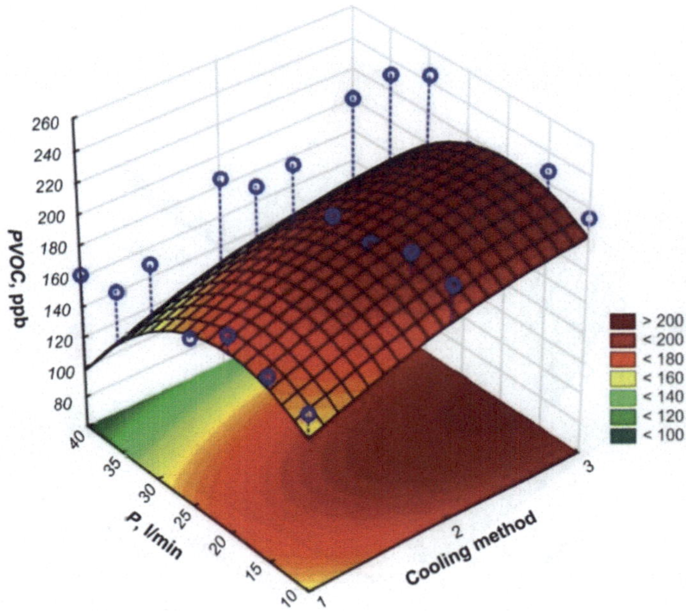

Fig. 8. Response surface chart of the total concentration of gaseous organic compounds (TVOC) depending on the applied cooling method and changes in the air volumetric flow rate (*P*) for: 1–MQL, 2–MQCL method, 3–MQL + Al_2O_3

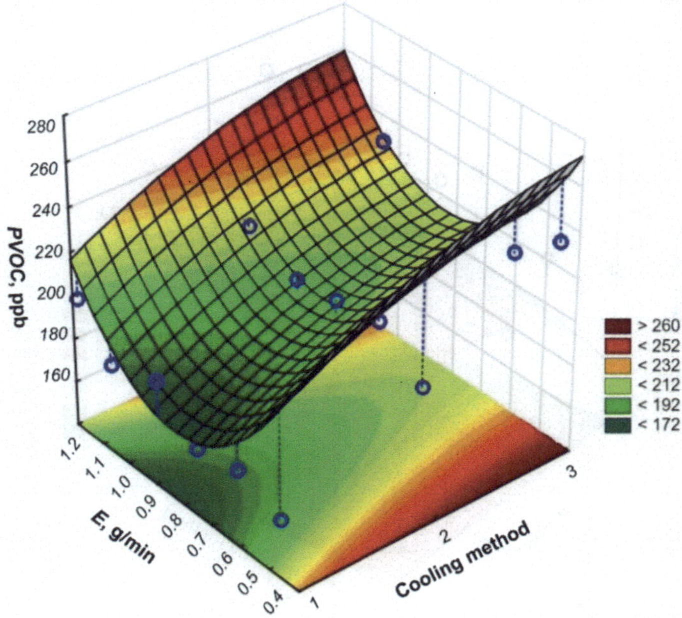

Fig. 9. Response surface chart of the total concentration of gaseous organic compounds (TVOC) depending on the applied cooling method and changes in oil mass flow rate (*E*) for: 1–MQL, 2–MQCL method, 3–MQL + Al_2O_3

3.2 Airborne Particulate Matter of PM2.5

Figure 10 presents the Pareto chart of standardized effects for harmful airborne particulate matter PM2.5. The absolute value of the standardized effect score for individual factors on the test result is $p = 0.05$. The factors analysed using the Pareto method were: cooling method, volumetric airflow rate *P*, and mass flow rate of the cutting fluid (CF) *E*. The value for each factor is lower than the absolute value of the standardized effect score. The closest to the absolute value was the change in the cooling method. Analysing the Pareto chart (Fig. 10), it was found that none of the independent variables have a statistically significant impact on the selected air pollution parameter.

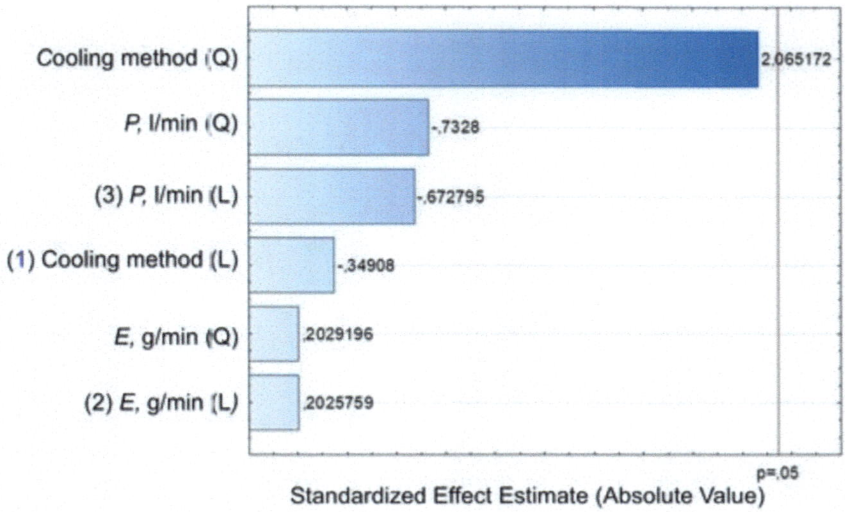

Fig. 10. Pareto charts of the standardized effects for particulate matter (PM2.5) in the air

Figure 11 presents the total concentration of PM2.5 particle emissions, depending on the cooling method used, along with the percentage changes compared to dry machining. The highest value of PM2.5 emissions was observed for MQL method, recording concentration of 6529 μg/m^3, which is 161% higher than generated during dry machining. These changes in PM2.5 values are attributed to temperature changes in the cutting zone, which promote the atomization of the CF due to thermal effects, leading to the formation of particles with a diameter smaller than 2.5 μm [31]. On the other hand, the implementation of Al_2O_3 nanoparticles into the CF enhances the lubrication properties of the nanofluid, improving lubricant penetration and reducing the contact length between the tool and the chip. As the result, the MQL+ Al_2O_3 method generates lower particle emission compared to MQL [29]. The use of the MQCL method leads to the tool temperature reduction compared to the MQL and MQL + Al_2O_3 methods, limiting the fluid evaporation and the formation of ultrafine particles. Additionally, due to the reduction in temperature, the viscosity of the CF changes, which also improves its lubricating properties [32].

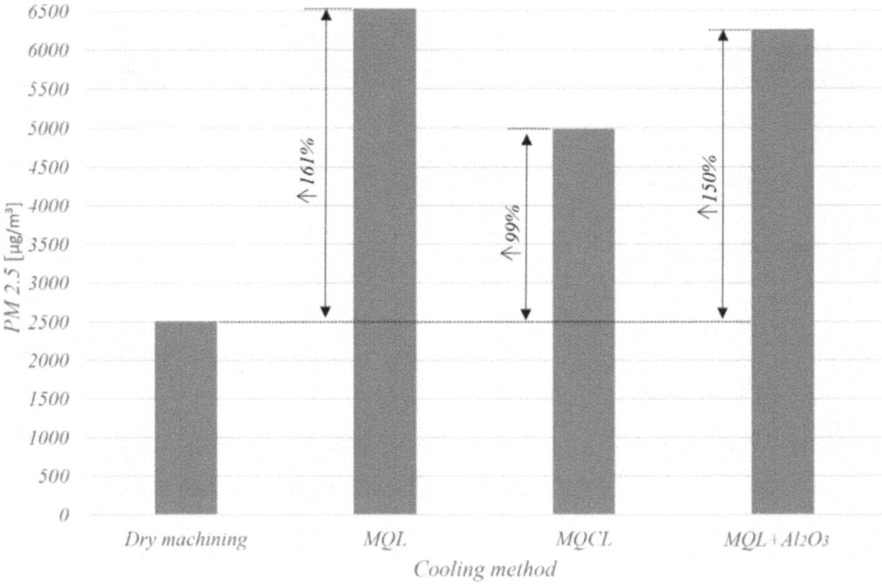

Fig. 11. Influence of the applied cooling method on the total concentration of particulate matter (PM2.5) and their percentage comparison to dry machining

Figure 12 presents the surface response chart for particulate matter PM2.5 as a function of the cooling method and variations in the volumetric airflow rate *P*. The characteristic curve (Fig. 12) is similar for all analysed cooling methods. In the initial range of airflow changes (< 25 l/min), a decrease in airflow rate results in an increase in the concentration of harmful PM2.5 particulate matter. However, as the volumetric airflow rate exceeds 25 l/min, the concentration of PM2.5 particulate decreases. During turning process, ultrafine particles are generated through the condensation of evaporated CF particles and through atomization in the tool-chip contact zone [31]. At lower airflow values, the temperature in the cutting zone may rise, leading to an increase in particle generation [33]. Additionally, in the MQCL method, as the droplet diameter decreases, the number of active components in the tribological film on the workpiece surface increases [34], which enhances both cooling and lubrication of the workpiece. The highest concentration of PM2.5 particles during turning occurs with MQL method and MQL with Al_2O_3 nanoparticles, in the airflow range of 15 to 25 l/min, where concentration of PM2.5 particles exceed 7000 µg/m^3. The lowest concentration of PM2.5 particles was observed during turning with MQCL method, in the airflow range between 35 and 40 l/min, where the PM 2.5 concentration was below 1000 µg/m^3.

Figure 13 presents the surface response chart for particulate matter PM2.5 as a function of the cooling method and changes in the mass flow rate *E* of the CF. The highest concentration of PM2.5 particles is observed for the minimum quantity lubrication (MQL) method at CF mass flow rates between 0.4 and 0.5 g/min, and between 1.1 and 1.2 g/min. The lowest concentration of PM2.5 particles is generated during turning with MQCL method at CF mass flow rate from 0.7 to 0.8 g/min. An increase in the CF flow rate leads to higher concentration of vapours, and thus to an increase in the number of suspended particles, which are one of its components [33].

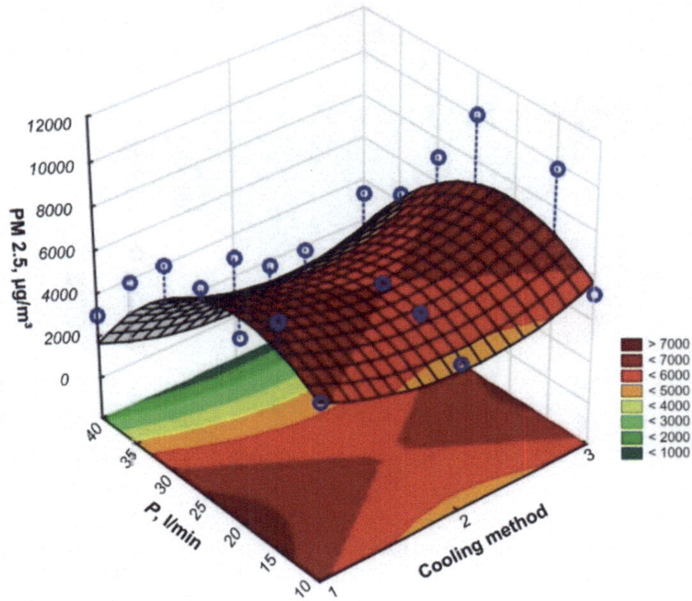

Fig. 12. Response surface chart of the total concentration of particulate matter (PM2.5) depending on the applied cooling method and changes in the air volumetric flow rate (*P*) for: 1–MQL, 2–MQCL method, 3–MQL + Al_2O_3

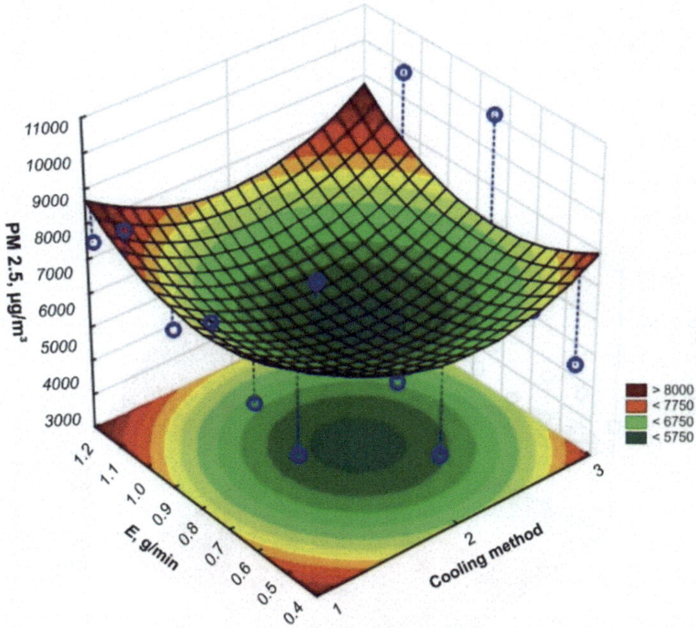

Fig. 13. Response surface chart of the total concentration of particulate matter (PM2.5) depending on the applied cooling method and changes in the oil mass flow rate (*E*) for: 1–MQL, 2–MQCL method, 3–MQL + Al_2O_3

4 Conclusions

This study examined the impact of changing the cooling method and aerosol formation parameters on air quality indicators: CO_2, TVOC, PM1.0, PM2.5, PM5.0, and PM10.0. Based on the conducted research, it should be concluded that:

1. The cooling method has the statistically greatest impact on air quality during the machining of the Ti6Al4V titanium alloy with the use of a cooling-lubricating medium for TVOC emissions, while the mass flow rate of the cutting fluid (CF) has the lowest impact.
2. For the PM2.5 air quality indicator, no statistically significant impact was found for the individual independent variables, so the conclusions for PM2.5 are trend-based.
3. The lowest levels of pollutants were observed for dry machining. However, when dry machining is not feasible, the MQCL method was found to be the most favourable alternative in terms of air quality.
4. Among the eco-friendly cooling methods, the MQCL method effectively reduces PM2.5 levels. The highest PM2.5 values were observed for MQL method, showing an increase of 161% compared to dry machining. Th increase may be attributed to the intense atomization and condensation of the oil aerosol at higher temperatures.

5. The implementation of Al_2O_3 nanoparticles into the MQL method increases TVOC levels compared to the use of pure oil – likely due to increased viscosity and changes in surface tension, which worsens atomization (larger SMD) and promotes the transfer of heavier organic fractions.
6. For all cooling methods, it was observed that the highest air quality was achieved for higher air flow rates P ($\geq$ 35–40 l/min) and mass flow rates E of CF in the range of 0.7–0.9 g/min.

Acknowledgement. This work was supported by the National Science Centre (Decision No. 2019/35/B/ST8/00552) "Modeling of problems related to hydrodynamics of nanoparticles-filled active medium formation during machining in minimum quantity cooling and lubricating conditions."

This work was also partially supported by a program of the Polish Ministry of Science under the title 'Regional Excellence Initiative', project no. RID/SP/0050/2024/1.

References

1. Rodriguez, I., Arrazola, P.J., Pušavec, F.: The impact of airborne emissions from coolants and lubricants on machining costs. CIRP Ann. **73**(1), 77–80 (2024). https://doi.org/10.1016/j.cirp.2024.04.056
2. Zhao, J., et al.: A comprehensive review of generating, monitoring, evaluating, and controlling particle emissions during machining process. J. Manuf. Syst. **70**, 395–416 (2023). https://doi.org/10.1016/j.jmsy.2023.08.007
3. Mokrani, H., Lounas R., Bennai, M.T., Salhi, D.E., Djerbi, R.: Air quality monitoring using IoT: a survey. In: IEEE International Conference on Smart Internet of Things (SmartIoT), pp. 127–134, Tianjin, China (2019). https://doi.org/10.1109/SmartIoT.2019.00028
4. Elsheikh, A., et al.: A review on sustainable machining: technological advancements, health and safety considerations, and related environmental impacts. Results Eng. **24**, 103042 (2024). https://doi.org/10.1016/j.rineng.2024.103042
5. Oláh, J., et al.: Impact of industry 4.0 on environmental sustainability. Sustainability. **12**(11), 4674 (2020). https://doi.org/10.3390/su12114674
6. McDuffie, E., Martin, R., Yin, H., Brauer, M.: Global burden of disease from major air pollution sources (GBD MAPS): a global approach. Res. Rep. Health Eff. Inst. **2021**(210), 1–45 (2021)
7. Hopf, N.B., et al.: Early effect markers and exposure determinants of metalworking fluids among metal industry workers: protocol for a field study. JMIR Res. Protocols. **8**(8), e13744 (2019). https://doi.org/10.2196/13744
8. European Chemicals Agency.: Scientific Report for Evaluation of Limit Values for Nitrosamines at the Workplace, Finland (2023)
9. Goldstein, B.D., Smith, M.T.: Formaldehyde. In: International Agency for Research on Cancer, IARC Monographs on the Evaluation of Carcinogenic Risks to Humans, vol. 88, pp. 1–436. IARC, Lyon, France (2006)
10. Henriks-Eckerman, M.L., Suuronen, K., Jolanki, R.: Analysis of allergens in metalworking fluids. Contact Derm. **59**(5), 261–267 (2008). https://doi.org/10.1111/j.1600-0536.2008.01438.x
11. Deng, W.J., Xia, W., Li, C., Tang, Y.: Ultrafine grained material produced by machining. Mater. Manuf. Process. **25**(6), 355–359 (2010). https://doi.org/10.1080/10426910902748024

12. Sen, B., Yadav, S.K., Kumar, G., Mukhopadhyay, P., Ghosh, S.: Performance of eco-benign lubricating/cooling mediums in machining of superalloys: a comprehensive review from the perspective of triple bottom line theory. Sustain. Mater. Technol. **35**, e00578 (2023). https://doi.org/10.1016/j.susmat.2023.e00578
13. Sterle, L., Krajnik, P., Pušavec, F.: The effects of liquid-CO2 cooling, MQL and cutting parameters on drilling performance. CIRP Ann. **70**(1), 79–82 (2021). https://doi.org/10.1016/j.cirp.2021.04.007
14. Ginting, Y.R., Boswell, B., Biswas, W.K., Islam, M.N.: Environmental generation of cold air for machining. Procedia CIRP. **40**, 648–652 (2016). https://doi.org/10.1016/j.procir.2014.07.087
15. Agrawal, C., et al.: Comprehensive analysis of tool wear, tool life, surface roughness, costing and carbon emissions in turning Ti–6Al–4V titanium alloy: cryogenic versus wet machining. Tribol. Int. **153**, 106597 (2021). https://doi.org/10.1016/j.triboint.2020.106597
16. Singh, R., et al.: Evaluating the sustainability pillars of energy and environment considering carbon emissions under machining of Ti-3Al-2.5 V. Sustain Energy Technol Assess. **42**, 100806 (2020). https://doi.org/10.1016/j.seta.2020.100806
17. Statnikov, R., Statnikov, A.: The Parameter Space Investigation Method Toolkit. Artech House, Boston/London (2011)
18. Park, K.-H., Olortegui-Yume, J., Yoon, M.-C., Kwon, P.: A study on droplets and their distribution for minimum quantity lubrication (MQL). Int. J. Mach. Tools Manuf. **50**(9), 824–833 (2010). https://doi.org/10.1016/j.ijmachtools.2010.05.001
19. Maruda, R.W., Szczotkarz, N., Wojciechowski, S., Gawlik, J., Królczyk, G.M.: Metrological relations between the spray atomization parameters of a cutting fluid and formation of a surface topography and cutting force. Measurement. **219**, 113255 (2023). https://doi.org/10.1016/j.measurement.2023.113255
20. Parida, A.K., Maity, K.: FEM and experimental analysis of thermal assisted machining of titanium base alloys. Measurement. **152**, 107292 (2020). https://doi.org/10.1016/j.measurement.2019.107292
21. Technical data sheet.: – Beruform STO 110, EuroFluid Poland Sp. z o.o.
22. Technical data sheet.: – WEDOLiT, Wilhelm Diets GmbH & Co. KG
23. Sharma, A.K., Singh, R.K., Dixit, A.R., Tiwari, A.K.: Characterization and experimental investigation of Al2O3 nanoparticle based cutting fluid in turning of AISI 1040 steel under minimum quantity lubrication (MQL). Mater. Today Proc. **3**(6), 1899–1906 (2016). https://doi.org/10.1016/j.matpr.2016.04.090
24. Dylewicz, W., Szczotkarz, N., Maruda, R., Gupta, M.: Influence of Al2O3 nanoparticle mass concentration and aerosol formation parameters on tool vibration during turning of Ti6Al4V titanium alloy. Int. J. Appl. Mech. Eng. **29**(2), 52–66 (2024). https://doi.org/10.59441/ijame/189470
25. Szczotkarz, N., Adamczuk, K., Dębowski, D., Gupta, M.K.: Influence of Aluminium oxide nanoparticles mass concentrations on the tool Wear values during turning of titanium alloy under minimum quantity lubrication conditions. Adv. Sci. Technol. Res. J. **18**(1), 76–88 (2024). https://doi.org/10.12913/22998624/175917
26. Tang, L., et al.: Biological stability of water-based cutting fluids: Progress and application. Chinese J. Mech. Eng. (English Edition). **35**(1), 3 (2022). https://doi.org/10.1186/s10033-021-00667-z
27. Jurina, F., Peterka, J., Vozar, M., Patoprsty, B., Vopat, T.: Design of Real-Time Monitoring System for cutting fluids. Designs. **7**(2), 43 (2023). https://doi.org/10.3390/designs7020043
28. Cieśliński, J.T., Ronewicz, K., Smoleń, S.: Measurement of temperature-dependent viscosity and thermal conductivity of alumina and titania thermal oil nanofluids. Arch. Thermodynamics. **36**(4), 35–47 (2015). https://doi.org/10.1515/aoter-2015-0031

29. Kim, G., et al.: Stability, surface tension, and thermal conductivity of Al2O3/water nanofluids according to different types of alcohol and their proportion. Case Stud. Therm. Eng. **28**, 101385 (2021). https://doi.org/10.1016/j.csite.2021.101385
30. Chideme, N., de Vaal, P.: Effect of liquid viscosity and surface tension on the spray droplet size and the measurement thereof. J. Appl. Fluid Mech. **17**(12), 2652–2672 (2024). https://doi.org/10.47176/jafm.17.2.2532
31. Iwasaki, M., et al.: Characterization of submicron oil mist particles generated by metal machining processes. Aerosol Air Qual. Res. **20**, 1469–1479 (2020). https://doi.org/10.4209/aaqr.2019.11.0607
32. Pervaiz, S., Rashid, A., Deiab, I., et al.: An experimental investigation on effect of minimum quantity cooling lubrication (MQCL) in machining titanium alloy (Ti6Al4V). Int. J. Adv. Manuf. Technol. **87**, 1371–1386 (2016). https://doi.org/10.1007/s00170-016-896
33. Liang, L., et al.: A comprehensive review of industrial workshop oil mist control technology based on electrostatic collection. Atmos. **16**(3), 242 (2025). https://doi.org/10.3390/atmos16030242
34. Maruda, R.W., et al.: A study on droplets sizes, their distribution and heat exchange for minimum quantity cooling lubrication (MQCL). Int. J. Mach. Tools Manuf. **100**, 81–92 (2016). https://doi.org/10.1016/j.ijmachtools.2015.10.008

Navigating the Energy Transition: Conditions for Effective Renewable Energy Adoption in Polish Enterprises

Daria Motała(✉), Julia Gumińska, Katarzyna Ragin-Skorecka, and Agnieszka Krugiełka

Politechnika Poznańska, ul. Rychlewskiego 1, 61-131 Poznań, Poland
daria.motala@put.poznan.pl

Abstract. The energy transition is one of the key challenges facing today's economy and environmental policy. The implementation of renewable energy sources (RES) in enterprises requires not only access to modern technologies, but also organizational changes, transparent regulations, and access to investment financing. The aim of the study was to identify the key factors influencing the process of RES implementation in Polish enterprises. The research had an exploratory character and was based on a qualitative method, individual in-depth interviews (IDI) conducted with three respondents representing both expert and implementation perspectives. Qualitative data analysis using MAXQDA software made it possible to identify four groups of factors determining the adoption of RES: financial-organizational, technological and quality-related, regulatory, and strategic-environmental. The results confirm that the effectiveness of the energy transition in Poland depends equally on access to financial instruments, the quality of investment planning and energy management, as well as stable legal frameworks and effective institutional support. The study provides an in-depth insight into the conditions shaping the implementation of renewable solutions at the enterprise level and constitutes a starting point for further analyses on the role of organizational competencies and environmental awareness in the process of green transformation.

Keywords: Renewable Energy Adoption · Energy Transition · Polish Enterprises

1 Introduction

The energy transition is currently one of the key challenges of contemporary economic and environmental policy. The gradual shift from fossil fuels to renewable energy sources (RES) is not only a technological transformation but also requires a systemic approach and organizational change. Growing engagement in decarbonization, increasing energy prices, and the rising importance of environmental, social, and governance (ESG) criteria encourage enterprises to adopt renewable technologies as part of their sustainable development strategies [1–3]. However, the pace and scope of this transition vary significantly across countries and sectors, depending on local institutional frameworks, financial instruments, and cultural contexts [4, 5].

B. Gapiński et al. (Eds.): MANUFACTURING 2026, LNME, pp. 316–328, 2026.
https://doi.org/10.1007/978-3-032-22943-4_22

In the Polish context, the transition toward renewable energy remains a complex and evolving process. Despite significant progress in aligning regulations with EU energy policy and the increasing availability of financial incentives, Polish enterprises still face numerous economic, administrative, and infrastructural barriers [6, 7]. The national energy system continues to rely heavily on coal, and the deployment of decentralized renewable solutions often encounters challenges related to grid capacity, regulatory complexity, and insufficient coordination among institutional actors [8, 9]. As a result, the diffusion of renewable energy technologies among enterprises remains uneven, and implementation is often limited to larger entities or those with access to external financing and technical expertise.

Existing research on RES implementation in Poland has focused mainly on technological and economic aspects, while social, organizational, and regulatory dimensions remain underexplored [10–13]. The limited number of empirical studies addressing these issues in the Polish context, particularly those incorporating the perspective of enterprises themselves, indicates a significant research gap. Qualitative research is especially valuable in this regard, as it enables the identification of emerging factors, relationships, and perceptions that are difficult to capture using quantitative methods [1, 4].

This study employed an exploratory qualitative design aimed at identifying key factors influencing the implementation of renewable energy sources in Polish enterprises. In-depth individual interviews (IDIs) were conducted to gain detailed insight into the attitudes, experiences, and motivations of respondents representing both expert and practitioner perspectives [14, 15]. This approach provides a deeper understanding of the determinants of RES adoption within a specific national and organizational context in which renewable energy initiatives are still in the process of development.

The findings presented here contribute to expanding the knowledge base on the energy transition by offering insights into the interactions among regulatory, financial, and organizational factors shaping the effectiveness of RES implementation in Polish enterprises.

2 Literature Review

The growing importance of renewable energy sources (RES) in the corporate sector reflects global trends related to economic decarbonisation, regulatory pressure, and rising energy costs. In recent years, an increasing number of firms have chosen to invest in photovoltaic, wind, and biomass installations, driven both by economic considerations and by environmental commitments stemming from ESG (Environmental, Social and Governance) policies [16–20]. The adoption of RES enhances firms' competitiveness, reduces energy expenditures [21, 22], and strengthens their image as sustainable entities in the context of global climate change [23–27].

In the Polish industrial sector, photovoltaic and wind energy projects dominate, offering relatively quick returns on investment and high operational efficiency. The dynamic expansion of photovoltaics is driven by the declining cost of the technology [27] and widespread access to support schemes [14, 28, 29], such as Mój Prąd or Energia Plus. At the same time solutions based on biomass, biogas, and heat pumps are gaining significance, enabling enterprises to increase energy self-sufficiency and reduce CO_2

emissions [30–32]. As highlighted in the World Energy Outlook [32], these technologies constitute an essential component of the energy transition in countries with a high share of coal in their energy mix.

The energy transition in Poland is both an economic and social process. Its goals extend beyond increasing the share of clean energy in the mix to include labour market restructuring and the development of local employment opportunities [33]. In this context, the modernisation of transmission infrastructure and advancements in energy storage technologies are crucial for ensuring the stable operation of the electricity system. Research indicates that a successful energy transition requires simultaneous development of technological and regulatory frameworks [34].

According to the Polish Energy Policy 2040 [35], by 2030 the share of RES in final gross energy consumption is expected to reach at least 32.6%, and in the electricity sector, 56%, with a further increase to 70% by 2040. The long-term objective is the near-complete elimination of coal from the energy mix. This trajectory aligns with the European Green Deal [36] and the Fit for 55 package, which promotes the transition towards a low-carbon economy. As REN21 [37] notes, Poland is currently among the countries with a moderate pace of transition to renewables, presenting a challenge to intensify efforts in the coming decade.

High electricity prices remain one of the most significant factors influencing investment decisions related to RES. In the first three quarters of 2025, energy prices for enterprises in Poland were approximately 10–15% higher than the EU average [38]. This situation compels companies, especially in energy-intensive sectors such as metallurgy, chemicals, and cement, to seek alternative energy sources [39]. At the same time, the Global Status Report on Renewables [37] indicates that the increasing competitiveness of RES technologies is gradually reducing these cost disparities.

Despite dynamic growth, the RES sector in Poland faces numerous barriers. The most significant include high investment costs [40–42], regulatory instability [43, 44], and lengthy administrative procedures [45–47], all of which generate investment uncertainty and delay project implementation [47]. Additionally, outdated grid infrastructure limits the capacity for new installations [48, 49]. The absence of coherent regulations on energy storage and heterogeneous technical standards adopted by grid operators further hinder development [45, 50–52]. Social barriers, including the NIMBY effect and low environmental awareness, also affect the pace of transition [53].

Auction mechanisms and support systems such as FIT (Feed-in Tariff) and FIP (Feed-in Premium) represent important tools for promoting the development of renewable energy in Poland, providing revenue stability for investors [18, 54]. The Power Purchase Agreement (PPA) market is also growing, enabling enterprises to purchase energy directly from producers [32]. According to Kubiczek and Przedworska [55], the future of Poland's energy transition will depend on the capacity to further integrate RES with transmission infrastructure, advance energy storage technologies, and ensure regulatory stability [54].

3 Materials and Methods

The study adopts an exploratory, qualitative research design aimed at identifying the key factors influencing the implementation of renewable energy sources in Polish enterprises. This approach was chosen due to the need for an in-depth understanding of a phenomenon that has not yet been extensively examined in the Polish economic context. It is grounded in inductive logic and serves primarily to uncover new ideas, concepts, and empirical generalisations. Reasoning from the particular to the general enabled the discovery of new relationships and patterns based on empirical data, without relying on predefined hypotheses. The results are general in nature, and applying them to a broader population would require in-depth research.

The applied research method was the in-depth individual interview (IDI), which allows for obtaining detailed information about respondents' attitudes, experiences, and emotions [14, 15]. The interviews were guided by a pre-prepared scenario that nonetheless allowed for flexibility in the order and formulation of questions, supporting a natural flow of conversation and enabling rich, authentic responses. The purposive sample consisted of three respondents: two experts from the energy sector (system-level and techno-regulatory perspectives) and one entrepreneur implementing a renewable energy investment (implementation perspective). This diversity made it possible to compare practical and expert viewpoints, enriching the interpretation process.

The collected research material was subjected to qualitative analysis using MAXQDA software, which enabled systematic coding, structuring, and visualisation of the data. The coding process followed a hybrid approach, incorporating both deductive codes (derived from the literature) and inductive codes (emerging from respondents' statements). Tools such as code matrices, word clouds, and frequency analysis were applied, allowing for the identification of recurring themes and relationships across thematic categories.

The choice of a qualitative analytical approach facilitated a deeper interpretation of the attitudes and contextual factors accompanying the implementation of renewable energy solutions in Polish enterprises. Combining a logically structured interview design with an analytically flexible approach yielded multifaceted insights that form a basis for further conceptualisation and comparison with findings from other studies in the field of sustainable energy transition.

4 Results

The qualitative data gathered through the in-depth interviews were analyzed using the MAXQDA software, which enabled the coding of respondents' statements and their visualization in the form of code clouds. These clouds illustrate the most frequently occurring concepts, thus, the more often a given term appeared in the responses, the larger its representation in the visualization. Two separate code clouds were developed for the purposes of this study: a deductive one and an inductive one.

In the deductive code cloud (Fig. 1), created on the basis of the literature review and theoretical knowledge, the dominant concepts included renewable energy sources, energy, costs, environment, efficiency, implementation, and energy management. Their frequent occurrence reflects the key thematic areas of the study, the economic, environmental, and organizational motivations that drive firms to invest in renewable energy technologies.

Fig. 1. A word cloud of deductive codes illustrating the most frequently occurring concepts related to the motivations and benefits of implementing renewable energy sources. Source: authors' own elaboration.

By contrast, the inductive code cloud (Fig. 2), generated through the analysis of respondents' actual statements, emphasized terms such as investment quality, Energy Management System, energy balancing, financial support, technical problems, and sustainable development. The visualization assigned the greatest prominence to practical aspects, such as the execution of the investment, the risks associated with selecting a contractor, and the importance of proper energy management. These visual outputs confirm that the interviewees focused primarily on implementation quality as well as the long-term economic and organizational outcomes of renewable energy investments.

Fig. 2. The inductive code cloud illustrating categories related to investment quality, energy management, and techno-organizational barriers. Source: authors' own elaboration.

A key element of the interview analysis is the code matrix developed in MAXQDA, which enables a clear presentation of the frequency of individual codes and the identification of the most significant research themes. Figure 3 shows the matrix with coded terms, color, coded according to thematic groups, and the interviews in which the frequency of each topic is indicated numerically. To facilitate interpretation, a color scale was applied, from red for high frequency to blue for low frequency, and the last column sums the values, allowing the most prevalent subcodes to be identified.

Coding System		Interview_3	Interview_2	Interview_1	Sum
Inductive codes					
	Quality	1	2	4	7
	Energy storage		2	3	5
	Hydrogen systems		1	2	3
	Social barriers	1		2	3
	Method of billing electricity	1		2	3
	Energy management systems		1	3	4
	Energy efficiency		1	2	3
	Energy security	3	3	1	7
	Balancing energy demand		4	8	12
Deductive codes					
Innovativeness:					
	Competitiveness of enterprises	1	1	2	4
	Perspectives	2		6	8
	Energy transformation		2	6	8
The use of Renewable Energy					
	Solar energy	1	4	2	7
	Wind energy		4	1	5
	Hydro energy				
	Biomass		4		4
Society and environment					
	Impact on society	1		2	3
	Impact on the environment	2	3	1	6
	Sustainable development	3	1	4	8
Barriers to Renewable Energy implementation					
	Legal and administrative barriers	1	2	5	8
	Technological and technical barriers		1	6	7
	Infrastructural barriers	1	1	2	4
	Financial barriers		2	5	7
Benefits of Renewable Energy Implementation in the enterprise					
	Financial benefits	4	5	8	17
	Legal and administrative benefits	3	1	2	6
	Infrastructural benefits			1	1
	Technological and technical benefits		1	1	2
TOTAL		25	46	81	152

Fig. 3. Code Matrix developed based on type and frequency of occurrence. Source: authors' own elaboration

Based on the code matrix and the compilation of the frequency and nature of the occurring codes, sets of codes were developed (Fig. 4). Their purpose is to support further inference regarding solutions that are significant for the development of renewable energy.

The analysis of the conducted interviews made it possible to distinguish four main sets of factors related to the implementation of renewable energy sources (RES) in Polish enterprises (Fig. 4). **The first set** includes financial benefits, perspectives, and the balancing of energy demand. This results from the impact of modern technologies on energy efficiency, which translates into long-term savings. Respondents indicated that, although RES investments require time, they help reduce operational costs and improve the energy balance of the enterprise. The necessity of rational planning of installations

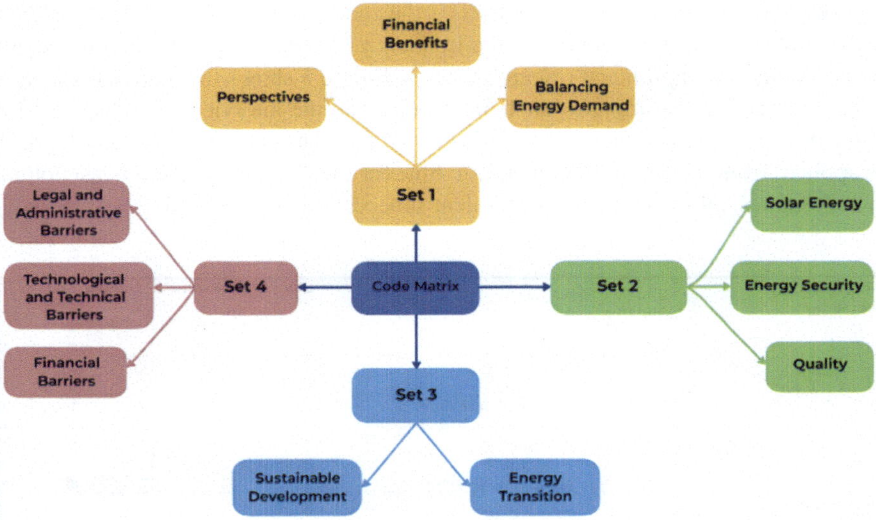

Fig. 4. Compilation of Factors Related to the Implementation of Renewable Energy Sources. Source: authors' own elaboration

was emphasized, as overproduction of energy without storage capabilities may lead to losses.

The second set relates to solar energy, energy security, and quality. Photovoltaics was identified as the most commonly chosen technology among enterprises, offering independence from the grid and the ability to maintain production continuity. The quality of components and professional installation are crucial for system reliability, directly affecting energy security. Participants highlighted that cutting corners on quality leads to failures and losses, while high technological standards ensure the durability of the investment.

The third set encompasses sustainable development and energy transition, which are closely interrelated. Energy transition concerns changes in the structure of energy generation towards renewable sources, whereas sustainable development integrates economic, environmental, and social aspects. Respondents stressed that implementing RES is not only a way to reduce costs but also an expression of ecological responsibility and a component of brand building. Increasingly, compliance with ESG standards has become important, motivating enterprises to pursue environmentally friendly and transparent practices.

The fourth set includes legal and administrative, technological and technical, and financial barriers. **Legal and administrative barriers** primarily concern the complexity of regulations, ambiguous interpretations, and lengthy formal procedures required to obtain necessary permits. Respondents noted that the investment process requires navigating multiple administrative stages, significantly extending project timelines, especially for larger installations. Procedures related to obtaining environmental decisions proved particularly burdensome, as their issuance can be time-consuming and dependent on local administrative conditions. One interviewee described the waiting time for a permit as "absurdly long," confirming that current regulations do not favor the rapid development of RES. On the other hand, institutional support, e.g., from Regional Funds for Environmental Protection and Water Management (WFOŚiGW) and local municipal authorities, was positively assessed, as it can accelerate procedures and assist in preparing project documentation.

Technological and technical barriers mainly arise from infrastructure limitations and the lack of adaptation of the national power system to modern renewable technologies. In many regions, the Polish power grid is not adapted for bidirectional energy flow, which complicates the integration of multiple RES installations. As a result, energy overproduction, which the grid cannot fully absorb, often leads to overloads or losses. Respondents also highlighted the need to develop energy storage technologies to balance supply and demand and stabilize system operation. Technical challenges also include the need to adapt transmission infrastructure, the quality of connections, and efficient energy management within the enterprise. The implementation of modern Energy Management Systems (EMS) addresses these issues but requires significant investment and technical expertise.

Financial barriers are among the most frequently mentioned by respondents and constitute a real obstacle to implementing RES projects, particularly for small and medium-sized enterprises. The main challenge is the high initial costs associated with purchasing, installing, and integrating renewable systems such as photovoltaic installations or energy storage. Despite available forms of support, including grants, preferential loans, and EU programs, many entrepreneurs still perceive RES investments as requiring substantial capital and a long payback period. At the same time, rising conventional energy prices make renewable energy investments increasingly cost-effective in the long term. It was also noted that implementing EMS can generate savings of 10–30%, providing an additional incentive to invest in efficient energy solutions.

5 Discussion

A comparison of the present study's findings with a literature review conducted via ScopusAI shows a clear alignment in the identification of key factors driving the implementation of renewable energy sources in enterprises (Fig. 5). Across both approaches, three main groups of determinants emerge: regulatory (set 3 vs. Policy and Regulation), organizational (sets 1 and 2 vs. Drivers), and economic-technological (set 4 vs. Barriers).

The literature emphasizes the importance of regional energy policy, support mechanisms, and stakeholder engagement, which foster the development of the renewable energy market [5, 6, 11, 12, 54]. **The findings of the** present qualitative **study** confirm this relationship, respondents indicated that institutional forms of support, such as initiatives by the Regional Funds for Environmental Protection and Water Management (WFOŚiGW) and local government programs, play a key role in reducing administrative barriers and facilitating investment implementation. Similar conclusions appear **in the literature**, where government grants and fiscal instruments are identified as primary drivers of RES adoption [1, 6, 7, 56]. **Empirical research** also shows that micro and small enterprises (SMEs) derive relatively greater benefits from direct financial support compared to larger companies [57].

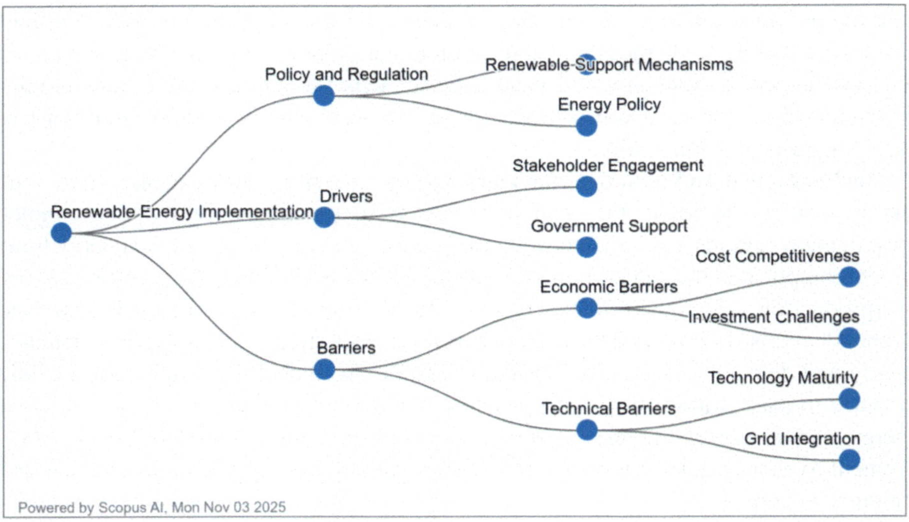

Fig. 5. Thematic Structure of Research on RES Implementation in a Bibliometric Approach. Source: ScopusAI

The results of in-depth **interviews** allow for a more comprehensive understanding of the barriers highlighted in the literature. ScopusAI (**the literature**) **analysis** mainly emphasizes economic obstacles, such as investment costs and price competitiveness, and technical challenges, including technological maturity and grid integration [5, 6, 47]. In the Polish context (**in both types of analyses**), procedural and organizational barriers are also evident. The complexity of administrative processes, lack of clear legal interpretations, and time-consuming environmental procedures represent significant constraints for RES investments [8, 9]. It is recommended (**in both types of analyses**) that energy transition efforts in Poland address both the development of financial mechanisms and the improvement of regulatory frameworks and decision-making processes [8, 11].

Both the literature and the present study confirm the growing role of social and strategic factors in the implementation of RES [13]. **Respondents** emphasized that adopting renewable solutions is not only a way to optimize costs but also a means of building the image of a responsible and modern enterprise, which is corroborated by **other studies** (the literature) [2, 8, 10]. **In both types of analyses** aspects of sustainable development, ESG reporting, and pro-environmental organizational strategies clearly align with global energy transition trends [1, 10]. In this context, the development of energy management technologies (EMS) and investment in high-quality installations can be regarded as evidence of Polish firms' growing maturity in participating consciously in the green economy [6, 12, 41]. Global research trends also indicate a shift in which sociocultural, political, and management factors are gaining increasing importance relative to financial determinants [13]. Comparing the findings of the present study with the literature suggests that the effective adoption of RES requires the simultaneous strengthening of the institutional environment, organizational capabilities, and ecological awareness within enterprises.

6 Conclusion

Based on the data analysis, it can be concluded that the key factor for the successful implementation of renewable energy is the quality of the investment undertaken. This does not refer solely to technical parameters or the quality of the materials used, but primarily to careful project planning and precise definition of the objectives the enterprise aims to achieve. Respondents emphasized that the implementation of renewable energy sources should be based on a thorough analysis tailored to the specific characteristics of the business, rather than limited to the mere purchase of technology.

It was also noted that the industry is seeing an increasing number of companies offering installation services at low prices; however, cooperation with such providers can carry significant risks. Excessive cost-cutting during the investment phase often leads to reduced service quality, resulting in system failures, energy losses, and interruptions in electricity supply. From the enterprise perspective, this translates into lower efficiency and extended payback periods for the investment.

Proper investment planning and energy management, on the other hand, allow for effective balancing of production and demand. The use of Energy Management Systems (EMS) enables optimization of system operation, monitoring of consumption, and reduction of losses. According to the respondents, energy management is often a critical stage, as only the proper utilization of data and process automation allows for real savings to be realized.

Respondents also emphasized that investments in renewable energy are a response to rising energy prices and increasingly stringent regulatory requirements. Implementing renewable energy sources enhances enterprise independence, improves cost efficiency, and supports market competitiveness. Companies that adopt such measures additionally gain reputational benefits, being perceived as modern, responsible, and innovative.

The interview findings also point to a growing awareness of the importance of corporate social responsibility in the context of the energy transition. Enterprises investing in renewable energy sources build stronger relationships with customers and business partners, respond to societal expectations, and comply with EU and national energy efficiency regulations. Furthermore, access to EU and national funds supporting the green transition accelerates the return on investment and enables further technological development of the companies.

Acknowledgments. This research was funded by the Poznan University of Technology, Faculty of Engineering Management - project No. 0812/SBAD/4240.

References

1. Pandey, A.K., Gupta, A., Bijalwan, P., Sayal, A.: A review of the determinants and barriers to renewable energy utilization in driving economic growth. In: Rethinking Resources: Strategies, Policies, and Partnerships for a Circular Future, pp. 315–333. Springer (2025)
2. Liu, S., Thurasam, R., Mohamad, M.N.: Towards a holistic understanding of green innovation adoption: a systematic review of determinants, barriers, and outcomes. J. Logist. Inform. Serv. Sci. **11**(2), 1–24 (2024)

3. Ragin-Skorecka, K., Motała, D.: Industry 4.0 technologies as a catalyst for lean transformation: bridging the gap between declaration and practice. In: European Lean Educator Conference, pp. 46–62. Springer, Cham (2025)
4. Akil, M., Senyapar, H.N.D., Bayindir, R.: Culturally driven renewable energy adoption: cross-regional insights and strategies for inclusive energy transitions. In: In: 2025 13th International Conference on Smart Grid (icSmartGrid), pp. 571–578. IEEE (2025)
5. Aparisi-Cerdá, I., Ribó-Pérez, D., García-Melón, M., D'Este, P., Poveda-Bautista, R.: Drivers and barriers to the adoption of decentralised renewable energy technologies: a multi-criteria decision analysis. Energy. **305**, 132264 (2024)
6. Bachanek, K.H., Drożdż, W., Kolon, M.: Development of renewable energy sources in Poland and stability of power grids-challenges, technologies, and adaptation strategies. Energies. **18**(8), 2036 (2025)
7. Zbroński, D., Otwinowski, H., Górecka-Zbrońska, A., Urbaniak, D., Wyleciał, T.: Analysis of changes in electricity generation from renewable energy sources after Poland's accession to structures of the European Union. Energies. **16**(12), 4794 (2023)
8. Cholewa, M., Mammadov, F., Nowaczek, A.: The obstacles and challenges of transition towards a renewable and sustainable energy system in Azerbaijan and Poland: the obstacles and challenges of transition towards a renewable and sustainable energy system in Azerbaijan and Poland. Miner. Econ. **35**(1), 155–169 (2022)
9. Kurowska, K., Kryszk, H., Bielski, S.: Location and technical requirements for photovoltaic power stations in Poland. Energies. **15**(7), 2701 (2022)
10. Awamleh, F.T., Shwawreh, S., Al-Kharabsheh, S.A.I., Alzghoul, A.: The integration of renewable energy adoption in sustainability practices for sustainable competitive advantage in Jordanian SMEs. Chall. Sustain. **13**(1), 135–145 (2025)
11. Brodzińska, K., Błażejowska, M., Brodziński, Z., Łącka, I., Stolarska, A.: Energy cooperatives as an instrument for stimulating distributed renewable energy in Poland. Energies. **18**(4), 838 (2025)
12. Drożdż, W., Mróz-Malik, O., Kopiczko, M.: The future of the Polish energy mix in the context of social expectations. Energies. **14**(17), 5341 (2021)
13. Zhindon-Almeida, R.G., Ruiz-Carrillo, J.A.: Factors influencing the adoption of renewable energies in developing countries. Sustain. Dev. (2025)
14. Berkhout, T., Rowlands, I.H.: The voluntary adoption of green electricity by Ontario-based companies: the importance of organizational values and organizational context. Org. Environ. **20**(3), 281–303 (2007)
15. Brooks, S., Sahaym, A., Datta, A., Srivastava, S.: Risky combination: the role of managerial perceptions of social media use and entrepreneurial orientation on SME innovation. Manag. Decis. **61**(1), 33–56 (2023)
16. Sovacool, B.K.: Deploying off-grid technology to eradicate energy poverty. Science. **338**(6103), 47–48 (2012)
17. Wüstenhagen, R., Menichetti, E.: The influence of energy policy on strategic choices for renewable energy investment. In: The Handbook of Global Energy Policy, pp. 373–387 (2013)
18. De Villiers, C., Maroun, W.: Introduction to sustainability accounting and integrated reporting. In: De Villiers, C., Maroun, W. (eds.) Sustainability Accounting and Integrated Reporting. Routledge, Abingdon (2017)
19. Maroun, W., Prinsloo, A.: Drivers of combined assurance in a sustainable development context: evidence from integrated reports. Bus. Strateg. Environ. **29**(8), 3702–3719 (2020)
20. Senyapar, H.N.D., Ayik, S., Colak, I., Bayindir, R.: Financial incentives for renewable energy: impact, effectiveness, and practices. In: 2025 13th International Conference on Smart Grid (icSmartGrid), pp. 51–58. IEEE (2025)
21. Ergun, S.J.: Does higher income lead to more renewable energy consumption? Energy Econ. (2023)

22. Usman, O., Jahanger, A., Radulescu, M.: Effects of domestic material consumption, renewable energy, and financial development on environmental quality. Renew. Sust. Energ. Rev. **158**, 112113 (2022)
23. Longstaff, H., Secko, D.M., Capurro, G., Hanney, P., McIntyre, T.: Fostering citizen deliberations on the social acceptability of renewable fuels policy: the case of advanced lignocellulosic biofuels in Canada. Biomass Bioenergy. **74**, 103–112 (2015)
24. Ofori, C.G., Aboagye, A.Q., Afful-Dadzie, A., Bokpin, G.A.: A real options approach to assessing the cost savings potential of renewable energy adoption among SMEs in Ghana. Clim. Pol. **22**(7), 851–864 (2022)
25. Sitompul, H., Saifi, M., Hutahayan, B., Sunarti.: Use of renewable energy to enhance firm performance: a systematic review. Sustainability. **16**(21), 9157 (2024)
26. Hussain, M.M., Altıparmak, S.O., Hatipoglu, E.: Collaborative currents: the promise of conflict resolution in public acceptance of utility-scale renewable energy projects. Land. **14**(5), 1056 (2025)
27. International Renewable Energy Agency: World Energy Transitions Outlook 2023. IRENA, Abu Dhabi (2023)
28. Jia, Y., et al.: Powering China's sustainable development with renewable energies: current status and future trend. Electr. Power Compon. Syst. **43**(8–10), 1193–1204 (2015)
29. Chen, Y., Lin, B.: Decomposition analysis of patenting in renewable energy technologies: from an extended LMDI approach perspective based on three five-year plan periods in China. J. Clean. Prod. **269**, 122402 (2020)
30. Dovichi Filho, F.B. Santiago, Y.C., Lora, E.E.S., Palacio, J.C.E., del Olmo, O.A.A.: Evaluation of the maturity level of biomass electricity generation technologies using the technology readiness level criteria. J. Clean. Prod. **295**, 126426 (2021)
31. International Energy Agency: Renewable Energy Market Update: June 2024. IEA, Paris (2024)
32. International Energy Agency: World Energy Outlook 2023. IEA, Paris (2023)
33. Ivanovski, K., Marinucci, N.: Policy uncertainty and renewable energy: exploring the implications for global energy transitions, energy security, and environmental risk management. Energy Res. Soc. Sci. **82**, 102292 (2021)
34. Abdelaziz, E.A., Saidur, R., Mekhilef, S.: A review on energy saving strategies in industrial sector. Renew. Sust. Energ. Rev. **15**(1), 150–168 (2011)
35. Ministry of Climate and Environment: Polityka Energetyczna Polski do 2040 roku (PEP2040). Government of Poland, Warsaw (2024)
36. European Commission: REPowerEU: 2024 Progress Report. European Commission, Brussels (2024)
37. REN21: Global Status Report on Renewables 2024. Paris: REN21 (2024)
38. Eurostat: Electricity prices for non-household consumers. Data extracted in October 2025 (2025)
39. World Bank: Energy Prices and Transition Pathways in Central and Eastern Europe. World Bank, Washington, D.C. (2023)
40. Fleiter, T., Schleich, J., Ravivanpong, P.: Adoption of energy-efficiency measures in SMEs-AN empirical analysis based on energy audit data from Germany. Energy Policy. **51**, 863–875 (2012)
41. Mishra, R., Naik, B.K.R., Raut, R.D., Kumar, M.: Internet of Things (IoT) adoption challenges in renewable energy: a case study from a developing economy. J. Clean. Prod. **371**, 133595 (2022)
42. Shi, X., Liu, H.: How does digital supply chain transformation enhance sustainable performance of renewable energy enterprises? Int. Rev. Econ. Finance, 104460 (2025)
43. Chebotareva, G.S.: Impact of state support mechanisms on the cost of renewable energy projects: the case of developing countries. WIT Trans. Ecol. Environ. **217** (2018)

44. Jiang, W., Jiang, K., Zhang, Z., Martek, I., Ruan, C., Lu, G.: The impact mechanism of political risk on foreign renewable energy investment in developing countries: the mediating role of vulnerability. Energy Policy. **198**, 114467 (2025)
45. Berg, S.V.: Regulatory functions affecting renewable energy in developing countries. Electr. J. **26**(6), 28–38 (2013)
46. Gajdzik, B., Wolniak, R., Nagaj, R., Grebski, W.W., Romanyshyn, T.: Barriers to renewable energy source (RES) installations as determinants of energy consumption in EU countries. Energies. **16**(21), 8401 (2023)
47. Javed, A., Ashraf, J., Yong, L.: Political and financial risks in developing countries: Implications for energy security and the transition to renewable energy. J. Environ. Manag. **387**, 125961 (2025)
48. Juszczyk, O., Juszczyk, J., Juszczyk, S., Takala, J.: Barriers for renewable energy technologies diffusion: empirical evidence from Finland and Poland. Energies. **15**(2), 527 (2022)
49. Kryszk, H., Kurowska, K., Marks Bielska, R., Bielski, S., Eźlakowski, B.: Barriers and prospects for the development of renewable energy sources in Poland during the energy crisis. Energies. **16**(4), 1724 (2023)
50. Pollitt, M., Stern, J.: Human resource constraints for electricity regulation in developing countries: Developments since 2001. Util. Policy. **19**(2), 53–60 (2011)
51. Campos Inês, I., Luz, G., Marín González, E., Gährs, S., Hall, S.: Regulatory challenges and opportunities for collective renewable energy prosumers in the EU. Energy Policy. **138**, 111235 (2020)
52. Ernst, Young: Renewable Energy Country Attractiveness Index (RECAI). Ernst, Young (2024)
53. Wüstenhagen, R., Wolsink, M., Bürer, M.J.: Social acceptance of renewable energy innovation. Energy Policy. **78**, 673–684 (2021)
54. Satrianto, A., Ikhsan, A., Dirma, E.Y., Gusti, M.A.: The effect of institutional quality on renewable energy: evidence from developing countries. Int. J. Energy Econ. Policy. **14**(5), 678–686 (2024)
55. Kubiczek, J., Przedworska, J.: Dynamics of energy transition in the context of Poland. Front. Energy Res. **13**, 1034567 (2025)
56. Lu, Q., Shen, R.: A tale of dual institutional logics: the impact of government venture capital involvement on SMEs' green innovation performance. Ventur. Cap., 1–26 (2025)
57. Dvouletý, O., Blažková, I., Potluka, O.: Estimating the effects of public subsidies on the performance of supported enterprises across firm sizes. Research Eval. **30**(3), 290–313 (2021)

Integration of Hydrogen Technologies into Sustainable Energy Management Systems in the Automotive Industry

Katarzyna Barnuś(✉) and Dorota Stadnicka

Rzeszów University of Technology, Rzeszów, Poland
k.barnus@prz.edu.pl

Abstract. The automotive industry is facing increasing pressure to reduce energy consumption and greenhouse gas emissions while maintaining high operational efficiency. This article examines energy management system models proposed in the literature, with particular emphasis on their applicability to the automotive sector. A systematic research methodology was applied, including a literature review, an analysis of existing EMS models, an assessment of the requirements and guidelines of ISO 50001 and ISO 50004, as well as an evaluation of the energy-consumption profile of vehicle manufacturing processes. The results indicate limited integration of multiple energy carriers and highlight the growing need to incorporate sustainability principles into energy-related decision-making. Based on the identified gaps, a conceptual model of a Sustainable Energy Management System (S-EMS) dedicated to the automotive industry is proposed. The model integrates both the functional and infrastructural requirements of modern manufacturing plants, addressing high-energy-intensity processes such as paint shops, robotics, compressed air systems, and HVAC (Heating, Ventilation, Air Conditioning). Particular attention is given to the role of hydrogen as a complementary energy carrier, especially in high-temperature applications and future low-emission production scenarios. The proposed S-EMS framework emphasizes requirements related to measurement systems, digital infrastructure, and integration with operational strategies oriented toward sustainable development. The article concludes with a synthesis of the main findings, a discussion of technological, economic, organizational, and regulatory limitations, and an outline of future research directions concerning hydrogen integration and the development of standardized S-EMS architectures.

Keywords: Energy Management Systems · Sustainable Manufacturing · Automotive Industry · PN-ISO 50001 · Hydrogen Integration · Energy Efficiency · Multi-Carrier Systems · Sustainable Energy Management

1 Introduction

The transformation of industrial energy systems has become one of the central challenges in the global transition toward low-carbon manufacturing. Among all industrial sectors, the automotive industry is particularly exposed to increasing regulatory pressure, rising energy prices, and the need to reduce greenhouse-gas emissions while maintaining

B. Gapiński et al. (Eds.): MANUFACTURING 2026, LNME, pp. 329–342, 2026.
https://doi.org/10.1007/978-3-032-22943-4_23

high production efficiency. Automotive manufacturing is characterised by a combination of energy-intensive thermal processes, highly automated electrical systems, and complex auxiliary infrastructure, all of which together create a heterogeneous and dynamically changing energy environment. This structural complexity underscores the need for advanced and systematic approaches to energy management.

Energy Management Systems (EMS), especially those aligned with ISO 50001, provide a well-established framework for monitoring, analysing, and improving energy performance across industrial facilities. These systems emphasise structured data collection, performance indicators, continuous improvement cycles, and organisational integration. However, industrial energy systems are undergoing rapid evolution driven by digitalisation, decentralisation, and the emergence of new energy technologies. As a result, traditional EMS architectures—designed primarily around single-carrier systems, each of which focuses exclusively on one energy carrier, such as electricity, natural gas, or compressed air—face increasing limitations in addressing the needs of modern multi-carrier manufacturing environments.

A particularly important trend influencing the future of industrial energy management is the growing interest in hydrogen as an alternative and low-carbon energy carrier. Hydrogen offers potential applications in high-temperature processes and other areas difficult to electrify, positioning it as a promising complement to existing industrial energy infrastructures. Advances in electrolysis, storage systems, and hydrogen combustion technologies indicate that hydrogen may play a meaningful role in decarbonising selected industrial operations in the coming decades. However, the introduction of hydrogen also raises new challenges related to safety, measurement, system integration, energy balancing, and operational decision-making—areas that are not addressed in conventional EMS models.

Despite increasing technological readiness, the integration of hydrogen into manufacturing-level energy management remains insufficiently explored. The lack of methodologies, models, and conceptual frameworks capable of combining traditional energy carriers with hydrogen limits the ability of automotive manufacturers to plan and evaluate future low-carbon scenarios. This creates a need for harmonised, multi-carrier energy-management approaches that incorporate hydrogen alongside conventional energy flows and that reflect the operational realities of automotive production.

The purpose of this study is to address this need by analysing existing approaches to industrial energy management and proposing a conceptual Sustainable Energy Management System (S-EMS) tailored to the requirements of the automotive industry and capable of accommodating hydrogen as an energy carrier. The work aims to contribute to the ongoing discussion on the evolution of EMS architectures in light of emerging energy technologies and the transition toward sustainable manufacturing.

The following section of this paper presents an overview of the contemporary energy landscape in industrial manufacturing and the energy-consumption profile of the automotive sector. The next part outlines the research objectives and the methodology applied. Subsequently, the results of a systematic literature review are presented, integrating perspectives on energy management systems, the energy demand of vehicle manufacturing processes, and hydrogen technologies. An analysis of energy management systems, including the model proposed by ISO 50001, is carried out, followed by a discussion of the energy-consumption structure in automotive production facilities and the identification of processes with the highest potential for hydrogen integration. The final part of the paper introduces the conceptual model of the Sustainable Energy Management System (S-EMS), complemented by a discussion of its key assumptions and directions for future research.

2 Energy Landscape of Industrial Manufacturing and the Automotive Sector

The energy landscape of contemporary industrial manufacturing is characterised by increasing complexity, diversification of energy carriers, and growing demands for flexibility and decarbonisation. Industrial facilities have traditionally relied on electricity, natural gas, steam, and compressed air as their primary energy sources, with energy management focused on improving efficiency, reducing operational costs, and ensuring continuity of production. However, the ongoing technological, environmental, and regulatory transitions are reshaping the structure of industrial energy use, accelerating the need for integrated, data-driven, and multi-carrier energy-management approaches.

Across most industrial sectors, electricity consumption has risen steadily due to automation, digitalisation, robotics, and the proliferation of advanced control systems. At the same time, thermal energy remains indispensable in processes requiring high or stable temperatures, particularly in sectors such as chemicals, steel, ceramics, and automotive manufacturing. The coexistence of thermal and electrical energy systems creates complex, interdependent load profiles with varying temporal dynamics, ranging from continuous high-temperature demand to short, high-power electrical peaks. This heterogeneity presents substantial challenges for energy forecasting, balancing, and optimisation at the facility level.

Within this broader industrial context, the automotive sector stands out as one of the most energy-intensive and structurally diversified manufacturing environments. Automotive production integrates a wide range of processes with fundamentally different energy characteristics. Thermal processes—including pre-treatment, drying, and paint curing—account for a significant share of total energy consumption and are responsible for the largest and most stable heat loads within the factory. Conversely, electrically driven processes such as robotic welding, stamping, and automated assembly exhibit dynamic load profiles, high peak-power requirements, and strong dependence on production scheduling.

Auxiliary systems further contribute to the sector's energy complexity. Compressed-air installations, often characterised by low efficiency and high leakage rates, represent a critical area of energy losses across automotive plants. HVAC systems play a central role in ensuring environmental stability in painting, assembly, and quality-control zones, where precise humidity and temperature control are essential. The interaction of these systems with production equipment creates facility-wide energy dependencies that require integrated monitoring and control mechanisms.

In recent years, additional pressures have intensified the transformation of the automotive energy landscape. Decarbonisation policies, corporate sustainability commitments, and the shift toward electrified vehicle platforms have increased demand for low-carbon and renewable energy sources. Furthermore, the volatility of energy markets and the growing importance of energy security have prompted manufacturers to explore alternative energy carriers, local energy storage, and on-site generation technologies. Among these emerging solutions, hydrogen has attracted considerable interest as a potential complement to electricity, particularly for high-temperature and hard-to-electrify processes.

As a result, the automotive sector is transitioning from a traditionally single-carrier, electricity-dominated structure toward a more diversified, multi-carrier energy architecture. This evolution underscores the need for advanced energy-management systems capable of integrating heterogeneous energy flows, supporting predictive analytics, and enabling strategic decision-making in a technologically complex and sustainability-driven environment.

3 Research Objective, Questions and Methodology

The main objective of this study is to analyse the energy management system models presented in the literature and to propose a concept of a sustainable energy management system. To operationalise this objective, three research questions (RQs) were formulated, each addressing a key thematic area of the study:

RQ1: *What types of energy management system models are proposed in the literature for industrial applications, and how do they address sustainability aspects?*
RQ2: *How are hydrogen technologies incorporated into the energy management systems presented in the literature?*
RQ3: *What elements should be included in a sustainable energy management system?*

The research methodology includes:

Stage 1. Systematic literature review

The objective of the first research stage was to conduct a review of scientific literature concerning Energy Management Systems (EMS), the energy intensity of automotive manufacturing processes, and the industrial applications of hydrogen. The review was carried out exclusively using the Scopus database, in accordance with the assumptions of the study and the search criteria defined at the beginning of the research process. The search was performed within the fields *Article title*, *Abstract*, and *Keywords*, using combinations of keywords presented in Table 1. First, the aim was to identify publications describing the use of EMS in the automotive industry and the energy characteristics of automotive manufacturing processes. Second, publications related to hydrogen were selected, which made it possible to narrow the results to studies linking EMS with hydrogen technologies in the context of the automotive sector. Only English-language publications within the field of Engineering, available in full text and related to industrial energy systems, hydrogen technologies, or energy management in manufacturing environments were included in the review. The final selection process involved an analysis of titles, abstracts, and full-text articles.

Table 1 presents information on the number of publications resulting from the search procedure.

Table 1. Search keyword combinations applied in search query formulations for the systematic literature review.

No	Search keyword combinations	Number of papers
1	energy management system	123 604
2	energy management system AND industry	11 663
3	energy management system AND industry AND automotive	832
4	energy management system AND industry AND automotive manufacturing	137
5	energy management system AND industry AND automotive manufacturing AND hydrogen	5

A total of 137 publications related to energy management systems in the automotive industry were identified, along with 5 publications that additionally addressed hydrogen technologies.

Stage 2. Review of existing EMS models presented in the literature

In the second stage of the methodology, existing EMS models presented in the literature were analysed, with particular emphasis on their applications in industrial environments. The aim of this step was to determine the extent to which the EMS models available in the literature can be applied or adapted to the automotive manufacturing sector, and whether they include the possibility of integrating alternative energy carriers such as hydrogen.

Stage 3. Review of the requirements and guidelines of ISO 50001 and ISO 50004

The review of the standards was carried out in the context of modelling energy management systems, with the aim of identifying the system elements that need to be designed and any potential references to hydrogen as an energy carrier.

Stage 4. Analysis of energy consumption in automotive manufacturing

To identify the processes with the highest potential for integrating alternative energy carriers—including hydrogen—a targeted analysis of the energy profile of automotive manufacturing plants was conducted. Its objective was not only to indicate the most energy-intensive areas, but above all to determine which processes, due to their technological characteristics, may in the future utilise energy sources other than those currently in use. This approach made it possible to move beyond a simple comparison of energy-consumption volumes and instead focus on assessing the susceptibility of individual processes to a change in energy carrier. The key criterion for selecting the areas analysed in terms of hydrogen integration was the temperature profile of the processes and the extent to which they depend on thermal or electrical energy.

Stage 5. Development of the proposed S-EMS for application in the automotive industry

In the final stage of the methodology, a conceptual model of a Sustainable Energy Management System (S-EMS) dedicated to the automotive industry was developed. The model was constructed based on the gaps identified in the literature review, the analysis of existing EMS frameworks, the requirements of ISO 50001 and ISO 50004, and the energy-consumption analysis of automotive manufacturing processes.

4 Systematic Literature Review Results

4.1 Energy Management Systems (EMS) in the Industrial Sector

The analysis of the systematic literature review results revealed a distinct research gap. Out of 137 publications concerning Energy Management Systems (EMS) in the automotive industry, only five addressed both EMS and hydrogen-related topics, and just one publication focused directly on processes applied within the automotive sector [1]. The remaining studies primarily dealt with hydrogen energy systems, sustainability, or Industry 6.0, but failed to establish a systematic connection with energy management in manufacturing plants. This confirms that the integration of hydrogen technologies with energy management systems in the automotive industry remains an underexplored area in the scientific literature.

The most widespread energy management model remains the structure based on the ISO 50001 standard, which defines an EMS as a set of processes enabling the systematic monitoring, analysis, and optimization of energy consumption. The standard, grounded in the PDCA (Plan–Do–Check–Act) cycle, serves as a reference framework across various industrial sectors [2, 3].

The literature also describes EMS models integrated with Manufacturing Execution Systems (MES) and Supervisory Control and Data Acquisition (SCADA) systems, enabling multilayer real-time monitoring and control of energy use [4]. It is emphasized that effective energy management requires not only energy-related data, but also contextual information regarding production processes, logistics, and quality.

Another important area of EMS development is the use of data analytics and predictive methods. The integration of machine learning enables the forecasting of energy consumption, the identification of energy-related anomalies, and the optimization of energy use in highly dynamic environments such as stamping and welding lines [5].

Most existing EMS models, however, focus on conventional energy carriers such as electricity, natural gas, and thermal energy. Hydrogen technologies are predominantly discussed in the literature from a technical perspective—for example, in the context of hydrogen combustion, integration with high-temperature industrial processes, or economic feasibility assessments of green hydrogen production [6]. What is lacking, however, are studies that present hydrogen as an integral component of EMS architecture, including its measurement, balancing, and integration with the existing energy streams of an industrial facility.

In the automotive industry, this gap is particularly evident. Despite the growing body of research on the use of hydrogen in industrial processes, the available EMS models do not include mechanisms for integrating hydrogen with the energy systems of automotive manufacturing plants. This indicates the need to extend existing architectures with functionalities that enable the monitoring and management of hydrogen in the same manner as other energy carriers.

In summary, the literature confirms that energy management systems are evolving rapidly, particularly in the areas of digital monitoring and analytics. However, their current architecture is not designed to handle alternative energy carriers such as hydrogen. This highlights the need for further research on multi-carrier EMS models adapted to the integration of hydrogen and hydrogen-based energy systems into automotive manufacturing plants.

4.2 Overview of the ISO 50001 Requirements and Guidelines of ISO 50004

The ISO 50001 standard defines the requirements for an Energy Management System (EMS), enabling organizations to systematically plan, implement, and improve actions aimed at enhancing energy performance [7]. The document is normative and specifies mandatory system elements, including the energy policy, objective planning, performance monitoring and evaluation, and the energy review. For establishing the energy baseline (EnB) and energy performance indicators (EnPIs), ISO 50001 refers to ISO 50006 [8]. In contrast, ISO 50004 serves as an implementation guide [9], providing practical recommendations for establishing, maintaining, and improving an EMS in line with ISO 50001, without introducing additional requirements.

The EMS model defined in ISO 50001 is based on the High-Level Structure (HLS) and the PDCA (Plan–Do–Check–Act) cycle. It comprises several key functional areas, starting with the organizational context, including the definition of the EMS scope and the identification of interested parties' needs. Leadership is another core element, emphasizing top management's role in setting the energy policy and ensuring resources. The most extensive component is energy planning, which according to ISO 50004 includes the energy review, identification of significant energy uses (SEUs), development of EnPIs, establishment of the EnB, and planning of energy data collection. Further elements include operational energy management, performance evaluation through monitoring and audits, and improvement actions completing the PDCA cycle. The ISO 50001 EMS model is universal and applicable to all types of organizations, regardless of their energy profiles.

An analysis of both documents shows that neither ISO 50001 nor ISO 50004 treats hydrogen as a distinct energy carrier. ISO 50004 mentions hydrogen only incidentally, as an example of "other media" that may be included within the EMS scope, without classification or dedicated guidance. Likewise, the energy review section refers only to general energy groups—such as electricity, heat, biomass, fossil fuels, steam, and compressed air—without recognizing hydrogen separately. Consequently, the standards provide no system-level guidance on measuring, monitoring, or balancing hydrogen within an EMS, leaving its integration entirely to the discretion of the organization.

4.3 Energy Consumption Structure in Automotive Manufacturing

According to the literature, hydrogen is primarily applied in processes that require stable medium or high temperatures [6, 10]. For this reason, the analysis focuses on areas characterized by significant thermal energy consumption. In this context, paint shop operations represent a natural direction of investigation. These processes include surface preparation, drying, and curing, all of which are highly dependent on thermal energy and generate repetitive streams of low- and medium-temperature waste heat [11, 12] This makes paint shops a potential area for the use of alternative energy carriers, including hydrogen, particularly in auxiliary systems and heat-recovery configurations.

In contrast to thermal processes, areas that rely mainly on electricity, such as robotic welding, are characterized by highly repeatable load profiles and limited susceptibility to changes in the type of energy carrier [5]. The literature emphasizes that in welding shops the greatest potential lies in optimizing energy-use control, advanced monitoring, and integration with the EMS, rather than substituting electricity with alternative carriers [2, 3].

A distinct category comprises processes with highly variable power-demand profiles, such as stamping operations. Hydraulic and mechanical presses operate cyclically, generating steep load peaks, which limits the feasibility of using carriers other than electricity due to the requirement for high stability and substantial instantaneous power [13]. At the same time, the energy profiles of stamping facilities constitute valuable input for predictive systems within the EMS, supporting the stabilization of the plant's overall energy balance [5].

The analysis also included auxiliary systems such as compressed air and HVAC (Heating, Ventilation, Air Conditioning). Compressed-air installations are characterized by low efficiency and high losses resulting from leakages, making them an important area for optimization, although not a natural candidate for hydrogen supply [5]. HVAC systems, particularly in assembly and paint shop areas, require stable environmental parameters and large airflows, which creates opportunities for integration with local heat sources, potentially including hydrogen-based solutions [12].

Based on the conducted analysis, it was determined that the processes most susceptible to hydrogen integration are those that use thermal energy continuously and generate significant streams of waste heat. This applies primarily to: (1) paint booths, (2) drying tunnels, (3) surface-preparation processes, and (4) low-temperature waste-heat recovery systems (45–50 °C), which, according to the literature, are currently underutilized [11].

Electrical processes such as robotic welding or stamping presses remain areas of optimization within conventional EMS frameworks and do not represent a natural direction for hydrogen application [2].

The structure of energy consumption in automotive manufacturing plants results from the diverse nature of the processes, environmental requirements, and the operational intensity of equipment. OEM (Original Equipment Manufacturer) factories are among the most energy-complex industrial facilities, combining processes with extremely different energy-demand profiles—from high-energy thermal operations to stable electrical processes [14]. Therefore, energy management in the automotive sector must encompass both the technical infrastructure and the dynamic operation of production lines.

The paint shop represents by far the largest component of energy consumption—accounting for approximately 50–70% of an automotive assembly plant's total energy demand [11]. This results from high thermal inertia and the necessity to maintain stable process conditions and elevated temperatures. For this reason, paint shops exhibit the greatest potential for integrating alternative energy carriers, including hydrogen.

The second strategic area is the welding shop. It is characterized by high electricity consumption associated with the operation of robots, control systems, and auxiliary equipment. Although its proportional share is lower than that of the paint shop, the load profile is highly repeatable and stable, facilitating integration with predictive EMS modules. Welding operations also generate transient loads that influence the stability of the plant's electrical systems.

Stamping facilities constitute an area with extremely high instantaneous power requirements and sharp load fluctuations, which complicate the management of energy profiles. Although the potential for applying alternative energy carriers is lower, stamping remains a key element for load stabilization through prediction and modelling.

A significant portion of the energy profile of automotive plants is associated with compressed-air systems (10–20% of electricity consumption) [14]. Due to losses resulting from leakages (even 20–50%), these systems are attractive targets for optimization [2]. HVAC systems also contribute substantially to total energy use (10–15%), driven by the need to maintain stable environmental parameters in large building volumes [11].

Auxiliary infrastructure—including internal transport, lighting, process cooling, industrial pumps, and safety systems—although smaller in individual contribution, may collectively account for a considerable share of overall energy use. Modernizing these systems, particularly through integration with the EMS, operational regulation, and automation, enables significant energy savings [12].

In summary, the energy structure of automotive manufacturing facilities is highly diversified and results from the simultaneous operation of processes with different energy profiles. The greatest modernization potential and the highest feasibility of hydrogen integration relate to thermal processes, primarily paint shop operations, whereas electricity-based processes depend more on advanced EMS tools than on changes in the type of energy carrier. These findings form the basis for further analyses aimed at developing Sustainable Energy Management Systems (S-EMS) and integrating hydrogen into modern energy architectures of OEM plants.

4.4 Discussion and Conclusions from the Systematic Literature Review

The systematic literature review on energy management systems, the energy intensity of automotive production processes, and hydrogen technologies revealed substantial differences in the maturity of these research areas. In the field of classical energy management, the dominant approach is based on ISO 50001 and on concepts of systematic improvement of energy efficiency. Organizational structures, the PDCA cycle, and processes for monitoring and analysing energy use play a key role [2, 3]. These models are widely described and implemented across multiple industrial sectors, yet they focus primarily on traditional energy carriers such as electricity, heat, and compressed air. This answers **RQ1**, indicating that existing EMS architectures remain largely single-carrier and efficiency-oriented, with limited direct consideration of broader sustainability dimensions beyond incremental improvements in energy performance.

In parallel, research on the energy consumption characteristics of automotive manufacturing processes is developing rapidly. Studies by Giampieri (2020) and Sato and Nakata (2020) emphasize the dominance of thermal and material-intensive operations in automotive energy use, with particular relevance to surface preparation and painting

[12, 13]. Research on the energy intensity of paint shops [11] and on process-chain simulation [5] indicates that the greatest optimization potential lies in areas requiring high temperatures, high thermal inertia, and intensive use of auxiliary media.

A separate research stream concerns hydrogen technologies. Publications by Marghani et al. (2023) and Sokil, Podolchak, Stetsiv, Zuiev and Chepil (2025) focus on hydrogen production, storage, and application as an alternative energy carrier [6, 15]. Meanwhile, Małek, Karowiec and Jóźwik (2023) analyse hydrogen utilisation in the automotive sector, mainly in the context of vehicle infrastructure and fuel supply chains [16]. Although these studies confirm the technological maturity of hydrogen solutions and the growing interest in their industrial applications, they only marginally address the integration of hydrogen with EMS at the level of manufacturing facilities. This directly addresses **RQ2**, showing that hydrogen technologies are discussed predominantly in technological terms and are not systematically embedded within EMS frameworks, nor are they supported by methodological guidelines for their operational integration.

Altogether, the review results clearly indicate the presence of a systemic research gap. On the one hand, there are well-developed EMS concepts based on standards and organizational solutions; on the other, an extensive body of literature on hydrogen energy systems and detailed analyses of the energy intensity of automotive processes exists. However, research that combines these two domains into a coherent model—one that accounts for the multi-carrier nature of automotive factory energy systems and the infrastructural requirements necessary for hydrogen implementation—is lacking. Particularly absent are empirical case studies describing pilot integrations of hydrogen as an energy carrier in automotive manufacturing plants, including OEM facilities, and analysing the impact of such integration on the operation of existing EMS. These findings refer directly to **RQ3**, emphasising the need for a system architecture capable of managing multiple energy carriers, including hydrogen, and equipped with analytical and decision-support mechanisms suitable for modern, decarbonised manufacturing environments.

The findings of the review provide the foundation for developing a concept of a Sustainable Energy Management System (S-EMS) for the automotive industry. The results clearly indicate that future systems must go beyond traditional approaches to energy management by incorporating alternative energy carriers such as hydrogen, as well as advanced analytical tools that support decision-making in complex, multi-carrier energy systems. In response to **RQ3**, the proposed direction of development includes a multi-layered EMS architecture integrating monitoring, predictive analytics, and decision-making capabilities, together enabling comprehensive management of both conventional and emerging energy carriers.

5 Sustainable Energy Management Systems (S-EMS)

The development of sustainable energy management systems in the automotive industry requires moving away from the classical, single-carrier approach based primarily on electricity and natural gas, toward models that enable the integration of new energy sources and technologies, including hydrogen. The literature review shows that although EMS frameworks are extensively described in the context of normative and organizational requirements, there is a lack of solutions indicating how these systems should evolve toward multi-carrier, decentralized, and hybrid energy architectures characteristic of modern automotive factories, including OEM plants.

In response to this gap, a concept of a Sustainable Energy Management System (S-EMS) is proposed, aimed at providing a coherent approach to monitoring, controlling, and optimizing energy flows in a complex production environment. This model assumes the parallel handling of traditional energy carriers (electricity, natural gas, heat, compressed air) and alternative carriers such as hydrogen, treated as a potential element of the plant's future energy infrastructure. Unlike classical EMS solutions, the S-EMS concept also incorporates predictive functionalities, risk assessment, scenario analysis, and the capability for dynamic switching between energy carriers depending on operational or economic conditions.

The proposed model consists of three interdependent functional levels. The basic layer includes the measurement, monitoring, and reporting of energy consumption for all carriers in accordance with ISO 50001 [7], expanded to include hydrogen streams. Both operational data and energy-quality characteristics are recorded, enabling comparisons between different energy carriers. In addition, this layer incorporates structured databases and energy-quality registers as well as integration with MES, SCADA, and even BMS systems, ensuring consistent data acquisition from production and facility infrastructures [4, 6, 17].

In the case of hydrogen integration, particular attention must be paid to measurement accuracy and operational safety. The basic layer of the S-EMS should therefore be extended to include dedicated hydrogen measurement protocols, covering mass flow, pressure, purity, leakage detection, and energy content, in accordance with emerging standards and safety guidelines for industrial hydrogen systems. Continuous monitoring of hydrogen-related parameters enables not only energy balancing but also early detection of abnormal operating conditions, thereby supporting risk mitigation and compliance with safety requirements. Integrating hydrogen safety indicators into the EMS measurement framework allows hydrogen to be treated as a managed energy carrier rather than as an isolated technological subsystem.

The analytical–predictive layer encompasses energy-consumption modelling, load forecasting, cost analysis, identification of areas with the highest savings potential, and the assessment of the feasibility of substituting one energy carrier with another, for example with hydrogen. This layer also includes advanced data-analysis and visualization functions, machine-learning and AI-based predictive models, multi-carrier energy balancing, and quantitative evaluation of costs and emissions [5, 12]. Scenario assessment modules—such as for example evaluating the replacement of natural gas with hydrogen—are important elements at this stage [6]. In addition to technical and energy-performance analyses, this layer supports economic feasibility assessments, including cost–benefit analyses, estimation of operational expenditures, and comparative evaluation of different energy-carrier scenarios. This enables a preliminary assessment of return on investment for hydrogen-related infrastructure within the overall energy strategy of the plant (Fig. 1).

The decision layer integrates analytical results with plant-management processes. It enables the implementation of optimization strategies such as load control, energy-cost minimization, emission reduction, or the selection of the most efficient carrier

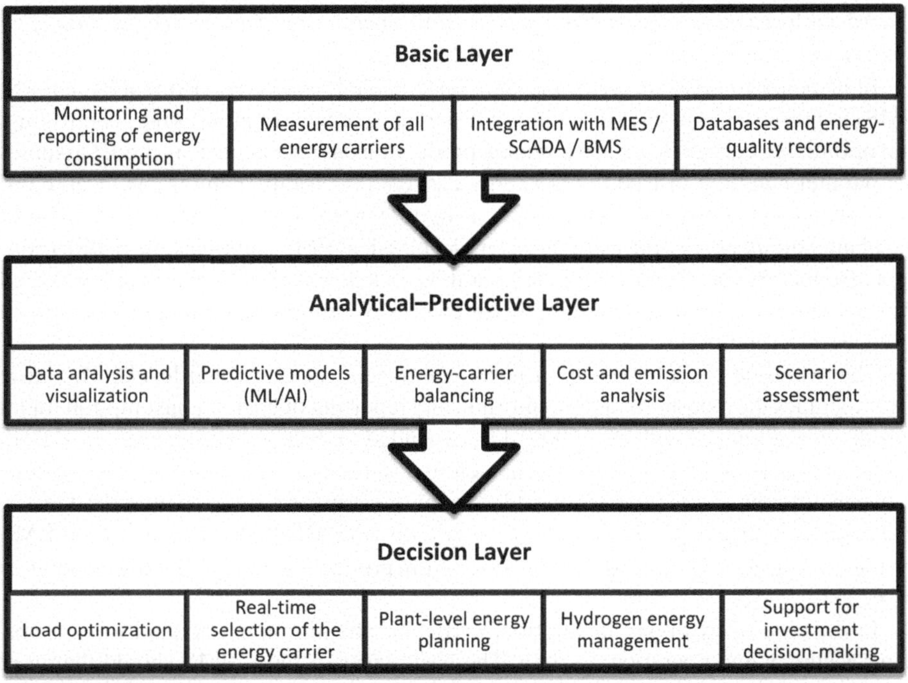

Fig. 1. Model of Sustainable Energy Management System (S-EMS); MES – Manufacturing Execution System, BMS – Building Management System, SCADA – Supervisory Control and Data Acquisition, ML – Machine Learning, AI – Artificial Intelligence.

under given operational conditions [2, 3]. Furthermore, this layer enables real-time selection of the energy carrier, plant-level energy planning, and dedicated hydrogen-energy management functions covering production, storage, and use [15]. It also provides decision-support tools for investment planning in energy technologies and infrastructure.

A key feature of the S-EMS is its flexibility, which allows the model to be adapted to diverse technological configurations found in automotive factories. It can be applied both in plants with traditional energy infrastructures and in facilities introducing elements of the hydrogen economy, such as local H_2 storage, electrolysers, or hydrogen-powered thermal processes [10]. This flexibility provides an advantage over classical EMS solutions, which do not account for the high variability of energy flows or the integration of new carriers.

The S-EMS is not a system competing with ISO standards, but rather an extension of them. It forms an additional layer built on top of ISO 50001, complementing it with analytical, predictive, and planning capabilities, as well as with hydrogen-carrier management. As a result, the model addresses future needs by combining improved energy efficiency with the decarbonization of manufacturing processes.

6 Summary

The EMS models review and the energy analysis of automotive manufacturing processes made it possible to formulate conclusions regarding both the current state of research and the future directions for the development of EMS in the automotive industry. The results clearly indicate that available EMS solutions focus primarily on traditional energy carriers and organizational structures aligned with ISO 50001. They only marginally account for the multi-carrier nature of contemporary energy systems in automotive factories, and, in particular, they lack solutions capable of integrating hydrogen technologies, despite hydrogen being a promising energy carrier for processes.

The analysis of the energy-consumption structure in automotive manufacturing plants confirmed that the greatest optimization potential lies in high–thermal-intensity processes and auxiliary systems. These areas are characterized by both load variability and high sensitivity to operating parameters, which means that effective energy management requires analytical tools extending beyond classical monitoring functions. Hydrogen, when used as an energy carrier in thermal processes, may complement electricity, provided that appropriate infrastructure and system-level integration are ensured.

In response to the identified research gap, a concept of a Sustainable Energy Management System (S-EMS) has been proposed. It consists of three functional levels: monitoring and reporting, analysis and prediction, and operational and investment-related decision-making. The model enables parallel management of multiple energy carriers, including hydrogen, and supports decision-making related to energy planning, cost optimization, emission reduction, and the assessment of technological alternatives. In practice, it represents an extension of classical EMS approaches with advanced analytical tools required in the complex energy environment of automotive factories.

Although the S-EMS framework incorporates analytical functionalities for cost evaluation and scenario comparison, a comprehensive economic assessment of hydrogen integration—including full life-cycle costs and return on investment—requires empirical operational data from industrial installations. Therefore, detailed economic validation should be treated as a direction for further research.

The findings suggest that the integration of hydrogen with EMS is technically feasible; however, its full implementation requires further efforts related to measurement standardization, operational safety, economic assessment, and analysis of the interdependencies between local hydrogen production and the plant's energy profile. Pilot implementations will be a crucial next step, enabling verification of the practical operation of the S-EMS.

A potential roadmap for pilot implementation of the proposed S-EMS may include the following steps: (1) selection of a representative production area with significant thermal energy demand; (2) deployment of extended hydrogen measurement and safety-monitoring infrastructure; (3) integration of hydrogen-related data streams into the S-EMS analytical layer; (4) execution of controlled operational scenarios comparing conventional and hydrogen-based energy supply; and (5) empirical validation of energy, economic, and safety performance indicators.

In summary, the proposed S-EMS concept responds to the growing demands for energy efficiency and decarbonization in the automotive industry. Its further development should involve empirical studies, simulation models, and analyses of energy integration using hydrogen, which will enable the creation of comprehensive tools supporting the energy transition of the automotive sector.

References

1. Takahashi, T., Ueno, N., Sakuma, D.: Development of hydrogen burners and vacuum insulated furnaces for zero CO_2 emissions. SAE Technical Paper 2018-01-1187 (2018)
2. Thollander, P., Palm, J.: Industrial energy management decision making for improved energy efficiency. Energies. **8**, 5694–5703 (2015)
3. May, G., Stahl, B., Taisch, M., Kiritsis, D.: Energy management in manufacturing: toward eco-friendly production. In: Advances in Production Management Systems, pp. 287–294. Springer (2017)
4. Abdelaziz, E.A., Saidur, R., Mekhilef, S.: A review on energy saving strategies in industrial sector. Renew. Sust. Energ. Rev. **15**(1), 150–168 (2011)
5. Herrmann, C., Thiede, S.: Process chain simulation to foster energy efficiency in manufacturing. CIRP J. Manuf. Sci. Technol. **1**(4), 221–229 (2009)
6. Sokil, O., Podolchak, N., Stetsiv, I., Zuiev, M., Chepil, B.: Hydrogen for heating: technologies, challenges, and opportunities. Open Res. Europe. **5**, 18 (2025)
7. International Organization for Standardization: ISO 50001: 2018 Energy Management Systems—Requirements with Guidance for Use. ISO (2018)
8. International Organization for Standardization: ISO 50006: 2023 Energy Management Systems—Measuring Energy Performance Using Energy Performance Indicators (EnPIs). ISO (2023)
9. International Organization for Standardization: ISO 50004: 2020 Energy Management Systems—Guidance for Implementing, Maintaining and Improving an Energy Management System. ISO (2020)
10. Chandra, R., Kaushik, N., Singh, H., Hammouch, Z.: Hydrogen energy systems for industrial applications. Int. J. Hydrog. Energy. **50**, 1122–1138 (2025)
11. Oktaviandri, M., Safiee, C.: Modelling electrical energy consumption in the automotive paint shop. Energy Procedia. **144**, 33–39 (2018)
12. Giampieri, A., Ling-Chin, J., Ma, Z., Smallbone, A., Roskilly, A.P.: A review of the current automotive manufacturing practice from an energy perspective. Appl. Energy. **263**, 114–146 (2020)
13. Sato, F.E., Nakata, T.: Energy consumption analysis for vehicle production through a material flow approach. Energies. **13**(21), 5619 (2020)
14. Cagno, E., Trianni, A., Worrell, E., Miggiano, F.: Evaluating industrial energy efficiency opportunities: a case study analysis. Energy Procedia. **61**, 2634–2637 (2014)
15. Marghani, M., et al.: Recent developments in hydrogen production and applications. Int. J. Hydrog. Energy. **48**, 17233–17250 (2023)
16. Małek, M., Karowiec, A., Jóźwik, I.: Hydrogen technologies in the automotive industry: challenges and opportunities. Sustainability. **15**(6), 5123 (2023)
17. Sayed, K., Gabbar, H.A.: Building energy management systems (BEMS). In: Smart Energy Grid Engineering, pp. 381–410. Academic Press (2018)

The Importance of Green Hydrogen for a Sustainable Economy

Michał Wieczorowski[1,2](✉), Jacek Bogusławski[3], Katarzyna Stec[4], Paweł Fuć[5], Adam Lewandowski[2], Wojciech Prus[2], and Rehan Khan[6]

[1] Faculty of Mechanical Engineering, Poznan University of Technology, Poznan, Poland
michal.wieczorowski@put.poznan.pl
[2] Polish Chamber of Hydrogen, Poznan, Poland
[3] Marshal Office of the Wielkopolska Region in Poznan, Poznan, Poland
[4] Łukasiewicz Institute of Ceramics and Building Materials, Gliwice, Poland
[5] Faculty of Civil and Transport Engineering, Poznan University of Technology, Poznan, Poland
[6] College of Electrical and Mechanical Engineering, National University of Sciences and Technology, Islamabad, Pakistan

Abstract. The aim of this paper is to analyse the role of green hydrogen in the energy transition process and its importance for the development of a sustainable economy. Green hydrogen is one of the most promising solutions in the context of decarbonising industry, transport, and the energy sector. The study provides an overview of current activities and trends as well as an assessment of various aspects of hydrogen economy development. Green hydrogen can significantly contribute to achieving climate neutrality goals by replacing fossil fuels in hard-to-abate sectors. It can also support the development of renewable energy sources, while naturally taking into account certain barriers, including high investment costs, limited availability of transmission infrastructure, and insufficient legal regulations. The authors present global directions of change and emerging projects related to the production of green energy, green hydrogen, and its derivatives. They also emphasise the need to intensify research on these technologies, as well as the development of certification and standardisation. The development of the hydrogen economy is hindered by high costs, technological limitations, underdeveloped infrastructure, slow and inconsistent regulations, and limited public acceptance.

Keywords: Green Hydrogen · Renewable Energy Sources · Transport · Energy Transition · Climate Neutrality

1 Introduction

In recent years - despite the pandemic and the armed conflict in Ukraine - a growing importance of the energy transition and climate neutrality has been observed, leading to a dynamic acceleration of efforts to transform economic systems toward low-emission models. Although these efforts are not always global in nature, and some countries strongly resist them, the global trend is very clear. Striving for a balance between greenhouse gas emissions and their absorption has become one of the main strategic objectives

B. Gapiński et al. (Eds.): MANUFACTURING 2026, LNME, pp. 343–357, 2026.
https://doi.org/10.1007/978-3-032-22943-4_24

for both the European Union and an increasing number of countries worldwide, as well as corporations. One of the most important elements of this process is the long-term transformation of the structure of energy production, distribution, and consumption, along with the reduction of harmful emissions [1]. Within this framework, humanity is primarily seeking to shift from a fossil fuel-based system to one dominated by renewable energy sources (RES), although improving energy efficiency and focusing on zero-emission technologies are also highly significant [2]. This transformation is also embedded in the sustainable development strategy - the 2030 Agenda [3] - which is universal and cross-cutting in nature, setting highly ambitious directions.

However, the transformation process involves numerous challenges [4]. These include, for example, ensuring energy security (which is particularly important in the context of an unstable political and economic situation) and maintaining the stability of power grids, which becomes problematic with the growing share of intermittent sources such as wind and solar. Equally important is the issue of social justice in the energy transition - namely, distributing its costs in such a way that the most economically vulnerable groups are not disproportionately burdened [5]. In this context, technologies that enable the decarbonisation of hard-to-abate sectors - such as heavy industry and large-scale maritime, aviation, and land transport - are gaining increasing importance. Energy storage and all related technologies are also crucial. As a result, new economic sectors are emerging, generating innovation and creating new jobs. Green hydrogen is starting to play a key role, and according to forecasts cited in the same report, it will account for 85% of the entire hydrogen market by 2050. For this reason, the 21st century is increasingly referred to as the era of green hydrogen. Therefore, this article is focused on its importance and the associated technological development. It aims to analyze the role of green hydrogen in the energy transition and its significance for the development of a sustainable economy by reviewing current trends, technologies, and global initiatives. It also examines key barriers to hydrogen economy development, including economic, technological, regulatory, and infrastructure-related challenges.

2 Hydrogen and Its Use

Hydrogen as a gas and chemical element has been known for several centuries, although it was not always correctly identified. As early as the 17th century, it was observed that some metals react with acids, releasing a gas, but it was not recognised until about 100 years later. It was described as "flammable air," and the most interesting discovery was that water was the product of its combustion. Hydrogen began to be used, among other things, in hydrogen balloons, and later - after mastering water electrolysis (controlled hydrogen production) - also in chemical syntheses (ammonia production, which revolutionised agriculture, as well as acids, fuels, etc.). The Apollo programme marked the use of hydrogen as rocket fuel in space missions. The concept of hydrogen as an environmentally friendly energy carrier emerged as early as the 20th century, giving rise to the term "hydrogen economy". Hydrogen is a versatile energy carrier and can be produced from various raw materials [6]. Its greatest advantage is the absence of carbon dioxide, sulphur oxide, and very small amount of nitrogen oxide emissions, because during use - whether in combustion or in fuel cells - the only by-product is water vapour.

From a production perspective, hydrogen has another major advantage: unlike fossil fuels that occur only in specific locations on Earth, hydrogen can be produced almost anywhere, regardless of natural resource availability [7]. It can store large amounts of energy over long periods, addressing the issue of variable energy production from wind and solar power (RES) and the limitations of electric battery storage. The universality of hydrogen as a fuel and energy carrier - usable in many fields - should also be emphasised. In the energy sector, hydrogen can be used to store surplus energy from renewables. In district heating, it can be combusted directly or blended with natural gas, and ultimately used as a pure gas [8]. Achieving high temperatures is another area of industrial application, particularly in processes requiring large amounts of heat, such as steel production, chemicals, and fertilisers. A sample furnace used as a research station for hydrogen combustion and co-combustion is presented in Fig. 1. A further key segment of the hydrogen economy is transportation, where fuel cells and internal combustion applications hold significant promise. Thanks to its versatility, hydrogen can simultaneously support the decarbonisation of multiple sectors.

Fig. 1. Furnace with hydrogen combustion capability at the Łukasiewicz Institute of Ceramics and Building Materials.

A few years ago, when widespread fascination with hydrogen in the context of the energy transition began, some claimed that it would soon replace all other fuels. Naturally, this idea lacked any real foundation and did not align with reality, including for economic and technological reasons [9]. Hydrogen will certainly not be a panacea for all the world's problems, but it does have every chance of playing an important role in the energy mix. This, however, requires investments in infrastructure development and ensuring safety (flammability, gas permeability, hydrogen embrittlement, etc.) [10].

The literature reviews show a very dynamic increase in the number of publications on green hydrogen, electrolysers, storage, and applications in energy and transport [11]. Despite this high level of interest, the implementation of hydrogen technologies encounters economic and technological barriers. Emphasis is placed primarily on the high costs of producing low-emission hydrogen, as well as challenges related to storage, transport, and safety [12]. In the long term, massive infrastructure expansion, regulatory standardisation, and mechanisms for creating demand for clean hydrogen will be necessary - factors that are currently lacking or still in pilot phases. It is also highlighted that current research focuses mainly on technical aspects (process efficiency, material parameters, system performance), while relatively less attention is devoted to analyses covering business models and social acceptance [13].

One of the key directions in the development of a sustainable energy transition is the integration of hydrogen technologies with the circular economy (CE). Hydrogen is used in closed industrial loops, where some waste gases can be reused. It also plays a special role in the closed carbon cycle - Carbon Capture and Utilization / Carbon Capture, Utilization and Storage (CCU/CCUS). Combining hydrogen with captured CO_2 enables the production of synthetic fuels, chemicals, and materials, creating a loop in which carbon dioxide becomes a raw material rather than waste. Such solutions help reduce emissions in sectors that are difficult to decarbonise, such as aviation, the chemical industry, and the petrochemical sector.

3 Hydrogen Production

Hydrogen can be produced from various sources and using different types of energy [14]. Depending on these factors, the resulting gas is categorised by different colours. Since not all colour definitions are consistent across the literature, we adopt the assumption that the most desirable green hydrogen is produced through water electrolysis powered by renewable energy, or through biogas reforming - which involves biochemical conversion of biomass. This renewable energy comes from sources such as the sun and wind. Nevertheless, a subtype of green hydrogen is sometimes distinguished - yellow hydrogen, produced by water electrolysis powered specifically by solar energy. Also considered environmentally friendly is white hydrogen, which comes from natural geological sources - found in natural caverns or mantle rocks. A related category is gold hydrogen, present in depleted oil wells and inactive mines. Another promising type is purple hydrogen, which, like green hydrogen, is produced via electrolysis, but using electricity generated in nuclear power plants, where energy comes from nuclear reactions in the production process (e.g., reactor heat). When electricity from a nuclear power plant powers an electrolyser, the product is referred to as pink hydrogen. Also viewed positively is turquoise hydrogen, produced through methane pyrolysis without oxygen or through the processing of waste plastics. This method emits no carbon dioxide, and the by-product is solid carbon, which is easier to store compared to the CO_2 capture and storage required for blue hydrogen. Blue hydrogen is produced using fossil fuels but supplemented with technologies for capturing, storing, or converting carbon dioxide. Hydrogen production technologies from the previous century were, to put it mildly, far from environmentally friendly. The hydrogen produced through them is most commonly classified as grey hydrogen, obtained from natural gas reforming or from other hydrocarbons derived from oil refining. In addition, there is black hydrogen, produced through hard coal gasification, and brown hydrogen, produced through lignite gasification.

Naturally, the most important role in the energy transition is played by green hydrogen, which is completely emission-free and relatively easy to obtain. To produce it, an electrolyser is required [15], that is, a device that uses electricity to split water (H_2O) into hydrogen (H_2) and oxygen (O_2). The water electrolysis process itself has been known for over 200 years; for green hydrogen, it is essential that the energy comes from renewable sources. Water molecules supplied to the device break apart in the presence of an electrolyte due to the flow of electric current, producing hydrogen - which accumulates at the cathode (−) and oxygen, which is released at the anode (+). Both gases are then collected in separate storage tanks. A very important element in the electrolysis process is the electrolyte, because it enables the flow of ions between the electrodes (the anode and the cathode), as pure water alone conducts electricity very poorly. Electrolysis requires a significant amount of energy, roughly 50–55 kWh of electricity to produce 1 kg of hydrogen. There are several technologies for producing electrolysers, differing in efficiency, operating temperature, and materials used. The oldest, chronologically, and at the same time relatively simple and proven over many years, is the Alkaline Electrolyzer (AEL), in which the electrolyte is a KOH or NaOH solution and the device operates at around 70–90 °C [16]. A newer solution is the Proton Exchange Membrane (PEM) electrolyser, which uses a polymer membrane (separating the anode from the cathode, enabling ion flow, and preventing hydrogen and oxygen from mixing) and operates at 50–80 °C [17]. It works faster and has a more compact design than the alkaline version. It is also more flexible in terms of operating conditions. Additionally, the membrane prevents gas mixing, allowing the production of very high-purity hydrogen. Its drawback, however, is the need for so-called rare earth materials, which are difficult to obtain in many parts of the world. A promising new technology similar to PEM is the Anion Exchange Membrane (AEM) electrolyser, operating at similar temperatures. The main difference lies in the polymer membrane itself (anionic instead of protonic), which uses a basic rather than acidic polymer - the operating environment is also alkaline, not acidic. This affects electrode materials: PEM requires acid-resistant (expensive, noble) materials, increasing the construction and operating costs. In contrast, AEM allows the use of cheaper metals (e.g., nickel), potentially reducing overall production and operating costs. Another difference concerns the type of ions conducted: in a proton electrolyser, positive ions (H^+) flow from the anode to the cathode, whereas in an anion electrolyser, negative ions (OH^-) move from the cathode to the anode. Electrolysis can also be conducted at much higher temperatures (600–800 °C) using oxide ceramics as the electrolyte; electrolysers based on this technology are known as Solid Oxide Electrolysis Cells (SOEC). While alkaline electrolysers typically achieve about 65% efficiency in industrial settings, and proton and anion electrolysers around 80%, oxide electrolysers can achieve even higher values, i.e. 80–100% or more, especially when waste heat is used in the process. This makes them a good choice in settings where large amounts of surplus heat are available, for example from industrial processes.

An important aspect is also the possibility of producing hydrogen from waste and industrial by-product streams [18], including waste gases, biomass, organic waste, and even sewage sludge or plastics. In many industrial sectors, waste gases rich in hydrogen or hydrocarbons are generated, and after appropriate processing they can become valuable energy feedstocks. Another significant direction is hydrogen production through microbiological conversion of plant waste, municipal waste, or sewage sludge. In parallel, biomass gasification is developing, in which wood or agricultural waste is converted into syngas, from which hydrogen is subsequently separated [19]. "Waste-to-hydrogen" technologies use plastic waste and mixed municipal waste to produce hydrogen. Pyrolysis of plastics makes it possible to break them down into gas fractions rich in hydrogen, while gasification of mixed waste enables the production of syngas from materials that cannot be recycled mechanically. Municipal waste can also serve as a hydrogen source through the processing of sewage sludge into an energy carrier, improving the efficiency of municipal systems [20].

When discussing white and gold hydrogen, the deposits found underground were mentioned. White hydrogen, unlike green or grey hydrogen, is natural, does not require energy-intensive processing, and does not generate carbon dioxide emissions. Furthermore, there is evidence suggesting that it may be possible to extract it from rock formations, as it forms through serpentinisation processes beneath the Earth's surface [21]. If the assumption is confirmed that billions of tonnes of geological hydrogen may exist within the Earth's crust, the outlook for hydrogen economy development will become significantly more favourable. Mountain ranges not only naturally produce large amounts of hydrogen but also enable its storage in rock reservoirs. Efficient exploration and resource-prospecting methods must be developed for this purpose. Underground, there is also gold hydrogen, trapped in rock formations beneath the surface. It forms naturally and often accumulates near other fossil fuels [22]. Its origin is again linked to iron-rich ophiolites—rocks that generate hydrogen when exposed to water. These are igneous rocks that originally formed beneath the ocean floor as parts of the lithosphere and were only later incorporated into continental crust. It is precisely in such formations that deposits of clean hydrogen are most promising.

4 Transport of Hydrogen and Its Compounds

A crucial link in the hydrogen value chain - that is, the delivery of hydrogen-related energy - is its transport, which significantly affects the economic aspects of the energy transition [23]. Transport connects production sites, such as electrolysers, with end-use locations (hydrogen hubs), because hydrogen is not always produced where it is needed. The places where hydrogen production is the cheapest and most environmentally friendly are often regions with large amounts of renewable energy (e.g., solar and wind - solar farms and onshore/offshore wind farms), which are typically located far from major industrial centres or cities, i.e., the main consumption hubs. Hydrogen produced in such regions must be delivered to end users, including primarily industry (especially heavy industry), but also the energy sector (including cogeneration and trigeneration, hydrogen power plants, and energy storage facilities), as well as vehicle refuelling stations. Hydrogen transport - both at the global, international scale (e.g., from countries in the southern hemisphere to Europe), at the national level (from coastal areas to more industrialised regions), and at the local level (from regional distribution centres to refuelling stations) - is therefore of great importance for the energy transition [24].

In terms of transport media, hydrogen can be delivered via pipelines or transported in mobile form - by land (vehicles), rail (trains), sea (ships), or air (airborne carriers). When considering pipelines, two types of installations can be distinguished: existing natural gas pipelines (which must be adapted for hydrogen transport) and dedicated hydrogen pipelines, designed and built specifically for this purpose. In general, pipelines are the most efficient and stable method of transmission at large scales and over long distances, with a lower energy footprint than other options (not counting construction-related impacts). It is also possible to partially use the existing gas infrastructure, although this requires additional adaptation work [25]. Safety and reliability issues are also critical. Hydrogen transport requires materials resistant to gas permeation and metal embrittlement because hydrogen's low molecular weight and high diffusivity lead to higher permeation levels, increasing the risk of leaks and material degradation [26]. Additionally, hydrogen has low density and is highly flammable, which significantly affects safety considerations. Firefighting and emergency personnel - regardless of region - have very limited or even no practical experience in handling hydrogen-related incidents. Each hydrogen pipeline requires, and will continue to require, appropriate monitoring using specialised measuring devices aligned with the latest metrology trends [27], as well as leak detection systems and regular maintenance to prevent failures. As with any pipeline, surface condition analysis is necessary - both in terms of wear [28] and mechanical defects or damage [29]. Non-destructive testing is also important for identifying internal material changes, for example through X-ray based techniques [30]. Unfortunately, regulatory frameworks in these areas are still very underdeveloped. It must also be taken into account that hydrogen's low density results in approximately three times higher pressure losses compared with natural gas, which requires higher inlet pressures and additional compression over long distances [31]. Long-distance hydrogen transport generally involves relatively high pressures (up to 12 MPa), requiring additional equipment and increasing the complexity and cost of the transport process. Pipelines must therefore be adapted to higher flow velocities and potential pressure drops, which affects compressor performance requirements and leak detection strategies. Lower pressures (up to 10 bar) are used in local distribution networks [32]. It is also significant that despite many years of hydrogen-related work worldwide, uniform standards for hydrogen transport and storage have still not been established. Even so, hydrogen transmission pipelines are viewed as the true backbone of large-scale hydrogen transport (and thus, indirectly, energy transport), connecting production hubs with demand centres.

At present, there is already a network of pipelines worldwide capable of transporting hydrogen, along with ambitious plans for its expansion. In Europe, there are discussions about connecting EU countries into a single transmission network and establishing additional links extending beyond Europe. The Gulf countries are also considered potential future exporters of hydrogen, which is driving plans for additional pipelines from that region. A similar situation is developing in other parts of the world, such as China, India, Australia, and Canada. These projects demonstrate how seriously the world views hydrogen transport via pipelines and how important a part of the economy it will become within the next decade or two. Despite the growth of pipeline projects, their network remains limited, which makes mobile transport - by road, rail, sea, or air - necessary. This mode involves much smaller quantities than pipeline transport and is intended for significantly shorter distances, similar to today's mobile fuel transport. Its key advantage is the flexibility of delivery routes and destinations, which pipelines naturally cannot provide [33]. In this approach, hydrogen can be transported in gaseous, liquid, or solid form, and the choice of method depends on the transport volume and delivery distance. The first option (gaseous hydrogen) is and will remain used in local distribution and to supply refuelling stations over short distances. The gas, compressed to around 30 MPa (a compromise based on tank weight and production cost), is transported in special high-pressure cylinders mounted on trailers (tube trailers) or in truck or rail tankers from hydrogen storage facilities over distances of up to several hundred kilometres - longer distances are economically unfeasible. Rail transport can cover greater distances, while long-distance options include maritime transport - where transporting the gas is possible, though not necessarily the best option.

Hydrogen is also transported in liquid form, in which it exists at temperatures between approximately −253 °C and −259 °C. In this state, it has higher energy density and a high purity level, but the liquefaction process and the subsequent return to the gaseous state (regasification) cause significant energy losses. Cryogenic tankers are used for its transport [34], and therefore, the economic analysis must account for both the cost of the cryogenic tanks and the phase-change processes. This method is more practical for long-distance transport, particularly international transport, and can also be used in maritime shipping, aviation, and even space applications.

The third option for mobile hydrogen transport is chemical compounds (hydrogen carriers), which include, among others, ammonia (NH_3), methanol (CH_3OH), and LOHCs (Liquid Organic Hydrogen Carriers). Ammonia has been transported worldwide on a large scale for decades as a feedstock for fertiliser production, meaning that mature infrastructure, technologies, and safety procedures already exist [35]. In the energy context, its importance has grown as an efficient hydrogen carrier - it is easy to liquefy (at about −33 °C at atmospheric pressure or at 8–10 bar at room temperature), has high energy density, and can make use of existing liquid-fuel and liquefied-gas infrastructure [36]. It can be transported in cryogenic or pressurised tanks similar to those used for LPG, which is logistically simpler than hydrogen transport. The most important method for transporting large quantities of ammonia is maritime shipping, though it is also transported by rail (pressurised steel tank cars) and road (ADR tank trucks). The biggest challenge associated with ammonia transport is its toxicity, as even small leaks can pose health hazards, causing respiratory and eye irritation. Ammonia can also be corrosive to certain metals and, at high concentrations, forms flammable mixtures with air. For these reasons, transport systems are equipped with special safety features, and strict loading, unloading, and technical inspection procedures are followed. Finally, ammonia can be decomposed into hydrogen and nitrogen, but may also be used directly as a fuel in engines (ammonia engines) and in fuel cells.

Transporting methanol is much simpler and less demanding than transporting hydrogen or ammonia [37]. Methanol is a liquid with relatively low toxicity compared to ammonia, easy to store and handle using standard liquid-fuel infrastructure. It does not require high pressures or low temperatures; as a stable liquid at room temperature, it can be stored and transported in steel tanks similar to those used for gasoline or diesel. There is no need to create new logistics systems, which significantly reduces costs and facilitates scalability. In maritime transport, methanol is shipped in chemical tankers in tanks with corrosion-resistant coatings and systems preventing evaporation. By rail, it is transported in chemical tank cars protected against corrosion, and by road in tank trucks commonly used for liquid fuels.

Hydrogen can also be transported using LOHCs - Liquid Organic Hydrogen Carriers - which are liquid organic substances capable of chemically binding hydrogen and subsequently releasing it on demand [38]. This allows hydrogen to be transported and stored at temperatures and pressures similar to conventional fuels. The main LOHC substances include benzyltoluene and dibenzyltoluene, which - unlike pure hydrogen - are neither highly flammable nor explosive, and additionally have significantly higher energy density (once hydrogen is bound, a relatively large amount can be stored per unit volume). Unfortunately, the processes of binding and releasing hydrogen require energy input; some carriers are expensive or difficult to recycle, and hydrogen release reactions may require high temperatures (e.g., 250–300 °C). Nevertheless, this is a very promising method not only for long-distance hydrogen transport (ambient-condition transport is highly preferred, e.g. in intercontinental shipping), but also for hydrogen storage and even for supplying refuelling stations and industrial installations.

A new transport and storage option is already emerging, namely the use of chemical compounds in solid form - metal hydrides [39]. These compounds absorb hydrogen, are stable, and allow it to be released as needed. In this case, there is no risk of leaks or explosions, as may occur with gases or liquids. Research on these compounds is ongoing, and the results allow for an optimistic outlook on future applications. It is also possible that drones will appear among hydrogen transport methods for airborne delivery, which will require solving numerous safety-related issues but could revolutionise short-distance deliveries.

Hydrogen transport also plays an indirect role, as it enables storage at various points within the energy network. This allows surplus renewable energy to be used later where it is needed, with hydrogen acting as an energy carrier that helps balance supply and demand across both time and location.

5 Further Technological Challenges

Although the economic and social mood, as well as perceptions and expectations regarding the hydrogen economy, fluctuate like a sine wave - at times entering phases of optimism (sometimes excessive) and at times pessimism (also excessive) - the future of hydrogen is clearly taking shape in optimistic tones. It will not replace all other energy sources, but it will have an important and significant place in the global economy. The geographical landscape of the emerging hydrogen world is also becoming increasingly clear, with defined areas of production and consumption. In many scenarios for the development of the hydrogen economy, it is assumed that hydrogen producers and exporters will be countries with high solar irradiation or strong winds, making hydrogen (or hydrogen-derived products) production there very cost-effective. The only major challenge to solve will be cheap, large-scale transport to importing countries. Countries expected to become hydrogen producers and exporters include Australia, Chile, the Gulf states, Morocco, Algeria, Saharan countries, Kazakhstan, Namibia, and Colombia. Importers, in turn, will be those with high energy demand but limited space for renewable energy deployment - such as the European Union, Japan, and South Korea. Large production projects are also emerging in Canada, though in this case supply may serve local needs or potentially those of neighbouring United States. The map of new centres of energy wealth, marked in green, is shown in Fig. 2.

Of course, this map does not account for all possibilities. Hydrogen electrolysis requires water resources, which means that countries with abundant water also have the opportunity to find their place in this emerging landscape. Large river dams and

Fig. 2. New zones of global energy wealth.

regulation systems (e.g., on the Congo River in Africa), as well as better management of areas affected by floodwaters from heavy rainfall (e.g., Bangladesh), may contribute to this. The new map of energy wealth will change how producer countries operate - some of which are currently considered poor. It is highly likely that a situation similar to the one observed in the 20th century will repeat itself, when regions where oil and natural gas were discovered underwent rapid economic development. Based on this, global trade and the hydrogen economy will continue to grow, in a way comparable to today's markets for crude oil and natural gas. Every part of the world will have its own concept for exporting or importing hydrogen. Europe is implementing systems for blending natural gas with hydrogen, but it is also intensively developing pipeline infrastructure - so-called hydrogen corridors. Examples include the Germany - Netherlands route, the Norway - EU route, and perhaps the most spectacular project: the Nordic-Baltic Hydrogen Corridor, planned to run from Finland through Estonia, Latvia, Lithuania, and Poland to Germany. Its length is expected to reach around 2,500 km, and once completed by 2040, it is projected to transport up to 2.7 million tonnes of hydrogen annually. A similar direction is outlined in the Baltic Sea Hydrogen Collector (BHC) project, which will also connect hydrogen production (especially from offshore wind energy) in Finland with the markets of Nordic and Central European countries. The planned route includes a Finland - Germany connection via the Åland Islands and Gotland, with potential extensions to Denmark, Poland, and other Baltic states. Both projects demonstrate the immense potential of the Baltic Sea region for producing offshore wind energy and how this potential is expected to be used to decarbonise hard-to-abate industrial sectors in Central Europe. The construction of subsea hydrogen pipeline infrastructure will undoubtedly be a major technological as well as regulatory and legal challenge. Another important European project is H2Med, aimed at creating a corridor for transporting green hydrogen from the Iberian Peninsula (Portugal and Spain) to north-western Europe. The plans include a subsea section connecting Spain (Barcelona) and France (Fos-sur-Mer / Marseille area). The target transmission capacity is approximately 2 million tonnes of hydrogen per year. In this case, green hydrogen will be generated using both wind energy (produced onshore) and solar energy. Pipelines to Europe are also being planned from North African countries, particularly Morocco and Algeria.

However, the European Union is fully aware that these projects may prove insufficient. Therefore, maritime transport from the southern hemisphere is being planned - either of liquefied hydrogen or, more likely, ammonia or methanol. In Africa, a key producer for the German economy will be Namibia, from which several million tonnes of hydrogen produced from onshore solar and wind energy are expected to be delivered annually starting in 2035. South America is also an interesting direction, with southern regions of Chile - characterised by excellent wind energy conditions - and northwestern Colombia, combining wind and solar energy, expected to play important roles. These regions are already cooperating with Europe on developing production capacity and maritime transport infrastructure. Canada is also investing in hydrogen transport to Europe, developing production in the Labrador Peninsula region, which is relatively close to potential buyers.

Australia, located in latitudes highly favourable for generating solar energy during the day and wind energy at night, is likewise interested in exporting hydrogen to Europe. Here, research and transport projects based on next-generation solid carriers and liquid carriers are developing rapidly, enabling safe and economical hydrogen transport. In terms of hydrogen export, Australia is also active on the Asian market - particularly in Japan and South Korea. Both countries already have, and continue to expand, fleets of tankers for transporting liquid hydrogen and ammonia. Both are also building hydrogen hubs on their coastlines to receive the carriers from ships and distribute them locally. In this region, no long-distance transmission pipelines are planned - only local distribution networks - due in part to mountainous terrain and, especially in Japan, frequent earthquakes and associated risks. Singapore, located along this transport route and recognising its strategic role in trade between East Asia and Europe, is testing the possibilities and challenges of hydrogen hubs for maritime transport.

Middle Eastern countries are also implementing their hydrogen plans, aiming to secure a position as key exporters in the global hydrogen market. Qatar is significantly expanding its production capacity, the United Arab Emirates are not far behind, and Saudi Arabia aims to become the world's largest hydrogen supplier. This ambition is supported by the region's vast solar energy potential and substantial accumulated capital resources. An added advantage is the possibility of using the existing pipeline infrastructure - naturally with the necessary modifications.

North America, similar to Europe, is expanding its pipeline network while also investing in terminals adapted for handling liquid hydrogen. Both the United States and Canada are already making use of existing infrastructure, which in the U.S. alone includes approximately 1,600 km of hydrogen pipelines. Considering the extensive natural gas pipeline network in the U.S., the HyBlend initiative has been launched to study the possibilities and challenges of blending hydrogen with the natural gas grid, specifically by injecting hydrogen into existing natural gas transmission pipelines. There is also the Regional Clean Hydrogen Hubs (H2Hubs) program, which aims to create regional clean hydrogen hubs connecting producers, end users, and hydrogen transport/distribution infrastructure. Its objective is to build a national-scale commercial hydrogen network, reduce risks associated with early-stage investments in clean hydrogen, develop a full value chain (production - transmission - end user), and support the standardisation of solutions.

Thinking long-term, the hydrogen market carries significant economic and social implications, affecting both production-related issues and broader structural changes in the economy. We can expect a gradual decrease in the cost of producing green hydrogen and the creation of new value chains and jobs in industry, logistics, and engineering services. Hydrogen also strengthens energy security. Among the negative aspects are the use of rare-earth materials and the high capital intensity of investments, which should encourage policymakers to actively develop public-private partnerships, support instruments, and stable regulatory frameworks.

In addition to technological and socio-economic considerations, political and institutional frameworks are also essential for the development of the hydrogen economy. These are defined primarily by European and national hydrogen strategies, which set emission reduction targets, forecast electrolyser capacity, define support mechanisms (investment grants, tax incentives, etc.), and regulate infrastructure and energy market rules. Within these frameworks, systems for certification and guarantees of origin for green hydrogen are also being developed to ensure transparency regarding the fuel's carbon footprint and environmental parameters, thereby enabling the trading of "green" volumes within a unified market.

The development of the hydrogen economy also faces numerous challenges and barriers that significantly influence the pace of technology commercialisation. They may lead to the classic dilemma in which investors hesitate to commit without guaranteed demand, while end users do not adopt hydrogen technologies without an operational supply system. Also, regulatory, standardisation, and certification issues are being resolved very slowly. Finally, public acceptance plays a crucial role. Hydrogen as an energy carrier still raises safety concerns, and its role in the energy transition is often underestimated by the public. Increasing awareness and providing appropriate education are essential elements for the successful implementation of hydrogen technologies.

6 Conclusions

The energy transition and the pursuit of climate neutrality are civilizational processes that are redefining the functioning of modern economies. Their growing importance results both from the need to counteract global warming and from the necessity to build economic resilience based on low-emission, innovative technologies. In this context, green hydrogen is a key element enabling the achievement of climate goals in a manner consistent with the principles of sustainable development.

The levelized cost of hydrogen (LCOH) for green hydrogen is expected to decrease substantially over time. Currently, in some countries it reaches even up to 19 USD/kg, depending on location and technology. By 2030, most projections indicate a reduction to approximately 2.5–5 USD/kg as a result of declining renewable electricity costs and technological progress. By 2050, it could fall further to around 1–2 USD/kg.

However, despite its growing potential, significant challenges remain. Hydrogen production requires large amounts of energy and water, which in some regions may conflict with local resource availability. Hydrogen technologies are also material-intensive and rely on critical raw materials for which recycling chains are still lacking. In hydrogen-related activities, a clear need emerges for an interdisciplinary approach that combines engineering expertise (technology development and optimisation), economic competence (cost analysis, business models, support mechanisms, investment risks), and environmental perspectives (LCA, carbon footprint, impact on natural resources). Only such an integrated approach will allow for a reliable assessment of the real economic potential of hydrogen technologies and the identification of the conditions under which hydrogen can become a lasting element of a sustainable energy transition - rather than merely a technological and political hope or promise.

References

1. Eaton, E.: Approaches to energy transitions: carbon pricing, managed decline, and/or green new deal? Geogr. Compass. **15**, e12554 (2021)
2. Popescu, C., Apostu, S.A., Radulescu, I.G., Muresan, J.D., Brezoi, A.G.: Energizing the now: navigating the critical landscape of today's energy challenges - an in-depth review. Energies. **17**, 675 (2024)
3. Transforming our World: The 2030 Agenda for Sustainable Development, A/Res/70/1 United Nations (2015)
4. Ahmad, H., Yaqub, M., Lee, S.H.: Global trends in carbon neutrality: a scientometric review on energy transition challenges, practices, policies, and opportunities. Environ. Dev. Sustain. (2025)
5. Radtke, J., Scherhaufer, P.: A social science perspective on conflicts in the energy transition: an introduction to the special issue. Util. Policy. **78**, 101396 (2022)
6. Yue, M., Lambert, H., Pahon, E., Roche, R., Jemei, S., Hissel, D.: Hydrogen energy systems: a critical review of technologies, applications, trends and challenges. Renew. Sust. Energ. Rev. **146**, 111180 (2021)
7. Beyazit, N.I.: Journey to a sustainable future: the importance of renewable hydrogen in industrial and domestic applications. Int. J. Hydrog. Energy. **99**, 836–851 (2025)
8. Abdelghany, M.B., et al.: Hydrogen energy systems for decarbonizing smart cities and industrial applications: a review. Renew. Sust. Energ. Rev. **226**(D), 116370 (2026)
9. Johnson, N., Liebreich, M., Kammen, D.M., Ekins, P., McKenna, R., Staffell, I.: Realistic roles for hydrogen in the future energy transition. Nat. Rev. Clean Technol. **1**, 351–371 (2025)
10. Khan, R., Wieczorowski, M., Mourad, A.-H.I., Seikh, A.H., Ahmed, T.: Evaluation of erosion of AISI 1045 carbon steel due to non-cohesive microparticles. J. Mater. Res. Technol. **28**, 1423–1432 (2024)
11. Agbadaola, O., Qadir, D., Ahmad, F., Suleman, H., Karousos, D.S., Favvas, E.P.: Hydrogen technologies and policies for sustainable future: a review. Chem. Pap. **78**, 4057–4073 (2024)
12. Abdellatif, M., et al.: Hydrogen energy systems: technologies, trends, and future prospects. Sci. Total Environ. **939**, 173622 (2024)
13. Sharma, G.D., Verma, M., Taheri, B., Chopra, R., Parihar, J.S.: Socio-economic aspects of hydrogen energy: an integrative review. Technol. Forecast. Soc. Chang. **192**, 122574 (2023)
14. Zhang, L., et al.: A comprehensive review of the promising clean energy carrier: hydrogen production, transportation, storage, and utilization (HPTSU) technologies. Fuel. **355**, 129455 (2024)
15. Akyuz, E.S., Telli, E., Farsak, M.: Hydrogen generation electrolyzers: paving the way for sustainable energy. Int. J. Hydrog. Energy. **81**, 1338–1362 (2024)
16. Schneider, L.P.C., et al.: Advancements in hydrogen production using alkaline electrolysis systems: a short review on experimental and simulation studies. Curr. Opin. Electrochem. **47**, 101552 (2024)

17. Carmo, M., Fritz, D.L., Mergel, J., Stolten, D.: A comprehensive review on PEM water electrolysis. Int. J. Hydrog. Energy. **38**(12), 4901–4934 (2013)
18. Yadav, S., Singh, D., Mohanty, P., Sarangi, P.K.: Biochemical and thermochemical routes of H2 production from food waste: a comparative review. Chem. Eng. Technol. **46**, 191–203 (2023)
19. Lukajtis, R., et al.: Hydrogen production from biomass using dark fermentation. Renew. Sust. Energ. Rev. **91**, 665–694 (2018)
20. Giakoumakis, G., Sidiras, D.: Production and storage of hydrogen from biomass and other sources: technologies and policies. Energies. **18**, 650 (2025)
21. Zwaan, F., et al.: Rift-inversion orogens are potential hot spots for natural H2 generation. Sci. Adv. **11**, 8 (2025)
22. Truche, L., et al.: A deep reservoir for hydrogen drives intense degassing in the Bulqizë ophiolite. Science. **383**(6683), 618–621 (2024)
23. Bogusławski, J., Wieczorowski, M., Lewandowski, A., Fuć, P., Khan, R., Prus, W.: Prospects and risks of green hydrogen use in transport - a review. In: Machado, J., Trojanowska, J., Antosz, K., Leão, C.P., Knapcikova, L., Sover, A. (eds.) Innovations in Industrial Engineering IV. ICIENG Lecture Notes in Mechanical Engineering. Springer, Cham (2025)
24. Faye, O., Szpunar, J., Eduok, U.: A critical review on the current technologies for the generation, storage, and transportation of hydrogen. Int. J. Hydrog. Energy. **47**(29), 13771–13802 (2022)
25. Abbas, A.J., Haruna, S.K., Burby, M., John, I.J., Yar'Adua, K.H.: Transitioning to a hydrogen economy: exploring the viability of adapting natural gas pipelines for hydrogen transport through a case study on compression vs. looping. Gases. **4**, 74–96 (2024)
26. Cristello, J.B., Yang, J.M., Hugo, R., Lee, Y., Park, S.S.: Feasibility analysis of blending hydrogen into natural gas networks. Int. J. Hydrog. Energy. **48**, 17605–17629 (2023)
27. Wieczorowski, M., Trojanowska, J.: Towards metrology 4.0 in dimensional measurements. J. Mach. Eng. **23**(1), 100–113 (2023)
28. Krolczyk, G., Kacalak, W., Wieczorowski, M.: 3D parametric and nonparametric description of surface topography in manufacturing processes. Materials. **14**(8), 1987 (2021)
29. Wieczorowski, M., Gapiński, B., Swojak, N.: The use of optical scanner for analysis of surface defects. Ann. DAAAM Proc. Int. DAAAM Symp. **30**(1), 76–85 (2019)
30. Gapinski, B., Wieczorowski, M., Grzelka, M., Alonso, P.A., Tomé, A.B.: The application of micro computed tomography to assess quality of parts manufactured by means of rapid prototyping. Polimery/Polym. **62**(1), 53–59 (2017)
31. Sayani, J.K.S., Wang, M., Ma, Z., Sharan, P., Mehana, M., Chen, B.: Techno-economic analysis of hydrogen transport via repurposed natural gas pipelines: flow dynamics and infrastructure tradeoffs. Int. J. Hydrog. Energy. **147**, 150033 (2025)
32. Khan, M.A., Young, C., Layzell, D.B.: The techno-economics of hydrogen pipelines. Transit. Accel. Tech. Briefs. **1**, 1–40 (2021)
33. Solomon, M.D., Heineken, W., Scheffler, M., Birth-Reichert, T.: Cost optimization of compressed hydrogen gas transport via trucks and pipelines. Energ. Technol. **12**, 2300785 (2024)
34. Ahluwalia, R.K., Peng, J.-K., Roh, H.-S., Papadias, D., Wang, X., Aceves, S.M.: Liquid hydrogen storage system for heavy duty trucks: capacity, dormancy, refueling, and discharge. Int. J. Hydrog. Energy. **48**(87), 34120–34131 (2023)
35. Menefee, A.H., Schwartz, B.A.: Comparing green hydrogen and green ammonia as energy carriers in utility-scale transport and subsurface storage. Energy Clim. Change. **5**, 100163 (2024)
36. Aziz, M., Wijayanta, A.T., Nandiyanto, A.B.D.: Ammonia as effective hydrogen storage: a review on production, storage and utilization. Energies. **13**, 3062 (2020)

37. Liu, L.: Study on route optimization of methanol safety transportation routing. Chem. Eng. Trans. **59**, 1177–1182 (2017)
38. Niermann, M., Beckendorff, A., Kaltschmitt, M., Bonhoff, K.: Liquid organic hydrogen carrier (LOHC) – assessment based on chemical and economic properties. Int. J. Hydrog. Energy. **44**(13), 6631–6654 (2019)
39. Nemukula, E., Mtshali, C.B., Nemangwele, F.: Metal hydrides for sustainable hydrogen storage: a review. Int. J. Energy Res. **2025**, 6300225 (2025)

Internet of Things

Digital Maturity in Poland: Investigating the Level of Digital Transformation and Corresponding Barriers

Julia Żmuda and Maria Rosienkiewicz(✉)

Department of Production Engineering and Management, Wroclaw University of Science and Technology, Lukasiewicza 5, 50-370 Wroclaw, Poland
maria.rosienkiewicz@pwr.edu.pl

Abstract. Digital transformation has become a key factor influencing the competitiveness and development of enterprises, particularly in the context of Industry 4.0. Despite growing interest in this topic, there is still a noticeable lack of comprehensive and sector-specific research on digital transformation and digital maturity, especially with regard to manufacturing enterprises in Poland. This paper aims to present the level of digitalization and digital transformation in Polish enterprises based on available data and to assess the availability of relevant studies. The study adopts a review-based research approach, relying on secondary data from national and European governmental sources and recognized institutional reports. The results indicate that most existing studies are general in nature, focus primarily on SMEs, and address individual technologies rather than holistic transformation strategies. Moreover, the adoption of advanced digital technologies decreases as their complexity increases. The paper highlights a significant research gap and emphasizes the need for further sector-specific research and systematic monitoring of digital maturity to support informed strategic decision-making.

Keywords: Digital Transformation · Digital Technology · Manufacturing · Digitalization

1 Introduction

Nowadays, technological development is rapid and dynamic. Each day, new IT solutions that replace the old and used ones, are introduced. As a result, the need to implement advanced technologies in companies is increasing. Currently, digital transformation is becoming a mandatory component of enterprise development rather than a stage to be considered [1]. Digital transformation is associated with introduction of innovative approaches that differ from the traditional ones. It relates to manufacturing and business models in production processes, as well as management supported by digital technologies [2]. The implementation of fast-growing solutions, such as artificial intelligence or Big Data analytics, provides numerous growth opportunities for enterprises. However, these solutions are also associated with various change-related challenges [2]. By

B. Gapiński et al. (Eds.): MANUFACTURING 2026, LNME, pp. 361–375, 2026.
https://doi.org/10.1007/978-3-032-22943-4_25

implementing eco-friendly production process, enterprises can achieve significant economic and environmental benefits [1]. Advanced technologies make it possible to reduce costs while maintaining high quality and efficiency. They also enable the introduction of new products and manufacturing services that are more competitive and attractive for consumers. As a result, companies become more recognizable and achieve a stronger position in the market [3]. Nevertheless, the outcomes are not limited solely to the benefits of digital transformation and implementation of advanced technologies. Enterprises also face numerous challenges and barriers related to these processes. One of the most important aspects of this area is that digital transformation involves the adoption of advanced technologies and digitization of processes. It affects the entire enterprise strategy [4] and requires adjustments across all business model elements and strategic objectives. Traditional solutions are no longer sufficient to withstand the ongoing technological revolution [5]. A particularly significant aspect of digital transformation is the high level of investment risk. The risk is related to the need for appropriate competencies required to implement organizational changes. In addition, digital transformation demands continuous learning from both failures and successes, as well as the development of effective risk management systems [6]. Publicly available data based on governmental research indicate that many industries experience significant failure during the digital transformation initiatives. Findings from 2022 suggest that only one in eight companies achieves satisfactory results with that implementations [7]. These observations indicate that similar projects may be ineffective and may lead to improper resource allocation, thereby increasing overall risk [8]. The advantages and disadvantages of digital transformation demonstrate that, although it has become necessity in modern enterprises, it is strongly associated with high risk of making incorrect decisions regarding both technology selection and human resources.

Digital transformation is closely related to Industry 4.0, which is commonly described as the Fourth Industrial Revolution. It represents a concept of enterprise transformation encompassing both technology and all organizational elements. Industry 4.0 relies on advanced technologies, such as Big Data, IT systems integration, cloud computing, augmented reality (AR), the Internet of Things (IoT), additive manufacturing (3D printing), autonomous robots, simulations and cybersecurity [9]. "Its aim is a more efficient, flexible and individual production, achieved through decentralized controls of production and completely digitally controlled or even self-organized value chains, and where automation, real-time and sensor technologies play a crucial role" [10]. Industry 4.0 also includes elements and activities that significantly influence the labor market and education systems [11], making it particularly important for manufacturing enterprises. The implementation of advanced technologies within the Industry 4.0 framework offers numerous benefits. It primarily enables cost reduction, improves production efficiency, and extends product lifespans. Moreover, it does not limit the enterprises to mass production, as it also supports single-unit and customized manufacturing [12]. Today, manufacturing business activity is highly dynamic, driven by constantly changing and unpredictable customer needs, as well as the necessity to continuously respond to market changes related to competition and technological development. Therefore, enterprises should implement innovative solutions and advanced technologies enabled by digital transformation and Industry 4.0 [13].

From the perspective of manufacturing enterprise, the most crucial objective is to optimize all stages of the manufacturing process to achieve the highest possible effectiveness. These stages should operate in a coordinated sequence to enable cost reduction and process shortening [14], while maintaining high quality and efficiency. At present, the most commonly used advanced technology in manufacturing process are production line automation and robotization. These technologies help eliminate human errors and accelerate production processes. In addition, simulation tools and artificial intelligence (AI) are increasingly being applied to support process optimization. Artificial intelligence (AI) and simulation tools are increasingly applied to support process optimization. They are also used to generate predictive reports, which are highly important in business decision-making. In manufacturing, these technologies are applied in vision systems that enable robot movement and response [15]. Currently, the level of digitalization in Poland remains relatively low, as confirmed by governmental statistics, industry reports and scientific studies. According to the available data, 33% of the surveyed, SME companies are not interested in implementing advanced technologies [16]. Although, the national economy recognizes the necessity of introducing digitalization into enterprise operations. However, it is a long-term and demanding process.

The main objective of this study was to assess and analyze the overall digital transformation landscape in Poland and to identify key barriers and challenges. A significant lack of research and data concerning the implementation of modern technologies in manufacturing enterprises can be observed. This gap may pose a challenge for companies planning to introduce digital transformation initiatives, as they lack sufficient data to support informed business decisions

2 Literature Review

To properly analyze the digital transformation of companies, it is necessary to clarify the concepts of digital transformation, digitization and digitalization. These concepts are often used interchangeably, despite their distinct meanings, and understanding their differences is essential. The literature provides several definitions of these terms; however, their number remains limited due to the relatively recent growth of interest in this area, particularly between 2014 and 2015 [17].

At the core of digital transformation lies the distinction between digitalization and digitization. According to Argilovski, Janovski and Minovski, digitization refers to the process that enables access to digital information. In contrast, digitalization is understood as the broader process of automation through the use of advanced digital technologies [18]. From the perspective of Gartner, digitization involves converting analogue information to the digital form, whereas digitalization is defined as the use of digital technologies to implement changes in business models and to create new value propositions [19]. Other approaches describe digitization as the automation of processes using advanced IT technologies [20, 21], while digitalization focuses on leveraging digital technologies to generate new revenue streams and improve business activities and processes [22]. The literature thus presents diverse perspectives on these concepts, yet a common understanding of their core principles can be identified.

The concept of digital transformation has gained significant popularity and has been widely discussed in the literature since the early 20th century [17]. As highlighted by

Stolterman and Fors [23], the digital transformation encompasses all changes that various aspects of people's lives. As reported by Martin, digital transformation involves the use of IT and communication technologies without a strong focus on automation in the process of creating new opportunities [24]. Gartner [19], in turn, defines digital transformation as a comprehensive process of leveraging digital technologies to fundamentally transform business models, organizational structures, and value creation mechanisms. Digital transformation has also been defined as a comprehensive and holistic process in which enterprises use digital technologies to transform their operations, develop new business models, and create value propositions. This perspective emphasizes the necessity of changes across the entire organizational strategy, including customer relationship and competitive positioning, rather than focusing solely on technological aspects [25]. All of the above definitions, along with others available in literature, converge on the core idea of digital transformation as a process of change affecting business processes and business models through the use of advanced technologies. At present, digital transformation has become increasingly prominent, as enterprises have recognized the rapid pace of market development and the need to implement continuous adjustments in order to maintain their competitive positions.

Digital transformation is strongly connected with the Industry 4.0, which provides its technological and organizational foundation. Therefore, it is essential to understand this concept in the context of transformation within manufacturing enterprises. The term, Industry 4.0, was first introduced in Hanover in 2011 [26]. There is no single, universally accepted definition of Industry 4.0. It is commonly described as the Fourth Industrial Revolution and refers to the transformation of traditional manufacturing processes into digital driven production systems through the use of advanced technologies [27]. From the standpoint of various researchers [28], Industry 4.0 represents a set of advanced digital technologies that deliver new value to consumers. It is also defined as a group of concepts and technologies that enable real-time connectivity between people, machines and production systems [29]. Overall, the definitions presented emphasize the core essence of Industry 4.0, namely the application of advanced digital technologies to improve industrial activities and processes [26]. The literature identifies several key principles of Industry 4.0, including interoperability, virtualization, real-time capacity, service orientation, modularity, decentralization and system integration [26]. The Fourth Industrial Revolution is closely associated with the use of advanced technologies such as data analytics and Big Data, autonomous robots, simulations, information systems integration, cybersecurity, cloud, additive manufacturing and augmented reality [30]. Some sources also highlight the role of the Internet of Things (IoT), Artificial Intelligence [31] and digital twins. Each of these technologies has the potential to deliver significant benefits for enterprises. Advanced data analytics and Big Data technologies enable the processing of very large datasets and the identification of non-obvious correlations that may be overlooked by humans analysts [32]. These technologies support managerial decision-making by providing insights into historical trends and customer preferences. Autonomous robots are described as intelligent machines capable of sensing and scanning their environment, performing assigned tasks, and making decisions. Their application contributes to improved production line efficiency and product quality, primarily through the reduction of errors and employee workload [26]. Simulations applied in manufacturing processes enable the testing of proposed improvements and allow for predictive analyses that support management in business decision-making [9]. The primary objective of information systems integration is to connect and consolidate enterprise data across the organization. Such systems can be accessed from various locations and used by employees across the organization [9]. At the same time, the increasing number of cyberattacks and fraudulent activities has made cybersecurity one of the most critical technological areas. Due to the growing need to protect key systems and production lines from cyber threats, secure and reliable communication has become essential, along with advanced identity and access management solutions for both machines and users. he primary objective of cybersecurity systems is to ensure the protection of data and production infrastructure [9]. Cloud computing enables data to be shared and processed across multiple locations by connecting various devices to centralized storage systems [30]. Additive manufacturing, commonly referred to as 3D printing, is primarily applied in job-lot or batch production, where a certain degree of product customization is required [30]. Augmented reality plays an increasingly important role in manufacturing by supporting production processes and responding to the growing demand for faster and more efficient solutions in today's dynamic industrial environment. The use of these technologies enables real-time information sharing among employees, which enhances decision-making processes [30]. The technologies discussed above represent the most important components of Industry 4.0 and are crucial for manufacturing enterprises, as they significantly improve production processes. Among the potential organizational benefits, the most notable include increased revenues and cost reduction, as well as improved production efficiency and flexibility. Moreover, the Industry 4.0 offers numerous environmental and social benefits, such as enhanced competitiveness, more effective decision-making, improved cooperation and communication, and greater personalization as added value for the products [26].

One additional and highly significant concept in this area the digital transformation strategy. It represents a structured approach to digital transformation that relies on organizational adaptation models [33]. A digital transformation strategy marks the initial phase of the transformation process, as it defines business and strategic objectives, sets the direction, and identifies the most appropriate areas for the implementation of advanced technologies [34]. From the point of view of various researchers, a digital transformation strategy encompasses all aspects of an organization, including its structure processes, and capabilities, enabling coordination across operations at every level [35]. It addresses changes implemented by the enterprise with the aim of creating new value propositions and building a sustainable competitive advantage [5]. Initiating digital transformation requires strong managerial commitment and readiness to address potential challenges and risks that may arise during the process [36]. It is important to note that implementation of advanced technologies should not be equated with a digital transformation strategy or digital transformation itself. Such initiatives require substantial resources, as well as advanced skills and competencies [37]. The scope of activities necessary to initiate transformation is associated with numerous challenges, particularly for small and medium-sized enterprises (SMEs). One of the most significant barriers is limited financial capacity, which often prevents enterprises from undertaking transformation initiatives. Additionally, digital strategies may encounter resistance from employees due to their complexity and time-consuming nature, potentially slowing down implementation processes. A lack of specialized, technical expertise further constitutes a major obstacle [38]. Nevertheless, a well-designed digital transformation strategy represents a significant opportunity for enterprises to achieve long-term development and improvement, provided that potential challenges and limitations are appropriately addressed.

The final concept closely related to digital transformation that warrants explanation is digital maturity. It reflects the level of digitalization within an industry and indicates an enterprise's position in the context of digital transformation. Digital maturity refers to the stage at which an enterprise is positioned in the process of digital transformation, taking into account both the operational changes implemented and the organizational capabilities that enable effective navigation of this process [39, 40]. Assessing an enterprise's digital maturity remains particularly challenging. The literature does not provide a universally accepted set of dimensions or clearly defined areas that should be measured for this purpose [41]. Moreover, the rapid pace of technological development and continuous innovation can quickly affect the relevance of existing technologies; consequently, previously assessed maturity levels may become outdated over time [42].

While the literature offers extensive insights into digital transformation, including its principles and frameworks, there is a noticeable lack of specialized empirical studies focused on manufacturing enterprises. In particular, limited attention has been given to the processes of implementing digital transformation and systematically measuring digital maturity within this sector.

3 Research Methodology

The presented research on digital transformation and the digital maturity of enterprises in Poland has a review character. It is based on publicly available data, industry reports, and governmental statistics. The results primarily address SMEs (manufacturing-oriented SMEs in particular), as they constitute the majority of companies in Poland. However, the availability of data providing detailed and country-specific information on Poland is limited. The overall research methodology consisted of the three steps is presented in Fig. 1. The research began with the selection of appropriate sources to ensure their timeliness, reliability, and relevance to the topic of digital transformation. Most of the sources considered were published within the last five years to ensure the timeliness of the data; however, older sources were also included to present different perspectives and approaches. The next stage involved the analysis of selected sources, aimed at assessing the availability of such studies in Poland and identifying their scope. Step three of the above-presented methodology focused on identifying key barriers and challenges in the digital transformation of Polish companies.

The results of this study are limited by the nature of the data sources used. The analysis is based exclusively on publicly available sources, which restricts the level of detail of the findings. Consequently, it was not possible to tailor the research scope or refine the study area according to specific analytical needs. This limitation may affect the depth and granularity of the results and should be considered when interpreting the findings.

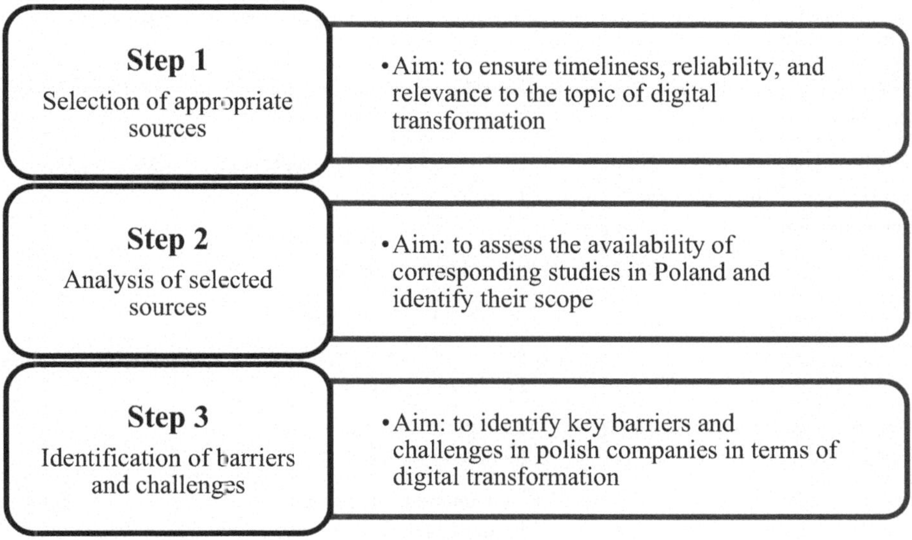

Fig. 1. Scheme presenting the overall research methodology

4 Results and Discussion

This section presents the results of the analysis of available data concerning the level of digital maturity among enterprises in Poland, with a particular focus on small and medium-sized enterprises (SMEs). The analysis addresses the level of digital maturity in Polish SMEs, the scope of digital transformation initiatives, as well as the identification of barriers and challenges related to the implementation of advanced technologies. The research focuses on assessing the availability and scope of relevant data. The data used in the analysis originate from national and European governmental sources, as well as industry reports, which provide insights into the current state of digital maturity and digital transformation in Polish enterprises. They consider a small research group, so they are not as much accurate.

The research was conducted in Poland and focused primarily on SMEs. Key findings related to the assessment of the level of digitalization in Polish companies, selected from the analyzed sources, are presented in Table 1.

SMEs constitute the vast majority of enterprises operating in Poland, accounting for approximately 99.8% of all companies in the country [16]. As a result, the available data are not disaggregated by economic sectors, such as manufacturing. According to DESI – Digital Economy and Society Index, which is measured annually by European Commission, Poland ranked 24th among the 27 Member States of the European Union in 2022. The DESI assesses four main dimensions: human capital, connectivity, integration of digital technology and digital public services. The results for each dimension indicate that Poland performs below the European Union average. This is particularly evident in the area of human capital, where the proportion of the population possessing essential digital skills remains relatively low. Furthermore, the results indicate that Poland has a relatively low number of specialists in advanced digital technologies. Compared to

Table 1. Assessment of the level of digitalization in Polish companies - key findings.

Measured/ analyzed aspect (in Poland)	Result	Year	Source
The most effective way to ensure access to specialists with the required competencies	54% - via improving employees' knowledge and skills	2020	[43]
	26% - via recruitment of highly skilled professionals		
	14% - via outsourcing		
	6% - via changes in formal employee qualifications		
Investment plans related to digital technologies	44% plan to maintain their current level of digitalization		
	31% intend to scale up the implementation of digital technologies		
	25% prefer to limit further technological investments		
Planned changes in business models	56% do not intend to introduce new business model		
	38% plan to implement changes		
Digital Enterprise Index (average)	1,8 out of 4	2021	[44]
How do companies assess the level of production digitalization in their firms?	80% or more – 6,0%; 60–79% – 14,0%; 40–59% – 16,7%; 20–39% – 30,0%; 19% or less – 26,7%; don't know / hard to say – 6,6%		
Digital Economy and Society Index (DESI)	Ranking: 24 out of 27 EU member states	2022	[45]
Amount of households with permanent internet access	70%		
Percentage of enterprises using cloud computing in their operational processes	19%		
Percentage of enterprises reporting the use of electronic information exchange	32%		
Adoption rates of cloud computing, big data, and artificial intelligence	3% to 19%		

(*continued*)

Table 1. *(continued)*

Measured/ analyzed aspect (in Poland)	Result	Year	Source
Digital Intensity Index (DII)	Ranking: 21 out of 27 EU member states	2022	[46]
Percentage of surveyed enterprises classified as traditional companies	73%		
Percentage of surveyed enterprises adopting digital technologies	16%		
SMEs grouped by level of digitalization to highlight differences in digital maturity	Uninterested (33%), Perceiving no benefits (5%), Ready to act (13%), Ambitious but lacking knowledge (17%), Requiring support (20%), Digital champions (9%)	2024	[16]
Percentage of surveyed companies using websites	68%		
Percentage of companies utilizing online business cards	65%		
Percentage of companies engaging in social media activities	52%		
Percentage of companies using online sales channels to offer services	26%		

previous years, an increase in the number of households with permanent internet access can be observed – 70% compared to 65% [45]. However, this result remains below the EU average. Another relevant indicator is the Digital Intensity Index (DII), which assesses the level of digitalization of enterprises, particularly in areas such as artificial intelligence, Big Data, electronic invoicing, and cloud computing. As indicated by DII, Poland was ranked 21st among the 27 Member in EU in 2022. The data indicate that 73% of surveyed enterprises are classified as traditional companies, while 16% use selected digital technologies. Only 4% of enterprises apply more advanced and innovative digital solutions [46].

The Digi Index report by Siemens provides insights into the current level of digitization among Polish enterprises, indicating an average Digital Enterprise Index score of 1.8 out of 4 [44]. Additionally, the European project *"Advancing the Digital Transformation of Polish Enterprises"* [16] analyzes the degree of digitalization in Poland in order to identify the reasons for its relatively low level. As part of this analysis, SMEs are grouped according to their level of digitalization to highlight differences in digital maturity. There are six groups with results: Uninterested (33%), Perceiving no benefits (5%), Ready to act (13%), Ambitious but lacking knowledge (17%), Requiring support (20%), Digital champions (9%). A comparison of the results indicates that nearly 40% of enterprises are not willing to consider the implementation of advanced technologies, while only 9% can be classified as digitally mature. In contrast, over 50% of companies plan to implement digital technologies or express a need for support in this area. These findings help explain the relatively low overall level of digitalization observed in Poland. Research conducted within the European project includes an analysis of the types of technologies used by enterprises. The results show that most companies rely primarily on basic digital tools, such as computers (98%), smartphones (89%) and business e-mail systems (96%). However, as the level of technological sophistication increases, the proportion of enterprises adopting such solutions declines. Only 68% of surveyed companies use websites, 65% utilize online business card, 52% engage in social media activities, and 26% use online sales channels to offer services [16]. This trend is further confirmed by DESI data, which indicate limited adoption of advanced digital technologies among SMEs in Poland. For example, only 19% of Polish enterprises use cloud computing in their operational processes. Additionally, 32% of enterprises report the use of electronic information exchange. The adoption rates of cloud computing, Big Data and Artificial Intelligence range between 3% and 19% [45]. Overall, these findings confirm the low level of digital maturity among Polish enterprises.

In 2020, a study was conducted to examine organizational approaches to digital transformation. To obtain insights from enterprises, researchers surveyed digital experts and individuals with significant influence on digital transformation initiatives. The questions focused on the implementation of digital transformation strategies and companies' future plans in this area [43]. The results highlight the importance of developing employee skills in the context of digital transformation. Among the surveyed companies, more than half (54%) indicated that the most effective way to ensure access to specialists with the required competencies is through improving employees' knowledge and skills. Additionally, 26% of respondents emphasized the recruitment of highly skilled professionals, while 14% pointed to outsourcing as a viable solution. Only 6% of enterprises indicated changes in formal employee qualifications as a key approach. These findings suggest a growing awareness of the importance of human-oriented digitalization. Focusing on employee development plays a crucial role in supporting digital transformation and has a positive impact on workforce productivity. The 2020 study also examined enterprises' investment plans related to digital technologies. The results indicate that 44% of surveyed companies plan to maintain their current level of digitalization over the next twelve months. In contrast, 31% intend to scale up the implementation of digital technologies, while 25% prefer to limit further technological investments. These findings suggest that the level of digital maturity in Poland is gradually increasing. It is important to emphasize that a digital transformation strategy encompasses all aspects of an organization, rather than focusing solely on technology. For this reason, respondents were also asked about planned changes in business models. The results show that 56% of enterprises do not intend to introduce new business model solutions, whereas 38% plan to implement such changes.

The project-based research – according to the step 3 of the research methodology – further enables the identification of barriers and challenges that may hinder the implementation of digital transformation initiatives. In total, more than 20 distinct barriers were identified at this stage of the analysis. The identified barriers to digital transformation include the following [16]: a lack of awareness of areas that could be improved (7%), insufficient knowledge of available solutions (6%), limited time to explore digital opportunities (8%), cooperation with partners that do not use modern solutions (4%), a lack of perceived need to improve customer service (9%), limited interest from company owners (6%), low confidence in the success of transformation initiatives (8%), inadequate infrastructure (5%), excessive financial investment requirements (14%), the absence of a digital strategy (7%), technological complexity (5%), insufficient employee competencies (7%), fear of change (8%), security concerns (4%), and low employee skill levels (6%). The most significant barrier identified is the lack of perceived need for digital transformation within enterprises, indicated by 80% of respondents. This finding suggests a low level of awareness regarding the benefits of digital transformation and the necessity of continuous development in order to maintain a competitive position in the market. According to Kowal et al., the main obstacles to digitalization include a lack of financial support (34%), insufficient knowledge on how to develop a plan, strategy, and business goals (16%), difficulties in integrating systems from different suppliers (16%), and a lack of knowledge regarding the use of collected data (16%) [47].

In summary, the analysis of data from multiple sources indicates that the level of digital maturity in Poland remains relatively low, particularly with regard to the adoption of advanced digital technologies. This situation results largely from the presence of numerous barriers that hinder the initiation and implementation of digital transformation initiatives. Research findings indicate that the barriers to the digital transformation of Polish enterprises are primarily organizational and mental, not solely technological. The most significant problem is the low level of awareness of the need for digitalization, which significantly limits companies' willingness to undertake transformational activities. Furthermore, financial factors, the lack of a coherent digital strategy, and insufficient employee competencies remain significant constraints, confirming the need for systemic support for enterprises in planning and implementing digital processes. The analysis also reveals limitations in the available data, which are often insufficiently detailed to allow for an accurate assessment of digital maturity across all economic sectors, including manufacturing. Consequently, there is a clear need for further research and continuous monitoring of digital transformation processes in order to obtain more precise and sector-specific insights.

5 Conclusions

In the context of rapid technological development, digital technologies have become a necessity for enterprises seeking to remain competitive. There is a growing awareness of the importance of enhancing digital maturity within organizations. The aim of this research was to present the level of digitalization and digital transformation among enterprises in Poland based on available data, as well as to assess the availability of relevant studies in this area and identify key barriers and challenges in terms of digitalization. The level of digital maturity among Polish enterprises remains low, as the results of all analyzed indices fall below the European Union average. Consequently, there is a clear need for continuous development and the implementation of advanced digital technologies. The analysis indicates that, although some enterprises recognize the necessity of undertaking digital transformation, such initiatives involve multiple barriers and challenges that many organizations are unable to effectively address. These barriers occur across various organizational processes, significantly hindering transformation efforts. At the same time, a substantial number of enterprises remain uninterested in implementing digitalization initiatives or advanced technologies, as they do not perceive clear benefits or an immediate need for change. This lack of engagement negatively affects national performance in terms of economic digitalization and efforts aimed at increasing enterprise-level digital maturity. A key factor in successful transformation is the presence of a well-defined digital transformation strategy, which remains insufficiently developed or overlooked in many organizations.

A significant challenge in defining and presenting clear findings on digital maturity in Poland lies in the strong generalization of existing research and its predominant focus on the SME sector. Although SMEs constitute the majority of active enterprises in Poland, there is a clear need to structure research according to economic sectors.

Comprehensive and detailed data on digital transformation in manufacturing enterprises remain scarce, which results in limited knowledge in this area. This constitutes a serious problem, particularly for the economy and the manufacturing sector, as enterprises are often forced to pursue digital transformation initiatives without adequate guidance or evidence-based recommendations. In the absence of sector-specific data, it is difficult to effectively plan digital transformation strategies, as there is no reliable basis for comparison with similar cases. Consequently, the lack of comprehensive data significantly constrains the development and implementation of effective digital transformation initiatives. This conclusion in itself constitutes an important outcome of the preliminary research conducted within the framework of this article, namely that studies of this type are needed and that further research in this direction is planned by the authors, including studies based on empirical data and/or case studies.

This situation represents a significant challenge, particularly for the manufacturing sector, as organizations are often required to operate independently, without adequate institutional support or clear guidance. Therefore, the implementation of systematic monitoring and repeated empirical research is strongly recommended. The analysis demonstrates that rapid technological development can quickly shift digital maturity thresholds, highlighting the need for regular and cyclical evaluation. Expanding research on digital maturity may play a critical role in increasing both the scope and effectiveness of digital transformation initiatives in Polish enterprises.

References

1. Wang, W., Cao, Q., Li, Z., Zhu, J.: Digital transformation and corporate environmental performance. Finance Res. Lett. **76**, 106936 (2025). https://doi.org/10.1016/j.frl.2025.106936
2. Nahum, N., Larsson-Olaison, U., Uman, T., Achtenhagen, L.: Corporate governance for digital transformation: the role of ownership and the board of directors. Technol. Forecast. Soc. Change. **223**, 124453 (2026). https://doi.org/10.1016/j.techfore.2025.124453
3. Liu, J., Xian, X.: Impact of digital transformation on high-quality economic development: the mediating role of human capital investment. Finance Res. Lett. **86**, 108657 (2025). https://doi.org/10.1016/j.frl.2025.108657

4. Warner, K.S.R., Wäger, M.: Building dynamic capabilities for digital transformation: an ongoing process of strategic renewal. Long Range Plan. **52**, 326–349 (2019). https://doi.org/10.1016/j.lrp.2018.12.001
5. Kringelum, L.B., Holm, C.G., Holmgren, J., Friis, O., Jensen, K.F.: Digital transformation: strategy comes first to lay the groundwork. J. Bus. Strateg. **46**, 3–10 (2025)
6. Cavalcante, M., Varajão, J., Silva Rodrigues, L.: Digital transformation initiatives: motivations, objectives, and strategies. Telemat. Inform. Rep. **19**, 100246 (2025). https://doi.org/10.1016/j.teler.2025.100246
7. Karabag, S.F., Simonsson, J., Berggren, C., Andreasson, M., Eriksson, R.: Digital transformation as a multi-phase process: a longitudinal study of corporate strategy and business unit adaptation. J. Bus. Res. **202**, 115796 (2026). https://doi.org/10.1016/j.jbusres.2025.115796
8. Cao, Z., Tang, C., Li, Z.: The impact of digital transformation on firm risk: effort by talk or effort by walk. Pac.-Basin Finance J. **94**, 102971 (2025). https://doi.org/10.1016/j.pacfin.2025.102971
9. Klasa, P.: Industry 4.0: digitalization of the economy and society. In: Zakrzewska-Półtorak, A. (ed.) Oddziaływanie megatrendów na gospodarkę i społeczeństwo, Debiuty studenckie - Uniwersytet Ekonomiczny we Wrocławiu, pp. 22–36. Wydawnictwo Uniwersytetu Ekonomicznego we Wrocławiu, Wrocław (2024)
10. Priffti, L., Knigge, M., Kienegger, H., Kremar, H.: A competency model for "industrie 4.0" employees. In: Presented at the Proceedings der 13. Internationalen Tagung Wirtschaftsinformatik, St. Gallen (2017)
11. Poszytek, P., Lis, M., Jefmański, B., Fila, J., Jeżowski, M., Kotelska, J.: Digital Transformation of Enterprises in the Era of Industry 4.0. Akademia WSB, Dąbrowa Górnicza (2024)
12. Götz, M.: Industry 4.0 and International Economic Cooperation. (2018)
13. Yang, Y., Yee, R.W.Y., Lee, P.K.C., han Zhu, J.: When digital technologies meet lean production: a socio-technical system perspective for operational efficiency and innovativeness. Inf. Manag. **63**, 104238 (2026). https://doi.org/10.1016/j.im.2025.104238
14. Wicej-Janka, E., Pawlicki, J., Walkowski, P.: An example of process improvement implementation in production processes: a case study. Zesz. Nauk. Politech. Pozn., 1–12 (2018). https://doi.org/10.21008/j.0239-9415.2018.076.20
15. Kaczmarek, W., Panasiuk, J.: Robotization of Production Processes. Wydawnictwo Naukowe PWN (2017)
16. Digital transformation.: Ministry of Development and Technology. https://www.gov.pl/web/rozwoj-technologia/transformacja-cyfrowa. Last accessed 7 Dec 2025
17. Kim, S., Choi, B., Lew, Y.K.: Where is the age of digitalization heading? The meaning, characteristics, and implications of contemporary digital transformation. Sustainability. **13** (2021). https://doi.org/10.3390/su13168909
18. Argilovski, A., Jovanoski, B., Minovski, R.: An overview of the digital transformation and industry 4.0 technologies implementation frameworks. In: Presented at the IS'23 - 19th International Scientific, Novi Sad, Serbia (2023). https://doi.org/10.24867/IS-2023-T1.1-7_03841
19. Gartner.: In IT Glossary. https://www.gartner.com/en/information-technology/glossary?glossaryletter=D. Last accessed 8 Dec 2025
20. Horváth, D., Szabó, R.Z.: Driving forces and barriers of industry 4.0: do multinational and small and medium-sized companies have equal opportunities? Technol. Forecast. Soc. Change. **146**, 119–132 (2019). https://doi.org/10.1016/j.techfore.2019.05.021
21. Hess, T., Benlian, A., Matt, C., Wiesböck, F.: Options for formulating a digital transformation strategy. MIS Q. Exec. **15**, 123–139 (2016)
22. Kraus, S., Durst, S., Ferreira, J.J., Veiga, P., Kailer, N., Weinmann, A.: Digital transformation in business and management research: an overview of the current status quo. Int. J. Inf. Manag. **63**, 102466 (2022). https://doi.org/10.1016/j.ijinfomgt.2021.102466

23. Stolterman, E., Fors, A.C.: Information systems research: relevant theory and informed practice. In: Information Technology and the Good Life. Kluwer Academic Publishers, London (2004)
24. Martin, A.: Digital literacy and the "digital society". In: Digital Literacies: Concepts, Policies, and Practices. Peter Lang Publishing, New York (2008)
25. Westerman, G., Bonnet, D., McAfee, A.: Leading Digital: Turning Technology into Business Transformation. Harvard Business Review Press, Boston (2014)
26. Elnadi, M., Abdallah, Y.O.: Industry 4.0: critical investigations and synthesis of key findings. Manag. Rev. Q., 1–34 (2023). https://doi.org/10.1007/s11301-022-00314-4
27. Ahmmed, M.S., Isanaka, S.P., Liou, F.: Promoting synergies to improve manufacturing efficiency in industrial material processing: a systematic review of industry 4.0 and AI. Machines. **12** (2024). https://doi.org/10.3390/machines12100681
28. Pacchini, A.P.T., Lucato, W.C., Facchini, F., Mummolo, G.: The degree of readiness for the implementation of industry 4.0. Comput. Ind. **113**, 103–125 (2019). https://doi.org/10.1016/j.compind.2019.103125
29. Pagliosa, M., Tortorella, G., Ferreira, J.: Industry 4.0 and lean manufacturing: a systematic literature review and future research directions. J. Manuf. Technol. Manag. (2021). https://doi.org/10.1108/JMTM-12-2018-0446
30. Saurabh, V., Prashant, A., Santosh, B.: Industry 4.0 – a glimpse. In: Presented at the 2nd International Conference on Materials Manufacturing and Design Engineering, India (2018)
31. Meindl, B., Ayala, N.F., Mendonça, J., Frank, A.G.: The four smarts of industry 4.0: evolution of ten years of research and future perspectives. Technol. Forecast. Soc. Change. **168**, 120784 (2021). https://doi.org/10.1016/j.techfore.2021.120784
32. Jędrzejczyk, W.: Industry 4.0 as a source of innovative change in industrial enterprises: the Polish context. In: Innowacje w dobie technologii IT: obszary, koncepcje, narzędzia. Oficyna Wydawnicza Politechniki Wrocławskiej, Wrocław (2020)
33. Chanias, S., Myers, M.D., Hess, T.: Digital transformation strategy making in pre-digital organizations: the case of a financial services provider. J. Strateg. Inf. Syst. **28**, 17–33 (2019)
34. Bharadwaj, A., El Sawy, O.A., Pavlou, P.A., Venkatraman, N.: Digital business strategy: toward a next generation of insights. MIS Q. Spec. Issue Digit. Bus. Strategy. **37**, 471–482 (2013)
35. Saeed, A.: Developing digital transformation strategy for manufacturing. Procedia Comput. Sci. **170**, 664–671 (2020)
36. Tekic, Z., Koroteev, D.: From disruptively digital to proudly analog: a holistic typology of digital transformation strategies. Bus. Horiz. **62**, 683–693 (2019)
37. Cyfert, S., Chwilkowska-Kubala, A., Malewska, K., Chomicki, M.: Reframing the digital strategy–organisational culture relationship: a contingency-theoretic study of digital transformation in Polish energy SMEs. J. Strateg. Manag. **18**, 849–872 (2025)
38. Omowole, B.M., Olufemi-Phillips, A.Q., Ofodile, O.C., Eyo-Udo, N.L., Ewim, S.E.: Barriers and drivers of digital transformation in SMEs: a conceptual analysis. Int. J. Sch. Res. Sci. Technol. **5**, 19–36 (2024)
39. Thordsen, T., Bick, M.: A decade of digital maturity models: much ado about nothing? Inf. Syst. E-Bus. Manag. **21**, 947–976 (2023). https://doi.org/10.1007/s10257-023-00656-w
40. Waara, Å.: Examining digital government maturity models: evaluating the inclusion of citizens. Adm. Sci. **15** (2025). https://doi.org/10.3390/admsci15030073
41. Van Tonder, C., Bossink, B., Schachtebeck, C., Nieuwenhuizen, C.: Key dimensions that measure the digital maturity levels of small and medium-sized enterprises (SMEs). J. Technol. Manag. Innov. **19**, 110–130 (2024)
42. Bąk, S., Jedynak, P.: Assessment of enterprise digital maturity. Organ. Kier., 69–86 (2024)
43. PricewaterhouseCoopers.: 13 Facts About Digital Transformation. PwC Report. https://www.pwc.pl/pl/publikacje/13-faktow-o-transformacji-cyfrowej.html. Last accessed 9 Dec 2025

44. Digiindex.: https://www.siemens.com/pl/pl/o-firmie/raporty-siemens/digi-index-2021.html (2021). Last accessed 17 Dec 2025
45. Poland in the Digital Economy and Society Index (DESI). https://digital-strategy.ec.europa.eu/pl/policies/desi-poland. Last accessed 9 Dec 2025
46. The Digital Condition of Polish Enterprises and the Economy: Report Summary. https://pfr.pl/artykul/jaka-jest-cyfrowa-kondycja-polskich-firm-i-gospodarki-podsumowanie-raportu. Last accessed 9 Dec 2025
47. Kowal, D., Radzik, M., Domaracká, L.: Assessment of the level of digitalization of Polish enterprises in the context of the fourth industrial revolution. Sustainability **16** (2024). https://doi.org/10.3390/su16135718

Physics-Informed Neural Networks for Anomaly Detection: A Mechanical Proof-of-Concept Towards IIoT-Enabled Mechatronic Systems

Michał Kobielski[2] and Andrzej Loska[1(✉)]

[1] Silesian University of Technology, Faculty of Mechanical Engineering, 18a Konarskiego st, 44-100 Gliwice, Poland
andrzej.loska@polsl.pl

[2] APA Sp. z o.o, 251 Tarnogórska st, 44-100 Gliwice, Poland

Abstract. This paper investigates the use of physics-informed neural networks (PINNs) for anomaly detection in IIoT-enabled mechatronic systems. Classical data-driven diagnostic approaches based on artificial neural networks (ANNs) often require large amounts of labelled data and may not explicitly enforce the underlying physical laws, which limits their interpretability and robustness in complex industrial environments. In contrast, PINNs embed prior knowledge in the form of governing equations and boundary conditions directly into the learning process. As a proof-of-concept study, we consider a physical pendulum with multiple damage scenarios modelled as changes in geometry and moment of inertia. Synthetic time series of angular position are generated from the analytical model and used to train a PINN that enforces the nonlinear equation of motion and energy conservation. The trained network acts as a physics-based filter and supports the construction of a dictionary of damage states, enabling the assignment of observed trajectories to candidate models. We discuss methodological aspects of the approach, including the definition of the loss function, the role of normalisation and the sensitivity to hyperparameters. Finally, we outline how analogous PINN-based diagnostic methods can be developed for selected subsystems of the CTTP4.0 industrial testbed, which integrates mechatronic transport and manipulation units with an Industrial Internet of Things platform.

Keywords: Physics-Informed Neural Networks (PINN) · Failure Detection · Industrial Internet Of Things (Iiot)

1 Introduction

The transformation of industry towards the concepts of Industry 4.0 and the Industrial Internet of Things (IIoT) is leading to the increasing integration of machines, control systems, and analytical platforms into coherent, networked production environments [1–3]. A key element of this transformation is the ability to continuously monitor the condition of devices and processes, particularly systems responsible for transport, manipulation, and the implementation of technological operations. In such conditions, the

B. Gapiński et al. (Eds.): MANUFACTURING 2026, LNME, pp. 376–389, 2026.
https://doi.org/10.1007/978-3-032-22943-4_26

importance of diagnostic systems is growing, as they can detect symptoms of irregularities early, supporting maintenance, maintenance planning, and reducing unplanned downtime [4–6].

In recent years, classical diagnostic methods based on signal analysis, mathematical models, and machine learning techniques have been extensively researched and developed. Deep neural networks have gained particular importance, used, among others, for operating state classification, fault detection, and component remaining life prediction. Despite many successful implementations in the IIoT environment, these approaches often require large sets of labeled data, and their models remain largely "black boxes," not explicitly reflecting the physical laws governing the systems under consideration. This limits the interpretability of results and hinders the transfer of trained models between different machine configurations [7–9].

One direction of diagnostics development is the integration of data models with physical knowledge, achieved through, among other methods, Physics-Informed Neural Networks (PINNs). In this approach, equations of motion, boundary conditions, and other physical relationships are incorporated directly into the neural network's loss function, allowing for the training of models that, by definition, respect the structural properties of the system being described. This approach opens the possibility of developing diagnostic solutions that, while retaining the flexibility of deep learning methods, better reflect the dynamics of the systems under consideration and are more resilient to interference and limited data sets.

For example, with a short and noisy measurement of a vibrating system, it can be used to determine the entire waveform so that, on the one hand, it passes close to the measurement points, and on the other, its curvature and energy are consistent with the laws of dynamics. In effect, PINN acts as a physical filter: it rejects nonphysical trajectories (e.g., those that violate the conservation of energy or the equations of motion) and retains those that are both consistent with the data and the domain model. This approach opens the possibility of developing diagnostic solutions that—while retaining the flexibility of deep learning methods—better reflect the dynamics of the systems under consideration and are more resistant to interference and limited data sets.

In parallel, industrial IIoT platforms are being developed (e.g., the Polish Nazca 4.0 platform), which combine condition monitoring, data acquisition, and real-time analytics. Their ecosystem often includes research and demonstration stations, such as CTTP4.0 – the Industrial Technology Testing Center 4.0, established by APA at the Silesian University of Technology in Gliwice. It integrates typical transport and manipulation systems with sensor infrastructure, industrial communication, and an analytical layer [10]. In this context, diagnostic methods are needed that simultaneously utilize physical knowledge about the object and can be directly embedded in the IIoT architecture and scaled between different device configurations.

The aim of this article is to explore the potential of PINN networks to identify anomalies in transport and manipulation systems, in the context of their future integration with industrial IIoT platforms. A controlled, well-understood physical pendulum is used as a test object, for which a series of "damage" scenarios modeled as changes in geometry and moment of inertia are considered. Based on this, a dictionary of damage states is constructed, and the PINN network acts as a physical filter that adapts motion trajectories to a nonlinear equation and the energy conservation condition. In this POC, we assess the extent to which the PINN approach enables reliable reconstruction of the pendulum trajectory and the differentiation of variants m1–m4 with limited data. The case study on the pendulum serves as a mechanical proof-of-concept, and its results provide a basis for designing analogous diagnostic methods for selected CTTP4.0 station systems.

2 Overview of Solutions for the Use of PINN Networks in Machine Diagnostics

In this paper, machine exploitation anomalies are defined as any deviations from proper operation, resulting from both catastrophic failures (e.g., sudden breakage of a structural element or power cable) and parametric failures (e.g., gradual wear, corrosion, or misalignment of drive components) [11–13]. The task of identifying anomalies is broadly defined as a sequence of interconnected diagnostic steps: detecting a deviation from the reference state, locating it within the object's structure, identifying the most probable cause, and predicting the further evolution of the object's state over time [14]. This approach is particularly important in the case of transport and handling systems, where an anomaly can lead to a production line stop, a reduction in product quality, or a threat to operator safety.

The literature analysis indicates that despite numerous works on the application of machine learning methods in machine diagnostics and the development of industrial IIoT platforms [2, 10], there is no coherent solution aimed at identifying anomalies in complex systems related to transport and manipulation, taking into account both the dynamics aspects and integration with communication infrastructure and supervision systems. This gap is the direct motivation for the research undertaken in this work. Industrial experience and a literature review also indicate the limitations of data-based approaches that hinder the interpretability and reliability of predictions [5, 8]. For this reason, methods combining data with domain knowledge about the system dynamics (e.g., PINN) are considered as a natural direction for the development of diagnostics for target mechatronic systems.

Applications of PINNs in machine diagnostics and condition monitoring are primarily focused on systems with well-defined dynamics, particularly rotating elements. These studies demonstrate that incorporating equations of motion and physical conditions into the learning process improves model consistency with real-world dynamics and robustness to limited and noisy measurement data [15, 16]. For example, in rotating shaft monitoring, PINNs are used to estimate state variables and parameters, supporting the detection of deviations indicative of damage. In parallel, solutions combining PINNs with the concept of digital twins (PHM) are being developed, where the physical model and operational data co-create a coherent picture of the object's condition and predict its degradation [17].

In bearing and drive diagnostics, the "physics + features" trend is observed, where physics-informed components amplify important diagnostic signals and limit the impact of interference. PINN and related methods are used to weight features and reconstruct waveforms consistent with the laws of motion, which increases the reliability of fault detection in industrial environments [15].

In robotics, PINNs are used for both modeling and control of systems with flexible links and complex nonlinear dynamics. They are used, among others, for input waveform shaping and vibration damping in flexible manipulations (single-link/flexible link), where enforcing physical constraints stabilizes the system's behavior [9], and for inverse dynamics of manipulators, where a physics-informed component helps maintain compliance with the equations of motion and reduces susceptibility to overfitting [10].

Among the current possibilities and solutions, three advantages of physics-informed approaches in diagnostics and robotics can be distinguished:

- Consistency of predictions with the dynamics of the system (respecting the equations of motion, boundary conditions) [15, 18, 19].
- Greater resistance to limited data sets and noise (regularization by physical knowledge) [15–17].
- Better interpretability and transferability between configurations (physical parameters as a knowledge carrier) [17, 18].

Bibliographic analysis indicates that despite numerous works on the application of machine learning methods in machine diagnostics and the development of industrial IIoT platforms, there is no coherent solution aimed at identifying anomalies in complex systems related to transport and manipulation, taking into account both the dynamics aspects and integration with communication infrastructure and supervision systems. This gap is the direct motivation for the research undertaken in this work. Industrial experience and a literature review also indicate the limitations of data-based approaches that hinder the interpretability and reliability of predictions [5, 20]. For this reason, methods combining data with domain knowledge about the system dynamics (e.g., PINN) are considered as a natural direction for the development of diagnostics for target mechatronic systems.

3 The Concept of the PINN Network Application Method for Anomaly Detection

The proposed method involves physically based reconstruction of signal trajectories and diagnostic inference using quantities consistent with the system's dynamics.

The input is a time series of measurements (and optionally controls), and the output is the reconstructed waveforms and, optionally, estimated model parameters. We assess quality simultaneously with respect to the data and the "residual" of physical equations (and, where appropriate, invariants, e.g., energy). We formulate an anomaly decision by comparing the reconstruction/parameters with a nominal reference and assessing compliance with the laws of physics.

In practical terms, we reconstruct a directly diagnostically relevant quantity (e.g., angle, velocity, or other observable), while we obtain the measurement y(t) from the sensors. The C operator maps the network output into the measurement space (selecting the appropriate component, rescaling the units). In the simplest case, the reconstructed quantity is identical to the measured quantity.

The diagram of the proposed method is presented in Fig. 1.

The diagram synthesizes this flow: input data → PINN reconstruction (residuals: Physics, IC/BC, Data, optionally Invariants/Energy) → feature determination (differences and residual measures) → decision rule.

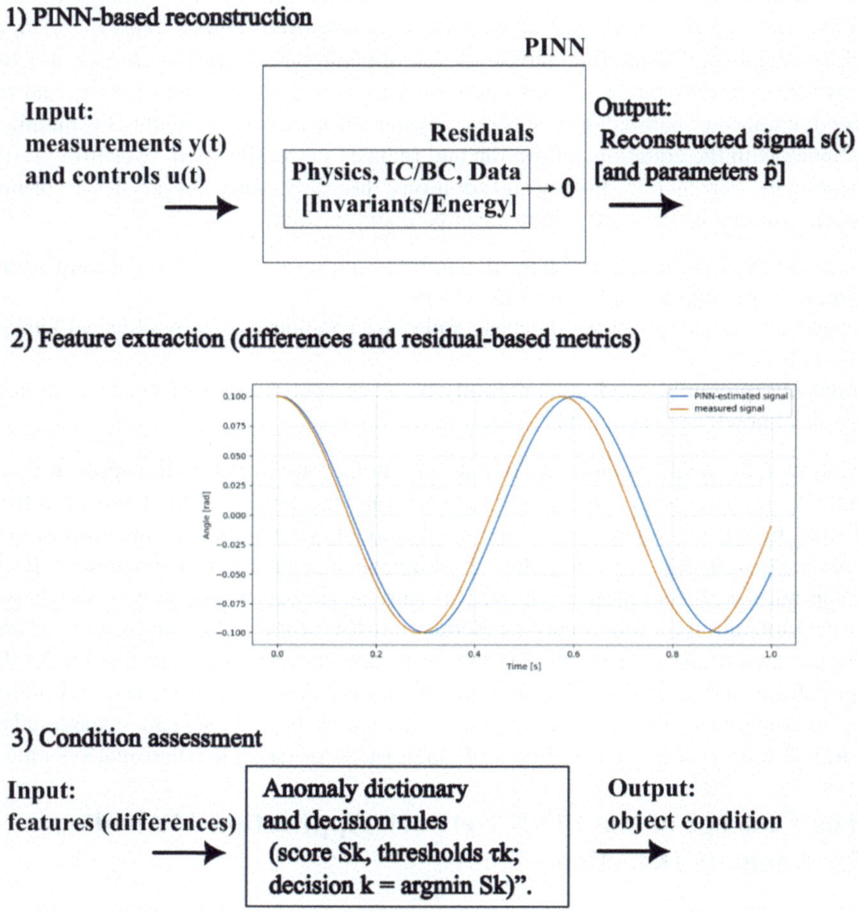

Fig. 1. The assumptions and elements of the proposed method.

To interpret the sequence of actions shown in Fig. 1, we first obtain the time series of observations y(t) and, if available, the forcing u(t). We normalize and time-synchronize the data set to ensure consistency of the model input. The PINN network then reconstructs the curves of diagnostically relevant quantities s_θ (t) in a manner consistent with the system dynamics. The cost function simultaneously enforces three types of consistency: with the laws of physics (residue of equations), with the initial and boundary conditions (IC/BC), and with the data (Data). Where necessary, we also include invariants, such as energy. This ensures that the reconstruction preserves the physical properties, and the features calculated from s_θ (t) are more stable than those calculated from the raw signals. In the final step, we determine the features and compare them with the state dictionary. One example feature is the relative period change ΔT/T. Based on the feature vector, we calculate the matching score S_k for k variants and make a decision using thresholds τ_k. In industrial applications, we combine several features and report the decision uncertainty.

In practical terms, we reconstruct a directly diagnostically relevant quantity (e.g., angle, velocity, or other observable), while we obtain the measurement y(t) from the sensors. The C operator maps the network output into the measurement space (selecting the appropriate component, rescaling the units). In the simplest case, the reconstructed quantity is identical to the measured quantity. PINN "filters" solutions to trajectories that obey the equations of motion (and invariants), so features computed from reconstruction are more stable and less susceptible to noise than those computed directly from the raw data.

In the general formulation, the state of the system $x(t) \in R^n$ fulfills the ordinary equation:

$$\dot{x}(t) = f\left[x(t), u(t), p\right], \quad x(0) = x_0 \tag{1}$$

where u(t) denotes the forces/controls and p groups the parameters (including variants of the technical condition). We measure the following characteristics:

$$y(t) = h\left[x(t), p\right] + \varepsilon(t) \tag{2}$$

which are a function of the state and parameters with the addition of noise. In many mechatronic systems, a second-order approach (Lagrange/Euler–Lagrange) is natural, e.g.:

$$M(q, p) \cdot \ddot{q}(t) + C\left[q(t), \dot{q}(t), p\right] + g\left[q(t), p\right] = \tau(t) \tag{3}$$

However, this does not determine the specific model – the key is that we know (at least an approximate) physical operator that we can use in the learning process.

The network $s_\theta(t)$ serves as an approximator of the course of a specific variable (state component or observable). With the help of automatic differentiation, we determine the derivatives $s'_\theta(t)$, $s''_\theta(t)$ and construct the residue:

$$F(s, \dot{s}, \ddot{s}, u; p) = 0 \quad R_\theta(t; p) = F\left[s_\theta(t), \dot{s_\theta}(t), \ddot{s}_\theta(t), u(t); p\right] \tag{4}$$

which determines the consistency of the reconstruction with the equations of motion. The parameterization is subject to physical constraints (e.g., positive definiteness of M or positivity of selected parameters).

The PINN training process minimizes a composite loss that balances fidelity to measured data with enforcement of physical equations. In practice, network parameters are optimized so that the reconstructed trajectory both matches observations and reduces residuals of the governing dynamics and initial/boundary conditions (and, when needed, invariants such as energy). Automatic differentiation is used to compute derivatives of the network output, and training proceeds with gradient-based optimization and carefully balanced loss weights. The result is a reconstruction that is consistent with both the data and the system physics.

PINN training is performed on two types of time points: T_y (where we compare to the data) and T_c (where we execute the equations). Initial/boundary conditions are included in the T_b subset. The cost function has a modular structure – each term has a clear role. Data component:

$$L_{data} = \frac{1}{|T_y|} \cdot \sum_{t_i \epsilon T_y} C \cdot s_\theta(t_i) - y_i{}_2^2 \tag{5}$$

maintains consistency with measurements, physical component:

$$L_{phys} = \frac{1}{|T_C|} \cdot \sum_{t_j \epsilon T_C} R_\theta(t_j; p)_2^2 \tag{6}$$

keeps track of the equations of motion, and the IC/BC:

$$L_{IC/BC} = \sum_{t \epsilon T_b} \left[s_\theta(t) - \bar{s}(t)_2^2 + \dot{s}_\theta(t) - \bar{\dot{s}}(t)_2^2 \right] \tag{7}$$

ensures consistency with initial/boundary conditions (in practice often "softly", i.e. co-learned). If an invariant is known (e.g. energy), we use an additional term.

$$L_{inv} = \frac{1}{|T_C|} \cdot \sum_{t_j \epsilon T_C} I\left[s_\theta(t_j), \dot{s}_\theta(t_j), p\right] - k_2^2 \tag{8}$$

which stabilizes the behavior of the reconstruction. We combine the whole thing, as:

$$L(\theta, p) = \alpha_{data} \cdot L_{data} + \alpha_{phys} \cdot L_{phys} + \alpha_{\frac{IC}{BC}} \cdot L_{\frac{IC}{BC}} + \alpha_{inv} \cdot L_{inv} + \alpha_{reg} \cdot L_{reg} \tag{9}$$

ensuring dimensionlessness of the components and a reasonable choice of weights (e.g., by uncertainty weighting or gradient balancing). We conduct training on normalized variables, choosing T_c on a grid or randomly (with adaptive resampling where the residual is large), and for optimization we use a combination of first-order methods (e.g., Adam) and L-BFGS. We base stopping criteria on stabilization of the objective function and monotonic improvement of the physical component with an acceptable fit to the data.

The diagnostic decision can take two forms. In parametric estimation mode, we treat p as a co-learned vector and solve $\min_{\theta,p} L(\theta,p)$ (which requires observability with respect to $h(\cdot)$ and sufficient stimulation u(t)). In dictionary mode, we define a set of variants $\{p_k\}$ and compute the matching score for each of them:

$$S_k = \beta_{data} \cdot L_{data}^{(k)} + \beta_{phys} \cdot L_{phys}^{(k)} + \beta_{inv} \cdot L_{inv}^{(k)} \tag{10}$$

then we choose $k = \arg\min_k S_k$ and compare it with the anomaly threshold (allowing for a rejection zone).

The selection and reporting of metrics is optional and depends on the application requirements; in practice, the S_k result is combined with simple indicators derived from the reconstruction and with information about uncertainty.

4 Application of the Proposed Method: Case Study I – Physical Pendulum

To illustrate the method's operation with a simple, controlled example, we consider a physical pendulum described by the angle θ(t). Geometric variations translate into parameters p = (I,m,r), and thus into the natural frequency of the system. Figure 2 visualizes the notations of the considered pendulum models before and after damage.

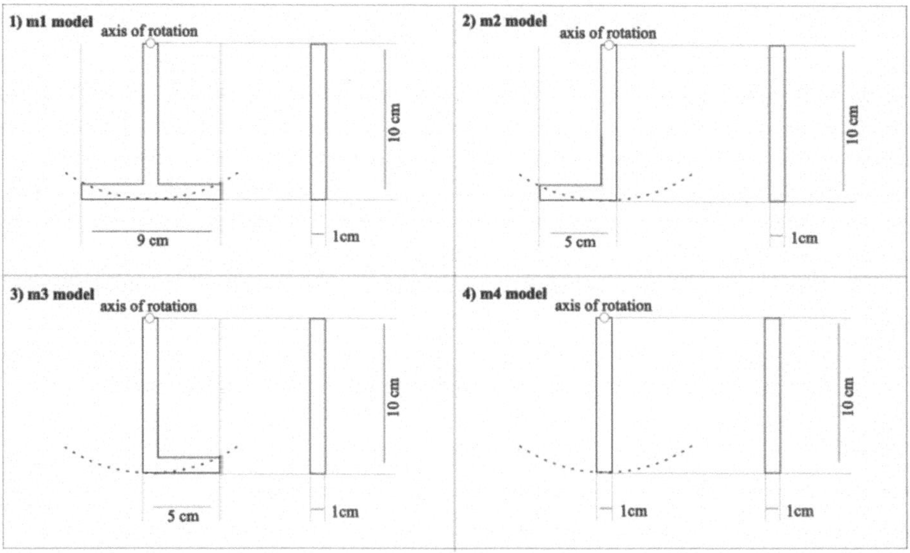

Fig. 2. Pendulum models for the m1–m4 cases discussed in the text.

The pendulum moves planarly in the sketch plane; the axis of rotation is perpendicular to this plane ($I=I_k$). The observed quantity is y(t) = θ(t), hence C=I and $s_\theta(t) = \theta(t)$.

In the model without damping, the following equation applies:

$$I\Big(p, \ddot{\theta}(t) + m(p) \cdot gr(p) \cdot \sin[\theta(t)] = 0 \quad (11)$$

and the mechanical energy remains constant:

$$E(t) = \frac{1}{2} \cdot I(p) \cdot \dot{\theta}^2(t) + m(p) \cdot gr(p) \cdot \{1 - \cos[\theta(t)]\} = const \quad (12)$$

PINN is trained on samples $y_i \approx \theta(t_i)$ (noise is treated additively), and collocation points are chosen within the analysis window, densifying them where the residual equations are largest. In this study, we directly approximate θ(t) (i.e., $s_\theta(t) = \theta(t)$ and C=I), and the cost function includes components: fit to the data, residual equations, initial conditions, and - as an invariant - energy. The configuration used was simple but sufficient for demonstration purposes: duration 1.0 [s], sampling step 0.01 [s], initial conditions

$\theta(0) = 0.1$ [rad], weights $\lambda_{IC} = 1.0$, $\lambda_{energy} = 1000.0$, and calculations in mixed-precision mode.

We formulate the diagnostic decision using a dictionary. First, we calibrate the nominal state p_{nom}. For a new observation, we estimate p^ and determine the vector of relative deviations:

$$\Delta p = |p - p_{nom}| \oslash p_{nom} \tag{13}$$

and then we evaluate compliance with the dictionary rules: for each class l_k we count:

$$S_k = w_k \odot (\Delta p)_{J_k l} = \sum_{j \in J_k} w_{k,j} \cdot \Delta p_j \tag{14}$$

and we find an anomaly when $S_k \geq \tau_k$ (in case of multiple matches, we choose the class with the largest S_k). In simple situations – as here – a single, robust feature derived from the reconstruction is sufficient to make a decision, namely the relative change in period $\Delta T/T$, which directly reflects the changes in I,m,r through the relationship $\omega = \sqrt{mgr/I}$. We define the τ_k thresholds as safe margins around the dictionary values (e.g., ±1 pp), calibrated on short validation sequences to avoid arbitrariness.

The summary of scenarios (indicative values) and their interpretation are presented in the Table 1.

Table 1. The summary of scenarios (indicative values) and their interpretation

Variant	Expected	Intuitive interpretation
nominal	0%	no parameter deviation (reference)
m2 (single-arm)	−3,56%	a slight change in inertia/effective length (e.g., loss of one "arm")
m4 (no-arms)	−13,19%	greater change in inertia/effective length (e.g., loss of both "arms")

The analysis was performed on pairs of series: "Real" (reference trajectories of variants) and PINN (reconstruction trajectory for the nominal state), using materials in the results/ directory (diagnosis_m2_m3.md for the single-arm scenario and diagnosis_m4.md for no-arms). For the decision, a single, robust feature derived from the reconstruction was used - the relative change in period $\Delta T/T$ - which is a surrogate for the deviations of the Δp parameters, since $\omega = \sqrt{mgr/I}$ directly relates I,m,r to the period.

The applied configuration proved to be sufficient for the demonstration purpose. Small differences from the dictionary values are typically due to the smoothness of the PINN reconstruction ("pseudo-damping" effect), finite sampling, and measurement noise; the stability of the decisions was achieved by tolerance bands around the dictionary values.

The classification results are shown in Table 2. In both cases, the decisions were consistent with the dictionary expectations, and the deviations were within fractions of a percentage point.

Table 2. The classification results

Case	Real T [s]	PINN T [s]	ΔT/T [%]	Class	Dev vs. dict
m2/m3	0.5756	0.5970	−3.4335	smaller_fault	≈0.13 pp
m4	0.5186	0.5970	−13.1274	larger_fault	≈0.065 pp

Anomaly detection visualization for case study I is shown in Fig. 3.

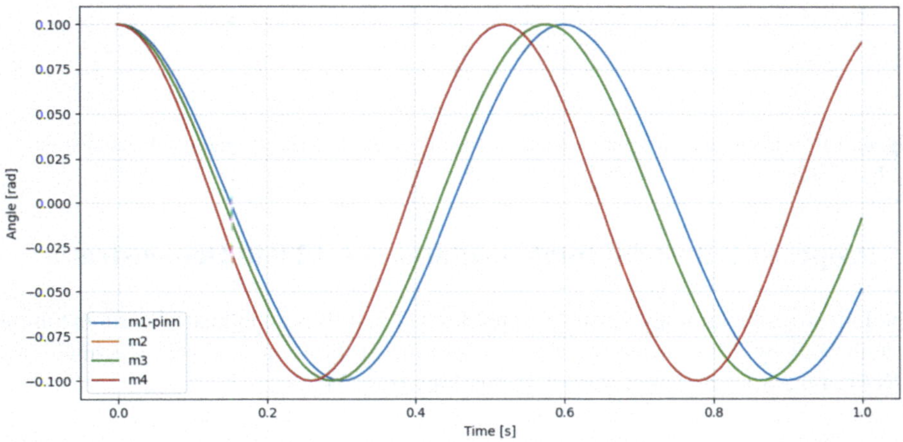

Fig. 3. Anomaly detection visualization for case study. Comparison of waveforms (Real vs. PINN) and the ΔT/T index used for variant classification.

PINN reconstruction acts as a "physical filter": the period determined from s_θ (t) is more stable than that from the raw data, which limits the influence of noise on ΔT/T and the classification result. This is illustrated by the plots of training convergence and waveform consistency in the key areas for period estimation.

PINN learning convergence (trend of cost function components and residual equations) – the basis of reconstruction reliability is presented in Fig. 4.

In POC, we do not yet estimate uncertainty in the formal sense; stability is ensured by the PINN structure (enforcing equations and invariants), and a full analysis (impact of noise, sensor drift, synchronization errors) is planned as future work. In this simple system, a single ΔT/T feature is sufficient to distinguish variants, but for more complex systems, the use of the full Δp vector and additional time-energy features is recommended.

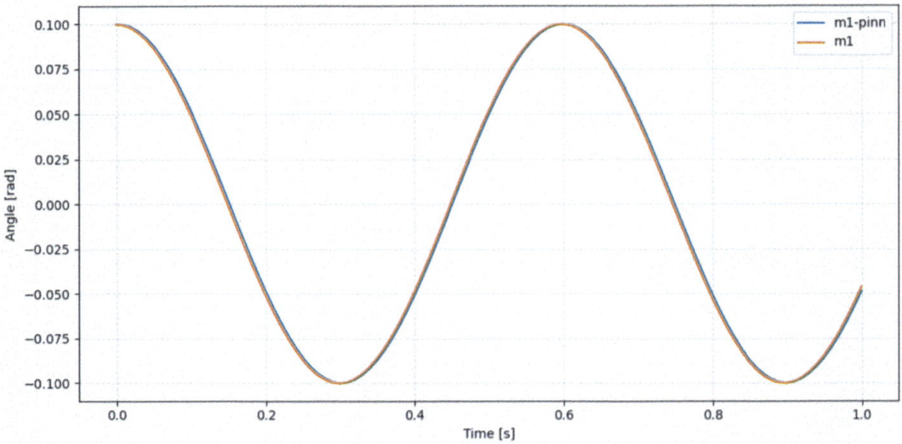

Fig. 4. PINN learning convergence (trend of cost function components and residual equations) – the basis of reconstruction reliability.

5 Proposed Industrial Implementation - CTTP40 Demonstrator

The Industrial Technology Testing Center 4.0 (CTTP4.0) is an industrial demonstration system built by APA Group Sp. z o.o., operating at the Center of New Technologies (CNT) of the Silesian University of Technology (Fig. 5).

a. b.

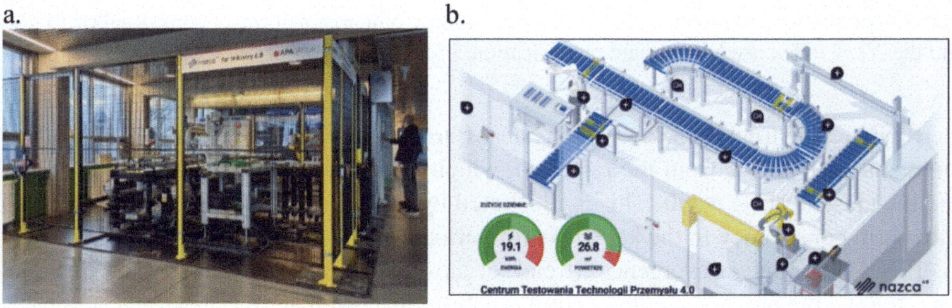

Fig. 5. View of the station of the Industry 4.0 Technology Testing Center (CTTP40).

The CTTP40 station presents the implementation of a typical production task, therefore it is an excellent environment for testing elements of this process, but also aspects of the "behavior" of machines and devices, taking into account the set parameters and constraints, or the impact of the industrial installation on the environment.

Structurally, CTTP40 is a complex industrial environment consisting of two main subsystems:

- The physical subsystem, which is a set of production machines and devices implementing a given production process (Fig. 5a).
- The virtual subsystem, which is the NAZCA 4.0 IT data processing platform (Fig. 5b).

The primary purpose of the material subsystem is to implement the production process, which in this case involves the permanent assembly of container components with pre-defined parameters. The main production process is complemented by the transport of components and the final product. This subsystem is fully automated and equipped with a large set of sensors enabling monitoring of the production and operational processes, as well as measuring the impact of line components on the environment.

The NAZCA 4.0 subsystem collects data from measurement sensors and implements analytical procedures related to the operation of the production line. The NAZCA 4.0 system is characterized by an open structure, which allows for a wide range of analyses and the development of new evaluation and optimization models using artificial intelligence methods and tools.

Figure 6 presents the CTTP4.0 station topology as an environment for studying physics-informed methods. Experience with the shuttle's POC suggests that selected station subsystems (transport, manipulation, and propulsion) will enable the formulation of physical operators and initial/boundary conditions analogous to those used in case study I.

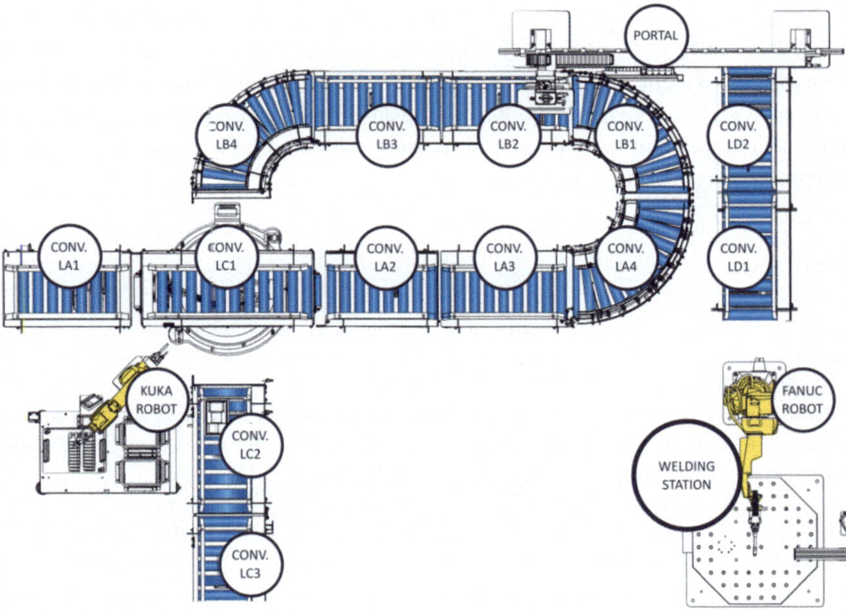

Fig. 6. CTTP4.0 station topology as an environment for research on physics-informed methods.

The station provides access to multimodal observations and controlled operating scenarios, which facilitates the design of PINN reconstructions and the definition of diagnostic features. This configuration allows for the planning of repeatable experiments, comparison of model variants, and verification of decision rules in industrial-like conditions.

At this stage, we are only describing the concept of method transfer: we base our observations from the POC, indicate potential directions for adaptation (selection of the physical operator, feature definition, uncertainty reporting), and defer hardware and system decisions to the research project phase.

6 Summary

In this work, we demonstrated that PINNs effectively combine measurement data with knowledge of the system's dynamics and provide stable, interpretable diagnostic features even with limited and noisy observations. The cost function construction enforces consistency with the equations of motion, initial/boundary conditions, and invariants, which increases the reliability of the reconstruction and facilitates decision-making.

Using a simple study of a physical pendulum, we confirmed that a single dynamic feature—the relative change in period $\Delta T/T$—allows us to distinguish between the analyzed variants, while the parameter deviation vector Δp further enhances the interpretability of the result. The stability of the reconstruction results from enforcing the laws of physics, which limit susceptibility to noise and measurement artifacts.

This paper is conceptual in nature. We limited the scope of our results to a demonstration of the PINN → feature → decision mechanism. Further work should include a systematic assessment of robustness to noise and drift, uncertainty calibration, and comparisons with baseline methods. We see the development perspective in transferring the method to more complex objects and in using the potential of the CTTP4.0 station as a research environment for PINN methods.

References

1. Janik, A., Ryszko, A.: Open innovation and environmental sustainability–A systematic literature review from the firm-level perspective. Technol. Forecast. Soc. Chang. **218**(124196), 1–31 (2025). https://doi.org/10.1016/j.techfore.2025.124196
2. Ragin-Skorecka, K., Motała, D.: Industry 4.0 technologies as a catalyst for lean transformation: bridging the gap between declaration and practice. In: European Lean Educator Conference, pp. 46–62. Springer Nature Switzerland, Cham (2025)
3. Rzydzik, S., Kroczek, P.: Time parameters verification of a numerical simulator of an automated store warehouse. Bull. Pol. Acad. Sci., Tech. Sci. **72**(3), 1–11 (2024). https://doi.org/10.24425/bpasts.2024.149816
4. Marcin, A., Krzysztof, P., Tomasz, R.: Diagnostic shell expert systems. Diagnostyka. **17**(1), 33–40 (2016)
5. Halenar, I., Halenarova, L., Tanuska, P., Vazan, P.: Machine condition monitoring system based on edge computing technology. Sensors. **25**(1), 180 (2025). https://doi.org/10.3390/s25010180
6. Loska, A., Moczulski, W., Wyczółkowski, R., Dąbrowski, A.: Integrated system of control and management of exploitation of water supply system. Diagnostyka. **17**(1), 65–74 (2016)
7. Paszkowski, W., Loska, A.: The use of neural network model in the assessment of annoyance of the industrial noise sources. In: Burduk, A., Mazurkiewicz, D. (eds.) Intelligent Systems in Production Engineering and Maintenance, ISPEM 2017 Advances in Intelligent Systems and Computing, vol. 637, pp. 428–439 (2018). https://doi.org/10.1007/978-3-319-64465-3_41

8. Pollak, A., Temich, S., Piąsiński, W., Kucharczyk, J., Gąsiorek, D.: Prediction of belt drive faults in case of predictive maintenance in industry 4.0 platform. Appl. Sci. **11**(21), 1–9 (2021). https://doi.org/10.3390/app112110307
9. Przystałka, P.: Model-based fault detection and isolation using locally recurrent neural networks. In: Rutkowski, L., Tadeusiewicz, R., Zadeh, L.A., Zurada, J.M. (eds.) Artificial Intelligence and Soft Computing – ICAISC 2008 Lecture Notes in Computer Science, vol. 5097, pp. 113–122. Springer, Berlin/Heidelberg (2008). https://doi.org/10.1007/978-3-540-69731-2_13
10. Pollak, A.: Wdrożenie polskiej platformy Nazca 4.0 w celu implementacji koncepcji przemysłu 4.0 w procesach produkcyjnych. Politechnika Śląska, Wydział Mechaniczny Technologiczny, Gliwice (2022)
11. Jasiulewicz-Kaczmarek, M., Waszkowski, R., Piechowski, M., Wyczółkowski, R.: Implementing BPMN in maintenance process modeling. In: International Conference on Information Systems Architecture and Technology, pp. 300–309. Springer International Publishing, Cham (2017)
12. Loska, A.: Variant assessment of exploitation policy of selected companies managing technical network systems. Manage. Syst. Prod. Eng. **3**(19), 179–188 (2015). https://doi.org/10.12914/MSPE-13-03-2015
13. Sun, B. et al.: Failure-based sealing reliability analysis considering dynamic interval and hybrid uncertainties. Eksploatacja i Niezawodnosc – Maint. Reliab. **23**(2), 278–284 (2021). https://doi.org/10.17531/ein.2021.2.7
14. Przystałka, P., Kalisch, M., Timofiejczuk, A.: Genetic optimization of meta-learning schemes for context-based fault detection. In: Timofiejczuk, A., Łazarz, B., Chaari, F. (eds.) Advances in technical diagnostics, Applied Condition Monitoring, vol. 10, pp. 287–297. Springer (2018). https://doi.org/10.1007/978-3-319-62042-8_26
15. Lu, H. et al.: A physics-informed feature weighting method for bearing fault diagnostics. Mech. Syst. Signal Process. **191**, 110171 (2023). https://doi.org/10.1016/j.ymssp.2023.110171
16. Parziale, M., Lomazzi, L., Giglio, M., Cadini, F.: Physics-informed neural networks for the condition monitoring of rotating shafts. Sensors. **24**(1), 207 (2024). https://doi.org/10.3390/s24010207
17. Barimah, A.K., Onu, O.P., Niculita, O., Cowell, A., McGlinchey, D.: Scalable data transformation models for physics-informed neural networks (PINNs) in digital twin-enabled prognostics and health management (PHM) applications. Computers. **14**(4), 121 (2025). https://doi.org/10.3390/computers14040121
18. Deng, W., Ardiani, F., Nguyen, K.T.P., Benoussaad, M., Medjaher, K.: Physics informed machine learning model for inverse dynamics in robotic manipulators. Appl. Soft Comput. **163**, 111877 (2024). https://doi.org/10.1016/j.asoc.2024.111877
19. Li, T., Xiao, T.: Physics-informed neural network-based input shaping for vibration suppression of flexible single-link robots. Actuators. **14**(1), 14 (2025). https://doi.org/10.3390/act14010014
20. Latif, S. et al.: Deep learning for the industrial Internet of Things (IIoT): a comprehensive survey of techniques, implementation frameworks, potential applications, and future directions. Sensors. **21**(22), 7518 (2021). https://doi.org/10.3390/s21227518

Innovative PLC-Based Architecture for Acquisition and Storage of Analog Sensor Data in Industrial Environments

Andrzej Chmielowiec[1(✉)], Adam Błachowicz[2], and Katarzyna Antosz[1]

[1] Rzeszow University of Technology, Rzeszow 35-959, Poland
achmie@prz.edu.pl

[2] Doctoral School of the Rzeszow University of Technology, Rzeszow 35-959, Poland

Abstract. This paper presents an innovative architecture for the acquisition and storage of data from industrial analog sensors, based on Programmable Logic Controllers (PLC) operating within Operational Technology (OT) networks. Examples from the automotive industry are provided, covering both process parameters and signals used in predictive maintenance. The study discusses a multilayer data acquisition architecture integrating remote I/O modules, PLCs, Supervisory Control And Data Acquisition (SCADA) systems, and data servers. The experimental section analyses the efficiency of various encoding and compression methods for measurement data (float32, int16, differential representation) using the Deflate algorithm and Huffman coding. The results demonstrate that replacing direct float32 data storage with integer-based scaling enables a reduction in memory requirements by a factor of three to nine. The proposed approach reduces the load on SCADA systems, lowers database maintenance costs, and improves the processing of signals used in predictive maintenance applications.

Keywords: Digitization · Process Monitoring · Data Acquisition · Data Storage · Data Compression · PLC

1 Introduction

In the times of widely implemented "Industry 4.0" mechanisms and methods, question of production data acquisition and storage, is one of the most important issue. Every production line, installation, or particular machine, is now equipped with plenty of analog sensors. To improve process control and performance, we need to collect and analyze signals from them. To build proper model of our process, and in the same time, improve its accuracy, we need more and more data. Increasing data collection sampling frequency is one of the method, but has a strong impact for available storage disc space. For this reason, proper design of data acquisition architecture is a crucial foundation of every efficient, production parameters, collection system.

According to [1] data acquisition systems mainly base on conversion one or more analog signals to digital form, and process it by a logic and computer systems. Choosing proper: analog-digital converters, sampling methods and signal conversion algorithm

B. Gapiński et al. (Eds.): MANUFACTURING 2026, LNME, pp. 390–403, 2026.
https://doi.org/10.1007/978-3-032-22943-4_27

allows us to build measurement channel with sufficient accuracy and processing speed. For different physical quantities, like: volume, flow, temperature, pressure, speed, vibration etc. we can find plenty of very good developed and optimized data collection methods. They are described, for example in [2–5], where all industry solutions based on PLC controllers, as a data conversion unit, connected with SCADA systems where data are stored.

In this elaboration, we want to focus on analog sensors, which measures physical machine parameters, connected with programmable logic controllers (equipped with remote I/O modules), working in industrial OT network. This type of network must be designed in proper, multilevel way to assure efficient traffic, according to increase frequency of acquired data sampling periods. Example descriptions of different industry collection architectures we can find in [6–8]. Of course approach and chosen solution will depend on scale and size of plant, where machines or installations are located. It will look different in small concentrated areas and for wide, large plants, for example in energy industry. For this kind of systems we can follow information from papers [9, 10].

Still increasing number of industry applications, used for analyze and process monitoring, require continuous extend of infrastructure and quantity of available measurement channels. This can be a problem for plants with centralized wired network topology, which requires cost and a lot of time to rebuild. Because of it, wired infrastructure start to be replaced with wireless systems, using IoT technology. Examples and description of those solutions we can find in [11, 12].

Numerous recent works focus on advanced data-driven diagnostics and predictive maintenance based on analog sensor signals, particularly in transportation and propulsion systems [13, 14]. However, much less attention is paid to the underlying data acquisition and storage architectures, which strongly affect the long-term costs and feasibility of such solutions.

New approach for the industry data acquisition systems, should allows to design architecture of devices, which will be able to collect data from wired and wireless networks in the same time. Just on the edge between OT (automation devices) network and business Information Technology (IT) environment, over SCADA systems and firewall, should be located Live Data Server (LDS), storing data using Open Platform Communications (OPC) or customized software libraries. It must allow us to read data from different types of controllers, using different communication protocols. This server must be able to collect data from all required sensors, machines and control devices, and store it according to quality system and customers' requirements rules, depend on type of production and technology.

This paper highlights on, how to operate with acquisition and storing of analog sensors data, based on examples from automotive industry, where are collected: process parameters and predictive maintenance signals. In such factories, a big quantity of data, with still increasing collection resolution, start to be a big challenge for SCADA systems and simple OT networks [15]. To keep systems reliability and robustness there is a necessity to introduce new: architectures, storage and compression methods, scaling signals algorithms and machine learning analysis. New architectures should be applicable in places with centralized systems – with central SCADA or Distributed Control Systems (DCS), and also in factories where digitalisation tools are specifically customized for technology. According to [16], Industry 4.0 tools and mechanism should promote decentralization, interoperability, modularity and real-time capability in connection with process customization.

In this paper, the authors focus on presenting a proposed architecture for the acquisition and storage of measurement data in industrial systems. The main emphasis is placed on data compression methods and mechanisms, whose effectiveness and efficiency are illustrated using real industrial use cases. The authors are aware that many other data acquisition and storage mechanisms are not addressed in this study; therefore, the presented solutions should be regarded as a proposal of a particular approach to efficient management of measurement data rather than a comprehensive or exhaustive framework.

2 Materials and Methods

The method by which a PLC acquires data from individual sensors varies according to the design of the machine or industrial installation. At the lower system level (beneath the PLC), data are transmitted either through standard electrical connections or via industrial networks connected to remote I/O modules. Depending on the type of industrial network in use, communication may rely on Ethernet/IP, Common Industrial Protocol (CIP), or Modbus/TCP protocols [17]. Across all measurement systems described above, a common feature is present: the data transmission pathway leading from the PLC to the upper layers of the acquisition architecture, including the Live Data server, the Historian database, and ultimately the Manufacturing Execution System (MES). To retrieve these data from controllers, collection servers must have appropriately installed and configured OPC server software or dedicated applications equipped with customized libraries capable of reading tag values directly from PLC memory and storing them in a database. To ensure system reliability and interoperability with a wide range of industrial devices, it is advisable to apply a unified acquisition architecture, as illustrated in (Fig. 1), along with a single communication protocol for higher-level data transmission. At present, Ethernet constitutes the most prevalent networking standard within industrial and factory automation environments [18], particularly in its Ethernet/IP implementation. At the uppermost layer of the proposed architecture, an advanced MES system or an industrial web portal may be incorporated, enabling the dissemination of measured data across geographically distributed locations via Ethernet using the HTTP protocol [18]. This approach is particularly recommended for global enterprises lacking an MES system and relying on web-based technologies, such as SharePoint or comparable platforms, for the presentation of production data within corporate networks.

The focus of MES systems on collecting and continuously analyzing data originating from a very wide range of sensors results in integration becoming the principal challenge, particularly due to the diversity of environments provided by different manufacturers. Issues related to the optimal storage of the collected data are generally treated as secondary. It may seem that in an era of readily available and easily deployable large-scale storage resources, such considerations can be neglected. Unfortunately, the increasing

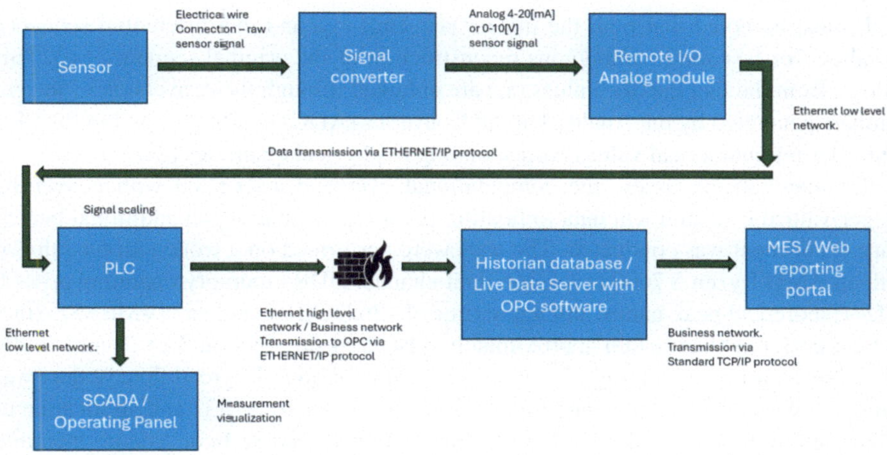

Fig. 1. Recommended acquisition architecture for analog sensor data

number of sensors, their measurement accuracy, and the frequency at which data are acquired clearly indicate that data storage systems are becoming a significant cost factor in the operation of an enterprise information infrastructure [19]. It should be emphasized that storing data in an uncompressed form and in user-readable formats requires substantially larger storage resources than would be implied by the actual informational content of the stored data. A commonly adopted measure of such informational weight is the information entropy defined by Shannon [20]:

$$H(X) = \sum\nolimits_{x \in X} -p(x)\log_2(p(x)), \tag{1}$$

where X is a random variable and $p(x)$ denotes the probability of occurrence of the symbol x.

An analysis of data storage methods commonly employed in industrial environments raises the question of whether such approaches are optimal from the perspective of storage resource utilization. To investigate this aspect in greater detail, data from four representative sensors were subjected to examination. The objective of the study was to determine the extent to which archived data may occupy disproportionately large storage resources. Assuming that (x_i) represents a sequence of measurement data, the following cases were analyzed:

1. the memory footprint of measurement data encoded in a 32-bit floating-point format (float32), both before and after compression using the ZLIB Deflate method (ZLIB implementation, compression level 9 – maximum compression),
2. the memory footprint of measurement data (y_i) converted to integer values using an inverse scaling transformation $y_i = ax_i + b$, both before and after Huffman compression [21],
3. the memory footprint of measurement data (z_i) expressed in differential form $z_i = y_{i+1} - y_i$, both before and after Huffman compression.

It must be noted that both the integer sequence (y_i) and the differential sequence (z_i) allow for fast and unambiguous reconstruction of the original sequence (x_i). This follows from the fact that the values (x_i) are obtained through the conversion of integer numbers generated by the Analog Digital Converter (ADC) module located in the PLC controller into numerical values corresponding to physical quantities [22].

To quantitatively assess the computational overhead associated with converting integer-valued measurement data to floating-point representation, an additional benchmark experiment was conducted. The tests were performed on a workstation equipped with an AMD Ryzen 5 7600 processor, dual-channel DDR5 memory operating at 4800 MT/s (theoretical peak bandwidth approximately 76.8 GB/s), and an NVMe SSD (theoretical peak read bandwidth approximately 6 BG/s), using Python 3.13.2 and NumPy 2.2.2. For an array containing 10^7 samples, the time required to read float32 data was compared with reading int16 data followed by conversion to float32 and application of a linear affine transformation $(x - u)/s$. The combined read-and-conversion operation required a median time of 0.0274 s, whereas direct reading of float32 data took 0.0090 s, corresponding to an approximately threefold increase in processing time.

The measurement data were collected from four sensors responsible for monitoring:

1. temperature,
2. vacuum level,
3. overpressure in the clean-room,
4. water conductivity.

The characteristics of the data acquisition process for these four selected sensors are presented below.

Temperature Measurement. is one of the most commonly applied control processes in industrial environments. Temperature can be measured using a variety of techniques, one of which is non-contact measurement with a laser pyrometer. This method is preferred in situations where direct contact with the measured object is not feasible due to high temperature levels, safety or environmental constraints, or product quality requirements. (Figure 2) presents a laser pyrometer sensor head together with its signal transducer, used for temperature measurement in the glass manufacturing industry.

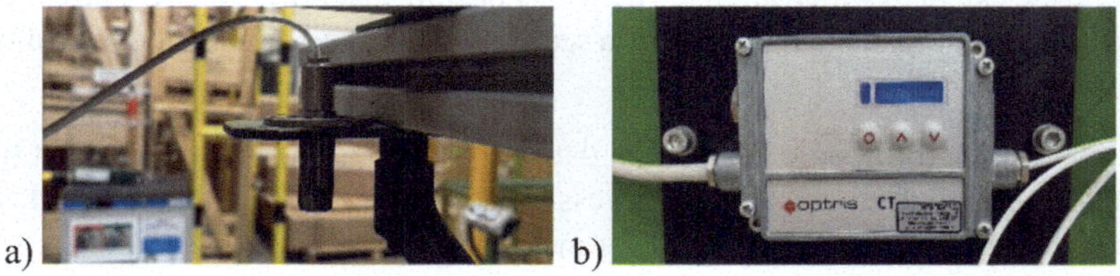

Fig. 2. Pyrometer for temperature measurement: a) sensor head, b) signal transducer

The signal produced by the measurement instrument (pyrometer head) must be converted to an automation-standard format—either 4–20 mA or 0–10 V — compatible with PLC input modules. The transducer is connected to a remote I/O input module, as shown in (Fig. 3) which in this case communicates with the PLC via a DeviceNet industrial network.

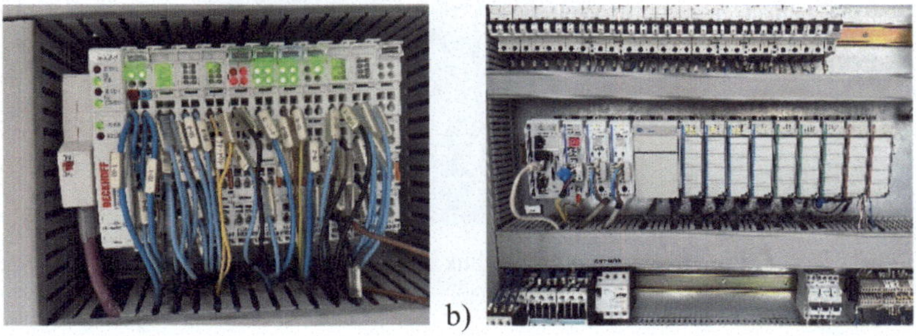

a) b)

Fig. 3. Temperature measurement connection: a) analog input module, b) PLC

Subsequently, an appropriate scaling procedure is implemented within the PLC program to convert the acquired current or voltage signal into physical units—degrees Celsius. Based on the information provided in the Optris device datasheet, the conversion function is defined within the PLC routine to translate the raw input signal into usable temperature values.

After the analog-to-digital conversion performed in the remote I/O module, the input signal is represented as an integer: 31206 for 20 mA and 6241 for 4 mA. For the pyrometer's measurement range of 0–200 ° C, the temperature is calculated using the following formula:

$$T = \frac{\text{Input} - \text{InputLowLimit}}{\text{InputRange}} \cdot \text{InstrumentMeasurementRange}. \quad (2)$$

Here, Input denotes the value received by the PLC from the analog input channel connected to the pyrometer transducer. The scaled signal is subsequently used for PLC logic operations, visualization on operator panels or SCADA systems, and ultimately for archival in the industrial database.

Vacuum Level. is an important parameter in lamination and gluing processes, as well as in hermetic testing and leakage detection within predictive maintenance procedures. Vacuum systems typically consist of a central buffer tank equipped with vacuum pumps and a network of pipelines connected to local vacuum-operated devices. (Figure 4) shows an example of a vacuum buffer tank with an integrated sensor and signal transducer.

The signal from the measurement instrument is transmitted directly to the remote I/O module as a standard 4–20 mA current signal. It is then converted from analog to digital form and forwarded to the PLC CPU. (Figure 5) presents the Beckhoff analog input module and the Rockwell PLC used in this setup, both connected to the Ethernet network.

Within the PLC ladder-logic routine the signal is converted from its raw digital representation into physical units, expressed in negative bar. The conversion is performed based on the sensor's reference documentation and the remote I/O module operating

Fig. 4. Vacuum buffer tank with sensor and transducer

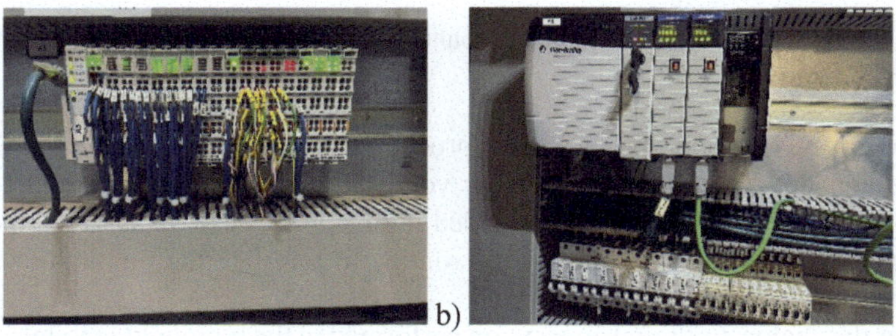

a) b)

Fig. 5. Vacuum measurement signal connection: a) remote I/O analog input module, b) PLC

manual, which specify memory allocation, signal resolution, scaling factors, and the conversion method:

$$\text{Vacuum} = \frac{\text{Input} - \text{ReferenceValue}}{\text{ScalingFactor}}. \tag{3}$$

Here, *Input* denotes the value acquired from the remote I/O analog channel connected to the vacuum sensor unit. The converted signal is then ready for subsequent processing steps. The integration of the sensor with the signal transducer eliminates one system layer and consequently reduces machine wiring. The remaining architecture layers are similar to those described earlier, featuring Ethernet/IP communication between the I/O module and the PLC CPU, and subsequently between the PLC and the Live Data server or Historian database. At the highest level, standard TCP/IP transmission is used between the data storage server and corporate IT systems (MES, web intranet portals, office workstations, etc.).

Overpressure in Clean-Room. is one of the key parameters monitored in assembly clean-rooms to prevent the presence of dust or other particulate contaminants. (Figure 6) shows an example of the DM-2000 overpressure sensor, which is used to control and display overpressure values to operators working in clean environments.

Similar to the vacuum measurement setup, the overpressure sensor also integrates a signal transducer and can be directly connected to standard automation input channels

Fig. 6. Overpressure signal sensor

operating at 4–20 mA or 0–10 V. (Figure 7) presents a Beckhoff remote I/O module and a Rockwell ControlLogix 5000 PLC unit, both connected to the same segment of the low-level Ethernet network. Within the PLC logic subroutine, a scaling procedure is implemented to convert the analog sensor signal into the corresponding overpressure values expressed in Pa

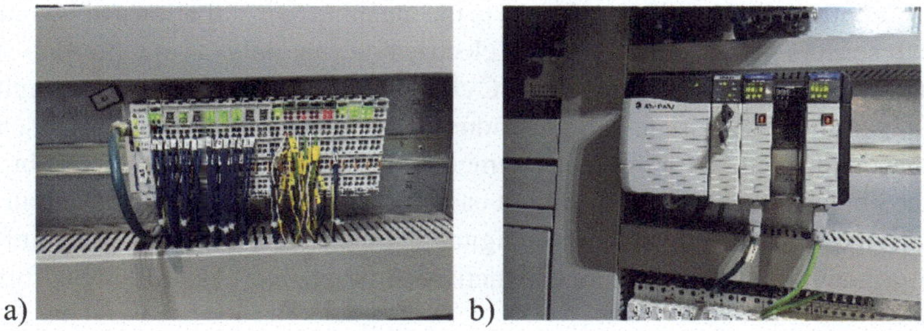

Fig. 7. Overpressure signal connection: a) remote I/O analog input module, b) PLC

The scaling method employs the embedded *SCL* function block from the predefined toolset available in Rockwell ControlLogix 5000 processors. Its configuration requires parameterisation according to the sensor datasheet and operating manual. In this case, the scaling is defined as a linear transformation between the raw input and the engineering units. According to the Rockwell PLC software documentation, the SCL instruction converts an unscaled value into a scaled value using the following relation:

$$\text{Out} = (\text{Input} - \text{InRawMin}) \cdot \left(\frac{\text{InEUMax} - \text{InEUMin}}{\text{InRawMax} - \text{InRawMin}}\right) + \text{InEUMin}. \quad (4)$$

The data acquisition architecture for this measurement is largely analogous to that used for vacuum measurement, with the distinction that one system layer is eliminated due to the integration of the sensor and transducer.

Water Conductivity. is a specific parameter associated with water purity and its level of mineralization, and it plays a crucial role in industrial water and steam installations. (Figure 8) presents an example of a Jumo conductivity sensor integrated into a water supply system.

Fig. 8. Water conductivity signal sensor

The signal generated by this sensor is not directly compatible with standard automation devices and therefore requires connection through a dedicated converter, which adapts the voltage and current levels to industrial standards. In this case, the converter used is the Jumo EcoTrans LF01 module, located in the same cabinet as the PLC controller (Fig. 9). This configuration does not rely on remote network I/O modules. The main PLC control cabinet is situated close to the machine or installation and is connected directly to sensors and actuators through electrical wiring only. This represents a more traditional solution, employing a large PLC chassis equipped with analog and digital I/O cards. Only a single network interface—mounted on the main PLC unit—is present. The signal conversion method for this measurement is implemented within a PLC subroutine. The value obtained from the analog input card corresponds to the 0–10 V output of the Jumo transducer and is converted into a digital PLC tag. This numerical value must then be transformed into the appropriate conductivity unit μ S using the following formula:

$$\text{CondValue} = \frac{200}{16000 \cdot (\text{InputValue} - 4000)}, \tag{5}$$

where CondValue represents the resulting conductivity measurement in μ S, and InputValue is the signal received by the PLC analog input card, stored as a floating-point variable.

a) b)

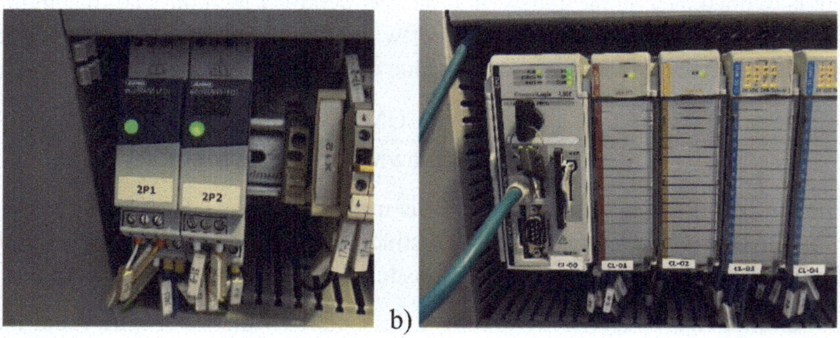

Fig. 9. Conductivity signal connection: a) sensor signal transducer, b) PLC

The data acquisition architecture for this analog signal reflects the design of traditional pre-Industry 4 0 machines. It is based primarily on the physical electrical layer rather than on network-based communication. Only one Ethernet interface, located on the PLC unit, is available, connecting directly to the higher-level network. This type of solution requires a large PLC chassis with numerous expansion cards and extensive electrical wiring. Data transmission between the PLC and the Live Data server uses the Ethernet/IP protocol, while communication in the enterprise network layer relies on standard TCP/IP.

3 Results

For the four analyzed sensors, a total of 193 583 samples were recorded:

- Temperature sensor: 42 653 samples, corresponding to 1364.9 kbit of data in the float32 representation (Table 1).
- Vacuum level sensor: 80 563 samples, 2578.0 kbit of float32 data (Table 2).
- Clean-room overpressure sensor: 67 716 samples, 2166.9 kbit of float32 data (Table 3).
- Water conductivity sensor: 2651 samples, 84.8 kbit of float32 data (Table 4).

In total, this amounts to 6194.6 kbit of raw measurement data, i.e. approximately 0.77 MB in the float32 format, prior to applying any compression method.

Based on Tables 1, 2, 3 and 4, the efficiency of different signal representations (original float32, rescaled int16, and differential representation) combined with ZIP Deflate compression and Huffman coding can be compared. In particular, we focus on the reduction relative to the original float32 representation.

For the temperature sensor (Table 1), the smallest data size was obtained for the int16 representation compressed using Huffman coding, yielding 202.6 kbit compared with 1364.9 kbit for float32. The compression ratio is:

$$\frac{202.6}{1364.9} \approx 0.15, \tag{6}$$

which corresponds to a reduction to approximately 15% of the original size.

Table 1. Compression results for Temperature sensor

Data type	Sample size	Unique values	Entropy [bit/sym]	Original size [kbit]	Deflate size [kbit]	Deflate Ratio	Huffman size [kbit]	Huffman Ratio
float32	42653	130	5.02	1364.9	269.6	0.197	221.7	0.162
int16	42653	105	4.63	682.4	208.7	0.306	202.6	0.297
int16-diff	42652	148	4.75	682.4	248.2	0.364	209.4	0.307

For the vacuum level sensor (Table 2), the most favorable result was achieved using the differential int16-diff representation compressed with Huffman coding, producing 287.7 kbit relative to 2578.0 kbit for float32. This yields:

$$\frac{287.7}{2578.0} \approx 0.11, \tag{7}$$

i.e. a reduction to approximately 11% of the initial data size. This is the best relative compression performance among the analysed sensors.

Table 2. Compression results for Vacuum level sensor

Data type	Sample size	Unique values	Entropy [bit/sym]	Original size [kbit]	Deflate size [kbit]	Deflate Ratio	Huffman size [kbit]	Huffman Ratio
float32	80563	33	4.01	2578.0	413.7	0.160	328.2	0.127
int16	80563	28	3.75	1289.0	351.8	0.273	306.8	0.238
int16-diff	80562	33	3.52	1289.0	376.9	0.292	287.7	0.223

For the clean-room overpressure sensor (Table 3), the lowest data size was obtained using int16 with Huffman coding (633.5 kbit). With the original size of 2166.9 kbit, the compression ratio becomes:

$$\frac{633.5}{2166.9} \approx 0.29, \tag{8}$$

corresponding to a reduction to approximately 29% of the original size. The high signal entropy (8.87 bit/symbol) markedly limits potential compression gains.

Table 3. Compression results for Overpressure sensor

Data type	Sample size	Unique values	Entropy [bit/sym]	Original size [kbit]	Deflate size [kbit]	Deflate Ratio	Huffman size [kbit]	Huffman Ratio
float32	67716	950	8.87	2166.9	936.8	0.432	648.7	0.299
int16	67716	950	8.87	1083.5	750.3	0.692	633.5	0.585
int16-diff	67715	1177	8.87	1083.4	773.8	0.714	642.0	0.592

For the water conductivity sensor (Table 4), the most efficient combination was the differential int16-diff representation compressed using Huffman coding, producing 11.2 kbit compared to 84.8 kbit in the float32 format. This yields:

$$\frac{11.2}{84.8} \approx 0.13, \tag{9}$$

Table 4. Compression results for Water conductivity sensor

Data	Sample	Unique	Entropy	Original	Deflate	Deflate	Huffman	Huffman
type	size	values	[bit/sym]	size [kbit]	size [kbit]	Ratio	size [kbit]	Ratio
float32	2651	293	6.91	84.8	24.8	0.292	31.7	0.374
int16	2651	272	6.87	42.4	19.7	0.464	26.3	0.619
int16-diff	2650	61	3.54	42.4	14.4	0.334	11.2	0.265

i.e. a reduction to approximately 13% of the original data size. The significantly reduced number of unique values in the differential signal, substantially improves entropy-based compression.

In summary, for all analysed sensors, a substantial reduction in data size was achieved relative to the original float32 representation.

4 Conclusions

The first objective of this work was to review selected examples of analog sensor data acquisition methods used in industrial environments. Analog sensors were chosen due to their essential role in controlling complex production processes. As noted in [23], many current industrial systems still rely on analog architectures in which transducers are connected to acquisition units through long electrical cables prone to electromagnetic interference. The use of network-distributed remote I/O modules integrated with PLCs within OT networks constitutes an effective approach to digitalizing such circuits, reducing wiring complexity, and improving signal reliability. Based on the analyzed cases, a generalized data acquisition architecture was proposed, adaptable to various industrial applications.

The second objective was to analyze a model of plant-wide data collection from PLCs, transmitted through the OT network to a centralized database. Such a model must ensure efficient data handling and high communication reliability across all system layers, which is particularly critical in predictive maintenance applications.

It should be noted that the commonly used industrial architecture, in which measurement values are automatically scaled by the PLC, is not efficient in terms of the memory required for data storage. The experiments conducted and the methods applied clearly demonstrate that, depending on the type of data, substantial reductions in memory usage are achievable. The presented results indicate that, depending on the method employed, the memory requirements can be reduced by a factor of three to nine. Consequently, such an approach can generate significant savings in database maintenance and backup system costs.

It should also be emphasized that, in the analyzed examples, a much more efficient solution is to store integer values originating directly from the ADC converter—or their differential representation—together with the corresponding conversion function. This is a non-obvious observation that has a significant impact on the design and functionality of sensor data acquisition and storage systems.

In conclusion, the proposed acquisition architecture provides a flexible and scalable framework for modern industrial data collection. As industrial databases evolve toward higher sampling rates and increasing volumes of sensor data, the need for efficient data acquisition and storage methods becomes even more pronounced. Continued development in this area will remain essential to support advances in process optimization and predictive maintenance. It should be noted, however, that the high compression efficiency achieved in the presented examples is closely related to the characteristics of the collected data. Although the selection of signals was not intentional, it must be taken into account that there are certainly scenarios in which the proposed mechanisms will be less effective than in the analyzed cases. This limitation arises from the fact that the effectiveness of the presented methods strongly depends on the probability distribution of the measured values. Consequently, for signals with a more random nature and significantly higher entropy, the potential for achieving substantial compression gains may be limited.

References

1. Taylor, H.R.: Data Acquisition for Sensor Systems. Springer, New York (2010). https://doi.org/10.1007/978-1-4757-4905-2
2. Hasan, B., Mohani, S.S.-H., Hussain, S.S., Yasin, S., Alvi, W.A., Saeed, O.: Implementation of supervisory control and data acquisition - SCADA on a plc and VFD controlled digital mixing plant using TIA portal. In: 2019 4th International Conference on Emerging Trends in Engineering, Sciences and Technology (ICEEST), pp. 1–6 (2019). https://doi.org/10.1109/ICEEST48626.2019.8981705
3. Kondratenko, Y., Korobko, O.V., Kozlov, O.V.: PLC-Based Systems for Data Acquisition and Supervisory Control of Environment-Friendly Energy-Saving Technologies, pp. 247–267. Springer International Publishing (2017). https://doi.org/10.1007/978-3-319-44162-7_13
4. Li, R., Du, X., Li, P.: Design of low-cost acquisition system for multi-channel analog signal for plc and its application in temperature control system. In: Proceedings of the 10th World Congress on Intelligent Control and Automation, pp. 93–98 (2012). https://doi.org/10.1109/WCICA.2012.6357846
5. Feng, C., Xiong, Q., Huang, J., Zhou, H.: Plc-based flywheel battery data acquisition system. Int. Core J. Eng. **8**(7), 441–453 (2022). https://doi.org/10.6919/ICJE.202207_8(7).0059
6. Khatri, V.K., Ghangro, A.J., Kumar, J., Haque, S.J.U.: Industrial data acquisition and control system using two PLCs' networked over MPI network. In: 2009 IEEE Symposium on Industrial Electronics and Applications, vol. 1, pp. 399–402 (2009). https://doi.org/10.1109/ISIEA.2009.5356437
7. Endi, M., Elhalwagy, Y.Z., Hashad, A.: Three-layer plc/scada system architecture in process automation and data monitoring. In: 2010 The 2nd International Conference on Computer and Automation Engineering (ICCAE), vol. 2, pp. 774–779 (2010). https://doi.org/10.1109/ICCAE.2010.5451799
8. Błachowicz, A., Klich, L., Michajłyszyn, A.: Predykcja w układach mechanicznych i automatycznych 2020–transmisja danych pomiarowych (MTKE 2). Oficyna Wydawnicza Politechniki Rzeszowskiej (2020)
9. Kaittan, K.H., Mohammed, S.J.: Communication protocols in plc-based data acquisition for wastewater management. In: 2024 3rd International Conference on Advances in Engineering Science and Technology (AEST), pp. 60–65 (2024). https://doi.org/10.1109/AEST63017.2024.10959998

10. Puchun, A., Jinhai, L.: A plc-based data acquisition system with adaptive threshold algorithm for leak detection of oil pipeline. In: 2013 Chinese Automation Congress, pp. 801–805 (2013). https://doi.org/10.1109/CAC.2013.6775843
11. Liptak, B.G., Eren, H.: Instrument Engineers' handbook. In: Process Software and Digital Networks, vol. 3, 4th edn. CRC Press (2012). https://doi.org/10.1201/9781315217093
12. Mishra, A., Mohaparto, M.: An iot framework for bio-medical sensor data acquisition and machine learning for early detection. Int. J. Adv. Technol. Eng. Explor. **6**(54), 112–125 (2019). https://doi.org/10.19101/IJATEE.2019.650027
13. Baranovskyi, D., Vladov, S., Bulakh, M., Sokurenko, V., Muzychuk, O., Vysotska, V.: The method for training dataset forming recorded by analog sensors to determine the helicopter turboshaft engines efficiency. Mech. Syst. Signal Process. **239**, 113,368 (2025). https://doi.org/10.1016/j.ymssp.2025.113368
14. Vladov, S., Bulakh, M., Czyżewski, J., Lytvynov, O., Vysotska, V., Vasylenko, V.: Method for helicopter turboshaft engines controlling energy characteristics through regulating free turbine rotor speed and fuel consumption based on neural networks. Energies. **17**(22), 5755 (2024). https://doi.org/10.3390/en17225755
15. Chmielowiec, A., Klich, L., Woś, W., Błachowicz, A.: Global spare parts exploitation costs optymization and its reduction to rectangular knapsack problem. In: Proceedings of the 24th International Conference on Enterprise Information Systems - Volume 1: ICEIS, pp. 364–373. INSTICC, SciTePress (2022). https://doi.org/10.5220/0011067400003179
16. Maseda, F.J., López, I., Martija, I., Alkorta, P., Garrido, A.J., Garrido, I.: Sensors data analysis in supervisory control and data acquisition (SCADA) systems to foresee failures with an undetermined origin. Sensors. **21**(8), 2762 (2021). https://doi.org/10.3390/s21082762
17. Wang, N., Fan, X., Feng, F., Li, Y., Ye, H., Liu, Y.: An industrial communication system for plc. In: 2024 IEEE 4th International Conference on Electronic Technology, Communication and Information (ICETCI), pp. 125–129 (2024). https://doi.org/10.1109/ICETCI61221.2024.10594451
18. Antony, J., Mahato, B., Sharma, S., Chitranshi, G.: A web plc using distributed web servers for data acquisition and control: wed based PLC. In: 2011 International Conference on Information Science and Applications, pp. 1–4 (2011). https://doi.org/10.1109/ICISA.2011.5772427
19. Dutta, A.K., Hasan, R.: How much does storage really cost? Towards a full cost accounting model for data storage. In: International Conference on Grid Economics and Business Models, pp. 29–43 (2013). https://doi.org/10.1007/978-3-319-02414-1_3
20. Shannon, C.: A mathematical theory of communication. Bell Syst. Tech. J. **27**(3), 379–423 (1948). https://doi.org/10.1002/j.1538-7305.1948.tb01338.x
21. Huffman, D.: A method for the construction of minimum-redundancy codes. Resonance. **11**(2), 91–99 (2006). https://doi.org/10.1007/BF02837279
22. Chmielowiec, A.: Algorithm for error-free determination of the variance of all contiguous subsequences and fixed-length contiguous subsequences for a sequence of industrial measurement data. Comput. Stat. **36**(4), 2813–2840 (2021). https://doi.org/10.1007/s00180-021-01096-1
23. Pinardi, D., Arpa, L., Toscani, A., Manconi, E., Binelli, M., Mucchi, E.: A novel hybrid acquisition system for industrial condition monitoring and predictive maintenance. IEEE Access. **12**, 98,121–98 129 (2024). https://doi.org/10.1109/ACCESS.2024.3428313

Load Test of the Communication Between VR and an Industrial Computer Connected to the Cloud

Jakub Brna, Jozef Svetlík(✉), Marek Sukop, and Juraj Kováč

Faculty of Mechanical Engineering, Technical University of Košice, Košice, Slovakia
jozef.svetlik@tuke.sk

Abstract. This study investigates the communication performance of wireless systems emulating Virtual Reality (VR) workloads in industrial computing environments. Industrial VR applications increasingly rely on distributed architectures, where VR devices exchange high-frequency data with industrial computers and cloud or edge servers, placing strict requirements on latency and jitter. To analyze these constraints, a controlled experimental testbed was established using an industrial-grade computer, a dedicated server, and multiple parallel client processes transmitting VR-like tracking and interaction messages over a wireless network. A custom ZeroMQ-based measurement framework was developed to generate deterministic, high-frequency bidirectional traffic representative of real VR communication loops. Each experimental scenario produced more than 40,000 round-trip time (RTT) measurements across parallel message streams, enabling statistically robust analysis of latency, jitter, and percentile-based performance metrics. The evaluation focuses on mean latency as well as higher percentiles (P90–P99.9), which are critical for identifying worst-case delays that can disrupt VR immersion. The results show that average RTT values remain consistently low, typically around 4–5 ms, indicating that wireless links can support many industrial VR communication tasks. However, sporadic high-latency spikes were observed at upper percentiles, reflecting the non-deterministic nature of wireless networks under load. These findings provide a quantitative baseline for assessing the suitability of wireless communication in industrial VR systems and highlight the need for predictive mechanisms, edge processing, or hybrid architectures to ensure reliable performance in latency-sensitive VR applications.

Keywords: Virtual Reality (VR) · Network Latency · Jitter · Industrial Computing · Cloud Communication · Performance Evaluation

1 Introduction

1.1 Background of VR Communication Systems

Virtual Reality (VR) technologies have become increasingly relevant in industrial domains such as manufacturing, maintenance training, remote supervision, and human–machine interaction. Unlike entertainment-oriented VR applications, industrial VR systems often operate in hybrid computational environments where rendering, computation,

B. Gapiński et al. (Eds.): MANUFACTURING 2026, LNME, pp. 404–419, 2026.
https://doi.org/10.1007/978-3-032-22943-4_28

and sensor processing are distributed across standalone VR headsets, industrial PC (IPC), and cloud services. This architectural distribution places high demands on the underlying communication systems, which must deliver continuous, low-latency, and jitter-free data streams to preserve immersion and ensure operational reliability.

Modern VR platforms rely on real time bidirectional communication to synchronize user motion, scene updates, sensor data, and interaction logic. In industrial contexts, these systems frequently operate over wireless networks due to mobility requirements or spatial constraints. However, Wi-Fi networks introduce variability in latency, susceptibility to interference, and throughput fluctuations factors that directly impact the stability of VR experiences. Understanding how this communication links behave under realistic load conditions is therefore crucial for integrating VR into mission-critical industrial workflows. The motivation for this study arises from the need to quantify these effects and to understand how communication stress generated by realistic industrial workloads - translates into performance degradation during VR operation [1, 2].

1.2 Importance of Low Latency in VR

Low communication latency is essential for maintaining presence and preventing perceptual discomfort in VR environments. Research consistently indicates that motion-to-photon latency should remain below approximately 20 ms to avoid motion sickness, disorientation, and a breakdown of immersion. While rendering pipelines and display technologies contribute to this latency budget, communication delays introduced between VR devices and industrial PCs play an increasingly significant role in distributed VR architectures. Beyond average latency, jitter defined as the variability of latency in time is a major contributor to degraded VR performance. Sudden spikes in latency, even if infrequent, can cause frame inconsistencies, loss of tracking precision, and delays in the delivery of critical scene updates. These issues are further amplified when the VR device relies on an industrial PC that concurrently processes multiple computation intensive tasks, such as CAD model streaming, physics simulation, or machine telemetry fusion. In such cases, the communication link becomes not only a medium for transmission but a limiter of the entire VR interaction loop. These effects are especially pronounced in multi-user VR scenarios and collaborative industrial simulations, where multiple streams of real time data must be synchronized across devices. Consequently, evaluating not only mean latency but also percentile-based performance (e.g., P90, P95, P99) is essential for understanding communication reliability and identifying worst-case conditions that are most harmful to immersion [1, 2].

1.3 Objectives of the Study

The primary objective of this study is to analyze the performance and reliability of wireless communication between industrial computers and real time VR systems. Specifically, the research focuses on the following goals:

- Evaluate latency and jitter characteristics in a controlled wireless environment using a custom ZeroMQ-based communication framework.

- Assess the behavior of parallel message streams to simulate real world VR workloads that involve simultaneous sensor, tracking, and application data.
- Investigate the impact of wireless variability on high-percentile latency values, which represent worst-case conditions known to disrupt VR immersion.
- Examine how increased data transfer from the industrial PC influences communication stability and contributes to latency spikes during VR operation.
- Identify communication bottlenecks in industrial Wi-Fi systems and provide recommendations for optimizing low-latency VR communication pipelines.

2 Research Problem

2.1 Identifying Challenges in Wireless Communication

Despite significant progress in wireless networking technologies, maintaining reliable low-latency communication for Virtual Reality (VR) in industrial environments remains a substantial challenge. Industrial VR systems frequently operate within complex electromagnetic environments characterized by metallic structures, machinery, and multiple overlapping wireless networks, all of which contribute to signal degradation, multipath interference, and inconsistent throughput. Such conditions make wireless performance considerably less predictable than in consumer or office settings.

A key motivation for this research stems from the observation that industrial VR workflows increasingly depend on external computation performed on industrial PC (IPC). As VR headsets offload rendering tasks, spatial processing, or simulation logic to nearby IPC, the communication channel becomes an integral part of the real time loop. Any increase in latency directly affects the timeliness of state updates transmitted from the headset to the industrial computer and back. Thus, the wireless link is not merely a transport layer - it is a functional component of the VR interaction cycle whose instability can degrade task performance. A major limitation arises from latency variability, where instantaneous fluctuations in packet delivery times—caused by interference, channel contention, or hardware buffering—introduce jitter that disrupts VR responsiveness. Packet loss and retransmission events further contribute to temporal instability, making wireless links inherently less suitable for continuous real time streaming without additional optimization layers.

Another challenge is protocol fragmentation. Industrial VR solutions often integrate heterogeneous devices head-mounted displays, tracking modules, edge compute units, and IPCs each operating with differing communication standards. Ensuring interoperability among these components places additional strain on the communication stack, potentially introducing mismatches in buffering strategies, packet scheduling, and Quality of Service (QoS) priorities. Network scalability presents a third challenge. As VR systems become more collaborative and data-intensive, wireless links must simultaneously support sensor fusion data, spatial tracking, visual updates, and application logic. Increased device density can lead to congestion, variations in access point load, and degraded latency performance. Understanding how wireless links behave under realistic load conditions is therefore essential for VR deployments intended to function reliably within industrial infrastructures [3].

2.2 Impact of Latency on User Experience

Latency in VR communication pipelines directly affects the perceived responsiveness of the virtual environment. Even small communication delays between user motion and the corresponding system update can introduce temporal misalignment between intended and rendered interaction states. This desynchronization is especially critical in industrial VR applications, where precision, timing, and consistent feedback are central to task performance. High-percentile latency events (e.g., P95, P99) are of particular interest because even short disruptions can produce noticeable inconsistencies in controller tracking, delayed reactions in interactive simulations, or interruptions in collaborative VR tasks. These effects are amplified in wireless IPC coupled VR systems, where state updates depend on round-trip communication rather than purely on-device processing. Although user sensitivity to latency varies across applications, industrial VR tasks such as remote teleoperation, safety training, and spatial planning require consistently low and predictable delays. An unreliable link diminishes the sense of agency, reduces operator trust, and undermines the reliability of VR as a tool for precision-oriented workflows. Mitigating latency effects often requires predictive modeling, local caching, or high-frequency state replication. However, the underlying wireless channel remains a dominant factor influencing VR responsiveness, particularly when VR devices rely on external processing units for computationally demanding tasks [3].

2.3 Need for Robust Testing Mechanisms

Given the unpredictable nature of wireless networks in industrial spaces, there is a clear need for testing methodologies capable of revealing both average and worst-case communication behavior under realistic conditions. Traditional network evaluation tools are often insufficient for VR-focused analysis because they do not capture moment-to-moment latency variations, nor do they simulate the parallel data streams commonly produced by VR systems.

To address these limitations, VR communication research requires high-resolution latency measurement frameworks capable of:

- Capturing jitters and performances across thousands of messages.
- Simulating multiple concurrent data streams (tracking, input data, scene updates).
- Evaluating bidirectional messaging under load.
- Reproducing conditions such as interference, degraded signal quality, or increased network congestion.

Controlled testing environments allow researchers to systematically introduce variables such as signal attenuation, competing network traffic, or fluctuating bandwidth. This facilitates a deeper understanding of how wireless conditions affect VR responsiveness and what communication architectures remain resilient under adverse scenarios.

Furthermore, automated and repeatable test frameworks—such as the ZeroMQ-based parallel messaging system developed for this study—enable consistent long-term measurement and the generation of statistically rigorous data. Such tools are indispensable for evaluating whether wireless networks can satisfy the strict temporal requirements of industrial VR systems.

2.4 Research Questions

Based on the identified challenges, this study investigates the following research questions:

- How do variations in wireless signal quality such as interference, congestion, and attenuation affect latency and jitter in communication between industrial PCs and VR systems?
- What latency thresholds are acceptable for maintaining interaction quality across different industrial VR use cases?
- Which testing methodologies most accurately capture real time communication performance, especially with respect to percentile latency, jitter, and parallel message flow?
- How can insights from controlled communication experiments support the scalability of wireless VR systems as device density and data complexity increase?

Addressing these questions is essential for designing robust wireless communication architectures capable of supporting high-performance VR applications in industrial environments.

3 Methodology

3.1 Description of the Test Environment

The test environment was designed to evaluate the communication performance between an industrial computer and a simulated VR-oriented data pipeline under controlled wireless network conditions. Although no physical VR headset was employed, the communication patterns closely replicate those observed in real VR systems, which are characterized by rapid updates, high-frequency bidirectional messaging, and strict low-latency constraints. This approach allows the study to isolate communication performance without interference from rendering or sensor-fusion subsystems, thereby producing measurements directly attributable to the network layer.

The experiments were conducted on an industrial-grade computer connected to a controlled laboratory network. The environment was configured to reproduce conditions typically encountered in industrial mobile VR scenarios, including variable wireless signal strength, fluctuating throughput, and external interference. The controlled setup ensured repeatability and minimized unpredictable environmental effects, which is essential when analyzing latency-sensitive communication loops.

A dual-band wireless router (2.4 GHz and 5 GHz) provided network connectivity, enabling systematic variation of distance, interference level, and channel congestion across test sessions. The industrial PC was equipped with a high-performance CPU and sufficient RAM to ensure that computation time did not influence the measured communication latencies.

In addition to the client components, the testbed included a dedicated server process responsible for receiving messages and instantly returning them. This architecture models a typical VR-to-edge or VR-to-cloud interaction loop, in which positional tracking and interaction data must be exchanged with minimal delay. Together, these components reproduce the temporal characteristics of real VR communication while allowing precise measurement of low-level network behavior [4, 5].

3.2 Tools and Technologies Used

To capture detailed communication characteristics, the study employed a custom measurement framework built on the ZeroMQ (ZMQ) messaging library. ZMQ was selected for its low overhead, its ability to sustain high message throughput, and its suitability for real time and near real time research applications. The framework was developed in Python 3.11 and optimized to generate deterministic, high-frequency message streams representative of VR tracking and interaction updates.

Each test employed multiple parallel client processes, typically four executing concurrently and independently on the industrial computer. These clients continuously transmitted request messages to the server, which immediately returned the received payload. For every message, the client computed RTT with microsecond-level precision. More than 40 000 measurements were collected for each configuration, enabling statistically robust analysis.

The testing framework incorporated:

- ZeroMQ REQ/REP communication sockets.
- Python multiprocessing for parallel traffic generation.
- High-precision timers for RTT measurement.
- Automated logging to CSV for raw latency values.
- Built-in analytical modules for computing percentiles (P50–P99.9), jitter, and error rates.
- Matplotlib-based visualization capable of producing:
  - Raw latency plots.
  - Running average plots,
  - Percentile bar charts,
  - Percentile line charts.

Supplementary diagnostics were conducted using tools such as Wireshark during preliminary experiments to verify packet flow, confirm expected transport-level behavior, and inspect potential retransmissions. Final latency datasets, however, were collected directly through the ZeroMQ framework to maintain temporal precision.

This combination of tools established a controlled yet flexible environment capable of reproducing VR-like communication loads and enabled precise quantification of latency stability, jitter, and performance under parallel message streams.

3.3 Experimental Procedure

The experimental procedure was designed to generate high-frequency communication patterns analogous to those found in VR tracking loops. Each experiment consisted of multiple parallel client processes running on the industrial computer and continuously exchanging messages with a dedicated ZeroMQ server. The use of four parallel clients reflects realistic multisource VR conditions in which headsets, controllers, and environmental sensors transmit updates simultaneously.

Each client generated a fixed number of messages, typically 10 000—resulting in a combined dataset exceeding 40 000 RTT measurements per experiment. Every message contained a unique sequence identifier along with randomly generated coordinate values representing synthetic VR positional updates. These messages were transmitted through a strict ZMQ REQ/REP pattern, ensuring that each request was paired with a corresponding response, thereby isolating pure communication delay from processing time.

The server maintained no internal state; it returned each message immediately upon receipt. This design mirrors VR edge-processing architectures, where tracking updates must be validated or acknowledged with minimal delay, and ensures that communication overhead is the primary contributor to measured RTT. All experiments were executed using wireless connectivity to reflect the constraints of mobile VR in industrial settings. Conditions such as distance, interference, and signal degradation were systematically varied to simulate realistic operational scenarios [4, 5] (Fig. 1).

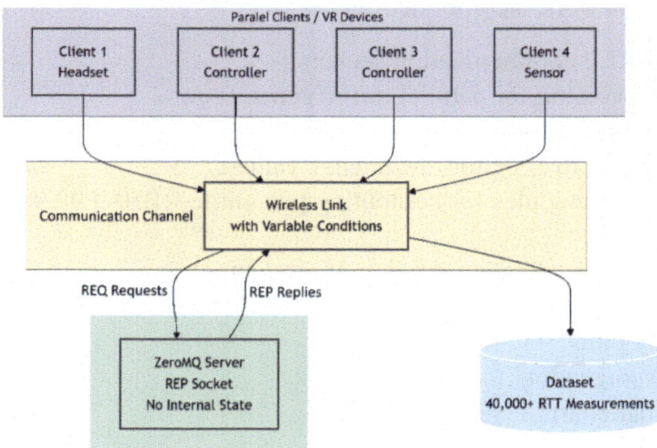

Fig. 1. A block diagram of the experimental communication loop

During execution, clients periodically reported progress (e.g., every 1 000 messages) to support monitoring and detection of anomalies, including packet loss or abnormal latency spikes. The testing scripts implemented precise scheduling and CPU-usage safeguards to reduce contention and improve reproducibility.

3.4 Data Analysis Pipeline

The collected latency measurements were processed through an automated data analysis pipeline developed specifically for this study. Each client produced detailed CSV logs containing:

- Message index.
- Raw latency in nanoseconds.

- Latency in milliseconds.
- Input parameters (X, Y).
- Metadata such as client identifier and timestamps.

After all streams were completed, the data were aggregated into a single unified dataset, enabling evaluation of both per-stream behavior and system wide communication performance [4, 5].

3.5 Statistical Metrics

For each parallel stream and for the aggregated dataset, the following metrics were computed:

- Mean latency.
- Minimum and maximum latency.
- Median (P50).
- Higher percentiles (P90, P95, P99, P99.9).
- Jitter, expressed as variance and absolute deviation.
- Error count, representing failed or unacknowledged messages.

These metrics correspond to thresholds commonly referenced in VR performance literature, where both average delay and delay variability strongly influence user perception and system stability.

3.6 Visualization

To support interpretation, the analysis pipeline automatically generated several types of visualizations:

- Raw latency plot. Displays the sequential RTT values across the entire experiment, highlighting jitter patterns and latency spikes.
- Running average plot. Utilizes a configurable sliding window (e.g., 200 messages) to reveal long-term trends while smoothing short-term fluctuations.
- Percentile bar chart. Enables side-by-side comparison of P50, P90, P95, and P99 values across the parallel streams (see Fig. 2).
- Percentile line chart. Illustrates percentile progression across streams using distinct visual markers for clarity (see Fig. 3).

These visualizations support identification of anomalies, assessment of latency stability, and evaluation of network behavior under varying conditions.

Interpretation Relevance for VR:

Because VR systems are highly sensitive to latency fluctuations, the analysis focused on:

- Upward drift in latency over time.
- Jitter values exceeding 10–15 ms.
- Infrequent but extreme latency spikes capable of disrupting immersion even when average latency remains low.

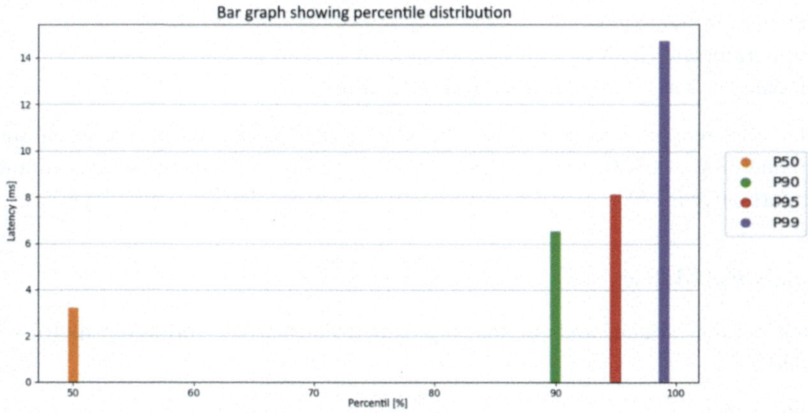

Fig. 2. Bar graph showing percentile distribution

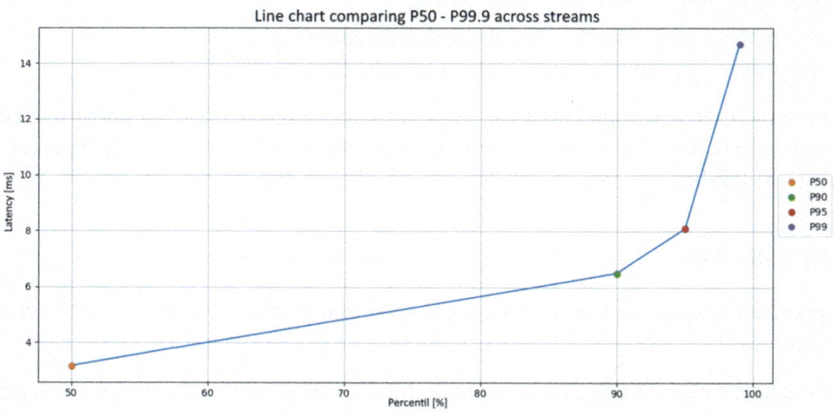

Fig. 3. A Line chart comparing P50 – P99.9 across streams

The resulting analytical framework offers a quantitative basis for assessing the suitability of wireless communication channels for industrial VR deployments, particularly in scenarios requiring consistent, low-latency bidirectional data exchange.

4 Results

4.1 Overview of Collected Latency Data

Across all test scenarios, the experimental setup generated more than 40 000 round-trip latency samples per run, distributed evenly across four parallel communication streams. All clients maintained uninterrupted bidirectional communication with the server, with no packet loss or stalled connections observed. Overall communication stability remained consistently high throughout the measurement sessions, confirming that the testing framework operated reliably under all examined network conditions.

Most raw latency measurements ranged from approximately 1.3 ms to 100 ms. The bulk of samples clustered around a mean of 4–5 ms, depending on the wireless conditions. Although the average latency was well within typical thresholds for industrial VR interaction, occasional spikes were detected—most prominently at the 95th and 99th percentile levels. These spikes are characteristic of short-term wireless interference or momentary bandwidth contention, phenomena expected in realistic industrial deployments.

For clarity, the following sections distinguish between aggregated system wide behavior and per-stream performance.

4.2 Performance Per Parallel Communication Stream

Each of the four parallel clients acted as an independent VR-like data source, conceptually representing different tracking and sensing channels (headset pose, controller pose, environmental mapping, and auxiliary sensor data). The latency results across these streams showed strong consistency.

Across all experiments, the following patterns were observed:

- P50: consistently 3–4 ms.
- P90: typically, 6–12 ms.
- P95: typically, 8–20 ms.
- P99: occasional spikes reaching 80–100 ms, consistent with Wi-Fi jitter.
- Maximum latency: rarely beyond 120–140 ms (Fig. 4).

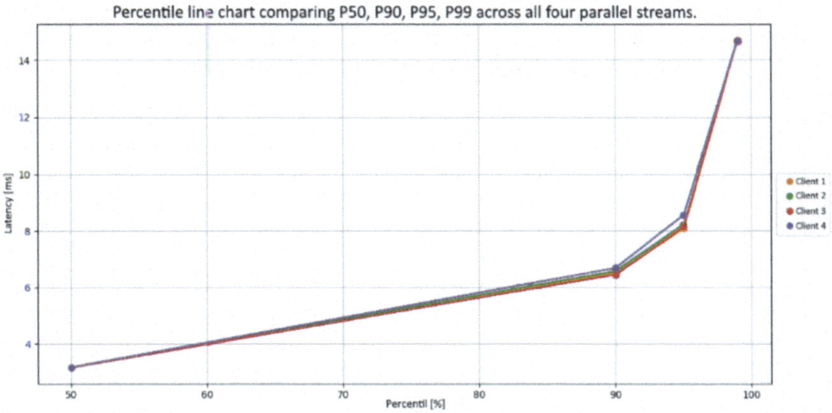

Fig. 4. Percentile line chart comparing P50, P90, P95, P99 across all 4 parallel streams

Although these spikes were infrequent, they are noteworthy because VR rendering loops are sensitive to isolated delays. Without mitigation (prediction layers, smoothing filters, or edge processing), even short anomalies may disrupt synchronization across VR subsystems.

Low variance across all streams indicates that the wireless network handled parallel load distribution effectively, with each client exposed to similar interference patterns.

This uniformity suggests that the observed anomalies are environmental rather than process specific.

4.3 Raw Latency Trends and Jitter Characteristics

The raw latency timeline exhibits micro fluctuations characteristic of wireless communication links. Short-term jitter (<5 ms) appeared consistently across all datasets. However, intermittent jitter bursts exceeding 20 ms were recorded at irregular intervals. These bursts align with expected industrial Wi-Fi behavior, where interference from machinery, metal structures, and competing devices can temporarily degrade signal quality (Figs. 5 and 6).

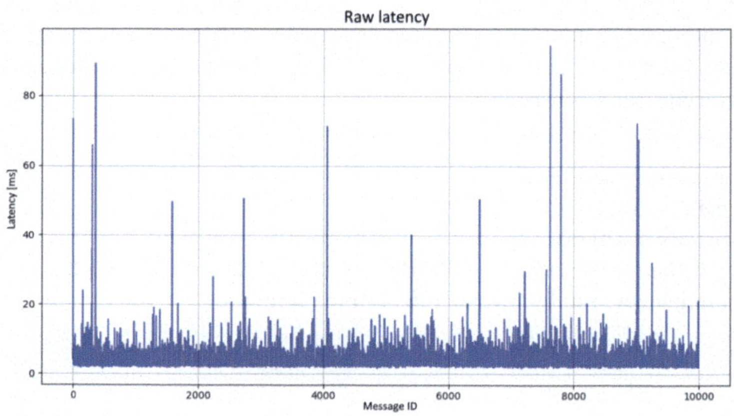

Fig. 5. Raw latency plot with highlighted jitter spikes

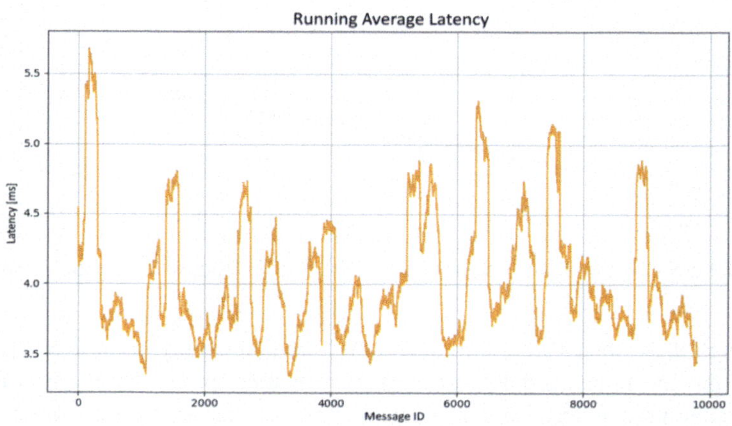

Fig. 6. Running average latency plot demonstrating long-term stability

Despite these fluctuations, long-term latency behavior remained stable. Using a moving-average filter with a 200-message window, all curves converged to approximately 3–6 ms. This indicates that while transient jitter exists, the underlying communication loop can sustain the responsiveness required for most VR interaction tasks.

4.4 Aggregate Statistical Results

Aggregated analysis across all parallel streams yields the following key statistics:

- Mean RTT: 4.0–4.5 ms.
- Median RTT: 3.5–4.0 ms.
- Minimum RTT: $\approx$1.4 ms.
- Maximum RTT: 90–140 ms.
- Error rate: 0% (no failed request–response events).
- Standard deviation 1.2–3.5 ms.

These results confirm that the communication channel can maintain low-latency operation for most messages. Percentile analysis shows that 99% of all samples remain below ~60–80 ms, which is acceptable for non-render-critical VR tasks such as cloud synchronization, telemetry reporting, or industrial analytics.

However, for motion-critical tasks—head tracking, eye tracking, and fast interaction loops—these spikes would require compensation mechanisms (Fig. 7).

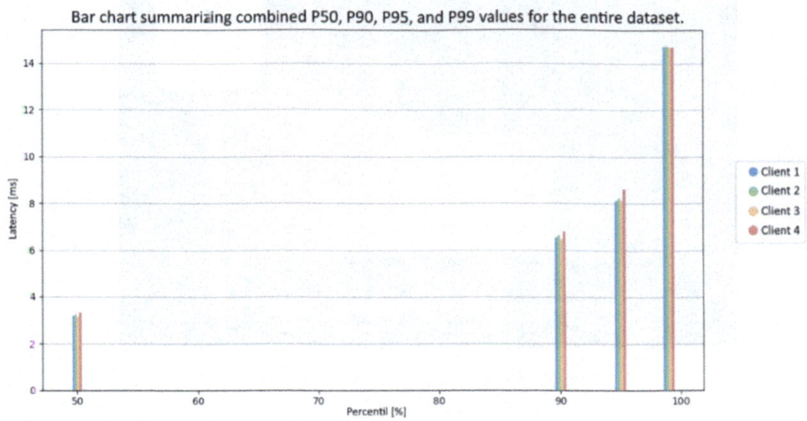

Fig. 7. A bar chart summarizing combined P50, P90, P95, P99 for the entire dataset.

4.5 Interpretation of Latency Spikes

Although the system exhibits low average and median latency, the presence of sporadic high-latency events is a critical observation for VR-industrial integration. These spikes are significant for several reasons:

- VR systems are hypersensitive to outliers.

  A single delayed frame can disrupt the temporal consistency of motion updates, affecting user comfort and degrading immersion.

- Industrial wireless environments amplify delay variability.

  Electromagnetic noise, reflective surfaces, and machinery introduce irregular interference patterns, causing unpredictable packet delays.

- Parallel data streams increase the probability of simultaneous spikes.

  VR systems typically involve multiple continuous data feeds. When spikes align across streams, cumulative degradation becomes more severe.

- These findings agree with existing research indicating that wireless communication, while capable of supporting auxiliary VR tasks, is inherently non-deterministic. To maintain stable interaction loops, VR systems require predictive rendering, smoothing layers, or local edge processing to absorb delay variability (Fig. 8).

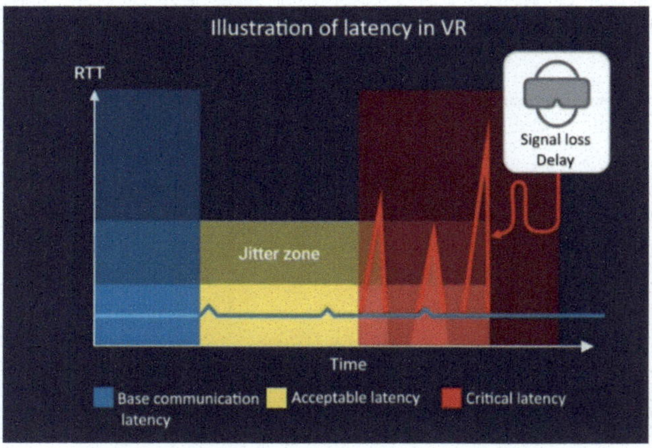

Fig. 8. Illustration of latency in VR

4.6 Implications for VR–Industrial Integration

The results show several important implications for the integration of VR into industrial wireless environments:

- Wireless links are sufficiently stable for non-render-critical VR operations, such as digital-twin updates, process monitoring, and supervisory functions.

- Real time rendering synchronization over pure wireless links remains challenging, unless supported by prediction, smoothing, or local fallback computation.
- Parallel communication does not significantly elevate average latency, but it increases the probability that multiple VR subsystems encounter simultaneous anomalies.
- Robust VR systems must include latency-tolerant design features, especially when dependent on cloud or edge services.

These findings reinforce the need for hybrid communication architectures, where latency-sensitive VR functions remain local while less time-critical processes leverage wireless or cloud resources.

5 Discussion

The results of this study provide a focused view on the performance limits of wireless communication between VR systems, industrial computers, and cloud environments. While the average latency remained consistently low across all test scenarios, the presence of irregular high-latency spikes reveals inherent constraints of wireless networks that must be considered in industrial VR deployments. Measured RTT values remained stable around 4 ms, indicating that wireless links are generally capable of supporting common VR data flows, such as sensor updates and cloud-assisted computations. However, the 95th and 99th percentile values showed occasional spikes exceeding 80–140 ms, observed across all parallel channels, suggesting that they stem from fundamental wireless network behavior, including interference, medium contention, and internal router scheduling. Such long-tail latency events align with findings from previous studies and represent a critical reliability challenge. Although they occur infrequently, their magnitude can significantly disrupt motion-sensitive VR applications. Even brief latency spikes undermine key VR experience components such as presence, agency, and motion stability. For industrial VR—where safety, precision, and uninterrupted immersion are essential—systems must incorporate mechanisms that reduce the perceptual impact of communication delays. Suitable techniques include client-side prediction and interpolation, local processing of motion-critical tasks, hybrid cloud/edge architectures, and adaptive prioritization of time-sensitive packets. These approaches allow VR systems to remain functional even when network conditions fluctuate. Parallel communication tests demonstrated stable behavior under moderate load, but real industrial setups often involve many devices operating simultaneously. Increasing the number of wireless nodes raises the risk of overlapping latency spikes. To support scalability, communication frameworks should incorporate QoS rules, network segmentation or slicing for VR traffic, dynamic bandwidth allocation, and local edge-processing nodes. These strategies help maintain predictable latency and ensure system reliability in dense network environments. The findings directly translate to typical industrial VR scenarios. Robotic teleoperation requires strict latency guarantees, and observed spikes make pure cloud operation impractical without prediction or local fallback. Digital twin visualization is less sensitive and can tolerate occasional delays. Training simulations require stable motion feedback but can incorporate smoothing techniques. Monitoring and diagnostics tolerate latency variability and can fully rely on wireless communication. Thus, applicability depends on the specific VR workload and its timing requirements. The controlled test environment cannot fully reproduce real industrial conditions, which often include stronger interference, heavy network traffic, and more diverse hardware configurations. Furthermore, mobility of VR headsets which affects signal quality was not included in this phase of testing. Future experiments should therefore incorporate realistic industrial settings and dynamic user movement. Overall, this study shows that wireless VR communication can achieve low average latency but remains vulnerable to unpredictable delays. These findings highlight the need for prediction mechanisms, edge computing, and adaptive communication protocols to ensure reliable industrial VR performance. They also provide a foundation for future work focused on testing in operational environments and optimizing communication architectures for deterministic behavior [6].

6 Conclusions

This study evaluated the communication performance of wireless connections emulating VR workloads between industrial-grade computers and a server environment. Through controlled experiments using multiple parallel client streams and a ZeroMQ-based measurement framework, over 40,000 round-trip latency samples per scenario were collected and analyzed. The results indicate that average latency remained consistently low (mean 4–4.5 ms, median 3.5–4 ms), demonstrating that wireless links can support high-frequency bidirectional communication.

However, occasional spikes in the 95th and 99th percentiles, reaching up to 120–140 ms, were observed. These outliers highlight the non-deterministic nature of wireless networks, particularly under interference, congestion, or multiple parallel streams. Jitter analysis confirmed that short-term fluctuations were generally small (<5 ms), though rare bursts exceeded 20 ms. Such deviations, if unmitigated, could disrupt real time applications requiring strict latency guarantees.

The findings emphasize that while wireless communication is viable for many non-motion-critical industrial VR interactions, high-precision tasks such as head-tracking or teleoperation require compensatory mechanisms. Strategies including client-side prediction, edge processing, and hybrid cloud architectures are recommended to maintain system responsiveness. This work provides a quantitative baseline for designing robust communication frameworks in industrial VR contexts and serves as a foundation for future studies under operational conditions with mobile devices and dynamic interference patterns.

Acknowledgement. This work was supported by the Slovak Research and Development Agency under the Contract no. APVV-18-0413, KEGA 043TUKE-4/2024 and VEGA 1/0294/24.

References

1. Elbamby, M.S., Perfecto, C., Bennis, M., Doppler, K.: Towards low-latency and ultra-reliable virtual reality. IEEE Netw. **32**(2), 78–84 (2018)
2. Yu, H., Zeng, P., Xu, C.: Industrial wireless control networks: from WIA to the future. Engineering. **8**(4), 18–24 (2022)

3. Fletcher, M., Paulz, E., Ridge, D., Michaels, A.J.: Low-latency wireless network extension for industrial internet of things. Sensors. **24**(7), 2113 (2024)
4. Kail, G. et al.: A highly reliable ultralow-latency wireless solution for industrial control loops: design and evaluation. In: Proceedings of the 55th Asilomar Conference on Signals, Systems and Computers, pp. 8–12. IEEE Computer Society (2021)
5. Tan, D.T., Kim, S., Yun, J.-H.: Enhancement of motion feedback latency for wireless virtual reality in IEEE 802.11 WLANs. In: Proceedings of the 2019 IEEE Global Communications Conference (GLOBECOM), pp. 1–6 (2019)
6. Havard, V., Jeanne, B., Lacomblez, M., Baudry, D.: Digital twin and virtual reality: a co-simulation environment for design and assessment of industrial workstations. Prod. Manuf. Res. **7**(1), 472–489 (2019)

Author Index

B. Gapiński et al. (Eds.): MANUFACTURING 2026, LNME, pp. 421–422, 2026.
https://doi.org/10.1007/978-3-032-22943-4

The manufacturer's authorised representative in the EU is Springer Nature Customer Service Centre GmbH, Europaplatz 3, 69115 Heidelberg, Germany. If you have any concerns regarding our products, please contact ProductSafety@springernature.com

Printed and bound by CPI Group (UK) Ltd, Croydon, CR0 4YY
21/09/2026
02219973-0012